T0185459

ROUTLEDGE HANDBOOK OF BIOSECURITY AND INVASIVE SPECIES

Edited by Kezia Barker and Robert A. Francis

Routledge
Taylor & Francis Group

LONDON AND NEW YORK

First published 2021
by Routledge
2 Park Square, Milton Park, Abingdon, Oxon OX14 4RN

and by Routledge
605 Third Avenue, New York, NY 10158

Routledge is an imprint of the Taylor & Francis Group, an informa business

British Library Cataloguing-in-Publication Data
A catalogue record for this book is available from the British Library

Library of Congress Cataloging-in-Publication Data
Names: Barker, Kezia, editor. | Francis, Robert A., editor.
Title: Routledge handbook of biosecurity and invasive species /
edited by Kezia Barker, Robert A. Francis.
Description: Abingdon, Oxon ; New York, NY : Routledge, 2021. |
Includes bibliographical references and index.
Identifiers: LCCN 2020049807 (print) | LCCN 2020049808 (ebook) |
ISBN 9780815354895 (hardback) | ISBN 9781351131599 (ebook)
Subjects: LCSH: Introduced organisms. | Introduced organisms—Social aspects. |
Biological invasions. | Biosecurity. | Ecosystem management.
Classification: LCC QH353 .R68 2021 (print) | LCC QH353 (ebook) |
DDC 333.95/23—dc23
LC record available at https://lccn.loc.gov/2020049807
LC ebook record available at https://lccn.loc.gov/2020049808

ISBN: 978-0-8153-5489-5 (hbk)
ISBN: 978-0-367-76321-3 (pbk)
ISBN: 978-1-351-13159-9 (ebk)

Typeset in Bembo
by Apex CoVantage, LLC

CONTENTS

Contents

CONTRIBUTORS

Editors

Kezia Barker is Lecturer in Geography in the Department of Geography at Birkbeck, University of London, UK. Her research considers public values and practices in relation to security futures, including biosecurity, health security and within prepping and survivalist subcultures. She co-edited *Biosecurity: The Socio-politics of Invasive Species and Infectious Diseases* (with Andrew Dobson and Sarah L. Taylor), also by Routledge.

Robert A. Francis is Reader in Ecology in the Department of Geography, King's College London, UK. His research focuses on urban ecology, freshwater ecology and nature and society interactions. He edited *A Handbook of Global Freshwater Invasive Species* and co-edited *Urban Landscape Ecology: Science Policy and Practice* (with James D.A. Millington and Michael A. Chadwick), also by Routledge.

Contributors

Jennifer Atchison lives and works in Dharawal Country. She is Senior Lecturer in the School of Geography and Sustainable Communities, University of Wollongong, where she teaches environmental crime and justice. Her research seeks to understand the place of plants in past, present and future lives; she has written about archaeobotany in Australia, agriculture and climate change and invasive plant management. She has had the privilege of learning about people, plants and environmental change in Gadjerong Country in the East Kimberley of Australia.

Enrique Baleriola is Lecturer in Social Psychology at the Open University of Catalonia (UOC). His main interests are related to public policies and non-experts' engagement, actor-network theory and biopolitics. He has a PhD in person and society in the contemporary world (Universitat Autònoma de Barcelona, Spain) and a MG in psychosocial research and intervention.

Luke Bergmann is Associate Professor and Canada Research Chair in the Department of Geography at the University of British Columbia. His research interests include critical-computational and social-theoretic geographies, often in intersection with economy, land and health.

Giuseppe Brundu is an agronomist and botanist, with a PhD on Monitoring Mediterranean Forest ecosystems. He is associate professor of environmental and applied botany at the University of Sassari, Italy. His main research interest are plant invasions in the Mediterranean regions, with a special focus on islands, invasive woody species, management and control of plant invasions, codes of conducts, and legislation applied to plant invasions. He has participated to several international scientific projects on biological invasions and collaborated as an expert or consultant with several international governmental and non-governmental organizations.

Mark A. Burgman is Professor of Risk Analysis and Environmental Policy, Director of the Centre for Environmental Policy at Imperial College London and Editor-in-Chief of the journal *Conservation Biology*. Previously, he was Head of the Department of BioSciences at the University of Melbourne, Australia. He works on expert judgement, environmental modelling and risk analysis. He has worked on environmental management and expert judgement in a range of settings from electrical landscape planning to species conservation. He received a BSc from the University of New South Wales (1977), an MSc from Macquarie University, Sydney (1981), and a PhD from the State University of New York at Stonybrook (1987). He worked as a consultant research scientist in Australia, the United States and Switzerland during the 1980s before joining the University of Melbourne in 1990. He joined CEP in February 2017. He has published over 250 refereed papers and book chapters and 7 authored books. He was elected to the Australian Academy of Science in 2006.

Thomas Campagnaro is assistant professor at Padova University, in the Department Agronomy, Food, Natural resources, Animals and Environment, Italy. His research regards forest management and planning, landscape ecology, biological invasions, nature protection and urban forestry. He has been involved in several international research projects concerning forest management, biodiversity conservation in protected areas and innovative use of alternative energy sources.

Jane A. Catford is a plant community ecologist with an interest in biological invasions, environmental change and biodiversity. She is fascinated in the causes, consequences and processes of plant invasions and loves to link the fundamental and applied aspects of invasion science. Working in terrestrial and freshwater ecosystems, she uses theoretical, quantitative and empirical approaches in her work.

Michael A. Chadwick is Senior Lecturer in Physical and Environmental Geography in the Department of Geography at King's College London. He is an aquatic biologist interested in exploring both applied and basic ecological questions. His work focuses on understanding how ecosystem structure and function, specifically related to macroinvertebrates, respond to changes in environmental conditions.

Kin Wing (Ray) Chan is a human geographer at the University of Exeter. He is currently working with Henry Buller and Steve Hinchliffe on a research project called 'Diagnostic Innovation and Livestock', which concerns understanding drivers for the uptake of rapid diagnostic technologies to reduce antimicrobial usage on farm animals in the UK. His research attempts to provide better understandings of farmers' and veterinarians' diagnostic practices in the UK and governance of animal health and antimicrobial usage on farm animals in Hong Kong and post-socialist China.

Evangeline Corcoran is a PhD candidate in quantitative ecology, with a background in conservation and biosecurity. Her current research focuses on improving the detection and modelling of pest species and species of conservation concern using drone platforms and computer vision. She also has experience in assessing the risk and managing the spread of insect pests in native Australian forests.

Elizabeth J. Cottier-Cook is Head of the United Nations University (UNU) and Scottish Association for Marine Science (SAMS)'s Associate Institute, Professor of Marine Biology, specialising in marine invasive species at the University of the Highlands and Islands (UHI), Programme Leader for the Erasmus Mundus Joint Master Degree in Aquaculture, Environment and Society (EMJMD ACES) and Fellow of the Royal Society of Biology.

Gareth Enticott is a reader in Human Geography at Cardiff University. His research examines farmers' understandings of animal disease, their biosecurity practices and their responses to disease control policies. His work covers bovine tuberculosis, antimicrobial resistance and vaccine use in England and Wales, Europe and New Zealand. He also examines the role of the veterinary profession and the production and dissemination of animal disease scientific knowledge.

Massimo Faccoli is associate professor at the Padova University, in the Department Agronomy, Food, Natural resources, Animals and Environment, Italy. His research activity mainly concerns biology, monitoring and systematics of wood boring insects, with special focus on the effects of climate change on forest pests of temperate regions. He published about 260 scientific papers in national and international journals, books and proceedings. He is member of the Italian Society of Entomology and Italian Academy of Forest Sciences.

Juliet J. Fall is a political and feminist geographer who has taught and written about invasive species, biosecurity, political borders and epistemologies of science. Her work explores how discourses of nature are invoked to depoliticise debates in domains as diverse as environmentalism, territorial conflicts and diplomacy.

Rebecca J. Giesler is a post-doctoral researcher in the Earth Institute, University College Dublin (UCD), Ireland. She specialises in marine invasive species in the UK and Ireland, with an interest in wider environmental policy. Her PhD from the Scottish Association of Marine Science and the University of Edinburgh focused on marine invasive species management in Scotland. Currently she is working on marine protected areas and the socio-economic and environmental impacts of marine invasive species in Ireland.

Nicholas Gill is a human geographer with interests in environmental management and rural cultures. His research focuses on rural areas, particularly on cultural and social aspects of land management, land-use change and environmental conflict. His research aims to bridge conventional natural resource management research and research on cultures of nature. He is interested in how people occupy landscapes and seek to inhabit, use, protect and conserve those landscapes, usually simultaneously and in ways that defy neat compartmentalisation of these activities.

Qinfeng Guo is a research ecologist at the US Department of Agriculture and is broadly interested in community ecology and biogeography. His current research includes (1) using life history traits to predict species invasiveness and (2) using information from invasive species' native ranges to predict the spread in introduced regions under climate and land-use change. He received a PhD from the University of New Mexico and was a post-doc fellow at UCLA. He worked at the University of Arizona (Ike Russell Fellow), the University of Tokyo (NSF-JSPS International Fellow) and the US Geological Survey before joining the USDA (Forest Service).

Grant Hamilton is a quantitative ecologist and Associate Professor in Ecology. With a focus on detection and risk in biosecurity and conservation, he has extensive experience and expertise in application machine learning and drone technology, risk assessment for agricultural systems and ecological modelling.

Timothy Hodgetts is an environmental geographer, formerly based in the School of Geography and the Environment at the University of Oxford, and also formerly Senior Research Fellow at the Wildlife Conservation Research Unit, Department of Zoology, University of Oxford. Past projects have investigated the political management of non-human life at a range of scales, from bacteria in domestic kitchens to the cross-border geopolitical institutions involved in large carnivore conservation. This work has particularly focused on the interplay of borders, biosecurity and transgression.

Tammi Jonas is a pastured pig and cattle farmer and meatsmith at Jonai Farms. Jonas is president of the Australian Food Sovereignty Alliance. She is undertaking a PhD at the University of Western Australia on the biodiverse practices of agroecological farmers, investigating the logistical, financial, social and legislative barriers to their efforts. She is active in the global fight for food sovereignty in the International Planning Committee for Food Sovereignty (IPC), advocating in UN governing bodies for the rights of Indigenous Peoples and peasants. She is co-author of *Farming Democracy: Radically Transforming the Food System from the Ground Up* (2019).

Richard Kock is a wildlife veterinary ecologist, infectious disease researcher and conservationist and has worked in the field of wildlife health since 1980 with a focus on African and Asian ecosystems. He holds a chair in Wildlife Health and Emerging Diseases at the Pathobiology and Population Sciences Department in the Royal Veterinary College. He is the co-chair of the IUCN SSC Wildlife Health Specialist Group, on the WHO IHR and OIE Crisis Management Committee expert list, and is an associate research fellow at Chatham House, past Council Member of the Wildlife Disease Association and an adjunct professor at Tufts University and Njala University.

Simon J. Lambert is Associate Professor of Indigenous Studies at the University of Saskatchewan, Canada, on Treaty Six territory and the Homeland of the Métis. He is originally from Aotearoa New Zealand, and his tribal affiliations are to Tūhoe and Ngāti Ruapani mai Waikaremoana. Simon's research is in Indigenous environmental management with particular interest in Indigenous disaster risk reduction and emergency management.

Ori Lev is Senior Lecturer at the Department of Public Policy and Administration, Sapir Academic College. Ori earned his doctorate in political theory from the London School of Economics (LSE) in 2006. From 2007–2010 he was a post-doctoral fellow in bioethics at the National Institutes of Health, US. After completing his fellowship, Ori became a science policy analyst in the Office of the Director at NIH. In that capacity, Ori worked on the regulation of dual-use research, the protection of human subjects in clinical studies and stem cell research.

Alex Liebman is a PhD student at Rutgers University, researching the turn towards big data science in agricultural research and its implications for climate change, peasant livelihoods and agroecological landscapes. He explores how agricultural data science reproduces particular forms of standardisation and homogeneity that constitute racialised and exclusionary forms of international development and environmental management across the Global South. He is also a researcher with the Agroecology and Rural Economics Research Corps, studying industrial agricultural drivers of land-use change and antibiotic resistant diseases.

Jamie Lorimer is an environmental geographer in the School of Geography and the Environment at the University of Oxford. His research explores the histories, politics and cultures of wildlife conservation. Past projects have ranged across scales and organisms – from elephants to hookworms. Jamie is the author of *Wildlife in the Anthropocene: Conservation after Nature* (2015). His current project draws together work on rewilding inside and beyond the human body to demonstrate that a 'probiotic turn' is underway in the management of both human and environmental health. *The Probiotic Planet: Using Life to Manage Life* was published in 2020.

Abigail L. Mabey is an invasive species ecologist with an interest in how traits contribute to species invasiveness. She has experience in marine intertidal environments, with a focus on the establishment and spread of invasive seaweeds. She has experience using quantitative and empirical research techniques and has familiarity with using systematic reviews and meta-analyses to summarise the existing literature.

Melanie Mark-Shadbolt is an indigenous environmental sociologist, CEO of Te Tira Whakamātaki, Kaihautū Chief Māori Advisor at the Ministry for the Environment and Kaihautū Ngātahi of New Zealand's Biological Heritage National Science Challenge. Her tribal affiliations include Ngāti Kahungunu ki Wairarapa, Ngāti Porou, Te Arawa, Ngāti Raukawa, Te Atiawa and clans Mackintosh and Gunn. Melanie's research focuses on traditional knowledge as it relates specifically to biosecurity and sustainable natural resource management, and her work has covered research in stakeholder values, attitudes and behaviours, social and cultural acceptability of management practices and the wider human dimensions of environmental health.

Damian Maye is Professor of Agri-Food Studies at the Countryside and Community Research Institute, University of Gloucestershire. He has a long-standing research interest in the geography and sociology of agri-food restructuring, including agricultural biosecurity, animal health governance and food security. He is Associate Editor of *Journal of Rural Studies* and *Frontiers in Veterinary Science* (*Veterinary Humanities and Social Sciences*) and has recently finished a major new textbook, *Geographies of Food*. His current biosecurity research examines animal health governance, particularly farmer understandings of bovine TB policy.

Shaun McKiernan is a human geographer with an interest in the social and cultural dimensions of natural resource management. Shaun earned his PhD in human geography in the School of Geography and Sustainable Communities, University of Wollongong. His research examines the discourses and practices of natural resource management, with a particular focus on how rural landholders and land managers develop the capacities to learn, manage and at times live with invasive plants.

Sebastián Moya is a PhD candidate in animal medicine and health at the Universitat Autònoma de Barcelona, Spain (UAB) and participates in a research project about biosecurity in dairy cattle farms (MINECO AGL2016–77269-C2–1-R). His main interests are related to biosecurity and participatory epidemiology in dairy farms and technologies that support this sector. He has a master's in psychosocial research and intervention (UAB).

John D. Mumford is Professor of Natural Resource Management in the Centre for Environmental Policy at Imperial College London. He works on invasive species risk analysis, risk assessment for genetic modification of insects, area-wide pest management and fisheries uncertainty. He received a BS from Purdue University (1974) and a PhD from the University of London (1978). He has worked extensively on pest management in Asia, Africa and Australasia and on fisheries in Europe. He chairs the Great Britain Non-native Species Risk Analysis Panel, which oversees the risk assessment and management processes for invasive species for the devolved governments. He has published over 250 articles, agency reports and book chapters.

Estibaliz Palma is an early career researcher with an interest in plant ecology and the mechanisms driving community assembly and species coexistence. She studies the role of functional traits in population and community dynamics, with a special focus on biological invasions. Her research combines a strong theoretical background, a functional perspective on ecological processes and quantitative analytical methods.

Ian D. Rotherham is an ecologist and landscape historian with a long-standing interest in invasive and non-native species. In recent years he has been developing ideas and concepts of recombinant ecologies formed by the fusion of natives and non-natives and how ecosystems ebb and flow through time and space. In this context he is working on the influences in human-nature interactions to generate eco-cultural systems. He is Professor of Environmental Geography and Reader in Tourism and Environmental Change in the Department of the Natural and Built Environment, Sheffield Hallam University.

Katarina Saltzman is Associate Professor in Conservation at the University of Gothenburg, Sweden. In her research she has investigated landscapes, nature/culture relations and heritage making from an ethnological point of view, often in transdisciplinary collaboration. Her research areas include vernacular practices such as gardening, recreational walking and rural landscape management. She has carried out field studies in rural, urban and semi-urban environments, including intensively tended private gardens and agricultural landscapes as well as transitory and temporarily leftover places at the urban fringe.

Limor Samimian-Darash is an anthropologist and Associate Professor at the Federmann School of Public Policy and Government at the Hebrew University. Her research focuses on preparedness for health and security threats, the governance of risk and uncertainty (in theory and practice) and scenario planning and the future. She received her PhD in anthropology and sociology from the Hebrew University (2010) and in 2013 was recognised as a promising early career social scientist in Israel (with the award of an Alon Fellowship for 2013–2016). She has been a visiting research scholar at the University of California, Berkeley (2006–2008); a post-doctoral fellow at the University of Illinois, Urbana-Champaign (2010); a post-doctoral fellow at Stanford University (2010–2012); and a visiting professor at UC Berkeley (2016).

Alberto Santini is a senior research fellow with the Institute of Sustainable Plant Protection of the Italian National Research Council. His work focuses on dynamics of arrival and spread of forest pathogens in Europe. His research encompasses pathogens early detection and diagnosis and forest tree species breeding and conservation. He is interested in the analysis and management of the risk of introducing plant pathogens through international trade and the ecological impacts of invasive alien species.

Tommaso Sitzia is associate professor at Padova University, in the Department of Land, Environment, Agriculture and Forestry, Italy. His primary research interests are in the management, use and conservation of forest and semi-natural habitats, subjects he teaches. His research spans a variety of ecosystems, with a particular focus on the Mediterranean and temperate regions of central Europe. He is experienced in planning for protected areas and forest lands, environmental impact assessment, and in drafting regulations and documents to implement nature and forestry related policies. He complements field-based research with modelling and practice.

Carina Sjöholm is Associate Professor in Ethnology at the Department of Service Management and Service Studies, Lund University, Sweden, and has a broad interdisciplinary profile. She has previously researched spatial aspects of popular culture and experience economy and has more specifically centred on social relations, identity formation, commodification of places and materiality, often by applying cultural analysis and ethnographic methods. She is currently focusing on tourist experiences such as gardening, hunting and lifestyle entrepreneurship in rural areas.

Kim B. Stevens is Senior Lecturer in Spatial Epidemiology within the Veterinary Epidemiology, Economics and Public Health (VEEPH) group, Royal Veterinary College, University of London. She completed a part-time PhD (University of London) in spatial epidemiology, focusing on spatial and ecological niche modelling methods and their use as risk-based decision-making tools for informing disease control and surveillance efforts. Previously, she worked for the University of Pretoria as Research Assistant in the Department of Veterinary Physiology (1995) and later as Senior Research Assistant for the Equine Research Centre (1996–2000). After moving to the UK, Kim joined the Epidemiology Division of the RVC as Research Assistant (2002) before being promoted to Assistant Lecturer (2008) and later Lecturer (2016).

Francisco Tirado is Senior Lecturer in Social Psychology at Universitat Autònoma de Barcelona, Spain (UAB), and the director of Barcelona Science and Technology Studies Group (STS-b). His main research interests cohere around four main topics: (1) science and technology studies, (2) biopower and biopolitics, (3) citizen participation in technoscientific controversies and (4) posthuman knowledge.

Peter A. Vesk is a plant ecologist who favours using functional traits and quantitative approaches to finding useful models for vegetation management.

Robert G. Wallace is an evolutionary epidemiologist at the Agroecology and Rural Economics Research Corps based in St Paul, Minnesota. He is author of *Big Farms Make Big Flu* and

Dead Epidemiologists: On the Origins of COVID-19. He is also co-author of *Neoliberal Ebola: Modeling Disease Emergence from Finance to Forest and Farm* and *Clear-Cutting Disease Control: Capital-Led Deforestation, Public Health Austerity, and Vector-Borne Infection*. He has consulted with the Food and Agriculture Organization and the Centers for Disease Control and Prevention.

Rodrick Wallace is a research scientist in the Division of Epidemiology of the New York State Psychiatric Institute at Columbia University. He received an undergraduate degree in mathematics and a PhD in physics from Columbia University, worked a decade as a public interest lobbyist and is a past recipient of Investigator Award in Health Policy Research from the Robert Wood Johnson Foundation. He is the author of numerous books and papers relating to public health and public order.

Bruce Webber is a principal research scientist at CSIRO, Australia's national science agency, Program Director of Processes and Threats Mitigation at the Western Australian Biodiversity Science Institute, and an Adjunct Associate Professor in the School of Biological Sciences at the University of Western Australia. His work focuses on the impacts of global environmental change on biological diversity and the role of plant-ecosystem interactions in shaping community composition. He leads projects that translate novel research findings into improved management solutions to address the biggest conservation challenges at the nexus of landscape change, species invasions and native species resilience.

David Weisberger is a PhD student at the University of Georgia. His research examines both biophysical and socio-economic dimensions of ecological weed management. His current work focuses on how cropping system diversification (via annual and perennial cover crops) alters weed population dynamics through resource competition and invertebrate seed predation. His associated sociological research uses participatory interviews with farmer, university and private industry stakeholders to explore narratives and conflicts around ecological weed management in the southeastern US.

Tina Westerlund is a gardener and Senior Lecturer at the Department of Conservation, University of Gothenburg, Sweden. In her research, she focuses on knowledge and skills in the craft of horticulture and how this knowledge can be analysed and communicated. She has specialised on this in relation to plant propagation, a knowledge needed both in the management and preservation of heritage gardens and plants. As a director of the Craft Laboratory, a national centre for craft in conservation, she works on building bridges between the academy and heritage institutions in Sweden.

Meike Wolf is a medical anthropologist interested in multispecies approaches. She explores how biomedical knowledge, technologies and practices are employed to shape and modify bodies and environments and to intervene in future scenarios. Building upon ethnographic fieldwork in Germany, the UK and France, her research focuses on the mutual entrapment of human and microbial life. As Assistant Professor at the University of Frankfurt, she has led projects on pandemic influenza preparedness and invasive tiger mosquitoes in Europe.

INTRODUCTION

Spatial tensions and divides in grappling with 'invasive life'

Kezia Barker and Robert A. Francis

'Making life safe is not . . . straightforward' (Maye and Chan, chapter 15, 7). Neither is defining exactly *what* the beguilingly simple terms 'biosecurity' and 'invasive species' actually entail. Biosecurity alone can be categorised by the threat to which measures are applied (e.g. 'microbial biosecurity'), the object it aims to protect ('animal biosecurity', 'plant biosecurity'), what it aims to achieve ('good biosecurity'), the action itself ('biosecuring health', 'to biosecure the nation'), the stage or site of intervention ('border biosecurity', 'farm or laboratory biosecurity'), or the practices entailed ('surveillant biosecurity'). As a set of governing frameworks, policies and practices, biosecurity extends from anticipatory practices of risk assessment (Mumford and Burgman, chapter 17) and preparedness (Samimian-Darash and Lev, chapter 19), to methods of surveillance (Corcoran and Hamilton, chapter 16; Stevens, chapter 6), to modes of emergency management (Tirado et al., chapter 18), to forms of enforcement. It refers simultaneously to both the mundane and the extraordinary: from precautions such as handwashing to the policing of national borders, from appropriate paper trails to the emergency culling of animal populations. What this semantic diversity and elasticity reveals is that the term 'biosecurity' has been affixed to a myriad of complex socio-technical assemblages and contested processes through which biological 'threats' are precipitated, mediated, made visible, interpreted, politicised and brought into the realm of significance by social, cultural, economic and political factors (Wilkinson et al., 2011).

Within these variations in meaning and application, concern for the protection of indigenous biota, agricultural assemblages and human health – and fears over bioterrorism and laboratory biosafety – mark national biosecurity regimes to a greater or lesser extent. These sectors do not operate in a void but coexist with multiple domains, things and intentions. As Hinchliffe et al. (2013) remind us, biosecurity regimes at an operational level are messy assemblages, as disease prevention is just one part of the entangled logics and practices that produce wider goods, such as food and capital.

Alongside this messiness, biosecurity can produce a range of contradictions and contrary effects. It can become a mode by which free trade and market penetration are enabled, as well as being seen as a form of trade protectionism (Maye and Chan, chapter 15). It can imply rigid categorical differences between native, non-native and invasive species, benign and disruptive, while operating in practice with subtleties our language cannot adequately convey (Fall, chapter 2; Rotherham, chapter 11). It can be mobilised to encourage economies of scale in animal husbandry that perversely increase biosecurity risk, removing natural heterogeneity and 'fire-breaks' (Wallace et al., chapter 12). It may be focused on protecting non-native agricultural eco-systems from non-native species invasions in colonial contexts where invasion science is a deeply painful term (Lambert and Mark-Shadbolt, chapter 3). It can rest on a snapshot of an ecological community that, with climate change, may soon become confined to the past (Rotherham, chapter 11). By drawing money and attention away from public health infrastructure and divert-ing public health surveillance systems (Parry, 2012), it can produce outcomes that exacerbate health and economic inequalities and increase disease risk for the most vulnerable and precarious lives (Sparke and Anguelov, 2012).

The academic sectors contributing to understandings of biosecurity and invasive species management are equally heterogeneous. For this collection alone, invasion ecologists, indig-enous scholars, geographers, anthropologists and epidemiologists, among others, have brought different perspectives to bear on the conceptualisation, research and critical response to biosecu-rity and invasive species management. But this is by no means a wholly collegiate enterprise. In practice, invasion ecologists may reject the language of biosecurity as too freighted with nationalisms. Critical biosecurity scholars may reject the language of invasion as overemphasising the production of disease as external to the systems biosecurity purportedly protects (Hinchliffe et al., 2013). Explorations of alternative discursive frameworks and recombinant ecologies can be shut out of academic space through accusations of denialism (Fall, chapter 2). And while biosecurity applies across issue areas from the microbial to the macro, from animals to plants, the development of theory in the social sciences has been driven by – and limited to – attention to global health, zoonoses and vector-borne diseases.

In producing this collection, we have come up against discursive barriers and chasms and have reproduced silences, omissions and assumptions. In places you will find incongruences between the theoretical premises advocated across different chapters, while a number reach across this divide and consider physical processes alongside management strategies. Further ten-sions have arisen in taking on both biosecurity and invasive species management in a way that implies that talking from one allows us to talk for both. Biosecurity and invasive species manage-ment have a history and a geography that touches and overlaps in times and places yet remains distinct in others. They are both more than just, and sometimes not at all, each other, and we risk limiting one to the other.

Did we take on too much 'mess' in attempting to consider both biosecurity and invasive species in all their guises from the micro to the macro, and from the sometimes-incongruent perspectives of the biological and critical social sciences, in one handbook? Epidemics and inva-sive events act like tracer dye in the network of global connectivity, demonstrating connections through circulations of biological, technological, cultural and political matter and highlighting health and economic inequalities. We can only hope that this collection acts like tracer dye in the network of meanings, issues and understandings of biosecurity and invasive species, highlighting connections but also incommensurability and contradiction. Gathered together, the book itself becomes a sort of meeting place where tensions become visible in a way that may yet draw our gaze across these disciplinary divides, widening the field of concern within all subdisciplines, posing alternative problem framings and challenging our theoretical silos and echo chambers.

Biosecurities' spatialities: circulation, exclusion and resilience

[B]oundary marking is far more complex cartographically than the traces it leaves on a map. Indeed, the techniques most central to it are not located at geographic borders at all.

(Law and Mol, 2008, 136)

Space – its differentiation through physical and conceptual borders and qualities of fixity or movement – has been central to early theoretical work seeking to understand biosecurity threats and responses. To paraphrase Enticott (2008, 433), resolving biosecurity problems 'has always been a matter of geography'. But exactly what spatial imagery comes to dominate imaginations of disease, invasive species and biosecurity has profound implications for understandings and practice. If important premises about the nature of security are embedded in the different spatial depictions of biosecurity (Rappert, 2009), these have material and political effects. This 'matter of geography' is neither simple nor resolved.

In the following, we draw out three ways in which the collection contributes to or opens up questions of the spatial tensions within biosecurity and invasive species management. We attend to circulations overlooked in traditional narratives, as industrial agriculture, invasion science and biosecurity regimes are themselves shown to be 'in motion' and imposed on other lifeworlds. This moves us from attention to invasive species to invasive economic production systems, from colonial perceptions of the diseased Other to invasive science as a colonial practice, and from knowledges of circulation to the circulation of knowledges. Finally, we point to alternative spatial concerns for understanding biosecurity as an issue not of convergence but of emergence, including attention to spatial inequalities and forms of cohabitation.

1 Species-space associations within 'fluid' terrains

The vision of ecological life that underpins much of conservation practice and by extension the understanding and management of invasive species seems to rest, as Hodgetts and Lorimer (chapter 20, 19) remind us, on a spatially fixed compositional logic: 'the idea that different locations each have ecological communities that are "composed" of a particular set of species . . . is the basis of the categories "native" (to a place), "non-native", and thus also "invasive"'. If a fixed set of species defines the true essence of place, the logical extension is that arrivals from outside the boundaries of this tight coupling of species and place threaten both health and identity.

Across the chapters in part 2, taking in temperate and tropical forest, island, marine and coastal and freshwater ecosystems, this compositional logic could be seen as an underpinning assumption. However, it is troubled in explicit and implicit ways. We learn, for example, that island systems serve as 'natural laboratories' for invasion studies (Guo, chapter 8). Boundary-dependent definitions of native, non-native and endemic species mean that, relative to studies on continental or mainland areas, 'non-native species can be better defined on islands due to their clear boundaries' (Guo, chapter 8, 3). Yet these ideal-type locations are losing their boundedness, with accelerated human traffic linking mainland and islands. Are they becoming more like their mainland counterparts and, it could be argued, more appropriate 'laboratories' for invasion studies?

Finding this fixed and boundary-dependent species-location coupling is also increasingly hard, Rotherham (chapter 11) points out, in contexts such as Britain where recombinant ecologies dominate in a fragmented landscape. This is creating ecological communities composed of species that have never coexisted in the same place before: 'Following centuries of fragmentation and transformation, many biological systems are essentially habitat patches in a sea of agroindustrial farming, industrial extraction and processing or urban sprawl' (Rotherham, chapter 11, 8). Rotherham draws on fluid metaphors to describe these ecological systems which 'have always fluxed as they ebb and flow' with changing environmental drivers and human-induced changes (Rotherham, chapter 11, 11).

This leads us to the fluid terrains historically overlooked in this traditional spatially fixed compositional narrative. As Francis and Chadwick (chapter 10) discuss, the fluid landscape characteristics of freshwater and marine environments themselves facilitate invasions through hydrological connectivity. Yet spatial metaphors are again turned on their heads as we learn that in marine and coastal ecosystems, it is *static* infrastructure, from shoreline modifications to offshore installations, which acts as stepping-stones, producing ecological connectivity that facilitates invasive alien species dispersal (Giesler and Cottier-Cook, chapter 9).

These fluid environments offer a metaphor for looking back at other terrains considered in this section. What does it mean to think of cities (Wolf, chapter 13), industrial agricultural environments (Wallace et al., chapter 12) or gardens (Saltzman et al., chapter 14) as 'fluid' terrains? Rewilding is explored as one avenue for envisioning a 'fluid' management approach to socio-natural difference and change, tending towards open-ended forms of experimentation and enabling connections in the landscape (Hodgetts and Lorimer, chapter 20). Finally, even the fixed identity of species might be better thought of in 'fluid' terms, with 'phenotypic plasticity and rapid evolution being key mediators of invasiveness' (Sitzia et al., chapter 7, pr; see also Mumford and Burgman, chapter 17). Together, this chips away at a simplified picture of predetermined individual species on the move, exploiting pathways and disrupting relations between otherwise static species-spaces associations.

2 Biosecurity's dangerous circulations

Next, circulation as a defining aspect of species' invasiveness and the basis of the biosecurity threat is examined and challenged. Following decades of research into functional traits likely to confer invasive ability, we learn from Palma et al. (chapter 1) that invasiveness is not an inherent property embedded in a species itself and stemming from particular biological characteristics. Instead, invasiveness has come to be understood as emergent, achieved by species' traits conferring with the specificities of the pathway on offer and the opportunities in the receiving environment. For example, Palma et al. (chapter 1) point to the pathway provided by the accidental introduction of seeds on clothing, which could favour traits of dormancy, small size and long attachment, or, for intentional introductions, traits desirable to humans including flowers, drought resistance or fast growth.

The stories written between species and pathway, the traveller and the journey, their contextual relations and contingent histories, begin to sort the benign arrival from the successful invader. The routes and opportunities for illicit species-pathway couplings are numerous, though often reduced to the catch-all of 'international trade and travel'. Opening this up to glimpse inside, we are told wild hybrid stories of boundary transgression and radical interconnectivity: stories of garden escapes (Saltzman et al., chapter 14), monastic aquaculture (Francis and Chadwick, chapter 10), colonial botanical gardens (Sitzia et al., chapter 7) and lucky bamboo moving tiger mosquitos through global city networks (Wolf, chapter 13). These relational stories challenge notions of human exceptionalism by emphasising the part played by the unruly agency and vitality of species which, so often regarded as an asset, becomes a problem that surprises when performed in excess (Saltzman, chapter 14; McKiernan et al., chapter 5). Invasive and non-native species are thus a living archaeology of human-non-human relations and co-dependent movements, exemplified by kiore, the Polynesian rat (*Rattus exulans*), who tells a story of Māori migrations to Aotearoa many hundreds of years ago (Lambert and Mark-Shadbolt, chapter 3).

Movement imprints on species identity, and not all movement is equal. The distinction between 'natural' and 'anthropic' is, as Fall (chapter 2) highlights, the defining distinction between safe and dangerous biological mobility and 'native' and 'non-native' species (also Rotherham, chapter 11). Circulation becomes loaded with further negative qualities in perceptions of the unnatural speed, scale and extent of species' translations (Fall, chapter 2). These qualities of motion are seen to be produced under the unique conditions brought about by globalisation, as expanding trade networks and rapid urbanisation dissolve 'traditional boundaries' of time, space and the everyday (Maye and Chan, chapter 15). The language of 'intensification' is frequently evoked, from exponentially intensified flows of people, goods and biota (Palma et al., chapter 1) to intensified urbanisation (Wolf, chapter 14) and from intensifying interactions between humans and non-humans to the intensification of processes of animal meat production (Wallace et al., chapter 12).

Standard transmission narratives of the dangerous circulation of undesirables across territories and bodies (Hinchliffe et al., 2013), from infected to disease-free spaces, can reinforce dualistic understanding of what constitutes the 'threat' and the 'threatened'. This produces simplified geographies of blame in the origin stories of biosecurity and invasion events (Maye and Chan, chapter 15). The spatialisation of risk within fears of invasion has a history and a geography, overlapping with fears of the contamination of the Global North by the dangerous 'Other'. As Lambert and Mark-Shadbolt highlight in their consideration of Indigenous biosecurity (chapter 3), this is tied to a colonial narrative of mastery of nature and biological superiority that drove both the circulation of Indigenous Peoples' lifeworld as a 'curiosity' alongside the colonial acclimatisation policy of biotic replacement through introduced species.

This spatial logic of disease as an external threat supports an understanding of biosecurity as the maintenance of static territorial integrity enforceable through procedures and practices – from risk management (Mumford and Burgman, chapter 17) to surveillance (Corcoran and Hamilton, chapter 16) to emergency modalities (Tirado et al., chapter 18) – that purify 'good' circulation from 'bad' circulation (Enticott, 2008). The success of these practices is measured by their ability to limit flow – *in*cursion or *in*vasion – of fully formed unwanted organisms from 'over there' to 'here' across territorial space, with biosecurity breaches a systematic boundary failure. As Maye and Chan (chapter 15) highlight, this impression that biosecurity threats are external to the systems that biosecurity is seeking to protect works to naturalise these systems. Yet, as Lambert and Mark-Shadbolt (chapter 3, 10) highlight, imperial expansion and colonial 'development', first and foremost, is the invasive system that has led to 'a catastrophic loss of biological heritage'. In the parts of the world where the effects of these colonial policies and the legacy of settler colonial resource extraction have been violently imprinted on landscapes, ecologies and cultures, the disciplinary label 'invasion science' is a tragically ironic term (Lambert and Mark-Shadbolt, chapter 3).

In their chapter on industrial agriculture, Wallace et al. (chapter 12) provide an aligned critical reading arguing that invasive species represent a second wave of invasion piggybacking on the first: that of industrial agricultural systems that impose their logic on local ecologies and human and non-human lives. The narrow attention within biosecurity to circulating risks, which demands protocols too costly for smallholders to implement, supports the further imposition of industrial farming as part of the solution. This super-scale 'coupling' between biosecurity and capitalist agricultural logics acts to shut out smallholders and alternate food regimes, compounding the operation of biosecurity as a form of trade protectionism and market penetration by the Global North (Wallace et al., chapter 12). Invasive species and disease therefore provide an opportunity for financialisation, extraction of value and (narrow) circulation of capital through particular conduits.

Our picture of biosecurity has developed from the imposition of boundaries to filter out the flow of unwanted life to the active *facilitation* of desired circulation (such as of capital). Rather than an irritant or impediment to friction-less global trade, travel and contemporary neoliberal life, biosecurity instead emerges as a practice that facilitates and optimises these flows. It does so not simply by removing their risky or negative elements but by operating as a badge of assurance for market penetration, a guard against the arbitrary imposition of trade barriers. Maye and Chan (chapter 15) argue that biosecurity has become crucial to the way in which globalisation and market penetration take place. International standards and lists of disease are applied by nation states to *enable* international trade.

Several chapters consider how intersecting biosecurity knowledges, practices and devices themselves circulate, embedded in heterogeneous materialities, relations and subjectivities. Tirado et al. (chapter 18) outline how emergency management techniques, including protocols, devices and technologies, are characterised by their mobility. They can, in principle, be deployed quickly, with low economic cost, anywhere in the world, regardless of the distinct character of particular settings. Enticott (chapter 4) demonstrates how through the rationalities, materialities and subjectivities of veterinary knowledges and veterinarians, New Zealand's neoliberal approach to cost and responsibility sharing has circulated beyond national boundaries, shaping how disease is understood and managed on a global stage. However, this account of policy mobilities shows how context-dependent biosecurity solutions in distant places are imperfectly translated into different national and regional settings. As Enticott (chapter 4) highlights, while the global mobility of veterinary expertise may be responsible for standardising approaches to governing disease, their local translation may result in a different set of geographies. McKiernan et al. (chapter 5) provide a detailed understanding of how local knowledge is formed and shared and influences wider social networks of people through close attention to landholders' relational learning about invasive plants.

Enabling privileged forms of knowledge to circulate and suppressing others is connected to modes of power (Fall, chapter 2; Lambert and Mark-Shadbolt, chapter 3; Enticott, chapter 4). Within 'invasion science' itself, Fall (chapter 2) outlines the contentious debates over its demarcation from other sciences and the active strategies that seek to curtail the circulation of alternative academic perspectives and knowledges within and across this field, policing the rhetorical boundaries of what constitutes 'sound science' (Bickerstaff and Simmons, 2004). As Wolf (chapter 13) argues, in the biomedical and political context of emerging infectious diseases, specific analytical frameworks determine what can be seen, known or said. The bordering practices that exclude non-dominant forms of knowledge or versions of reality mirror biosecurity attempts to 'wall out' invasive pests and diseases. Biosecurity emerges as a governing knowledge practice 'made in the Global North, . . . embedded and naturalized in northern knowledge practices, and exported to other domains through a mixture of the exceptionalism of western science and a triumphalism of the associated modernism' (Hinchliffe, 2015, 31).

Lambert and Mark-Shadbolt (chapter 3) highlight the circulation or invasion of Western scientific constructions of biosecurity, alongside the localisation or suppression of other forms of knowing and doing relational living. As Indigenous collectives increasingly bring their locally specific knowledges related to protecting the species and ecosystems on which Indigenous communities rely into existing and expanding biosecurity networks, they bring opportunities for efficient, effective and ethical biosecurity systems. Yet Lambert and Mark-Shadbolt urge a remaking of agroenvironmental management through attention to the whole of Indigenous Knowledge of biodiversity and agricultural techniques, not from distinct silos. Critically, this recognises the reassertion of Indigenous Knowledge to be indebted to a long history of Indigenous resistance, supported through acknowledgement of the rights of Indigenous Peoples.

Circulation has emerged as the focus of biosecurity, the basis for negative discourses and a problematic mirage deflecting attention from the contextual emergence of risk. Yet in Saltzman et al.'s (chapter 14, 24) detailed discussion of a surprisingly understudied form of circulation in the practice of sharing plant material between domestic gardeners, we are reminded of the emotional intensities within these supposedly dangerous circulations: 'The everyday practices by which private as well as professional gardeners divide, multiply and share plants are an important reminder that the spreading of species is closely connected to cultural contexts and social networks: practices that set plants in motion'. By attending to why people share – the motivations and drivers behind practices – they remind us that circulating species as biosocial beings hold together qualities of knowledge, love, belonging, remembrance and friendship, which come to be embodied and memorialised within plants and gardens.

3 Invasibility and resilience

Finally, we arrive back at the sites of invasion and disease events. The third of the tripartite relationship that promotes species invasions is strongly linked to ecological and historical context – the vulnerability to or resilience of an environment to invasion (Palma et al., chapter 1). This can relate to the physical features of the invaded community such as climate, topography or environmental and biological diversity. Island ecosystems, for example, are typically seen as vulnerable due to their isolation and naivety (Guo, chapter 8), whereas 'diverse communities are expected to be more resistant to invasion because fewer vacant niches are available' (Palma et al., chapter 1, 16).

Increasingly important in making sense of invasions is the human 'preparation' of landscape (Fall, chapter 2), where the conditions of certain human-modified landscapes favour the establishment and dispersal of invasive species. Palma et al. (chapter 1, 15–16) highlight that 'novel, usually disturbed, conditions created by human activity (e.g. novel niche space)' create opportunities for non-native species to colonise. Bio-*insecurity* can therefore be produced within, through monocultures, excessive use of fertilizers; in the pressured animal bodies in forms of poultry production; and through climatic or other environmental stresses, such as landscape fragmentation and habitat degradation. Freshwater ecosystems, for example, are among the most degraded in the world, with the vast majority impacted by hard engineering and flow modification, pollution, habitat loss and species invasions (Francis and Chadwick, chapter 10). The urban environment has been configured as a particular area of risk, a hospitable environment for undesirables such as invasive species and pathogens (Wolf, chapter 13). Invasive species may be considered passengers, riding into and exploiting these modified environments, or the very drivers of change themselves (Sitzia et al., chapter 7).

Despite this enlarged concern for environmental vulnerability or resilience, this remains a push-or-pull argument that retains attention to movement from the outside in (Hinchliffe, 2015). What is required is a recognition that these are not only sites of convergence but also of emergence. As Stevens (chapter 6, 5) highlights, rather than having a single cause, 'EIDs [emerging infectious diseases] often stem from complex interactions between several microbial, host and environmental factors together creating opportunities for pathogens to evolve into new ecological niches.' Due to the complexity of relationships among multi-host systems, environmental change and human populations, effective control measures require each EID event to be understood under its own merits (Stevens, chapter 6). This demands attention to the wider relations that make disease possible, including changing social and biophysical conditions required for disease emergence and spread.

As Wallace et al. (chapter 12, 17–18) argue, industrial agricultural systems not only produce space for invasive species but produce and 'train' pathogens and weeds themselves as 'integral by-products of operations' that select for their evolution, virulence and persistence through 'mechanisation, simplification, geographical re-ordering and incessant spatiotemporal movement'. This is a profoundly chilling picture of an industrial agriculture that produces its own vulnerability, which it then attempts to control through biosecurity lengths 'gothic in their gore' ('snatch harvesting'; in vitro harvesting; antibiotic overuse; Wallace et al., chapter 12, 6). Biosecurity practices become complicit both in producing viral vulnerability and in the production and training of pathogens and weeds.

Taken together, the chapters in the collection urge us to consider an approach to biosecurity that focuses less on circulation, borders and breach points, on movement in and out. Rather, they advocate attending to the qualities of (relational) space; the complexity of relationships between disease/invasive species, environment and host; and the practices and ecologies that work to produce spaces alongside vulnerability and health (Maye and Chan, chapter 15). This moves from a simplified biosecurity dependent on keeping things out to one that addresses factors that build internal health and resilience. Other impressions of geography that require attention include spatial inequalities produced by and compounding issues of ecological injustice, poverty, malnutrition, inadequate access to basic infrastructure and the breakdown of public health. While space matters in the production of these inequalities, this context is not spatially restricted but dispersed, relational and distributed. Taking this wider relational context into account reorientates our understandings of what biosecurity could mean. A diverse local food and local plant industry, agroecology, high standards of animal welfare, workers' rights including adequate sick pay and addressing socio-economic and health inequalities all emerge as contributing to a more expansive understanding of what 'good biosecurity' could entail.

Structure of collection

The collection draws together contributions from leading scholars, combining reflection on practices and policies with critical analysis of a range of theoretical issues. Our contributors follow biosecurity and invasive species around the disciplines, places and practices where they are found and enacted. They analyse their meanings and implications and the normative and practical challenges to which they give rise. The book is organised into three parts, 'Knowledges', 'Terrains' and 'Practices'. These incorporate chapter contributions from the social and physical sciences, attending to both invasive species issues and wider biosecurity concerns.

Part 1 Knowledges

This part attends to the question of how knowledge about biosecurity and invasive species is produced. As we cannot hope to review the contributions of all the diverse fields that make up this knowledge ecosystem, the six chapters that form the section attend to questions of knowledge conflicts and contestation, emerging understandings, gaps and omissions and research related to the formation or circulation of knowledges.

In chapter 1, Palma, Mabey, Vesk and Catford outline the history and contemporary research on the drive to characterise invasive species. This is significant in highlighting the importance of attending to traits, pathways and receiving environments in understanding the success of invasion. In chapter 2, Fall considers the debate over the definition and categorisation of alien invasive species from a critical social science perspective. She discusses how the demarcation of 'invasion science' as a distinct field of expertise separate from the rest of the life sciences is maintained and continues to be contentious. In chapter 3, Lambert and Mark-Shadbolt consider what constitutes 'best practice' biosecurity when it comes to Indigenous lands and waters. They consider the rights, knowledges and approaches of Indigenous communities that inform, challenge and expand contemporary biosecurity practices and detail the opportunities stemming from the inaugural International Indigenous Biosecurity gathering in New Zealand. In chapter 4, Enticott analyses the geography of veterinary experts and expertise across the United Kingdom,

Australia and New Zealand. He considers how certain neoliberal approaches to animal disease management, diagnostic techniques and practices, as well as the veterinarians themselves, have been mobilised and circulate globally, and he details the consequences for disease control and biosecurity. In chapter 5, McKiernan, Gill and Atchison provide a detailed understanding of how local knowledge is formed and shared and influences wider social networks of people through close attention to landholders' relational learning about a specific invasive plant, African lovegrass (*Eragrostis curvula*), on the southeast coast of Australia, in conditions of uncertainty and lack of effective formal institutional knowledge. This reveals how land managers are developing the capacity to manage and live with changing and unpredictable conditions. Finally, in chapter 6, Stevens considers what emerging infectious diseases are and why they occur and reviews opportunities for expanding our knowledge of the ecology and evolution of such diseases. This includes the use of internet-based surveillance to provide rapid identification of potential threats, cluster detection and predictive spatial modelling and molecular genome sequencing.

Part 2 Terrains

This part centres context in examining biosecurity and invasive species issues across a range of terrains, 'thresholds' (Hinchliffe et al., 2017) or environments. The eight chapters that make up the section consider mechanisms of invasion and emergence, contextual vulnerabilities, the impacts of invasion and biosecurity events and many review approaches to prevention and control. The chapters are authored by researchers across the social and ecological sciences and consider environments where human modifications are a contributing, or dominant, mode of ecosystem organisation.

In chapter 7, Sitzia, Campagnaro, Brundu, Faccoli, Santini and Webber focus on emblematic species groups in different forest ecosystems, giving temperate and tropical examples of plant, pathogen and arthropod invasions. These case studies are drawn on to highlight the diversity of factors that mediate and facilitate forest invasions. In chapter 8, Guo considers island ecosystems and their particular vulnerability and uniqueness for the study of invasions, comparing island invasions and conservation with that of mainlands. In chapter 9, Giesler and Cottier-Cook consider the impact of marine non-native species (NNS) on marine and coastal ecosystems, reviewing the ways human activity can move marine organisms around the globe through pathways including ballast water, hull fouling and marine urbanisation. They discuss the potential impact of NNS in recipient environments and the challenges of management and control in marine systems. In chapter 10, Francis and Chadwick provide a background to freshwater ecosystem invasions, arguing that intensive human use and ecological degradation of such ecosystems as well as landscape position and connectivity facilitate freshwater ecosystem invasions. They review the impacts of invasions and summarise key forms of non-native control utilised in rivers and lakes.

In chapter 11, Rotherham introduces the concept and history of recombinant communities and their ecological roles and conservation values. The chapter also considers the ways recombinant communities are perceived, and how they relate to the control of species invasions and biological conservation more broadly. In Chapter 12, Wallace, Liebman, Weisberger, Jonas, Bergmann, Kock and Wallace consider how industrial agricultural environments, including those of livestock and crops, produce biosecurity threats by selecting for pathogens and invasive plants. They argue that the standard narrative of global trade and species movement acknowledges the contribution of capitalism to facilitating pathogens and invasive species only through one form. They further detail how capital monetises the process of controlling those very same diseases that are of its own making. In chapter 13, Wolf draws from cultural anthropology and multispecies approaches to consider the complex interrelation between urban environments and infectious disease risks, critically scrutinising the material networks that entangle humans and non-humans, urban and other environments. Drawing from the example of the tiger mosquito and its associations with a series of vector-borne diseases, she argues that taking the manifold infectious disease entanglements in urban environments seriously means acknowledging that humans are embedded in material networks that exceed our control. Finally for this section, at the scale of the garden, in chapter 14, Saltzman, Sjöholm and Westerlund investigate the everyday practices of gardeners in Sweden to highlight the social and cultural aspects involved in the spread of invasive species. They point to the need to acknowledge the role of human as well as non-human agencies in order to understand the complexity of these interactions.

Part 3 Practices

The final part draws together six chapters that consider the practices involved in biosecurity and invasive species management. These form part of the complex socio-technical assemblages (natures, materialities and social practices) that make up management regimes, national policy and local practices.

Maye and Chan open the section in chapter 15 by considering how we can understand biosecurity at the national scale before profiling and comparing the national biosecurity policy regimes of the UK and China. For Maye and Chan, the political, socio-economic and cultural factors that shape national animal and plant biosecurity regimes represent forms of institutional 'blending', as ideal models of biosecurity play out differently in specific contexts, spaces and places. In chapter 16, Corcoran and Hamilton provide an overview of biosecurity surveillance, considering how surveillance approaches deal with uncertainty in data, before reviewing possible future developments. They argue that as innovative surveillance techniques and technologies are developed and implemented, attention is needed to reduce potential negative impacts on human freedoms. In chapter 17, Mumford and Burgman consider the contributions of and challenges faced in undertaking risk assessments. They explore quantitative and qualitative aspects and spatial and temporal scales in risk assessments, drawing in particular on the processes specified in international plant health standards and by the European Invasive Alien Species Regulation. They point to options for generating more statistically reliable risk assessments, with greater resource investment.

In chapter 18, Tirado, Baleriola and Moya consider the 'emergency modality' in biosecurity responses. They argue that the notion of emergency acquires a concrete and differentiated meaning that redefines the biological threat, according to the form of anticipatory technology applied – whether risk calculation, modern protocols or scenarios. Overall, they highlight the emergency modality as failing to attempt a long-term intervention in the cultural, social, political or economic dimensions associated with outbreaks of infectious diseases. In chapter 19, Samimian-Darash and Lev consider the intersection between biosecurity and the life sciences, outlining the emergence of bioterrorism threats as a political and technical problem for US national security. They consider the formation of the dual-use research concern, using the H5N1 controversy as a case study. They demonstrate that the US government and international bodies have been moving away from the notion that scientists should police dual-use research themselves towards an understanding that specific regulations are needed to protect national security and public health. Finally, closing the collection in chapter 20, Hodgetts and Lorimer consider the implications of 'rewilding' for biosecurity and the control of invasive species across rewilding's different interpretations and guises. They suggest that rewilding tends to diverge from biosecurity through its emphasis on the facilitation of ecological processes over and above the presence or absence of particular species in a specified location by its embrace of open-ended forms of experimentation rather than fixed goals or regimes of control and, in some cases, by enabling rather than restricting connections in the landscape.

In the midst of the multiple discussions, debates, divides and deadlines that criss-cross the story of this edited collection, SARS-CoV-2 emerged. The world was shuttered, the unthinkable enacted, the boundaries between reality and apocalyptic fiction seemed to momentarily collapse. We were reminded that the material world could reject all attempts at preparedness, surveillance and containment and 'object' to our definitional debates. The eternally repeated 'not if, but when', an almost proverbial reminder at the start of all pronouncements and publications about anticipatory biosecurity, became, for a time, 'now'. Not least with this context in mind, we would like to thank all the contributors for their hard work and patience. We hope you find the collection useful and meaningful, as we move from pre- into -post-COVID-19 worlds.

References

Bickerstaff, K. and Simmons, P. (2004) 'The right tool for the job? Modeling, spatial relationships, and styles of scientific practice in the UK foot and mouth crisis', *Environment and Planning D: Society and Space*, vol. 22, no. 3, pp. 393–412. doi: 10.1068/d344t

Enticott, G. (2008) 'The ecological paradox: Social and natural consequences of the geographies of animal health promotion', *Transactions of the Institute of British Geographers*, vol. 33, pp. 433–446. https://doi.org/10.1111/j.1475-5661.2008.00321.x

Hinchliffe, S. (2015) 'More than one world, more than one health: Re-configuring interspecies health', *Social Science & Medicine*, vol. 129, pp. 28–35.

Hinchliffe, S., Allen, J., Lavau, S., Bingham, N. and Carter, S. (2013) 'Biosecurity and the topologies of infected life: From borderlines to borderlands', *Transactions of the Institute of British Geographers*, vol. 38, pp. 531–543. https://doi.org/10.1111/j.1475-5661.2012.00538.x

Hinchliffe, S., Bingham, N., Allen, J. and Carter, S. (2017) *Pathological Lives: Disease, Space and Biopolitics*. Wiley, Oxford, UK.

Law, J. and Mol, A. (2008) 'Globalisation in practice: On the politics of boiling pigswill', *Geoforum*, vol. 39, pp. 133–143.

Parry, B. (2012) 'Domesticating biosurveillance: "Containment" and the politics of bioinformation', *Health & Place*, vol. 18, no. 4, pp. 718–725.

Rappert, B. (2009) 'The definitions, uses, and implications of biosecurity', in B. Rappert and C. Gould (eds.) *Biosecurity: New Security Challenges Series*. Palgrave Macmillan, London. https://doi.org/10.1057/9780230245730_1

Sparke, M. and Anguelov, D. (2012) 'H1N1, globalization and the epidemiology of inequality', *Health & Place*, vol. 18, no. 4, pp. 726–736.

Wilkinson, K., Grant, W.P., Green, L.E., Hunter, S., Jeger, M.J., Lowe, P., Medley, G.F., Mills, P., Phillipson, J., Poppy, G.M. and Waage, J. (2011) 'Infectious diseases of animals and plants: An interdisciplinary approach', *Philosophical Transactions of the Royal Society B*, vol. 366, no. 1573, pp. 1933–1942. https://doi.org/10.1098/rstb.2010.0415

PART 1

Knowledges

1

CHARACTERISING INVASIVE SPECIES

Estibaliz Palma, Abigail L. Mabey, Peter A. Vesk and Jane A. Catford

Prediction of species' invasiveness based on species' characteristics

The movement of biota worldwide has exponentially intensified in recent centuries as a result of increasing globalisation and expanding trade networks (van Kleunen et al., 2015; Seebens et al., 2018). Despite past (Drake et al., 1989) and recent (Pyšek and Richardson, 2010; Díaz et al., 2019) wake-up calls regarding the potential impacts of exotic species on the economy, the environment and human well-being, there is no sign that the number of species becoming established outside of their native range is decelerating (Seebens et al., 2017).

Invasion success is a function of propagule pressure, characteristics of invaders and biotic and abiotic characteristics of the receiving environments (Catford et al., 2009). For decades, invasion ecologists have adopted a functional perspective in order to understand the links between introduced species' traits and invasion success (Darwin, 1859; Baker, 1965; Macarthur and Levins, 1967; Callaway and Ridenour, 2004; Alpert, 2006). These studies investigate questions such as how traits relate to efficient transport into new areas, successful adaptation to novel local environments and high competitive ability within the recipient communities (Figure 1.1). The insight offered by functional assessments to understand invasion drivers increases when trait-based findings are contextualised against the broader background of introduction history and ecological features (Kueffer et al., 2013; Pearson et al., 2018).

Functional traits influence population fitness through their effect on individual growth, reproduction and survival (Violle et al., 2007). As such, differences among traits of co-occurring species translate into fitness inequalities (or lack thereof; i.e. neutrality) and influence the niche differences that are needed for species to coexist (Adler et al., 2007; Holt, 2009; HilleRisLambers et al., 2012). Functional traits are thought to operate in a similar manner across taxa, providing a 'common currency' to understand changes in community assembly resulting from biological invasions (McGill et al., 2006; Shipley, 2010). For this reason, the study of functional traits is expected to generalise the causes and mechanisms through which introduced species overcome barriers to invasion (Blackburn et al., 2011). Such generalisations may enable prediction of future invasion patterns (Moles et al., 2008) and their community-level impacts (Gibson et al., 2012), as well as inform management tools (Reichard and Hamilton, 1997; Pheloung et al., 1999; Barker et al., 2002).

Understanding links between species' invasiveness and their functional traits can inform biosecurity tools in multiple ways. In the particular case of plant species, a focus on functional

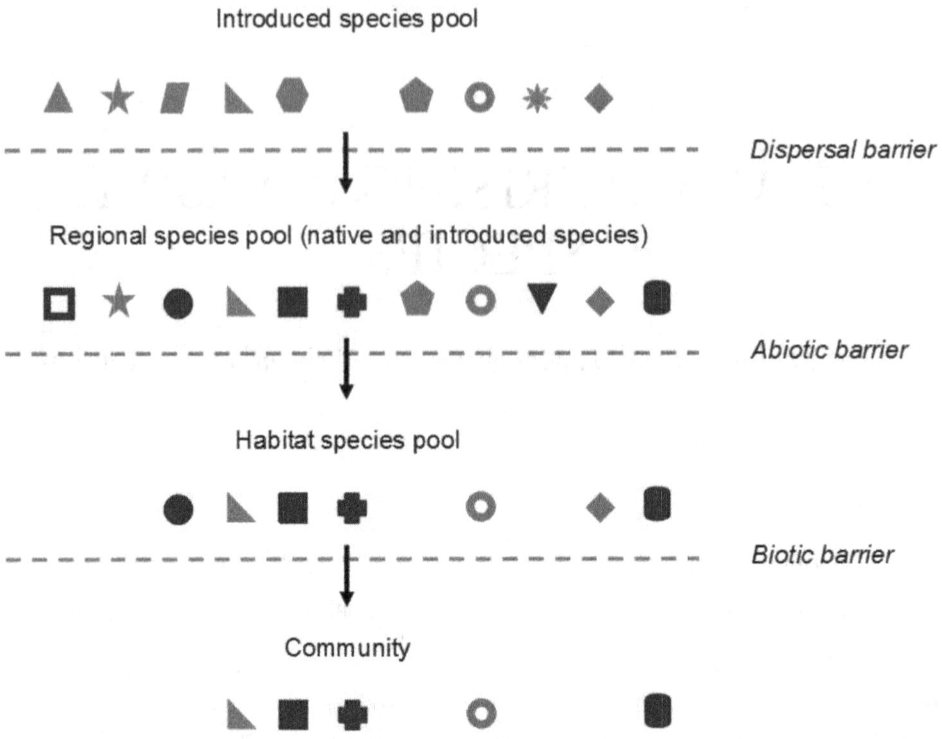

Introduced species pool

Regional species pool (native and introduced species)

Habitat species pool

Community

Figure 1.1 Following community assembly theory, introduced species need to overcome a series of dispersal, abiotic and biotic barriers to successfully invade the local community. Different shapes represent different species. Introduced species are in grey; native species in black.

Source: Adapted from Pearson et al. (2018).

traits offers two main benefits. First, traits can be correlated to expected risk of becoming invasive. Screening procedures can then be targeted to detect individuals with traits linked to high expected risk of becoming invasive across multiple species. Second, identifying the functional forms that cause severe ecological impacts in certain communities can support targeted management tools. For example, efforts can be focused on surveillance and prompt removal of exotic species suspected to become strong competitors within a particular habitat. While the introduction of species for commercial use is expected to continue (Cook and Dias, 2006; Seebens et al., 2015), accurate risk assessment is key to identifying and avoiding those most likely to become harmful. A trait-based biosecurity approach enables risk assessments to be applied to multiple species, and potentially different ecosystems, making it a cost-effective but powerful tool to prevent biological invasions (Yokomizo et al., 2012). An example of this is the Australian weed risk assessment (Pheloung et al., 1999), which screens imported plants for potential invasiveness using a combination of traits, habitat preferences and weed status elsewhere in the world.

In the next few sections, we present the state of the art of understanding on functional traits and invasiveness, with a particular emphasis on plants. We first focus on how invasive species, and their invasiveness, have been defined and quantified across the literature. We then move to how these definitions set limits on and impact the inferences made from trait-based comparative studies of invasiveness. After that, we review the current knowledge of links between commonly used plant functional traits and plant ability to become invasive before acknowledging that there are some context-specific factors that play an important role in shaping the outcome of biological invasions. Next, we reflect on some biosecurity tools currently in place and their success incorporating trait information within their risk assessment criteria. This section presents examples mostly from, but not restricted to, terrestrial plants. We finish the chapter with a brief summary on the prospect and future research needs for trait-based invasion studies.

Defining and studying invasiveness

Defining invasive species and species' invasiveness

There is no widely accepted, universal definition for species invasiveness, with different criteria being used to distinguish invasive from non-invasive species across the literature (Catford et al., 2016; van Kleunen et al., 2018a). However, in general, species invasiveness is defined as the ability of introduced species to survive, reproduce and disperse across multiple habitats (Richardson et al., 2000; Rejmánek, 2011). Occasionally, the impact of the species on the new range is also considered to characterise its invasiveness, especially by practitioners (Davis et al., 2011). In this work, we follow Blackburn et al. (2011) and define invasiveness as the ability to establish a self-sustaining population at a significant distance from the point of introduction.

Even though our working definition is a conceptually clear representation of the invasion process and is commonly accepted, its translation into an operational definition (i.e. which species can be categorised as invasive in the field) is unfortunately not straightforward. Frameworks of invasion often lean on ideas from community assembly theory to reflect that becoming invasive requires species to be capable of overcoming a hierarchy of ecological barriers, including transport, survival, reproduction and dispersal (Figure 1.2). Due to the multiple factors at play,

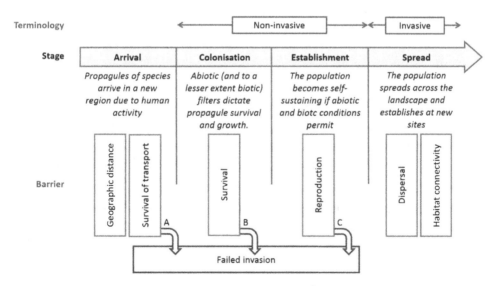

Figure 1.2 Description of the stages of invasion and associated barriers, terminology and reasons for failed invasions. Invasion may fail because: (A) species is not effectively transported beyond its native range; (B) individuals are unable to survive for prolonged periods of time; (C) individuals can survive but the population is not self-sustaining.

Source: Adapted from Theoharides and Dukes (2007) and Blackburn et al. (2011).

it is common that the criteria used to determine species invasiveness (and the metric used to categorise species as invasive) vary across studies; e.g. local relative abundance (Bernard-Verdier and Hulme, 2015) vs. frequency across the landscape (Thuiller et al., 2012). Without a strong focus on the particular demographic components underlying successful invasions, invasiveness studies risk classifying invasive species in a way that does not reflect any particular invasion mechanisms or strategy or that lumps several together (Catford et al., 2016; van Kleunen et al., 2018a). One of the fundamental ideas behind the multiple frameworks that have been proposed so far to summarise the multidimensionality of biological invasions (Catford et al., 2009; Blackburn et al., 2011; Gurevitch et al., 2011) is the fact that invasions are demographic processes. As such, process-based, mechanistic approaches may be better suited to unravel how functional traits modulate introduced species' success (Shea and Chesson, 2002; Drenovsky et al., 2012; Pearson et al., 2018).

Discrete categories of invasiveness

The most common way to quantify invasiveness is to create discrete categories that coincide with the main invasion stages presented on the introduction-naturalisation-invasion continuum (Blackburn et al., 2011). Following this approach, exotic species can be roughly classified as introduced (Blackburn et al., 2011, categories B and C0), naturalised (Blackburn et al., 2011, C1–C3) or invasive (Blackburn et al., 2011, D–E), or the multiple synonyms of these terms that abound across the literature (Richardson et al., 2000) – depending on their ability to overcome only some or all barriers to invasion. However, deciding which stage species sit in at the continuum presents some challenges. First, crossing survival, reproductive and dispersal barriers to move along invasion stages is not irreversible – such a framework of invasion is a simplification of multiple, interacting processes. Second, the specific stage that species occupy along the continuum changes through time, usually in a non-linear fashion (e.g. lag phase; Sakai et al., 2001; Gurevitch et al., 2011). Third, different populations of the same species can be at different points along the continuum, but enough information to evaluate each population is lacking.

Continuous metrics of invasiveness

An alternative approach to quantify species' invasiveness is to use continuous metrics, like species' economic cost or relative abundance. Continuous metrics are more nuanced in that they avoid the need to define boundaries of invasive/non-invasive (i.e. they embrace shades of grey rather than forcing species into categories of black or white) and enable more flexible and powerful analytical tests. They also have the potential to reflect individual demographic processes that promote invasion success – i.e. species growth, reproduction and spread (Richardson et al., 2000; Rejmánek, 2011) – that may be of special interest to practitioners.

Demographic metrics of invasiveness are well suited to reflect the continuous nature of plants' invasive ability at the same time that they isolate particular invasion mechanisms (Gurevitch et al., 2011; van Kleunen et al., 2018a). Plant invasiveness has been quantified through combined population performance measures (Colautti et al., 2014) and individual metrics of population size, frequency of occurrence, local abundance, spread rate, geographic range, niche breadth and impact (Moravcová et al., 2015; Carboni et al., 2016; Catford et al., 2016; McGeoch and Latombe, 2016). These different metrics reflect the fact that plants are able to invade natural vegetation through different mechanisms and likely rely on different traits to do so (Dawson et al., 2012; Lai et al., 2015; van Kleunen et al., 2018a).

Comparative studies of invasiveness

Invasiveness studies may take a different approach depending on the biological question they try to answer (van Kleunen et al., 2010b). Multispecies, trait-based studies of plant invasiveness mostly adopt one of two approaches according to their choice of study species (Figure 1.3). They may investigate differences in traits among introduced species with various degrees of success in becoming invasive across all established species (Figure 1.3 B; e.g. Bucharova and van Kleunen, 2009) or between invasive vs. non-invasive species (Figure 1.3 A1, A3; e.g. Moravcová et al., 2010). Alternatively, they may focus on trait differences between invasive exotic plants and native plants (Figure 1.3 C; e.g. Godoy et al., 2011). Some studies simultaneously present both comparisons (e.g. Dawson et al., 2009; Lavoie et al., 2016; Divíšek et al., 2018).

Studies restricted to exotic plants

Often, trait-based plant invasion studies present the comparison between two discrete groups of exotic species – those considered to be naturalised or invasive in a region and those considered not to be so. Differences in the two groups' invasiveness are then attributed to their functional differences (Kempel et al., 2013; González-Moreno et al., 2014; Novoa et al., 2016b;

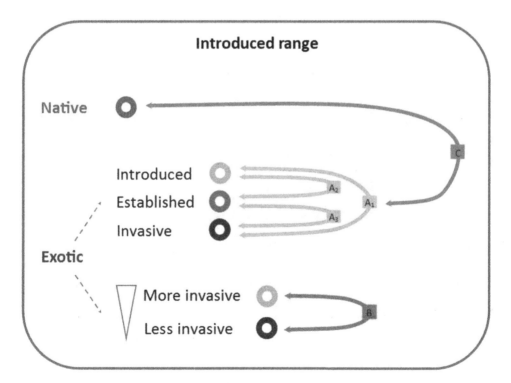

Figure 1.3   Common comparisons found on trait-based invasiveness studies. Comparisons may be made among discrete categories of invasiveness – e.g. invasive vs. non-invasive species (A1, A3), established vs. non-established species (A2) – or using a continuous metric of invasiveness (B). Some studies compare exotic, usually invasive, species and native ones (C).

Source: Adapted from van Kleunen et al. (2010b).

Lembrechts et al., 2018). Sometimes, multiple discrete groups of species with different probabilities to move along the introduction-naturalisation-invasion continuum are presented and compared in a similar fashion (Erskine-Ogden et al., 2016). As pointed out in the previous section, 'Discrete categories of invasiveness', there are some practical challenges associated with categorising species' invasiveness in discrete classes.

Many research studies overcome these issues by relying on regional lists of currently or potentially harmful species provided by government, managers, practitioners or experts (Figure 1.4). Even though these resources can be highly detailed and follow a comprehensive approach to describe individual species' invasiveness (Pyšek et al., 2012; White et al., 2018), trait-based invasiveness studies usually compact this knowledge to a few invasiveness classes into which the study species are lumped. Regardless of the methods used to decide which species belong to which invasiveness categories, from a practical point of view, comparing discrete levels of invasiveness risks overlooking the possibility that different ecological mechanisms enable plant invasions (Pearson et al., 2016) and that those mechanisms may relate to functional traits in different ways (Murphy et al., 2006).

Less common, but increasingly popular, are studies that investigate the correlation between functional traits and invasiveness of exotic plants using continuous metrics of invasiveness, as opposed to categorical classifications. These continuous metrics take diverse forms, including surrogates of geographic range size (Thuiller et al., 2012; Klinerová et al., 2018), regional frequency (Bucharova and van Kleunen, 2009; Speek et al., 2011), dominance (Moravcová et al., 2015) and niche breadth (Thuiller et al., 2012). The rationale behind this type of analysis is still

Figure 1.4   Examples of plant species considered to be highly invasive in Victoria, Australia (top row) and the UK (bottom row). Risk associated with introduced plants in Victoria is evaluated using a combination of five semi-quantitative criteria, including impact, potential distribution, potential for invasion, rate of dispersal and range of susceptible habitats (White et al., 2018). In the UK, plants that represent an environmental risk are listed under the Wildlife and Countryside Act 1981, schedule 9. Clockwise: *Disa bracteata* (by Andrew Massyn), *Cotoneaster pannosus* (by Forest and Kim Starr), *Gladiolus tristis* (by Vahe Martirosyan), *Poa bulbosa* (by Andreas Rockstein), *Heracleum mantegazzianum* (by Jean-Pol Grandmont), *Impatiens glandulifera* (by Maja Dumat), *Rhododendron ponticum* (by Ryan Somma) and *Crassula helmsii* (by Stéphane Delplanque).

Source: All images under Creative Commons licence.

the same as before; differences in species' traits should reflect the species' different degrees of invasiveness. Using continuous demographic quantifications of invasiveness, as opposed to categorical classifications, acknowledges that multiple processes can contribute to invasion and that traits may correlate with each of them in different ways.

Recent research on invasiveness (Dawson et al., 2012; Lai et al., 2015; Carboni et al., 2016; Catford et al., 2016; McGeoch and Latombe, 2016) has found that plants are able to invade native communities through several mechanisms. Failure to acknowledge the multidimensionality of invasiveness may limit our ability to identify traits related to invasion (Rejmánek, 2011; Pyšek et al., 2015; Catford et al., 2016; Catford et al., 2019). For example, different invasion metrics, measured at different spatial scales, can be driven by different processes; studies of invasion at regional scales often quantify invasion as the frequency of occurrence across the landscape (Thuiller et al., 2012) while local scale studies may define invasion ability as relative abundance in a sampled area (Bernard-Verdier and Hulme, 2015). As a result, dispersal traits and introduction features may appear more important to explain the former while resource acquisition traits and species competitive ability may be better suited to explain the latter.

Studies including native plants

Trait-based studies looking at differences between native and successful introduced plants assume that trait differences between the two groups allow introduced species to enter the community through either filling vacant niche space (Davis et al., 2000) or having some competitive advantage over native plants (Godoy and Levine, 2014), even if only temporarily or at particular sites. Comparisons between introduced plants and the native community usually have a mechanistic approach because trait differences are often framed within an ecological context and interrogated for a particular function. This includes linking differences between introduced and native species with resource availability (Daehler, 2003; Leishman et al., 2010; Bachmann et al., 2012; Dawson et al., 2012; Funk et al., 2016; Dyderski and Jagodzinski, 2019), competitive interactions (Dawson et al., 2012; Catford et al., 2019) and demographic performance (Daehler, 2003; Meiners, 2007; Mason et al., 2008; Flores-Moreno and Moles, 2013; Flores-Moreno et al., 2013).

Although comparisons between introduced and native species tell us about opportunities offered by the local community that introduced species can profit from to become invasive, only those introduced species with the right combination of traits will profit from them to become invasive. Hence, to understand invasiveness, the traits of successful (i.e. invasive), compared to non-successful (i.e. non-invasive or less invasive), introduced plants need to be identified (van Kleunen et al., 2018a). Identification of trait differences between native and invasive plants that are not evaluated along with how those traits correlate with the degree of invasiveness across naturalised plants may be subject to several caveats. Differences found between invasive and native plants may reflect differences in the environmental conditions they occupy (e.g. with invasive species being preferentially found in disturbed areas and native species in undisturbed remnants of vegetation; Leishman et al., 2010) or a potential introduction bias towards exotic species with specific traits (Knapp and Kühn, 2012; Driscoll et al., 2014; Pearson et al., 2018) or novel growth forms (Funk et al., 2016).

As with studies that are limited to exotic species, studies comparing invasive and native species may benefit from demographic metrics of success applied to both the native and the exotic plants. Native plants can also show different life-history strategies that allow them to persist in the community through different mechanisms (Elizabeth, 2007; Dawson et al., 2012).

The characteristics of successful invaders

Traits promoting invasiveness of exotic plants

The study of plant traits that correlate with the ability to become invasive began in the second half of the 20th century ('ideal weed' hypothesis; Baker, 1965, 1974) and identified characteristics such as rapid growth, high fecundity and general ecological requirements as important to promote exotic plant success, mostly focusing on disturbed environments. Many other hypotheses have been proposed since then to explain biological invasions from different perspectives, but it is generally agreed among invasion biologists that species' biological attributes are a key component in understanding the invasion process (Richards et al., 2006; Pyšek and Richardson, 2007).

Invasive plants have been consistently reported to have faster leaf economics than non-invasive plants (Matzek, 2012; Gallagher et al., 2015; Erskine-Ogden et al., 2016), as reflected by high specific leaf area. Plants with fast leaf economics are able to produce leaf biomass quickly through high assimilation rates, at the expense of leaf tissue quality and defence against herbivores. This makes it possible for plants to reach maturity and complete their life cycles in a short time (Westoby et al., 2002; Wright et al., 2004; Reich, 2014); however, the correlation between specific leaf area and growth rate varies through ontogeny (Gibert et al., 2016; Falster et al., 2018). High specific leaf area in invasive plants promotes quicker population growth rate (Pyšek and Richardson, 2007; Gallagher et al., 2015) and therefore species' local abundance (Hamilton et al., 2005). It has also been correlated with larger geographic range (Hamilton et al., 2005), likely as a consequence of rapid life cycles yielding frequent dispersal events (van Kleunen et al., 2010a; Kuester et al., 2014; Lavoie et al., 2016).

Regarding seed-related traits, both naturalised and invasive plants have been found to have smaller seeds and higher fecundity than their unsuccessful counterparts (casuals and non-invasive plants) (Hamilton et al., 2005; Dawson et al., 2009; Moravcová et al., 2010; Ordoñez, 2014; Chen et al., 2015; Moravcová et al., 2015). However, this pattern does not seem to be universal (Pyšek and Richardson, 2007; Castro-Díez et al., 2011; Gallagher et al., 2015). Although seed mass and seed size are linked with plant dispersal and fecundity (Moles, 2018), the role of these traits in plant demography, and thus invasion, is complex (Lembrechts et al., 2018) and reflects both a tolerance-fecundity trade-off (Muller-Landau, 2010) and the relative longevity of competing species (Leishman et al., 2000; Moles, 2018).

In general, plant height and plant size have been found to promote plant invasiveness (Pyšek and Richardson, 2007; van Kleunen et al., 2010a; Castro-Díez et al., 2011; Gallagher et al., 2011; Lockwood et al., 2013; Gallagher et al., 2015; Moravcová et al., 2015; Pyšek et al., 2015; Divíšek et al., 2018); but see Goodwin et al. (1999), Hamilton et al. (2005), Kyle and Leishman (2009) and Novoa et al. (2016b). Plant height correlates with multiple strategies for success; taller height may be a surrogate of slower growth rate, long lifespan and high investment in structural tissue or may represent faster growth rate among species with similar life forms (Westoby et al., 2002; Falster et al., 2018). Plant height and plant size are also good indicators of the competitive potential of the species through preferential sequestration of light (Westoby et al., 2002). Positive findings from invasiveness studies likely point to the ability of invasive plants to achieve rapid growth rates under disturbed conditions, where most invasions occur, which likely results from a bias on the types of introduced species, with short-lived plants being preferentially picked and moved around the globe (Pearson et al., 2018).

Plant invasiveness has also been linked to short longevity (Kuester et al., 2014), vegetative reproduction (Pyšek and Richardson, 2007; Rejmánek, 2011; Lockwood et al., 2013; Pyšek et al., 2015; Nunez-Mir et al., 2019), phenotypic plasticity and genetic polymorphism (Sakai et al., 2001; Rejmánek, 2011) and the ability to use multiple long-distance dispersal vectors (Nunez-Mir et al., 2019).

For the last few decades, trait-based invasion research has been a prolific area of science. However, the myriad of studies showing some correlation between plant invasiveness and multiple functional traits likely reflect the fact that the role of traits is highly context-dependent. Traits that are important to understand invasion patterns somewhere do not necessarily promote invasions everywhere. For example, functional traits linked to invasion in riparian corridors might include fragmentation and striking from detached fragments (Catford and Jansson, 2014); however, these same traits might not help in heathlands, where post-fire seed germination might be important (D'Antonio, 2000).

Are invasive plants different from native plants?

Many studies comparing exotic plants with the native community have found no differences between the two groups in their response to nutrient addition, survival rate or biomass production (Dawson et al., 2012), germination rate and seedling survival (Flores-Moreno and Moles, 2013), dispersal distance (Flores-Moreno et al., 2013) or even population dynamics and the general ecological strategies they use (Meiners, 2007; Leishman et al., 2010; Dawson et al., 2012). The studies that have found differences between native and exotic plant traits have generally concluded that these differences allow them to exploit novel, usually disturbed, conditions created by human activity (e.g. novel niche space) through trait values novel to the community (Daehler, 2003; van Kleunen et al., 2010a; Bachmann et al., 2012; Divíšek et al., 2018). Despite the apparent contradiction that these results present, according to community assembly theory, there are reasons to simultaneously expect both similarities and differences in the traits and behaviours of co-occurring exotic and native species. While exotic species may need to be similar enough to the native community to become established in the first place, exotic species with sufficiently distinct functional attributes may be strong competitors under particular conditions (e.g. underused niche space; Alpert et al., 2000). Therefore, the opportunities to become invasive depend on the degree of functional dissimilarity between the introduced plants and the recipient flora (Pyšek and Richardson, 2007; Carboni et al., 2016).

Biotic resistance is the ability of the resident species in a community to reduce the success of exotic species (Levine et al., 2004), which can be realised through interspecific competition, herbivory and lack of mutualists. These mechanisms will negatively impact introduced plants that are more closely related to species in the recipient community since these are expected to be more functionally similar to the native plants (Feng et al., 2019). However, the effect of native-exotic relatedness in invasion varies with spatial scale and invasion stage (Ma et al., 2016).

Closely related species are hypothesised to experience greater interspecific competition due to similar resource requirements (Mitchell et al., 2006). Diverse communities are therefore expected to be more resistant to invasion because fewer vacant niches are available for functionally novel species (Levine et al., 2004). Although some evidence suggests that interspecific competition can reduce establishment and individual performance of invaders (Suwa and Louda, 2012), the idea that limiting similarity is enough to competitively exclude exotic species is not broadly accepted (Levine et al., 2004; Mayfield and Levine, 2010; Kunstler et al., 2016). Positive correlations between some traits (e.g. height, specific leaf area) and invasion have been explained by the expected role of those traits in competition (Feng et al., 2019), suggesting they are important to overcome biotic resistance. In a similar way, closely related plants may undergo attack from the same pathogens and herbivores that consume the native species (Duncan and

Williams, 2002). Or, alternatively, their traits may pre-adapt them to the environment and the biotic interactions present in the invaded range (Duncan and Williams, 2002).

Trait syndromes and invasion strategies

Plant traits do not vary independently. Changes in one trait have been found to correlate with changes in other traits, as part of a coordinated trait economic spectrum (Reich, 2014; Moles, 2018). Well-known trade-offs among traits include the correlations between seed mass and seed number (Moles et al., 2004), between plant height and lifespan (Westoby et al., 2002) and between leaf longevity and photosynthetic rate (Wright et al., 2004). Moreover, particular combinations of traits provide competitive advantage – and therefore opportunities to become invasive – at different points along environmental gradients or at different successional stages (Grime, 1977; Westoby et al., 2002; Wright et al., 2004; Reich, 2014).

Areas with frequent disturbance events and communities at early successional stages are characterised by high availability of resources. Under these circumstances, those species capable of transforming light, nutrients and water into biomass before neighbouring plants do so will be the strongest competitors (Westoby et al., 2002; Falster et al., 2018). Such species will capitalise on resources and quickly occupy the space, both through quick vertical growth of individuals and rapid increase in population size (Davis et al., 2000). Quick resource acquisition relies on fast deployment of leaves, efficient resource acquisition by roots and high photosynthetic rate. These abilities are associated with short life cycles that can be completed between disturbance events. They often correlate with frequent release of large amounts of seeds too (often wind dispersed), with potential to colonise the surrounding areas (Salguero-Gómez, 2017). Species with rapid life cycles that also form seed banks (seed dormancy mechanisms) are further able to skip growth under suboptimal environmental conditions and capitalise on pulses of nutrients and water (Lavorel et al., 1997).

On the opposite side of the habitat templet (Southwood, 1977, 1988), highly stressful environments pose the challenge of survival under restricted light, nutrients or moisture. Under such limiting conditions, stronger competitors rely on a conservative resource acquisition strategy, which is based on slow growth rates, long lifespan and slow tissue turnover (Grime, 1977). Being conservative allows individuals to persist under resource shortages that temporarily prevent individual and population growth and show higher resistance against herbivores. Because they are better adapted to establish and survive during suboptimal environmental conditions, they do not rely on dispersal, at least in the short term (Moles et al., 2004), as much as species with quick resource acquisition (Muller-Landau, 2010). Despite their slower growth rate, in the long run, these plants reach higher stature than early successional species and sequestrate most of the light. As a result, they take over the community as ecological succession progresses. Having large seeds is also advantageous for plants facing stressful conditions, at least during establishment (Leishman et al., 2000; Muller-Landau, 2010; Moles, 2018).

Species with quick resource acquisition have often been found to have high invasion ability (Baker, 1974; Pyšek and Richardson, 2007). This is consistent with the fact that plant invasions have their origin in human-driven introductions, often around highly disturbed habitats (e.g. urban areas). However, distinct functional types may become invasive in different ecological and historical contexts (Lai et al., 2015; Pearson et al., 2017). Evidence is mounting that both acquisitive and conservative strategies can support invasions (Tecco et al., 2013; Funk et al., 2016; Dyderski and Jagodzinski, 2019). Therefore, currently harmless functional types may become invasive in the close future as introduction preferences evolve, e.g. new trade routes and ornamental fashion trends (Martin et al., 2009; Kueffer et al., 2013).

Traits and the temporal dimension of invasion

Even though the invasion process is a continuum, for practical purposes, it is usually divided into discrete stages (Blackburn et al., 2011), each of them characterised by particular barriers and ecological processes. The importance of species traits for invasion varies throughout the invasion process (Dawson et al., 2009) and is relative to the particular stage, or demographic process, that is being examined (Pyšek et al., 2009a).

Overall, species introduction, initial population size and early survival are promoted by similar factors worldwide. Propagule pressure, broad native ranges and economic interest have consistently been found to be good predictors of species' ability to become invasive (Krivanek et al., 2006; Bucharova and van Kleunen, 2009; Feng et al., 2016). Both human desirability – which selects for traits such as fast growth and particular aesthetics (Drew et al., 2010; Kendal et al., 2012) – and seed longevity have been found to increase the probability of being transported and surviving transport (Theoharides and Dukes, 2007). Wide abiotic tolerances and phenotypic plasticity can enhance colonisation success, as they can buffer the abiotic limitations the species may encounter in the introduced range (Higgins and Richardson, 2014).

On the other hand, the particular functional traits that promote survival and spread in the new range is context-dependent (Lloret et al., 2005), becoming more apparent towards later stages of invasion (Pyšek et al., 2009b; Pyšek et al., 2009a). After being introduced, species are met with the challenge of establishing self-sustained populations and spread across the landscape. Competitiveness has been linked with traits such as tall height, fast growth and efficient resource use (Theoharides and Dukes, 2007). Efficient dispersal, high fecundity, phenotypic plasticity and fast generation time are likely to boost quick spread across the landscape (Theoharides and Dukes, 2007).

Acknowledgement of the temporal dynamics of invasions allows researchers to test whether specific functional traits consistently promote invasion or, alternatively, play different (even contrasting) roles for particular demographic processes that underlie species' invasiveness. As an example, small seed size may promote plant dispersal but may hamper competitiveness (Catford et al., 2016; van Kleunen et al., 2018a; Palma et al., in press).

Context

The generalities gained through the functional study of invasions may be offset by the context-specificity of biological invasions. Acknowledgement of the introduction history and ecological conditions of the introduced area is key to understand and predict the path of any invasion (Lloret et al., 2005; Pearson et al., 2017). The successful ecological strategies in any place are going to be bounded by the species that get introduced to that place and the ecological conditions, including abiotic and biotic constraints, they experience after their introduction (Pearson et al., 2018). Clear definitions of the invasion mechanisms under investigation as well as information of the historical and ecological background can benefit trait-based studies. Without this information, trait characteristics of invasive species may just be reflecting an introduction bias towards species with those traits, rather than representing strategies to successfully cope with environmental and biotic constraints. Biosecurity protocols of risk assessment usually include an evaluation of the invasion context. For example, the Australian weed risk assessment uses trait data in conjunction with environmental preferences (Pheloung et al., 1999) to make predictions.

The following subsections briefly touch on which aspects of context-dependency may influence the relationships between functional traits and invasiveness found across invasion events.

Features of the native vs. novel ranges

Both abiotic and biotic conditions in the novel range, and their similarity to the introduced species' native range, will determine the extent of suitable habitat available for the introduced species. The interaction among these elements can either help or hinder the invasion by introduced species (Mitchell et al., 2006). It is logical to assume that species that have been moved across wide geographical areas are most likely to survive and thrive in locations with similar environmental conditions (e.g. climate) to those from which they came, as they possess traits that allow them to survive in these conditions. This logic has been used in risk assessments to predict which areas are likely to be invaded by novel species (Pheloung et al., 1999; Broennimann et al., 2007). However, invasive populations can sometimes occupy areas that are climatically distinct from their native range (Broennimann et al., 2007; Early and Sax, 2014), likely due to new ecological context (e.g. release from enemies) facilitating establishment and/or spread (Nuñez and Medley, 2011).

Spatial patterns in the landscape can also influence the invasion process. Spatio-temporal landscape heterogeneity can create windows of opportunity through dispersal, allowing invaders to move across the landscape (With, 2002), sometimes in spite of their traits. For example, fragmented habitat can create 'nascent loci', where an invasive population is split into several groups which are all dispersing (With, 2002), thereby creating multiple invasion fronts. Roads and rivers can also acts as dispersal corridors (Von Der Lippe and Kowarik, 2007). Likewise, the level of disturbance in the landscape can favour establishment and dispersal of species with a resource-acquisitive strategy, suited to take advantage of resource pulses (Davis et al., 2000).

Minimum residence time

Invasion success, measured as invasion range (Bucharova and van Kleunen, 2009) and dominance (Speek et al., 2011), is higher for species that have been introduced earlier on (Castro et al., 2005). Plant residence time is important to take into context when using traits to predict plant invasions because the species must have been in the environment long enough to either fail or succeed in the invaded range (Richardson and Pyšek, 2006). When predicting rates of spread of invasive alien plants in South Africa, including residence time and potential range increased the explanatory power and affected which factors were significant determinants of invasiveness (Wilson et al., 2007).

Propagule pressure

Introduced species need to be able to overcome demographic and environmental stochasticity early on in the invasion process to eventually become established. The stronger the propagule pressure of a given species (higher number of individuals released or quicker rate at which they are released; Simberloff, 2009), the higher the likelihood for that species to reach a minimum viable population size and eventually become invasive (Krivanek et al., 2006; Martin et al., 2009). If the amount of propagules arriving is not enough to overcome demographic stochasticity and reach a minimum viable population size, it is likely that the invasion will fail, even if the species has a suite of traits that would promote success in the invaded range (Zenni and Nuñez, 2013). Repeated release of propagules can function as a source pool of immigrants, thereby sustaining a population even if the initial propagule size was not sufficient (Lockwood et al., 2005) or the species' set of traits not well suited, thus increasing the genetic and phenotypic diversity of the established populations (Lockwood et al., 2005).

Introduction pathway

Finally, plant species are likely to be introduced non-randomly (Colautti et al., 2006; Martin et al., 2009), and their *introduction pathway* determines which traits are needed for surviving said pathway. For example, accidental introduction of seeds on clothing could favour traits of dormancy, small size and long attachment (Pyšek et al., 2011). In the case of intentional introductions, these species represent a non-random subsection of invasive species that have traits desirable to humans (van Kleunen et al., 2018b), such as aesthetic flowers, drought resistance or fast growth (Drew et al., 2010; Chrobock et al., 2011). International trade has created new introduction pathways across land, sea and air for both intentional and accidental introductions (Hulme, 2009). The type of introduction pathway can affect the traits of the propagules being imported, resulting in propagule bias and therefore creating misleading patterns between invasion success and the traits correlated with invasiveness (Colautti et al., 2006). Certain traits may be selected for during transport and introduction, or they may only be imported from a particular area – for example, along shipping routes (Colautti et al., 2006). In locations with a long history of naturalisation, similar species may be intentionally reintroduced by humans if they provide a benefit. For example, in Hawaii new species of grass genera that provide high-quality fodder (e.g. *Paspalum*) have been regularly introduced over the past 200 years for agricultural benefits (Daehler, 2001).

Functional traits: a tool for biosecurity

Biosecurity includes strategies and policies that aim to identify and prevent the release of invasive species into novel areas. Risk assessments are important biosecurity tools that use current knowledge to estimate the probability of future risk occurring, such as a species becoming invasive (Auer, 2008). Risk assessments can contribute to both pre-border and post-border biosecurity. Pre-border procedures screen taxa proposed for introduction to prevent the entrance of species expected to be harmful. Post-border assessments quantify the invasive potential of taxa already present in the environment to guide management and monitoring programmes (Keese et al., 2014). Risk assessments aim to prevent the release of invasive plants and reduce their associated management costs (Keller et al., 2007; Novoa et al., 2016a) while allowing for intentional introduction of plants for commercial purposes such as forestry, horticulture and agriculture (Nishida et al., 2009).

Risk assessment uses a range of information to make predictions about how likely a species is to become invasive. This can include information on weed status elsewhere (Pheloung et al., 1999; Daehler et al., 2004; Gordon et al., 2012), climate and environmental preferences (Pheloung et al., 1999) and species traits (Pheloung et al., 1999; Andreu and Vilà, 2010; Gordon et al., 2012). Traits usually evaluated in risk assessment include dispersal, reproduction and persistence (Table 1.1). Information of evaluation criteria is then combined and a score assigned to each taxon. Based on the score, taxa are assigned one of three outcomes: accept, reject or recommend further analysis or monitoring (Pheloung et al., 1999). This is the approach used by the Australian weed risk assessment (Pheloung et al., 1999), which is the most widely used approach (Hulme, 2012).

The Australian weed risk assessment protocol has been adopted and modified for use in many countries, including New Zealand (Gordon et al., 2010), Japan (Nishida et al., 2009), China (He et al., 2018) and Hawaii (Daehler et al., 2004). This suggests that the model of weed risk assessment can be applied successfully to multiple habitats (Chong et al., 2011).

After adjustments for New Zealand climate, this risk assessment model was adopted within the New Zealand Biosecurity Act of 1993 (Gordon et al., 2010). More recently, research

Table 1.1 Categories and descriptions used in the Australia Weed Risk Assessment. Categories including species traits in italics.

Category	Description
Domestication/cultivation	Is the species highly domesticated?
	Has the species become naturalised where grown?
	Does the species have weedy races?
Climate and distribution	Species suited to Australian climates
	Quality of climate match data
	Broad climate suitability
	Native or naturalised in regions with extended dry periods
	Does the species have a history of repeated introductions outside its natural range?
Weed elsewhere	Naturalised beyond native range
	Garden/amenity/disturbance weed
	Weed of agriculture/horticulture/forestry
	Environmental weed
	Congeneric weed
Undesirable traits	Produces spines, thorns or burrs
	Allelopathic
	Parasitic
	Unpalatable to grazing animals
	Toxic to animals
	Host for recognised pests and pathogens
	Causes allergies or is otherwise toxic to humans
	Creates a fire hazard in natural ecosystems
	Is a shade tolerant plant at some stage in its life cycle
	Grows on infertile soils
	Climbing or smothering growth habit
	Forms dense thickets
Plant type	Aquatic
	Grass
	Nitrogen-fixing woody plant
	Geophyte
Reproduction	Evidence of substantial reproductive failure in native habitat
	Produces viable seed
	Hybridises naturally
	Self-fertilisation
	Requires specialist pollinators
	Reproduction by vegetative propagation
	Minimum generative time
Dispersal mechanisms	Propagules likely to be dispersed unintentionally by people
	Propagules dispersed intentionally by people
	Propagules likely to disperse as a produce contaminant
	Propagules adapted to wind dispersal
	Propagules buoyant
	Propagules bird dispersed
	Propagules dispersed by other animals (externally)
	Propagules dispersed by other animals (internally)
Persistence attributes	Prolific seed production
	Evidence that a persistent propagule bank is formed
	Well controlled by herbicides
	Tolerates or benefits from mutilation, cultivation or fire
	Effective natural enemies present in Australia

Source: Adapted from Pheloung et al. (1999) and Weber et al. (2009).

institutions supported by the New Zealand government have developed a decision tool to assist with biosecurity enquires in the country. The novelty of this tool, called Biosecure, lies in the fact that the evaluation of the risk is based on the interaction between introduced species traits and the local environment (Barker et al., 2002). For its use in Hawaii, to the Australian risk assessment protocol was added a seven-question binary decision tree for species that required further analysis (Daehler et al., 2004). This additional step increased accuracy by 17.6% when both methods were tested on woody invasive and non-invasive species in central Europe (Křivánek and Pyšek, 2006).

For risk assessment to be successful and applicable, four main criteria must be met. Risk assessment must be transparent, i.e. follow clear criteria for rejection or acceptance of species; it must be based on scientific knowledge, it must minimise the use of subjective decisions and it must be repeatable (Daehler et al., 2004). Additionally, risk assessment procedures should be as cost-effective as possible and practical and quick to perform (Daehler et al., 2004). While risk assessments are limited by the data on which they are based, potential bias from risk assessors and a lack of consideration of hierarchal structures (Hulme, 2012), they provide a method for countries to screen against potential invasive species.

Risk assessments are most developed for terrestrial plants. Multiple attempts have been made to apply the terrestrial plant risk assessments to other taxa (Box 1.1). Risk assessments for aquatic plants have shown mixed results (Gordon and Gantz, 2011; Gordon et al., 2012) (Gordon and Gantz, 2011; Gordon et al., 2012). When the New Zealand weed risk assessment was modified for predicting aquatic plants imported to the US, major invaders were distinguished from minor and non-invaders with 91% accuracy (Gordon et al., 2012). When the Australian weed risk assessment was used, all the major invaders were identified. However, it also predicted that 83% of the non-invaders would be invasive (Gordon and Gantz, 2011). Such false positives could discourage use of the risk assessment by leading to some undesired economic impacts, e.g. the banning of some aquatic plants that would be safe to sell.

Applying weed risk assessment models to taxa other than plants has, unfortunately, been less successful. Emiljanowicz et al. (2017) investigated whether the Australian weed risk assessment could be used to predict invasiveness of insect pests, which can cause major commercial losses in agriculture and horticulture. While there was some overlap in the traits of invasive plants and insects (Table 1.2), it was found that traits related to social behaviour in insects strongly correlated with invasive potential (Emiljanowicz et al., 2017).

Table 1.2 Summary of trait categories used in risk assessment across taxa.

Taxa	Trait category					Reference
	Undesirable	Reproduction	Dispersal	Persistence	Diet	
Terrestrial plants	x	x	x	x		Pheloung et al. (1999)
Aquatic plants (seaweed)		x	x	x		Nyberg and Wallentinus (2005)
Vertebrate animals (fish)	x	x	x	x	x	Copp et al. (2005)
Vertebrate animals (mammals, birds, amphibians, reptiles)	x		x		x	Bomford (2006)
Invertebrate animals (marine invertebrates)	x	x	x			Drolet et al. (2016)

Box 1.1 Beyond weed risk assessment

Major invasive pests across the globe include not only terrestrial plants but also vertebrate and invertebrate animals and aquatic plants (Schaffelke and Hewitt, 2007; Harper and Bunbury, 2015; Russo, 2016). Risk assessments for these taxa often rely on criteria similar to that in the Australian weed risk assessment – e.g. invasiveness elsewhere, climate matching and biological traits (Table 1.1) – to provide a ranking of potential invasiveness. The traits used for different taxa vary as a reflection of the rules of community assembly that relate to the invasion potential of the group examined. The Australian weed risk assessment uses attributes that are grouped into four main categories (Table 1.1), namely undesirable traits, reproduction traits, dispersal traits and persistence traits (Pheloung et al., 1999). As reflected in Table 1.2, these categories may not always be relevant for risk assessment of other taxa.

Dispersal traits are deemed important across taxa, likely as the ability to spread across the landscape is key to determine species' invasiveness (Blackburn et al., 2011; Catford et al., 2016). Undesirable traits (e.g. carrying diseases or poison), reproduction traits (e.g. vegetative propagation) and persistence traits (e.g. propagule bank) are also used to evaluate invasiveness and assess potential species' risk in many groups. Risk assessment of invertebrates may require behavioural traits (e.g. social behaviour) to be considered. In extreme cases, invasive ants have formed super colonies, with many interconnected nests and multiple queens, which allows the ants to divert more resources to population growth and interspecific competition rather than intraspecific aggression (Tsutsui and Suarez, 2003). For vertebrates, consideration of their diet may be key (Copp et al., 2005). Feeding guild is particularly important for predator introductions since they can have severe impacts on native communities (Aloo et al., 2017).

While the Australian weed risk assessment is a valuable tool for predicting invasiveness of terrestrial plants, modifications that account for differences in behaviour, diet and other attributes relevant for the taxa under evaluation can increase assessment success.

Conclusion

Since Baker developed the theory of the ideal weed (Baker, 1965, 1974), ecologists have strived to generalise rules of invasion through species functional traits. Identifying the traits that promote invasiveness of non-native species not only has the potential to advance understanding of community ecology but also informs biosecurity and management tools aimed at biodiversity conservation (van Kleunen et al., 2010a). As research continued, however, it became clear that there was no simple answer to determine which traits conferred invasion success. Traits that confer invasive success are dependent upon context, such as environmental constraints (Broennimann et al., 2007), residence time (Castro et al., 2005) and the biological features of the invaded community (Lonsdale, 1999). As a consequence, future trait-based research of invasions could highly benefit from more increased mechanistic approaches (Kueffer et al., 2013). This includes the acknowledgement that population dynamics are key to understand invasion drivers and that invasiveness is a multidimensional concept, with different traits likely being associated with different forms of invasiveness (Catford et al., 2016). Using traits to predict potential species invasiveness will require an understanding of how context applies, which traits are appropriate for the mechanism of invasiveness and consistent accumulation of data that can be compared across replicate studies of various context.

References

Adler, P.B., HilleRisLambers, J. and Levine, J.M. (2007) 'A niche for neutrality', *Ecology Letters*, vol. 10, pp. 95–104.

Aloo, P.A., Njiru, J., Balirwa, J. and Nyamweya, C. (2017) 'Impacts of Nile Perch, *Lates niloticus*, introduction on the ecology, economy and conservation of Lake Victoria, East Africa', *Lakes & Reservoirs: Research & Management*, vol. 22, pp. 320–333.

Alpert, P. (2006) 'The advantages and disadvantages of being introduced', *Biological Invasions*, vol. 8, pp. 1523–1534.

Alpert, P., Bone, E. and Holzapfel, C. (2000) 'Invasiveness, invasibility and the role of environmental stress in the spread of non-native plants', *Perspectives in Plant Ecology, Evolution and Systematics*, vol. 3, pp. 52–66.

Andreu, J. and Vilà, M. (2010) 'Risk analysis of potential invasive plants in Spain', *Journal for Nature Conservation*, vol. 18, pp. 34–44.

Auer, C. (2008) 'Ecological risk assessment and regulation for genetically-modified ornamental plants', *Critical Reviews in Plant Sciences*, vol. 27, pp. 255–271.

Bachmann, D., Both, S., Bruelheide, H., Ding, B.-Y., Gao, M., Härdtle, W., Scherer-Lorenzen, M. and Erfmeier, A. (2012) 'Functional trait similarity of native and invasive herb species in subtropical China: Environment-specific differences are the key', *Environmental and Experimental Botany*, vol. 83, pp. 82–92.

Baker, H.G. (1965) 'Characteristics and modes of origin of weeds', in H.G. Baker and G.L. Stebbins (eds.) *The Genetics of Colonizing Species*. Academic Press, New York.

Baker, H.G. (1974) 'The evolution of weeds', *Annual Review of Ecology and Systematics*, vol. 5, pp. 1–24.

Barker, G.M., Stephens, A., Hunter, C., Rutledge, D., Harris, R.J., Lariviere, M.-C. and Gough, J.D. (2002) 'Biosecure: A model for analysis of biosecurity risk profiles', in S.L. Goldson and D.M. Suckling (eds.) *Defending the Green Oasis: New Zealand Biosecurity and Science*. New Zealand Plant Protection Society Inc.

Bernard-Verdier, M. and Hulme, P.E. (2015) 'Alien and native plant species play different roles in plant community structure', *Journal of Ecology*, vol. 103, pp. 143–152.

Blackburn, T.M., Pyšek, P., Bacher, S., Carlton, J.T., Duncan, R.P., Jarošík, V., Wilson, J.R.U. and Richardson, D.M. (2011) 'A proposed unified framework for biological invasions', *Trends in Ecology & Evolution*, vol. 26, pp. 333–339.

Bomford, M. (2006) *Risk Assessment for the Establishment of Exotic Vertebrates in Australia: Recalibration and Refinement of Models*. Department of the Environment and Heritage, Canberra, Australia.

Broennimann, O., Treier, U.A., Müller-Schärer, H., Thuiller, W., Peterson, A.T. and Guisan, A. (2007) 'Evidence of climatic niche shift during biological invasion', *Ecology Letters*, vol. 10, pp. 701–709.

Bucharova, A. and van Kleunen, M. (2009) 'Introduction history and species characteristics partly explain naturalization success of North American woody species in Europe', *Journal of Ecology*, vol. 97, pp. 230–238.

Callaway, R.M. and Ridenour, W.M. (2004) 'Novel weapons: Invasive success and the evolution of increased competitive ability', *Frontiers in Ecology and the Environment*, vol. 2, pp. 436–443.

Carboni, M., Münkemüller, T., Lavergne, S., Choler, P., Borgy, B., Violle, C., Essl, F., Roquet, C., Munoz, F., DivGrass, C. and Thuiller, W. (2016) 'What it takes to invade grassland ecosystems: Traits, introduction history and filtering processes', *Ecology Letters*, vol. 19, pp. 219–229.

Castro, S.A., Figueroa, J.A., Muñoz-Schick, M. and Jaksic, F.M. (2005) 'Minimum residence time, biogeographical origin, and life cycle as determinants of the geographical extent of naturalized plants in continental Chile', *Diversity and Distributions*, vol. 11, pp. 183–191.

Castro-Díez, P., Godoy, O., Saldaña, A. and Richardson, D.M. (2011) 'Predicting invasiveness of Australian acacias on the basis of their native climatic affinities, life history traits and human use', *Diversity and Distributions*, vol. 17, pp. 934–945.

Catford, J.A., Baumgartner, J.B., Vesk, P.A., White, M.D., Buckley, Y.M. and McCarthy, M.A. (2016) 'Disentangling the four demographic dimensions of species invasiveness', *Journal of Ecology*, vol. 104, pp. 1745–1758.

Catford, J.A. and Jansson, R. (2014) 'Drowned, buried and carried away: Effects of plant traits on the distribution of native and alien species in riparian ecosystems', *New Phytologist*, vol. 204, pp. 19–36.

Catford, J.A., Jansson, R. and Nilsson, C. (2009) 'Reducing redundancy in invasion ecology by integrating hypotheses into a single theoretical framework', *Diversity and Distributions*, vol. 15, pp. 22–40.

Catford, J.A., Smith, A.L., Wragg, P.D., Clark, A.T., Kosmala, M., Cavender-Bares, J., Reich, P.B. and Tilman, D. (2019) 'Traits linked with species invasiveness and community invasibility vary with time, stage and indicator of invasion in a long-term grassland experiment', *Ecology Letters*, vol. 22, pp. 593–604.

Chen, L., Peng, S. and Yang, B. (2015) 'Predicting alien herb invasion with machine learning models: Biogeographical and life-history traits both matter', *Biological Invasions*, vol. 17, pp. 2187–2198.

Chong, K.Y., Corlett, R.T., Yeo, D.C.J. and Tan, H.T.W. (2011) 'Towards a global database of weed risk assessments: A test of transferability for the tropics', *Biological Invasions*, vol. 13, pp. 1571–1577.

Chrobock, T., Kempel, A., Fischer, M. and van Kleunen, M. (2011) 'Introduction bias: Cultivated alien plant species germinate faster and more abundantly than native species in Switzerland', *Basic and Applied Ecology*, vol. 12, pp. 244–250.

Colautti, R.I., Grigorovich, I.A. and MacIsaac, H.J. (2006) 'Propagule pressure: A null model for biological invasions', *Biological Invasions*, vol. 8, pp. 1023–1037.

Colautti, R.I., Parker, J.D., Cadotte, M.W., Pyšek, P., Brown, C.S., Sax, D.F. and Richardson, D.M. (2014) 'Quantifying the invasiveness of species', *NeoBiota*, vol. 21, pp. 7–27.

Cook, G.D. and Dias, L. (2006) 'Turner review No. 12: It was no accident: Deliberate plant introductions by Australian government agencies during the 20th century', *Australian Journal of Botany*, vol. 54, pp. 601–625.

Copp, G., Garthwaite, R. and Gozlan, R. (2005) 'Risk identification and assessment of non-native freshwater fishes: A summary of concepts and perspectives on protocols for the UK', *Journal of Applied Ichthyology*, vol. 21, pp. 371–373.

Daehler, C.C. (2001) 'Darwin's naturalization hypothesis revisited', *The American Naturalist*, vol. 158, pp. 324–330.

Daehler, C.C. (2003) 'Performance comparisons of co-occurring native and alien invasive plants: Implications for conservation and restoration', *Annual Review of Ecology, Evolution & Systematics*, vol. 34, pp. 183–211.

Daehler, C.C., Denslow, J.S., Ansari, S. and Kuo, H.-C. (2004) 'A risk-assessment system for screening out invasive pest plants from Hawaii and other Pacific Islands', *Conservation Biology*, vol. 18, pp. 360–368.

D'Antonio, C.M. (2000) 'Fire, plant invasions, and global change', in H.A. Mooney and R.J. Hobbs (eds.) *Invasive Species in a Changing World*. Island Press, Washington.

Darwin, C. (1859) *On the Origin of Species by Means of Natural Selection, or the Preservation of Favoured Races in the Struggle for Life*. Murray, London.

Davis, M.A., Chew, M.K., Hobbs, R.J., Lugo, A.E., Ewel, J.J., Vermeij, G.J., Brown, J.H., Rosenzweig, M.L., Gardener, M.R., Carroll, S.P., Thompson, K., Pickett, S.T.A., Stromberg, J.C., Del Tredici, P., Suding, K.N., Ehrenfeld, J.G., Grime, J.P., Mascaro, J. and Briggs, J.C. (2011) 'Don't judge species on their origins', *Nature*, vol. 474, pp. 153–154.

Davis, M.A., Grime, J.P. and Thompson, K. (2000) 'Fluctuating resources in plant communities: A general theory of invasibility', *Journal of Ecology*, vol. 88, pp. 528–534.

Dawson, W., Burslem, D.F.R.P. and Hulme, P.E. (2009) 'Factors explaining alien plant invasion success in a tropical ecosystem differ at each stage of invasion', *Journal of Ecology*, vol. 97, pp. 657–665.

Dawson, W., Fischer, M. and van Kleunen, M. (2012) 'Common and rare plant species respond differently to fertilisation and competition, whether they are alien or native', *Ecology Letters*, vol. 15, pp. 873–880.

Díaz, S., Settele, J. and Brondízio, E. (2019) 'Summary for policymakers of the global assessment report on biodiversity and ecosystem services of the Intergovernmental Science-Policy Platform on Biodiversity and Ecosystem Services', www.ipbes.net, accessed 2 December 2019.

Divíšek, J., Chytrý, M., Beckage, B., Gotelli, N.J., Lososová, Z., Pyšek, P., Richardson, D.M. and Molofsky, J. (2018) 'Similarity of introduced plant species to native ones facilitates naturalization, but differences enhance invasion success', *Nature Communications*, vol. 9, p. 4631.

Drake, J.A., Mooney, H.A. and di Castri, F. (1989) *Biological Invasions: A Global Perspective SCOPE 37*. Wiley, Chichester, New York.

Drenovsky, R.E., Grewell, B.J., D'Antonio, C.M., Funk, J.L., James, J.J., Molinari, N., Parker, I.M. and Richards, C.L. (2012) 'A functional trait perspective on plant invasion', *Annals of Botany*, vol. 110, pp. 141–153.

Drew, J., Anderson, N. and Andow, D. (2010) 'Conundrums of a complex vector for invasive species control: A detailed examination of the horticultural industry', *Biological Invasions*, vol. 12, pp. 2837–2851.

Driscoll, D.A., Catford, J.A., Barney, J.N., Hulme, P.E., Inderjit, Martin, T.G., Pauchard, A., Pyšek, P., Richardson, D.M., Riley, S. and Visser, V. (2014) 'New pasture plants intensify invasive species risk', *Proceedings of the National Academy of Sciences*, vol. 111, pp. 16622–16627.

Drolet, D., DiBacco, C., Locke, A., McKenzie, C.H., McKindsey, C.W., Moore, A.M., Webb, J.L. and Therriault, T.W. (2016) 'Evaluation of a new screening-level risk assessment tool applied to non-indigenous marine invertebrates in Canadian coastal waters', *Biological Invasions*, vol. 18, pp. 279–294.

Duncan, R.P. and Williams, P.A. (2002) 'Darwin's naturalization hypothesis challenged', *Nature*, vol. 417, pp. 608–609.

Dyderski, M.K. and Jagodzinski, A.M. (2019) 'Functional traits of acquisitive invasive woody species differ from conservative invasive and native species', *NeoBiota*, vol. 41, pp. 91–113.

Early, R. and Sax, D.F. (2014) 'Climatic niche shifts between species' native and naturalized ranges raise concern for ecological forecasts during invasions and climate change', *Global Ecology and Biogeography*, vol. 23, pp. 1356–1365.

Elizabeth, J.F. (2007) 'Plant life history traits of rare versus frequent plant taxa of sandplains: Implications for research and management trials', *Biological Conservation*, vol. 136, pp. 44–52.

Emiljanowicz, L.M., Hager, H.A. and Newman, J.A. (2017) 'Traits related to biological invasion: A note on the applicability of risk assessment tools across taxa', *NeoBiota*, vol. 32, pp. 31–64.

Erskine-Ogden, J., Grotkopp, E. and Rejmánek, M. (2016) 'Mediterranean, invasive, woody species grow larger than their less-invasive counterparts under potential global environmental change', *American Journal of Botany*, vol. 103, pp. 613–624.

Falster, D.S., Duursma, R.A. and FitzJohn, R.G. (2018) 'How functional traits influence plant growth and shade tolerance across the life cycle', *Proceedings of the National Academy of Sciences*, vol. 115, pp. E6789–E6798.

Feng, Y., Fouqueray, T.D. and van Kleunen, M. (2019) 'Linking Darwin's naturalisation hypothesis and Elton's diversity: Invasibility hypothesis in experimental grassland communities', *Journal of Ecology*, vol. 107, pp. 794–805.

Feng, Y., Maurel, N., Wang, Z., Ning, L., Yu, F.-H. and van Kleunen, M. (2016) 'Introduction history, climatic suitability, native range size, species traits and their interactions explain establishment of Chinese woody species in Europe', *Global Ecology and Biogeography*, vol. 25, pp. 1356–1366.

Flores-Moreno, H. and Moles, A.T. (2013) 'A comparison of the recruitment success of introduced and native species under natural conditions', *PLoS One*, vol. 8, p. e72509.

Flores-Moreno, H., Thomson, F.J., Warton, D.I. and Moles, A.T. (2013) 'Are introduced species better dispersers than native species? A global comparative study of seed dispersal distance', *PLoS One*, vol. 8, p. e68541.

Funk, J.L., Standish, R.J., Stock, W.D. and Valladares, F. (2016) 'Plant functional traits of dominant native and invasive species in mediterranean-climate ecosystems', *Ecology*, vol. 97, pp. 75–83.

Gallagher, R.V., Leishman, M.R., Miller, J.T., Hui, C., Richardson, D.M., Suda, J. and Trávníček, P. (2011) 'Invasiveness in introduced Australian acacias: The role of species traits and genome size', *Diversity and Distributions*, vol. 17, pp. 884–897.

Gallagher, R.V., Randall, R.P. and Leishman, M.R. (2015) 'Trait differences between naturalized and invasive plant species independent of residence time and phylogeny', *Conservation Biology*, vol. 29, pp. 360–369.

Gibert, A., Gray, E.F., Westoby, M., Wright, I.J. and Falster, D.S. (2016) 'On the link between functional traits and growth rate: Meta-analysis shows effects change with plant size, as predicted', *Journal of Ecology*, vol. 104, pp. 1488–1503.

Gibson, M.R., Richardson, D.M. and Pauw, A. (2012) 'Can floral traits predict an invasive plant's impact on native plant: Pollinator communities?', *Journal of Ecology*, vol. 100, pp. 1216–1223.

Godoy, O. and Levine, J.M. (2014) 'Phenology effects on invasion success: Insights from coupling field experiments to coexistence theory', *Ecology*, vol. 95, pp. 726–736.

Godoy, O., Valladares, F. and Castro-Díez, P. (2011) 'Multispecies comparison reveals that invasive and native plants differ in their traits but not in their plasticity', *Functional Ecology*, vol. 25, pp. 1248–1259.

González-Moreno, P., Diez, J.M., Ibáñez, I., Font, X. and Vilà, M. (2014) 'Plant invasions are context-dependent: Multiscale effects of climate, human activity and habitat', *Diversity and Distributions*, vol. 20, pp. 720–731.

Goodwin, B.J., McAllister, A.J. and Fahrig, L. (1999) 'Predicting invasiveness of plant species based on biological information', *Conservation Biology*, vol. 13, pp. 422–426.

Gordon, D.R. and Gantz, C.A. (2011) 'Risk assessment for invasiveness differs for aquatic and terrestrial plant species', *Biological Invasions*, vol. 13, pp. 1829–1842.

Gordon, D.R., Gantz, C.A., Jerde, C.L., Chadderton, W.L., Keller, R.P. and Champion, P.D. (2012) 'Weed risk assessment for aquatic plants: Modification of a New Zealand system for the United States', *PLoS One*, vol. 7, p. e40031.

Gordon, D.R., Mitterdorfer, B., Pheloung, P.C., Ansari, S., Buddenhagen, C., Chimera, C., Daehler, C.C., Dawson, W., Denslown, J.S., LaRosa, A., Nishida, T., Onderdonk, D.A., Panetta, F.D., Pysek, P., Randall, R.P., Richardson, D.M., Tshidada, N.J., Virtue, J.G. and Williams, P.A. (2010) 'Guidance for addressing the Australian Weed Risk Assessment questions', *Plant Protection Quarterly*, vol. 25, pp. 56–74.

Grime, J.P. (1977) 'Evidence for the existence of three primary strategies in plants and its relevance to ecological and evolutionary theory', *The American Naturalist*, vol. 111, pp. 1169–1194.

Gurevitch, J., Fox, G.A., Wardle, G.M., Inderjit and Taub, D. (2011) 'Emergent insights from the synthesis of conceptual frameworks for biological invasions', *Ecology Letters*, vol. 14, pp. 407–418.

Hamilton, M., Hamilton, B., Murray, M., Cadotte, G., Hose, A., Baker, C., Harris, D. and Licari (2005) 'Life-history correlates of plant invasiveness at regional and continental scales', *Ecology Letters*, vol. 8, pp. 1066–1074.

Harper, G.A. and Bunbury, N. (2015) 'Invasive rats on tropical islands: Their population biology and impacts on native species', *Global Ecology and Conservation*, vol. 3, pp. 607–627.

He, S., Yin, L., Wen, J. and Liang, Y. (2018) 'A test of the Australian Weed Risk Assessment system in China', *Biological Invasions*, vol. 20, pp. 2061–2076.

Higgins, S.I. and Richardson, D.M. (2014) 'Invasive plants have broader physiological niches', *Proceedings of the National Academy of Sciences*, vol. 111, pp. 10610–10614.

HilleRisLambers, J., Adler, P.B., Harpole, W.S., Levine, J.M. and Mayfield, M.M. (2012) 'Rethinking community assembly through the lens of coexistence theory', *Annual Review of Ecology, Evolution, and Systematics*, vol. 43, pp. 227–248.

Holt, R.D. (2009) 'Bringing the Hutchinsonian niche into the 21st century: Ecological and evolutionary perspectives', *Proceedings of the National Academy of Sciences*, vol. 106, pp. 19659–19665.

Hulme, P.E. (2009) 'Trade, transport and trouble: Managing invasive species pathways in an era of globalization', *Journal of Applied Ecology*, vol. 46, pp. 10–18.

Hulme, P.E. (2012) 'Weed risk assessment: A way forward or a waste of time?', *Journal of Applied Ecology*, vol. 49, pp. 10–19.

Keese, P.K., Robold, A.V., Myers, R.C., Weisman, S. and Smith, J. (2014) 'Applying a weed risk assessment approach to GM crops', *Transgenic Research*, vol. 23, pp. 957–969.

Keller, R.P., Lodge, D.M. and Finnoff, D.C. (2007) 'Risk assessment for invasive species produces net bioeconomic benefits', *Proceedings of the National Academy of Sciences*, vol. 104, pp. 203–207.

Kempel, A., Chrobock, T., Fischer, M., Rohr, R.P. and van Kleunen, M. (2013) 'Determinants of plant establishment success in a multispecies introduction experiment with native and alien species', *Proceedings of the National Academy of Sciences*, vol. 110, pp. 12727–12732.

Kendal, D., Williams, K.J.H. and Williams, N.S.G. (2012) 'Plant traits link people's plant preferences to the composition of their gardens', *Landscape and Urban Planning*, vol. 105, pp. 34–42.

Klinerová, T., Tasevová, K. and Dostál, P. (2018) 'Large generative and vegetative reproduction independently increases global success of perennial plants from Central Europe', *Journal of Biogeography*, vol. 45, pp. 1550–1559.

Knapp, S. and Kühn, I. (2012) 'Origin matters: Widely distributed native and non-native species benefit from different functional traits', *Ecology Letters*, vol. 15, pp. 696–703.

Křivánek, M. and Pyšek, P. (2006) 'Predicting invasions by woody species in a temperate zone: A test of three risk assessment schemes in the Czech Republic (Central Europe)', *Diversity and Distributions*, vol. 12, pp. 319–327.

Krivanek, M., Pysek, P. and Jarosik, V. (2006) 'Planting history and propagule pressure as predictors of invasion by woody species in a temperate region', *Conservation Biology*, vol. 20, pp. 1487–1498.

Kueffer, C., Pyšek, P. and Richardson, D.M. (2013) 'Integrative invasion science: Model systems, multi-site studies, focused meta-analysis and invasion syndromes', *New Phytologist*, vol. 200, pp. 615–633.

Kuester, A., Conner, J.K., Culley, T. and Baucom, R.S. (2014) 'How weeds emerge: A taxonomic and trait-based examination using United States data', *New Phytologist*, vol. 202, pp. 1055–1068.

Kunstler, G., Falster, D., Coomes, D.A., Hui, F., Kooyman, R.M., Laughlin, D.C., Poorter, L., Vanderwel, M., Vieilledent, G., Wright, S.J., Aiba, M., Baraloto, C., Caspersen, J., Cornelissen, J.H.C., Gourlet-Fleury, S., Hanewinkel, M., Herault, B., Kattge, J., Kurokawa, H., Onoda, Y., Peñuelas, J., Poorter, H., Uriarte, M., Richardson, S., Ruiz-Benito, P., Sun, I.F., Ståhl, G., Swenson, N.G., Thompson,

J., Westerlund, B., Wirth, C., Zavala, M.A., Zeng, H., Zimmerman, J.K., Zimmermann, N.E. and Westoby, M. (2016) 'Plant functional traits have globally consistent effects on competition', *Nature*, vol. 529, pp. 204–207.

Kyle, G. and Leishman, M.R. (2009) 'Functional trait differences between extant exotic, native and extinct native plants in the Hunter River, NSW: A potential tool in riparian rehabilitation', *River Research and Applications*, vol. 25, pp. 892–903.

Lai, H.R., Mayfield, M.M., Gay-des-combes, J.M., Spiegelberger, T. and Dwyer, J.M. (2015) 'Distinct invasion strategies operating within a natural annual plant system', *Ecology Letters*, vol. 18, pp. 336–346.

Lavoie, C., Joly, S., Bergeron, A., Guay, G. and Groeneveld, E. (2016) 'Explaining naturalization and invasiveness: New insights from historical ornamental plant catalogs', *Ecology and Evolution*, vol. 6, pp. 7188–7198.

Lavorel, S., McIntyre, S., Landsberg, J. and Forbes, T.D.A. (1997) 'Plant functional classifications: From general groups to specific groups based on response to disturbance', *Trends in Ecology & Evolution*, vol. 12, pp. 474–478.

Leishman, M.R., Thomson, V.P. and Cooke, J. (2010) 'Native and exotic invasive plants have fundamentally similar carbon capture strategies', *Journal of Ecology*, vol. 98, pp. 28–42.

Leishman, M.R., Wright, I.J., Moles, A.T. and Westoby, M. (2000) 'The evolutionary ecology of seed size', in M. Fenner (ed.) *Seeds: The Ecology of Regeneration in Plant Communities*. 2nd edition. CABI Publishing, Wallingford.

Lembrechts, J.J., Rossi, E., Milbau, A. and Nijs, I. (2018) 'Habitat properties and plant traits interact as drivers of non-native plant species' seed production at the local scale', *Ecology and Evolution*, vol. 8, pp. 4209–4223.

Levine, J.M., Adler, P.B. and Yelenik, S.G. (2004) 'A meta-analysis of biotic resistance to exotic plant invasions', *Ecology Letters*, vol. 7, pp. 975–989.

Lloret, F., Medail, F., Brundu, G., Camarda, I., Moragues, E.V.A., Rita, J., Lambdon, P. and Hulme, P.E. (2005) 'Species attributes and invasion success by alien plants on Mediterranean islands', *Journal of Ecology*, vol. 93, pp. 512–520.

Lockwood, J.L., Cassey, P. and Blackburn, T. (2005) 'The role of propagule pressure in explaining species invasions', *Trends in Ecology & Evolution*, vol. 20, pp. 223–228.

Lockwood, J.L., Hoopes, M.F. and Marchetti, M.P. (2013) 'Predicting and preventing invasion', in J.L. Lockwood, M.F. Hoopes and M.P. Marchetti (eds.) *Invasion Ecology*. Blackwell Publishing Ltd, Somerset, UK.

Lonsdale, W.M. (1999) 'Global patterns of plant invasions and the concept of invasibility', *Ecology*, vol. 80, pp. 1522–1536.

Ma, C., Li, S.-P., Pu, Z., Tan, J., Liu, M., Zhou, J., Li, H. and Jiang, L. (2016) 'Different effects of invader-native phylogenetic relatedness on invasion success and impact: A meta-analysis of Darwin's naturalization hypothesis', *Proceedings of the Royal Society B: Biological Sciences*, vol. 283, p. 20160663.

MacArthur, R. and Levins, R. (1967) 'The limiting similarity, convergence, and divergence of coexisting species', *The American Naturalist*, vol. 101, pp. 377–385.

Martin, P.H., Canham, C.D. and Marks, P.L. (2009) 'Why forests appear resistant to exotic plant invasions: Intentional introductions, stand dynamics, and the role of shade tolerance', *Frontiers in Ecology and the Environment*, vol. 7, pp. 142–149.

Mason, R.A.B., Cooke, J., Moles, A.T. and Leishman, M.R. (2008) 'Reproductive output of invasive versus native plants', *Global Ecology and Biogeography*, vol. 17, pp. 633–640.

Matzek, V. (2012) 'Trait values, not trait plasticity, best explain invasive species' performance in a changing environment', *PLoS One*, vol. 7, pp. 1–10.

Mayfield, M.M. and Levine, J.M. (2010) 'Opposing effects of competitive exclusion on the phylogenetic structure of communities', *Ecology Letters*, vol. 13, pp. 1085–1093.

McGeoch, M.A. and Latombe, G. (2016) 'Characterizing common and range expanding species', *Journal of Biogeography*, vol. 43, pp. 217–228.

McGill, B.J., Enquist, B.J., Weiher, E. and Westoby, M. (2006) 'Rebuilding community ecology from functional traits', *Trends in Ecology & Evolution*, vol. 21, pp. 178–185.

Meiners, S.J. (2007) 'Native and exotic plant species exhibit similar population dynamics during succession', *Ecology*, vol. 88, pp. 1098–1104.

Mitchell, C.E., Agrawal, A.A., Bever, J.D., Gilbert, G.S., Hufbauer, R.A., Klironomos, J.N., Maron, J.L., Morris, W.F., Parker, I.M., Power, A.G., Seabloom, E.W., Torchin, M.E. and Vázquez, D.P. (2006) 'Biotic interactions and plant invasions', *Ecology Letters*, vol. 9, pp. 726–740.

Moles, A.T. (2018) 'Being John Harper: Using evolutionary ideas to improve understanding of global patterns in plant traits', *Journal of Ecology*, vol. 106, pp. 1–18.

Moles, A.T., Falster, D.S., Leishman, M.R. and Westoby, M. (2004) 'Small-seeded species produce more seeds per square metre of canopy per year, but not per individual per lifetime', *Journal of Ecology*, vol. 92, pp. 384–396.

Moles, A.T., Gruber, M.A.M. and Bonser, S.P. (2008) 'A new framework for predicting invasive plant species', *Journal of Ecology*, vol. 96, pp. 13–17.

Moravcová, L., Pyšek, P., Jarošík, V., Havlíčková, V. and Zákravský, P. (2010) 'Reproductive characteristics of neophytes in the Czech Republic: Traits of invasive and non-invasive species', *Preslia*, vol. 82, pp. 365–390.

Moravcová, L., Pyšek, P., Jarošík, V. and Pergl, J. (2015) 'Getting the right traits: Reproductive and dispersal characteristics predict the invasiveness of herbaceous plant species', *PLoS One*, vol. 10, p. e0123634.

Muller-Landau, H.C. (2010) 'The tolerance: Fecundity trade-off and the maintenance of diversity in seed size', *Proceedings of the National Academy of Sciences*, vol. 107, pp. 4242–4247.

Murphy, H.T., VanDerWal, J., Lovett-Doust, L. and Lovett-Doust, J. (2006) 'Invasiveness in exotic plants: Immigration and naturalization in an ecological continuum', in M.W. Cadotte, S.M. Mcmahon and T. Fukami (eds.) *Conceptual Ecology and Invasion Biology: Reciprocal Approaches to Nature*. Springer, Dordrecht, The Netherlands.

Nishida, T., Yamashita, N., Asai, M., Kurokawa, S., Enomoto, T., Pheloung, P.C. and Groves, R.H. (2009) 'Developing a pre-entry weed risk assessment system for use in Japan', *Biological Invasions*, vol. 11, p. 1319.

Novoa, A., Kumschick, S., Richardson, D.M., Rouget, M. and Wilson, J.R.U. (2016b) 'Native range size and growth form in Cactaceae predict invasiveness and impact', *NeoBiota*, vol. 30, pp. 75–90.

Novoa, A., Rodríguez, J., López-Nogueira, A., Richardson, D.M. and González, L. (2016a) 'Seed characteristics in Cactaceae: Useful diagnostic features for screening species for invasiveness?', *South African Journal of Botany*, vol. 105, pp. 61–65.

Nuñez, M.A. and Medley, K.A. (2011) 'Pine invasions: Climate predicts invasion success; something else predicts failure', *Diversity and Distributions*, vol. 17, pp. 703–713.

Nunez-Mir, G.C., Guo, Q., Rejmánek, M., Iannone III, B.V. and Fei, S. (2019) 'Predicting invasiveness of exotic woody species using a traits-based framework', *Ecology*, vol. 100, p. e02797.

Nyberg, C.D. and Wallentinus, I. (2005) 'Can species traits be used to predict marine macroalgal introductions?', *Biological Invasions*, vol. 7, pp. 265–279.

Ordoñez, A. (2014) 'Global meta-analysis of trait consistency of non-native plants between their native and introduced areas', *Global Ecology and Biogeography*, vol. 23, pp. 264–273.

Palma, E., Vesk, P.A., White, M., Baumgartner, J.B. and Catford, J.A. (in press) 'Plant functional traits reflect different dimensions of species invasiveness', *Ecology*. doi: 10.1002/ecy.3317

Pearson, D.E., Ortega, Y.K., Eren, Ö. and Hierro, J.L. (2016) 'Quantifying "apparent" impact and distinguishing impact from invasiveness in multispecies plant invasions', *Ecological Applications*, vol. 26, pp. 162–173.

Pearson, D.E., Ortega, Y.K., Eren, Ö. and Hierro, J.L. (2018) 'Community assembly theory as a framework for biological invasions', *Trends in Ecology & Evolution*, vol. 33, pp. 313–325.

Pearson, D.E., Ortega, Y.K. and Maron, J.L. (2017) 'The tortoise and the hare: Reducing resource availability shifts competitive balance between plant species', *Journal of Ecology*, vol. 105, pp. 999–1009.

Pheloung, P.C., Williams, P.A. and Halloy, S.R. (1999) 'A weed risk assessment model for use as a biosecurity tool evaluating plant introductions', *Journal of Environmental Management*, vol. 57, pp. 239–251.

Pyšek, P., Danihelka, J., Sádlo, J., Chrtek Jr, J., Chytrý, M., Jarošík, V., Kaplan, Z. and Krahulec, F. (2012) 'Catalogue of alien plants of the Czech Republic: Checklist update, taxonomic diversity and invasion patterns', *Preslia*, vol. 84, pp. 155–255.

Pyšek, P., Jarošík, V. and Pergl, J. (2011) 'Alien plants introduced by different pathways differ in invasion success: Unintentional introductions as a threat to natural areas', *PLoS One*, vol. 6, p. e24890.

Pyšek, P., Jarošík, V., Pergl, J., Randall, R., Chytrý, M., Kühn, I., Tichý, L., Danihelka, J., Jun, J.C. and Sádlo, J. (2009b) 'The global invasion success of Central European plants is related to distribution characteristics in their native range and species traits', *Diversity and Distributions*, vol. 15, pp. 891–903.

Pyšek, P., Křivánek, M. and Jarošík, V. (2009a) 'Planting intensity, residence time, and species traits determine invasion success of alien woody species', *Ecology*, vol. 90, pp. 2734–2744.

Pyšek, P., Manceur, A.M., Alba, C., McGregor, K.F., Pergl, J., Štajerová, K., Chytrý, M., Danihelka, J., Kartesz, J., Klimešová, J., Lučanová, M., Moravcová, L., Nishino, M., Sádlo, J., Suda, J., Tichý, L. and Kühn, I. (2015) 'Naturalization of central European plants in North America: Species traits, habitats, propagule pressure, residence time', *Ecology*, vol. 96, pp. 762–774.

Pyšek, P. and Richardson, D.M. (2007) 'Traits associated with invasiveness in alien plants: Where do we stand?', in W. Nentwig (ed.) *Biological Invasions*. Springer, Berlin, Heidelberg.

Pyšek, P. and Richardson, D.M. (2010) 'Invasive species, environmental change and management, and health', *Annual Review of Environment and Resources*, vol. 35, pp. 25–55.

Reich, P.B. (2014) 'The world-wide "fast-slow" plant economics spectrum: A traits manifesto', *Journal of Ecology*, vol. 102, pp. 275–301.

Reichard, S.H. and Hamilton, C.W. (1997) 'Predicting invasions of woody plants introduced into North America', *Conservation Biology*, vol. 11, pp. 193–203.

Rejmánek, M. (2011) 'Invasiveness', in D. Simberloff and M. Rejmanek (eds.) *Encyclopedia of Biological Invasions*. University of California Press, Berkeley, California.

Richards, C.L., Bossdorf, O., Muth, N.Z., Gurevitch, J. and Pigliucci, M. (2006) 'Jack of all trades, master of some? On the role of phenotypic plasticity in plant invasions', *Ecology Letters*, vol. 9, pp. 981–993.

Richardson, D.M. and Pyšek, P. (2006) 'Plant invasions: Merging the concepts of species invasiveness and community invasibility', *Progress in Physical Geography*, vol. 30, pp. 409–431.

Richardson, D.M., Pyšek, P., Rejmánek, M., Barbour, M.G., Panetta, F.D. and West, C.J. (2000) 'Naturalization and invasion of alien plants: Concepts and definitions', *Diversity and Distributions*, vol. 6, pp. 93–107.

Russo, L. (2016) 'Positive and negative impacts of non-native bee species around the world', *Insects*, vol. 7, p. 69.

Sakai, A.K., Allendorf, F.W., Holt, J.S., Lodge, D.M., Molofsky, J., With, K.A., Baughman, S., Cabin, R.J., Cohen, J.E., Ellstrand, N.C., McCauley, D.E., O'Neil, P., Parker, I.M., Thompson, J.N. and Weller, S.G. (2001) 'The population biology of invasive species', *Annual Review of Ecology and Systematics*, vol. 32, pp. 305–332.

Salguero-Gómez, R. (2017) 'Applications of the fast-slow continuum and reproductive strategy framework of plant life histories', *New Phytologist*, vol. 213, pp. 1618–1624.

Schaffelke, B. and Hewitt, C.L. (2007) 'Impacts of introduced seaweeds', *Botanica Marina*, vol. 50, pp. 397–417.

Seebens, H., Blackburn, T.M., Dyer, E.E., Genovesi, P., Hulme, P.E., Jeschke, J.M., Pagad, S., Pyšek, P., van Kleunen, M., Winter, M., Ansong, M., Arianoutsou, M., Bacher, S., Blasius, B., Brockerhoff, E.G., Brundu, G., Capinha, C., Causton, C.E., Celesti-Grapow, L., Dawson, W., Dullinger, S., Economo, E.P., Fuentes, N., Guénard, B., Jäger, H., Kartesz, J., Kenis, M., Kühn, I., Lenzner, B., Liebhold, A.M., Mosena, A., Moser, D., Nentwig, W., Nishino, M., Pearman, D., Pergl, J., Rabitsch, W., Rojas-Sandoval, J., Roques, A., Rorke, S., Rossinelli, S., Roy, H.E., Scalera, R., Schindler, S., Štajerová, K., Tokarska-Guzik, B., Walker, K., Ward, D.F., Yamanaka, T. and Essl, F. (2018) 'Global rise in emerging alien species results from increased accessibility of new source pools', *Proceedings of the National Academy of Sciences*, vol. 115, p. E2264–E2273.

Seebens, H., Blackburn, T.M., Dyer, E.E., Genovesi, P., Hulme, P.E., Jeschke, J.M., Pagad, S., Pyšek, P., Winter, M., Arianoutsou, M., Bacher, S., Blasius, B., Brundu, G., Capinha, C., Celesti-Grapow, L., Dawson, W., Dullinger, S., Fuentes, N., Jäger, H., Kartesz, J., Kenis, M., Kreft, H., Kühn, I., Lenzner, B., Liebhold, A., Mosena, A., Moser, D., Nishino, M., Pearman, D., Pergl, J., Rabitsch, W., Rojas-Sandoval, J., Roques, A., Rorke, S., Rossinelli, S., Roy, H.E., Scalera, R., Schindler, S., Štajerová, K., Tokarska-Guzik, B., van Kleunen, M., Walker, K., Weigelt, P., Yamanaka, T. and Essl, F. (2017) 'No saturation in the accumulation of alien species worldwide', *Nature Communications*, vol. 8, p. 14435.

Seebens, H., Essl, F., Dawson, W., Fuentes, N., Moser, D., Pergl, J., Pyšek, P., van Kleunen, M., Weber, E., Winter, M. and Blasius, B. (2015) 'Global trade will accelerate plant invasions in emerging economies under climate change', *Global Change Biology*, vol. 21, pp. 4128–4140.

Shea, K. and Chesson, P. (2002) 'Community ecology theory as a framework for biological invasions', *Trends in Ecology & Evolution*, vol. 17, pp. 170–176.

Shipley, B. (2010) *From Plant Traits to Vegetation Structure: Chance and Selection in the Assembly of Ecological Communities*. Cambridge University Press, Cambridge, UK.

Simberloff, D. (2009) 'The role of propagule pressure in biological invasions', *Annual Review of Ecology, Evolution, and Systematics*, vol. 40, pp. 81–102.

Southwood, T.R.E. (1977) 'Habitat, the templet for ecological strategies?', *Journal of Animal Ecology*, vol. 46, pp. 337–365.

Southwood, T.R.E. (1988) 'Tactics, strategies and templets', *Oikos*, vol. 52, pp. 3–18.

Speek, T.A.A., Lotz, L.A.P., Ozinga, W.A., Tamis, W.L.M., Schaminée, J.H.J. and van der Putten, W.H. (2011) 'Factors relating to regional and local success of exotic plant species in their new range', *Diversity and Distributions*, vol. 17, pp. 542–551.

Suwa, T. and Louda, S.M. (2012) 'Combined effects of plant competition and insect herbivory hinder invasiveness of an introduced thistle', *Oecologia*, vol. 169, pp. 467–476.

Tecco, P.A., Urcelay, C., Diaz, S., Cabido, M. and Perez-Harguindeguy, N. (2013) 'Contrasting functional trait syndromes underlay woody alien success in the same ecosystem', *Austral Ecology*, vol. 38, pp. 443–451.

Theoharides, K.A. and Dukes, J.S. (2007) 'Plant invasion across space and time: Factors affecting nonindigenous species success during four stages of invasion', *New Phytologist*, vol. 176, pp. 256–273.

Thuiller, W., Gassó, N., Pino, J. and Vilà, M. (2012) 'Ecological niche and species traits: Key drivers of regional plant invader assemblages', *Biological Invasions*, vol. 14, pp. 1963–1980.

Tsutsui, N.D. and Suarez, A.V. (2003) 'The colony structure and population biology of invasive ants', *Conservation Biology*, vol. 17, pp. 48–58.

van Kleunen, M., Bossdorf, O. and Dawson, W. (2018a) 'The ecology and evolution of alien plants', *Annual Review of Ecology, Evolution, and Systematics*, vol. 49, pp. 25–47.

van Kleunen, M., Dawson, W., Essl, F., Pergl, J., Winter, M., Weber, E., Kreft, H., Weigelt, P., Kartesz, J., Nishino, M., Antonova, L.A., Barcelona, J.F., Cabezas, F.J., Cárdenas, D., Cárdenas-Toro, J., Castaño, N., Chacón, E., Chatelain, C., Ebel, A.L., Figueiredo, E., Fuentes, N., Groom, Q.J., Henderson, L., Inderjit, Kupriyanov, A., Masciadri, S., Meerman, J., Morozova, O., Moser, D., Nickrent, D.L., Patzelt, A., Pelser, P.B., Baptiste, M.P., Poopath, M., Schulze, M., Seebens, H., Shu, W.-s., Thomas, J., Velayos, M., Wieringa, J.J. and Pyšek, P. (2015) 'Global exchange and accumulation of non-native plants', *Nature*, vol. 525, pp. 100–103.

van Kleunen, M., Dawson, W., Schlaepfer, D., Jeschke, J.M. and Fischer, M. (2010b) 'Are invaders different? A conceptual framework of comparative approaches for assessing determinants of invasiveness', *Ecology Letters*, vol. 13, pp. 947–958.

van Kleunen, M., Essl, F., Pergl, J., Brundu, G., Carboni, M., Dullinger, S., Early, R., González-Moreno, P., Groom, Q.J., Hulme, P.E., Kueffer, C., Kühn, I., Máguas, C., Maurel, N., Novoa, A., Parepa, M., Pyšek, P., Seebens, H., Tanner, R., Touza, J., Verbrugge, L., Weber, E., Dawson, W., Kreft, H., Weigelt, P., Winter, M., Klonner, G., Talluto, M.V. and Dehnen-Schmutz, K. (2018b) 'The changing role of ornamental horticulture in alien plant invasions', *Biological Reviews*, vol. 93, pp. 1421–1437.

van Kleunen, M., Weber, E. and Fischer, M. (2010a) 'A meta-analysis of trait differences between invasive and non-invasive plant species', *Ecology Letters*, vol. 13, pp. 235–245.

Violle, C., Navas, M.-L., Vile, D., Kazakou, E., Fortunel, C., Hummel, I. and Garnier, E. (2007) 'Let the concept of trait be functional!', *Oikos*, vol. 116, pp. 882–892.

Von Der Lippe, M. and Kowarik, I. (2007) 'Long-distance dispersal of plants by vehicles as a driver of plant invasions', *Conservation Biology*, vol. 21, pp. 986–996.

Weber, J., Dane Panetta, F., Virtue, J. and Pheloung, P. (2009) 'An analysis of assessment outcomes from eight years' operation of the Australian border weed risk assessment system', *Journal of Environmental Management*, vol. 90, pp. 798–807.

Westoby, M., Falster, D.S., Moles, A.T., Vesk, P.A. and Wright, I.J. (2002) 'Plant ecological strategies: Some leading dimensions of variation between species', *Annual Review of Ecology and Systematics*, vol. 33, pp. 125–159.

White, M., Cheal, D., Carr, G.W., Adair, R., Blood, K. and Meagher, D. (2018) *Advisory List of Environmental Weeds in Victoria: Arthur Rylah Institute for Environmental Research Technical Report Series No. 287.* Department of Environment, Land, Water and Planning, Victoria.

Wilson, J.R.U., Richardson, D.M., Rouget, M., Procheş, Ş., Amis, M.A., Henderson, L. and Thuiller, W. (2007) 'Residence time and potential range: Crucial considerations in modelling plant invasions', *Diversity and Distributions*, vol. 13, pp. 11–22.

With, K.A. (2002) 'The landscape ecology of invasive spread', *Conservation Biology*, vol. 16, pp. 1192–1203.

Wright, I.J., Reich, P.B., Westoby, M., Ackerly, D.D., Baruch, Z., Bongers, F., Cavender-Bares, J., Chapin, T., Cornelissen, J.H.C., Diemer, M., Flexas, J., Garnier, E., Groom, P.K., Gulias, J., Hikosaka, K.,

Lamont, B.B., Lee, T., Lee, W., Lusk, C., Midgley, J.J., Navas, M.-L., Niinemets, U., Oleksyn, J., Osada, N., Poorter, H., Poot, P., Prior, L., Pyankov, V.I., Roumet, C., Thomas, S.C., Tjoelker, M.G., Veneklaas, E.J. and Villar, R. (2004) 'The worldwide leaf economics spectrum', *Nature*, vol. 428, pp. 821–827.

Yokomizo, H., Possingham, H.P., Hulme, P.E., Grice, A.C. and Buckley, Y.M. (2012) 'Cost-benefit analysis for intentional plant introductions under uncertainty', *Biological Invasions*, vol. 14, pp. 839–849.

Zenni, R.D. and Nuñez, M.A. (2013) 'The elephant in the room: The role of failed invasions in understanding invasion biology', *Oikos*, vol. 122, pp. 801–815.

2

WHAT IS AN INVASIVE ALIEN SPECIES? DISCORD, DISSENT AND DENIALISM

Juliet J. Fall

Introduction: aliens and experts

There are different names for describing species that proliferate, create environmental problems and come from distant locations: aliens, invasives, exotics, invasive alien species. . . . Many of these species bewitched people in the past with colourful flowers, vigorous blooms or fancy plumage, encouraging enterprising travellers to love them and move them. From the middle of the 19th century, adding chosen species to particular landscapes was encouraged in some parts of the world through state-sanctioned colonial projects and institutionalised within acclimatisation societies. The first of these learned societies were founded in European countries when moving species around the globe had fundamentally positive connotations. Scientists enthusiastically embraced schemes for improving colonial landscapes deemed defective and rendering nature more cosmopolitan within metropolises. Less than 200 years later, much has changed. Today, protecting, improving and restoring harmony in nature involves not the addition but the subtraction of certain established species deemed problematic. Among the many animal, plant and insect species that have naturalised, scholars identify a limited number of invasive alien species (IAS) that have negative ecological and economic impacts. These are taken as being both in the wrong place and dangerous: belonging far from the place in which they pose a problem. They are accused of displacing native species, hybridising inappropriately with desirable species, carrying diseases or threatening the balance of whole ecosystems. But despite the apparent solidity of this category, what invasive alien species are remains a subject of vigorous scientific debate.

New domains of knowledge have emerged around IAS, crafting specialised scholarly and applied expertise to describe, understand, predict and control them. Recurrent concerns are where such species came from originally and what their native range was before they started travelling. To where do they *really* belong? What right do they have to settle in new places? To what extent are these ecological, cultural or political questions? And why might it matter that controversies still exist in defining the category as a whole when some species obviously create massive local problems?

This chapter has two objectives: it first shows how IAS have been defined and categorised, answering the question of what invasive aliens species are; second, it discusses how this demarcation of a field of scholarship separate from other ecological sciences continues to be contentious. Because hostile debates continue to flare up, there is a need to understand how the object of study has been historically constructed as part of a distinct field of expertise variously called 'invasion biology', 'invasion ecology' or 'invasion science'. This involves understanding how this scholarship emerged and has been maintained as a bounded discipline by a specific cohort of scholars and examining how new individuals are enrolled in maintaining it. Making sense of the remaining contentious aspects of dissent and discord helps unpack misunderstandings at a time when accusations of denialism, xenophobia and racism continue to cloud important debates on ecological change in what has been called the Anthropocene.

Constructing objects of knowledge: from loving the exotic to stigmatising it

Objects of knowledge – the things people study – do not emerge by themselves fully formed within the life sciences. They are identified, chosen, crafted and constructed by groups of people who make sense of the complexity of the world by creating categories of living things. As Francis Bacon's conception of the scientific method took hold in the 17th century, modern science became rooted in notions of prediction and control (Tiles, 1996), philosophically connected to a specific conception of the relation between humans and nature. This way of thinking about the world determined what constituted useful scientific knowledge, including defining how evidence and empiricism were enrolled in producing knowledge. Using the language of the time, Tiles noted that 'the scientific, intellectual interest in prediction and control presupposes a view of Man on which such an interest is worthy of his dignity, an expression and fulfilment of a distinctively human potential, which raises man above the animals and the rest of nature' (Tiles, 1996, 228). She argued that knowledge that conferred power was primarily knowledge of laws of action and the contexts in which they were applicable, leading to a high value being placed on knowledge that not only observed the natural flow of events (i.e. 'watching nature take its course') but directly served to shape them (i.e. 'to not only imitate nature but dominate it'). This interest in exercising a mastery over nature, and controlling, manipulating, dominating and conquering nature, was used as further evidence of the success of the process of constituting knowledge.

Defining certain species as invasive aliens relies on rhetorical acts of category-making that stem from this conception of the scientific method, a reflection of what has been called the uniquely Western classifying imagination (Ritvo, 1997). These classificatory practices emerged within the natural sciences in the 18th and 19th centuries and continued to be refined in the 20th century. During this time, European societies progressively looked upon exotic natures with increasing suspicion, and a slow cultural shift took place towards maintaining order perceived as valuable in itself (Cooper, 2003; Smout, 2003; Hall, 2003; Olwig, 2003; Osborne, 2000; Crosby, 1986). The specific emergence of invasive exotic species as a scientific object of knowledge is usually traced back to the 1950s and the work of Charles Elton (Elton, 1958). Elton's work – in particular, his book *The Ecology of Invasions by Animals and Plants*, a short book of popular science – has attained almost mythical status among scholars working on invasive alien species. Elton's progressive turn towards isolating alien invasions from other successional processes may have been the result of his experiences during World War II, when British people were preoccupied by German invasion (Davis et al., 2001), reflecting a change in personal and societal values. Scholars are increasingly referencing him (Ricciardi and MacIsaac, 2008), but few actually quote direct extracts or, perhaps, even read it.

There have been other attempts to suggest longer filiations, and scholars have also flagged up the work of William Turner in the 16th century in the British Isles (Preston, 2009). Still others have unearthed the work of Hewett Cottrell Watson, who worked in Britain in the mid-19th century (Thompson, 2014). Watson, for instance, distinguished naturalised alien plants 'with just as much claim as the descendants of Normans or Saxons to be considered part of the British nation' (Watson, 1835, 38) from those relying on human action to multiply who 'are no more entitled to be called Britons, than are the Frenchmen or Germans who occasionally make their homes in England' (Watson, 1835, 39). Scholars have also noted that non-Anglophone authors are more likely to be overlooked in the writing of such hegemonic histories, something Kowarik and Pyšek have sought to redress by shedding light on the work of the Swiss botanist Albert Thellung, who was botanising at the beginning of the 20th century (Kowarik and Pyšek, 2012), although this does little to make these tales of origins any less Eurocentric.

Such histories are often written with ulterior contemporary motives in mind: either to show how indisputable contemporary knowledge is by identifying ancestors and pioneers made out to be 'ahead of their times' or, on the contrary, to demonstrate how the decision to separate aliens from other species was a political and cultural choice, not an ecological one. These birth stories of invasive alien species all make implicitly clear the male genealogies: by focusing on the male master–male disciple relationships, authors unwittingly flesh out which gender is the apparent authentic recipient of this cultural and biological heritage. Identifying these distant forefathers participates in the creation of a noble, if exclusively male, lineage. Words and ideas often have long histories, so writing such cultural histories of emerging categories must be done cautiously, in clear reference to how terms change over time. Some of these studies display what is called presentism – that is, uncritical and unquestioned adherence to present-day mindsets that have a tendency to interpret past practices in terms of modern concepts, classificatory practices and values. The label 'native', for instance, meant wild, untamed, uncultivated or undomesticated until the end of the 18th century (Thompson, 2014), with no reference to original location, so tracing its historical appearance inevitably needs to situate terms clearly. Other scholars have tried to take the cultural values of past times seriously to explain why alien invasions captured the imagination, although attributing causation to political contexts should be done cautiously, notwithstanding interesting parallels with the prevalence, for instance, of military and bellicose vocabularies in contexts far removed from war (Uekötter, 2007; Rémy and Beck, 2008).

What are invasive alien species? Making a new category

The term 'invasive alien species' does not reflect an everyday, obvious and well-known category of species with widespread vernacular recognition across the world. It does not describe biological characteristics but rather stems from how certain species dispersed (i.e. their dispersal history). It needs explaining because it relies on the belief in an order of nature – an ontology or worldview – in which individual species are uniquely associated with specific places. Although this makes sense in relation to human lifespans and history, environmental change is more the norm than the exception over longer geological time frames. Ice flows have covered continents repeatedly, requiring species to move to survive, with related changes in rainfall, sea levels and continental shapes. Species composition and genetic make-up have been subsequently modified in the resulting landscapes, and nowhere is fully isolated or stable. The earth has always been a dynamic and unpredictable place, with extreme geological and climatic changes selecting species that can adapt to survive.

Thus the apparently straightforward idea of locating species rests on specific assumptions about time, space and order (Thompson, 2014) and on the uniqueness of human agency. Defining certain species as alien implies the existence of a sort of mythical state of grace: an imaginary time and place before humans purposefully or inadvertently assisted inappropriate long-distance species dispersals. If such a moment of rupture or global change took place, it must be identified. This is often conveniently associated with the European discovery of the New World in 1492 or, for island ecosystems, with the time preceding first contact with Europeans. While these are obviously Eurocentric definitions, in many cases they do make empirical sense in terms of understanding the causes of rapid environmental change and species composition. This idea of a sudden change fits neatly with monotheistic conceptions of a fall from a state of grace or a rupture in a metaphysical covenant and might be one reason for this argument's cultural resonance in Western contexts. It also fits neatly with the Western dichotomy between nature and culture. Humans, in this view, are distinct from nature, and their actions and impacts are to be considered accordingly.

For the category of invasive alien species to make sense in this context, spatial boundaries (*what* is *where*) and temporal states (*when* something is *somewhere*) are taken as natural only in the absence of human perturbations. Species introduced by humans, rather than by other forces (a storm, a floating log, a tsunami), are considered different because humans are somehow not part of nature, existing instead as cultural, social or unnatural beings governed by rules distinct from the rest of the living world. This is rooted in an ontological view of nature that identifies human agency in moving living species as distinct and singular (Blackburn et al., 2011; Bellard et al., 2016), rather than distinctions made only according to distance travelled or speed of dispersal.

Some scholars explain that it is not so much the nature or origin of ecological change that is the problem with invasive alien species but the scale, extent and speed of corresponding subsequent species loss. Precise sets of arguments have been used historically to justify the singularity of invasive alien species as a specific and global metataxonomic category: one that transcends the usual classificatory grid of biological nomenclature (Helmreich, 2005). Constructing such a category has involved finding ways of distinguishing invasive aliens species from others – including others that pullulated but were considered native (Nackley et al., 2017). As a consequence, proponents of differentiating IAS concluded that the effects of these species' movements were also distinct and uniquely unnatural (Pereyra, 2016; Munro et al., 2019). Notwithstanding the apparent solidity of this category, it was intensely debated from the start. Did the conduct of these species reflect the intrinsic and essential characteristics of nature, or were they somehow fundamentally unnatural, as the term 'alien' seemed to imply? Despite the category IAS gaining recognition among scholars, there continue to be many fundamental and heartfelt debates on ambiguities connected to whether species should be categorised according to place of origin (native vs. alien), mode of transportation (natural vs. anthropic) and/or behaviour (pioneer vs. coloniser) (Cresswell, 1997; Gould, 1997; Davis and Thompson, 2001; Davis et al., 2001; Colautti and MacIsaac, 2004; Robbins, 2004; Willis and Birks, 2006; O'Brien, 2006; Coates, 2007; Davis, 2011; Kull, 2018; Munro et al., 2019). There are many well-researched reviews of the nature and basis of the scientific arguments that explain what distinctions and discords remain (Humair et al., 2014; Head, 2017; Guiaşu and Tindale, 2018; Kull, 2018; Shackleton et al., 2019). Some of these reviews note that while a consensus on categories might be needed for action, concepts and risk assessments continue to be controversial among stakeholders, and major public disagreements among experts continue to exist.

In response to what has become an increasingly tense debate both from within and outside invasion science over how and whether invasive aliens species should be categorised, and how existing categories reflect specifically situated cultural values and histories, scholars have referred to alternative categories for classifying species, including post-colonial (Subramaniam, 2001; Cardozo and Subramaniam, 2013) or indigenous classifications (Helmreich, 2005), rather than endlessly rehearsing the same dichotomies. One purpose of diversifying sources of knowledge about these species might be to develop parataxonomies that would see scientific and indigenous taxonomies coexist, reflecting broader critical movements to rethink how expertise and legitimacy are attributed and appropriated. In this context of decentring and decolonising knowledge, it seems particularly ironic that much of the fieldwork on invasive alien species is taking place in former colonies where issues of 'indigenousness and belonging are discussed in contexts where settler human populations are still coming to terms with their own belonging' (Head and Muir, 2004, 203) and by authors focusing largely on former European colonies (Barker, 2010; Crosby, 1986; Clark, 2002; O'Brien, 2006; Robbins, 2004). Because many of these countries – Australia, Canada, the US, New Zealand and South Africa in particular – are English speaking and the studies produced by local researchers are widely read, these forms of post-colonial guilt and anxiety about identity end up orienting ecological debates in ways that still need to be fully examined in contexts with very different ecological and social histories (Fall, 2013).

Regardless of how human agency is defined precisely and differentiated from other ecological processes, it is clear that mass human occupation of the earth has led to fantastically effective networks for non-human species dispersal. Many species now use the global infrastructures that humans have spread across the world, hopping on and off container ships from Europe to North America and back again, catching rides on lorries and spreading along roads or moving into marginal urban spaces with disturbed land. These invasive species are now seen to be rhetorically doubly perverse: not only are they spreading in new and unexpected ranges, but they are also inhabiting spaces considered removed from nature – derelict train stations, industrial zones and other abandoned margins of human activities. Authors have argued that understanding what has been called the human preparation of landscape is a key factor in making sense of invasions (Robbins, 2004). These happen most when and where the invasiveness of certain species uniquely combines with the ecological conditions of certain human-modified landscapes. Furthermore, as anthropogenic climate change continues to modify our responsibility and role in global ecological change, redefining what Western thought has understood as 'natural' is further complicated (Preston, 2009; Fall, 2014; Rotherham, 2017; Fuentes, 2018).

Stabilising and institutionalising a new field of knowledge

The scientific biological knowledge produced on invasive alien species is now extensive, yet some of it remains deeply contentious and contested, with ongoing vivid debates taking place within scientific journals on the fundamental nature and impacts of these exotic invasive species. These debates continue to question the contours not only of the objects under discussion – the invasive species – but also the coherence of 'invasion science' as a stand-alone field of knowledge tasked with studying them. Historians and sociologists of science have shown how academics have demarcated science from non-science, defining the changing landscapes of academic scholarship as new knowledge and methods give rise to new fields of expertise (Gieryn, 1983; Roth, 2005; Waterton, 2002). These studies of science as a social practice have provided tools and concepts to understand the shifting processes of knowledge and expertise production and the positioning of experts within these landscapes. These also help scientists understand how epistemological and institutional processes are also deeply personal, connected to the construction and maintenance of individual professional reputation, identity, status and sense of belonging.

Although institutional processes have been different in different places, according to national and linguistic traditions – including how teams and laboratories operate, what roles top professors hold within them, how new staff are appointed and so on – as well as specific disciplinary locations and practices, it is almost inevitable that people associate strongly with the scientific communities they choose to belong to as well as with the objects they study. Chew (2015) has provided a careful and contextualised unpacking of some of this recent history. He focuses on the circulation of the claim first suggested by Edward O. Wilson in 2002 that invasive alien species were the second cause of global biodiversity loss after habitat destruction. Through a meticulous reading of texts appearing in different versions that had been quoted extensively but selectively, he showed how what was initially a general boast became a core belief, despite clear evidence that this was originally little more than a bad extrapolation from partial and highly specific data (see also Kull, 2018).

Versions of the terms 'invasion biology' and 'invasion ecology' have been in use by scholars since the mid-1980s, denoting a distinct field of knowledge grounded purportedly in the biological characteristics of the species and not only on their effects on specific environments. The increasing stabilisation of this field as distinct from other forms of ecology or conservation biology was further promoted on the cusp of the 21st century by the founding of specialised scientific journals such as *Biological Invasions* (1999), *Diversity and Distributions* (1998), *NeoBiota* (2002) and *Management of Biological Invasions* (2010). On the one hand, this reflected the well-known international boom in specialised scientific journals and the particular economic models that underpinned them (Larson et al., 2005; Bornmann and Mutz, 2015). But it also heralded the apparent stabilisation and institutionalisation of an academic field.

Because the contours of invasion science remain unstable and frequently challenged, it is a fascinating example for studying how professional and personal statuses are conferred in a competitive context in which individual career management and professional position are grounded in the successful maintenance of particular paradigms (Chew, 2015). The strategic coining around 2003 of the name 'invasion science' for the field tasked with understanding the characteristics and impacts of these species reflected this attempted stabilisation. This is seen as a distinct subfield of conservation biology, itself a discipline grounded in the idea of crisis and the need for urgent action. The specific term 'invasion science' was apparently first used in writing within the title of a paper by Costa-Pierce (2003). By 2011, it seemed to have gained widespread visibility, perhaps thanks to its definition, use and promotion by Richardson and co-authors (Richardson et al., 2011), reflecting the rebranding of many subfields with the epithet 'science' as a strategic way of asserting authority (Kull, 2018). Rather than simply describing the effects of certain species on specific ecosystems, invasion science scholars advocated that these should be viewed as negative (Sagoff, 2018), notwithstanding continuing vigorous debates (Schlaepfer et al., 2011; Schlaepfer, 2018). This lexical creativity also exists within scholars working on social science approaches to IAS, as Kull recently coined the title 'critical invasion science' (Kull, 2018) to denote approaches concerned with questions of power, inspired by methodologies within critical geography and political ecology.

The following sections map out some of the remaining debates around the coherence of this field of knowledge to show what fundamental points remain contested and to illustrate how these challenges are performed and maintained.

Contested objects of knowledge: performing dissent in academic journals

In 2011, Mark Davis and 18 chosen colleagues published a paper in *Nature* arguing that the existing orthodoxy around invasive species was unhelpful. 'Increasingly', they wrote, 'the practical value of the native-versus-alien species dichotomy in conservation is declining, and even becoming counterproductive. Yet many conservationists still consider the distinction a core guiding principle' (Davis et al., 2011, 153). They argued that, notwithstanding a number of examples where invasive alien species had caused ecological harm, the need to conserve nature within a fast-changing planet implied that this overwhelming preoccupation with the native-alien divide be ditched in favour of more pragmatic and dynamic approaches to all forms of species movements, thus dropping the distinction between invasion and natural colonisation. This, they argued, would allow inputs from ecology in general, island biogeography and epidemiology. Suggestions that IAS were some sort of apocalyptic threat were, they argued, simply not backed by data (Davis et al., 2011). Having received many angry individual responses, the *Nature* journal editors collected these and apparently suggested that Daniel Simberloff (2011) lead the response, presumably based on his reputational standing, asking other respondents to co-sign it. The core of their response was that biological invasions are not just a biological phenomenon: the human dimension of invasions is a fundamental component of the system in which these must be understood. The journal crafted and shaped their collective voice and chose a specific male ambassador as lead author. Although none of these arguments were particularly new – Davis and his colleagues had made similar points in previous publications (for example Davis et al., 2001) – this at the very least led to increased visibility for all the participants and a repeated carving out of battle lines.

Such flare-ups often materialise in interventions and responses in which key individuals or like-minded collectives symbolically and concretely reassert their authority either within the field of invasion science or outside as more or less sympathetic observers. Examples of other triggering starting points include: Gröning and Wolschke-Bulmahn (2003), Schlaepfer et al. (2011), Valéry et al. (2013), Hoffmann and Courchamp (2016), Russell and Blackburn (2017) and Ricciardi and Ryan (2018). Each of these, and the related responses, are subsequently much quoted. In these heartfelt exchanges, colleagues variously accuse each other not only of bad science but also of xenophobia, racism, relativism or even denialism. Against the suggestions that terms should be chosen with caution, some invasion biologists claim that loaded metaphors simply reflect what is really happening: 'In invasion science, as in other contexts, researchers talk about 'invasions' because what is observed really is reminiscent of armies moving. Media reports on biological invasions often attract readers' attention with military metaphors' (Simberloff et al., 2013, 63). This assimilation of scientific facts with metaphors and the confusion of scientific and popular modes of discourse reflects the frequent storying of these issues not only in lay terms but also among scientists. Notwithstanding much critical research on framing and rhetoric that flags up how and why terms and labels should be used with caution (Subramaniam, 2001; Larson et al., 2005; Barker, 2009; Warren, 2007, 2008; Chew and Hamilton, 2011; Fall, 2014; Chew, 2015; Ernwein and Fall, 2015), using emotions and metaphors to opportunistically 'get the message out' continues to be thought appropriate by some invasion scientists (Begon, 2017; Saul et al., 2017).

It is worth looking in more detail at one other such debate, which helps show how dissent is performed and managed and how loaded terms carry different connotations across disciplines. This episode started with Benjamin Hoffmann and Franck Courchamp's publication of a 'Discussion Paper' in the open-access journal *Neobiota* (Hoffmann and Courchamp, 2016). The authors suggested that the term 'colonisation' should be used instead of the concept of 'invasion' to describe what takes place when exotic species pullulate in new places. In their understanding, 'colonisation' was a more neutral term, better suited to describing ecological processes. It was chosen to disrupt the apparent orthodoxy that invasive alien species display inherently different characteristics than other species due to their dispersal history caused by humans, questioning the ontological existence of 'invasion science' as a distinct field of scholarship. The 'current polarisation of the respective sciences based on human mediation versus natural colonisation is hindering the progression of our understanding' (Hoffmann and Courchamp, 2016), they wrote. Their point was similar to that made repeatedly by Mark Davis and colleagues (Davis and Thompson, 2001; Davis, 2011) – i.e. it doesn't matter where problematic species come from, what matters is the effect they have. Nevertheless, suggesting using the term 'colonisation', a word with a deep and loaded imperial history, was baffling, particularly as this was apparently in complete ignorance that it might be problematic to some audiences.

Hoffmann and Courchamp relied on chosen examples to ground and illustrate their argument, showing how the processes underpinning how species change their range were the same regardless of whether humans were responsible. Their critique was polite but fundamental: colleagues clinging to the unity and distinction of invasion science were misguided, showing 'great resistance' (Hoffmann and Courchamp, 2016, 3) to acknowledging that their ontology rested upon what they termed an artificial separation between ecological processes concerning IAS and other species. Upholding invasion science, they argued, simply impeded collaboration among what could be a broader coalition including 'biologists, invasion biologists, restoration ecologists, island biogeography biologists, community assembly ecologists and epidemiologists' (Hoffmann and Courchamp, 2016, 9). They rejected the theoretical basis for the institutionalisation of invasion science: that human influence over species distribution was ontologically so distinct that it followed a completely different set of ecological rules from those relevant to other species.

The response was quick, collective and severe: John Wilson, Pablo García-Diaz, Phillip Cassey, David Richardson, Petr Pyšek and Tim Blackburn published a response a few months later in the same journal (Wilson et al., 2016; see also Kühn et al., 2017). 'There is something in the way humans move species that moves them like no others' (Wilson et al., 2016, 88), they stated unambiguously, reusing one of the Wilson paper's titles (Wilson et al., 2009). These six scientists suggested that the debate was no longer relevant. Hoffmann and Courchamp's idea, they argued, had already been repeatedly rebutted by the majority voice (Wilson et al., 2009): 'this has been acknowledged many times before' (p. 91); 'has been accepted by the majority of researchers' (p. 92); 'most of us' (p. 93); 'a series of multi-author collaborations' (p. 93). These arguments, they pointed out, were fallacious and unscientific (Wilson et al., 2009): 'stretch the concept of degree beyond breaking point' (p. 89); 'a distinction which is lost in Hoffmann and Courchamp's unhelpful edits' (p. 91); 'reductio ad absurdum' (p. 93). For these authors, invasion science had to be singular, since they were as different from other biologists as doctors were from vets (Wilson et al., 2016). They inscribed their scientific discipline in the intrinsic nature of its object of knowledge, maintaining both the unity and singularity of invasion science and their own position within it. At no point did they point out that the term 'colonisation' might be in any way problematic or politically sensitive: their point was instead to uphold the unity of the field of invasion science and their own positions as experts within it.

The next section steps back from these examples to understand the changing professional and institutional contexts in which such ritualised exchanges take place.

Constructing collectives and managing dissent within invasion science

Emerging critical labour studies of academic capitalism (Gill, 2014) have shown the extent to which massification, marketisation and internationalisation have profoundly changed how scientists work in contemporary academia (Barcan, 2013). In a parody of market capitalism, scientists who were once socialised to cooperate and exchange for the greater good of knowledge production have simultaneously and often unproblematically adopted the idea that competition should be disproportionately prized. This newly engrained culture of competition brings about a self-reinforcing process that rewards those already recognised by their peers with disproportionate visibility and increased scientific citation while further excluding the contributions of less-visible others. What is perverse is that this happens whatever the actual merits of the scientific results, thereby challenging the idea that the best ideas always get picked up. Merton suggested calling this the 'Matthew effect' (Merton, 1968), using a biblical reference to express the idea that the rich always get richer while the poor lose what little they had. Regardless of what one chooses to call the self-reinforcing effect of reputation, it is important because the myth of market efficiency as applied to scholarship assumes that the best ideas rise to the top and survive due to their intrinsic qualities (just as customers are assumed to choose the best quality goods for the right price) and not because of existing power differentials between scientists. The connected idea for individual careers is meritocracy: i.e. the belief in the survival of the fittest in a competitive system.

The joint myths of efficiency and meritocracy obscure what is in fact a power-infused game aimed at maintaining high status through ongoing processes of micropolitics (Morley, 2003). Maintaining acquired status requires constant vigilance from the key players who have risen to the top, particularly within the context of broader neoliberal changes in labour markets (Morley, 2003; Smyth, 2017) and continued issues of lack of diversity of women and visible minorities (Despret and Stengers, 2012). Invasion science, notwithstanding the increasing presence of women scientists, still reflects broader patriarchal structures that reward caricatural competitive masculine traits: 'because the academy is premised on the CV and individual competitiveness it gives rise to particular sorts of performative hyper-masculinities, often thinly disguised in egalitarian rhetoric' (Clegg, 2008, 219). The endlessly rehearsed scientific debates, making collective but distinct positions loud and visible, thus appear to be as much about claiming and/or maintaining status and legitimacy as advancing new knowledge. Additionally, it might be that because applied knowledge – such as invasion science – is often considered lower in the pecking order of academic subjects (Becher and Trowler, 2001), being able to attain high status in the first place relies even more on stabilising the community of reference and rising within it.

There are many other such exchanges, including around suggestions that critiquing aspects of invasion science is simply 'invasive species denialism' (Ricciardi and Ryan, 2018). As further recent papers have picked up the term 'denialism', explicitly linking this to climate change denialism, there have also been excellent critiques, including disputing the assertion that so-called denialism is quantitatively increasing (Boltovskoy et al., 2018). Other authors have argued that considering that all critique is simply denialism is not only unfair but also intellectually unsound. Some of these are simply unsubstantiated and 'unjustifiable accusations of science denialism . . . [that] ostracise novel, minority or outside perspectives' (Munro et al., 2019, 6). Munro and his colleagues have shown that such knee-jerk reactions come from scholars who refuse to

> entertain the possibility that others, sometimes with different values and from different disciplines or cultures, would reach a different conclusion given the same scientific information. They present their science and proceed to apply their own unique values to prescribing how it should be interpreted and applied.
>
> *(Munro et al., 2019, 11)*

These exchanges often appear to be about positioning and the micropolitics of laying claims to space as important authors within a competitive field. They are as much about exercising individual and collective power – through strategic co-authorship and knowledge community creation – as they are about debating facts.

Taking debates beyond academia: practice-based knowledge

Similar though less formalised debates also take place in various non-peer-reviewed spaces, such as within online specialist forums. These are usually made up of individual responses to a contribution – a article posted to the group, a question asked or an opinion voiced – deemed controversial. One of the best known is the *Aliens-1 listserv* belonging to the International Union for Conservation of Nature Invasive Species Specialist Group (aliens-l@list.auckland. ac.nz), hosted by the University of Auckland in New Zealand. This connects 1,500 members – a mix of practitioners, policy specialists and scientists – with weekly messages and exchanges. One frequent point of contention appears between those thinking that 'science knows best' and those willing to engage with the complexities of transforming scientific knowledge into workable environmental policies and management strategies. The two quotes that follow are good examples of the 'scientists-know-best' paradigm in response to an article about the extermination of feral cats, illustrating the extent to which invasion science frequently sees itself as purely reason-based expertise:

> Alien invasion is an ecological phenomenon, now well-documented by science. To 'muddy the waters' with human emotions is to devalue this science.
>
> *(May 2019, Aliens-l listserv)*

> Love and hate vis-à-vis of plant, animal or fungi species are irrelevant when it comes to preserve ecosystems or community livelihoods for future generations. . . . I would recommend IAS ecology to remain in the field of science and reason.
>
> *(May 2019, Aliens-l listserv)*

Both these quotes claim legitimacy for the field of invasion science by situating it within rational, objective and universal scientific practices as a way of gaining legitimacy through an appeal to the authority of science. But understanding how expertise is received in the field is about more than imposing knowledge from a position of apparent power. Instead, translating knowledge and transporting it to the wider public has been shown to be surprisingly fraught, particularly when some aspects of the impacts of IAS remain uncertain. A study of the biosecurity regime in New Zealand showed a pragmatism and acknowledgement of the need to accommodate 'non-native' species (Barker, 2008) but also that there was a keen sense of needing to grab public attention. In many such environmental issues, communicating uncertainty has been shown to be complex since the reception of uncertainty is as diverse as the types of uncertainties expressed, as illustrated within a study carried out in Switzerland (Ernwein and Fall, 2015). By drawing upon such emotions as the fear of illness and death, discourses promoting individual action against IAS have been deployed strategically in many places, assuming that fear of uncontrollable and threatening nature can be enrolled productively and somehow transformed into action to control undesirable species in public or private land. Nevertheless, rather than leading unproblematically to action, expert categories that come across as uncertain or incoherent end up having their legitimacy questioned, resisted and transformed in the field (Ernwein and Fall, 2015). Scholars have thus flagged up the need to democratise decision-making around the biological threats posed by IAS and to explain what uncertainties remain by making it clear how this is connected to the need for an inclusive politics that questions the ecological good life more broadly (Barker, 2010).

Thinking through objectivity, values and norms in invasion science

Social studies of science have shown how all scientific practices are situated social practices (Callon, 2006; Bucchi, 2002). The modern scientific enterprise has historically been characterised by two competing and contradictory images: on one side the objective, dispassionate investigators grounded in a belief in detached, rational objectivity and on the other heroic and strong competitors willing to sacrifice personal comfort and spurred on to action by emotion. As such, scientific practices have always been shot through with specific norms, including gendered ones, such as recurrent calls within the natural sciences to 'unveil the secrets and penetrate the depths of nature' (Oreskes, 1996, 102). Invasion science is also confidently normative: it is committed to the development of policies to control and, at times, eradicate the species it studies. Rather than being only about generating knowledge about the living world, it is fundamentally connected to the idea of taking action in a context of crisis and loss. As Takacs (1996) has shown ethnographically, conservation biology as a whole is uniquely traversed by these two competing and paradoxical figures: claims to authority based on detachment, rationality and objectivity yet personal motivations that are shot through with deeply felt emotional investment and a desire to act to save that which is seen as threatened.

The biological sciences have customarily defended themselves against the accusation of being shot through with values by arguing that these play no significant role in how they operate. Proponents state that the purpose of biology is simply the search for truth. Science studies scholars have argued that this defence is doomed to failure since it quickly turns into a circular argument: theory is defended in reference to method, and method to theory (Tiles, 1996). Yet science relies on persuasion and rhetoric: 'a series of efforts to persuade relevant social actors that one's manufactured knowledge is a route to a desired form of very objective power' (Haraway, 1988, 577). Admitting this does not mean giving up objectivity. Instead, there is a need for 'enforceable, reliable accounts of things not reducible to power moves and agonistic, high-status games of scientistic, positivist arrogance' (Haraway, 1988, 580). Social studies of science, like much feminist critique, have suggested that flagging and explaining more clearly from what social position one is speaking – including trying to voice what are often implicit norms and values held – helps make scientific results more, not less, objective (Harding, 1991, 2015). Taking seriously how objects of knowledge such as IAS are constructed in relation to claims about the 'real world' means assuming they are neither passive nor inert since such an assumption risks becoming little more than a mask for dominating interests, however worthy intentions may be.

One value that has often been suggested as needing to be made explicit is nativism, an ideological position and value that implicitly runs through many debates on invasive alien species (Fall, 2014). This is the idea of a discrepancy between the interests and rights of certain established inhabitants of an area or state as compared to claims of newcomers or immigrants. This might not be a problem in the natural sciences if, as is sometimes claimed, political and ecological domains were fundamentally different, with environmental concerns determined by value-free and science-led paradigms. Yet in the case of invasion ecology, this assumption is revealed for the fallacy (or myth) that it is. Ecological policies are far from being a politics-free zone since they reflect in multiple complex ways the underpinning social values of the societies that give rise to them (Robbins, 2004; Barker, 2010). Biological research, like all other scientific practice, is never innocent. Just as nature should not be considered the original sanctuary set apart from human observation, scientific practices should not be considered detached from their immersion in the natural and social world (Gardey, 2014).

Conclusion: policing and challenging the contours of invasion science

This chapter has shown how invasive alien species have been defined and categorised and has answered the question of what invasive alien species are. It has also discussed how this demarcation of a field of scholarship separate from the rest of biological or life sciences continues to be contentious. It has made explicit how important it is to give continued attention to understanding how 'invasion science' has been constructed historically as a distinct field of expertise and how it continues to evolve. By showing how this emerged and how it is continuously maintained as a bounded discipline by a specific cohort of scholars, this chapter has illustrated how new individuals are continuously enrolled in maintaining it within an academic landscape of increased competition in which visibility and participation in debates are seen to confer symbolic capital. Power relations and hierarchies within this collective structure play a role in the continued institutionalisation of the field. Furthermore, by unpicking recent heartfelt scholarly debates around loaded terms such as 'colonisation' and 'denialism', this chapter has sought to make sense of the remaining contentious dissent and discord, in particular surrounding the role of humans as agents of rapid ecological change. Through chosen examples of academic debates, it has also shown that despite a number of careful attempts to unpack the remaining misunderstandings and identify divergent positions and possibilities for dialogue, accusations of xenophobia and denialism continue to cloud and shut down exchanges. This narrowing of the possibility for constructive engagement with critical perspectives is worrisome. Invasion science needs to continue to engage with questions about where species came from, where they live now and how landscapes are changed by their presence. But it also needs to continue to be able to ask fundamental questions, including whether its perimeter and project are appropriate to deal with life in the Anthropocene, without its scholars fearing that sceptical engagement will always be labelled denialism and construed as giving potential ammunition to the wrong people. For unless this happens, invasion science risks turning into a populist caricature of itself: exterminating those whose voices are deemed unworthy at a time of change when diversity just might ensure survival.

References

Barcan, R. (2013) *Academic Life and Labour in the New University: Hope and Other Choices.* Ashgate, Farnham.

Barker, K. (2008) 'Flexible boundaries in biosecurity: Accommodating gorse in Aotearoa New Zealand', *Environment and Planning A*, vol. 40, no. 7, pp. 1598–1614.

Barker, K. (2009) 'Garden terrorists and the war on weeds', in K. Dodds and A. Ingram (eds.) *Spaces of Security and Insecurity: New Geographies of the War on Terror*, Ashgate, Farnham.

Barker, K. (2010) 'Biosecure citizenship: Politicising symbiotic associations and the construction of biological threat', *Transactions of the Institute of British Geographers*, vol. 35, no. 3, pp. 350–363.

Becher, T. and Trowler, P.R. (2001) *Academic Tribes and Territories: Intellectual Enquiry and the Culture of Disciplines.* The Society for Research into Higher Education and The Open University Press, Buckingham and Philadelphia.

Begon, M. (2017) 'Winning arguments: Coexistence not competition: A reply to Saul et al. trends', *Trends in Ecology & Evolution*, vol. 32, pp. 723–724.

Bellard, C., Cassey, P. and Blackburn, T.M. (2016) 'Alien species as a driver of recent extinctions', *Biology Letters*, vol. 12, p. 20150623.

Blackburn, T.M., Pyšek, P., Bacher, S., Carlton, J.T., Jarosik, V., Wilson, J.R.U. and Richardson, D.M. (2011) 'A proposed unified framework for biological invasions', *Trends in Ecology & Evolution*, vol. 26, pp. 333–339.

Boltovskoy, D., Sylvester, F. and Paolucci, E.M. (2018) 'Invasive species denialism: Sorting out facts, beliefs, and definitions', *Ecology and Evolution*, vol. 8, pp. 11190–11198.

Bornmann, L. and Mutz, R. (2015) 'Growth rates of modern science: A bibliometric analysis based on the number of publications and cited references', *Journal of the Association for Information Science and Technology*, vol. 66, no. 11, pp. 2215–2222.

Bucchi, M. (2002) *Science in Society: An Introduction to Social Studies of Science.* Routledge, London and New York.

Callon, M. (2006) 'Quatre modèles pour décrire la dynamique de la science', in M. Akrich, M. Callon and B. Latour (eds.) *Sociologie de la traduction: textes fondateurs.* Mines Les Presses, Paris.

Cardozo, K. and Subramaniam, B. (2013) 'Assembling Asian/American Naturecultures: Orientalism and invited invasions', *Journal of Asian American Studies*, vol. 16, no. 1, pp. 1–23.

Chew, M.K. (2015) 'Ecologists, environmentalists, experts, and the invasion of the "second greatest threat"', *International Review of Environmental History*, vol. 1, pp. 7–40.

Chew, M.K. and Hamilton, A.L. (2011) 'The rise and fall of biotic nativeness: A historical perspective', in *Fifty Years of Invasion Ecology: The Legacy of Charles Elton*. Blackwell Publishing Limited, Oxford. pp. 35–48.

Clark, N. (2002) 'The Demon-Seed: Bioinvasion as the Unsettling of Environmental Cosmopolitanism', *Theory, Culture & Society*, vol. 19, no. 1–2, pp. 101–125.

Clegg, S. (2008) 'Femininities/Masculinities and a Sense Self: Thinking Gendered Academic Identities and the Intellectual Self', *Gender and Education*, vol. 20, no. 3, pp. 209–221.

Coates, P. (2007) *American Perceptions of Immigrant and Invasive Species: Strangers on the Land*. University of California Press, Berkeley.

Colautti, R.J. and MacIssac, H.J. (2004) 'A neutral terminology to define "invasive" species', *Diversity and Distributions*, vol. 10, pp. 135–141.

Cooper, A. (2003) 'The indigenous versus the exotic: Debating natural origins in early modern Europe', *Landscape Research*, vol. 28, p. 51.

Costa-Pierce, B. (2003) 'Rapid evolution of an established feral tilapia (Orechodromis spp.): The need to incorporate invasion science into regulatory structures', *Bioinvasions*, vol. 5, pp. 71–84.

Cresswell, T. (1997) 'Weeds, plagues, and bodily secretions: A geographical interpretation of metaphors of displacement', *Annals of the Association of American Geographers*, vol. 87, pp. 330–345.

Crosby, A. (1986) *Ecological Imperialism*. Cambridge University Press, Cambridge.

Davis, M.A. (2011) 'Do native birds care whether their berries are native or exotic? No', *BioScience*, vol. 61, pp. 501–502.

Davis, M.A., Chew, M.K., Hobbs, R.J., Lugo, A.E., Ewel, J.J., Vermeij, G.J., Brown, J.H., Rosenzweig, M.L., Gardener, M.R., Carroll, S.P., Thompson, K., Pickett, S.T.A., Stromberg, J.C., Tredici, P.D., Suding, K.N., Ehrenfeld, J.G., Philip Grime, J., Mascaro, J. and Briggs, J.C. (2011) 'Don't judge species on their origins', *Nature*, vol. 474, p. 153.

Davis, M.A. and Thompson, K. (2001) 'Invasion terminology: Should ecologists define their terms differently than others? No, not if we want to be of any help', *Bulletin of the Ecological Society of America*, p. 206.

Davis, M.A., Thompson, K. and Grime, J.P. (2001, January–March) 'Charles S. Elton and the dissociation of invasion ecology from the rest of ecology', *Diversity and Distributions*, vol. 7, pp. 97–102.

Despret, V. and Stengers, I. (2012) *Les Faiseuses d'Histoire: Que Font les Femmes à la Pensée?* La Découverte, Paris.

Elton, C. (1958) *The Ecology of Invasions by Animals and Plants*. Methuen, London.

Ernwein, M. and Fall, J.J. (2015) 'Communicating invasion: Understanding social anxieties around mobile species', *Geografiska Annaler: Human Geography*, vol. 97, pp. 155–167.

Fall, J.J. (2013) 'Biosecurity and ecology: Beyond the nativist debate', in K. Barker, A. Dobson and S. Taylor (eds.) *Biosecurity: The Socio-Politics of Invasive Species and Infectious Diseases*. Earthscan/Routledge, Abingdon, pp. 167–181.

Fall, J.J. (2014) 'Governing mobile species in a climate-changed world', *Governing the Climate: New Approaches to Rationality, Power and Politics*, pp. 160–174.

Fuentes, M. (2018) 'Biological novelty in the anthropocene', *Journal of Theoretical Biology*, vol. 437, pp. 137–140.

Gardey, D. (2014) 'The reading of an oeuvre. Donna Haraway: The poetics and politics of life', *Zeitschrift für interdisziplinäre Frauen- und Geschlechterforschung*, vol. 32, pp. 86–100.

Gieryn, T.F. (1983) 'Boundary-work and the demarcation of science from non-science: Strains and interests in professional ideologies of scientists', *American Sociological Review*, pp. 781–795.

Gill, R. (2014) 'Academics, cultural workers and critical labour studies', *Journal of Cultural Economy*, vol. 7, pp. 12–30.

Gould, S.J. (1997) 'An evolutionary perspective on strengths, fallacies and confusions in the concept of native plants', in J. Wolschke-Bulmahn (ed.) *Nature and Ideology: Natural Garden Design in the Twentieth Century*. Dumbarton Oaks Research Library and Collection, Washington, DC.

Gröning, G. and Wolschke-Bulmahn, J. (2003) 'The native plant enthusiasm: Ecological panacea or xenophobia?', *Landscape Research*, vol. 28, p. 75.

Guiaşu, R.C. and Tindale, C.W. (2018) 'Logical fallacies and invasion biology', *Biology and Philosophy*, vol. 33, no. 34, pp. 1–24.

Hall, M. (2003) 'Editorial: The native, naturalized and exotic: Plants and animals in human history', *Landscape Research*, vol. 1, no. 1.

Haraway, D. (1988) 'Situated knowledges: The science question in feminism and the privilege of partial perspective', *Feminist Studies*, vol. 14, pp. 575–599.

Harding, S. (1991) *Whose Science? Whose Knowledge?: Thinking from Women's Lives*. Cornell University Press, Ithaca.

Harding, S. (2015) *Objectivity and Diversity: Another Logic of Scientific Research*. University of Chicago Press, Chicago.

Head, L. (2017) 'The social dimensions of invasive plants', *Nature Plants*, vol. 3, pp. 1–7.

Head, L. and Muir, P. (2004) 'Nativeness, invasiveness, and nation in Australian plants', *Geographical Review*, vol. 94, no. 2, pp. 199–217.

Helmreich, S. (2005) 'How scientists think about "natives", for example: A problem of taxonomy among biologists of alien species in Hawaii', *Journal of the Royal Anthropological Institute*, vol. 11, pp. 107–128.

Hoffmann, B.D. and Courchamp, F. (2016) 'Biological invasions and natural colonisations: Are they that different?', *NeoBiota*, vol. 29, p. 1.

Humair, F., Edwards, P.J., Siegrist, M. and Kueffer, C. (2014) 'Understanding misunderstandings in invasion science: Why experts don't agree on common concepts and risk assessments', *NeoBiota*, vol. 20, p. 1.

Kowarik, I. and Pyšek, P. (2012) 'The first steps towards unifying concepts in invasion ecology were made one hundred years ago: Revisiting the work of the Swiss botanist Albert Thellung', *Diversity and Distributions*, vol. 18, pp. 1243–1252.

Kühn, I., Pyšek, P. and Kowarik, I. (2017) Seven years of NeoBiota: The times, were they a Changin'?, *NeoBiota*, vol. 36, pp. 57–69. Pensoft Publishers. doi: 10.3897/neobiota.36.21926

Kull, C.A. (2018) 'Critical invasion science: Weeds, pests and aliens', in R. Lave, C. Biermann and S. Lane (eds.) *The Palgrave Handbook of Critical Physical Geography*. Palgrave, Cham.

Larson, B.M.H., Nerlich, B. and Wallis, P. (2005) 'Metaphors and biorisks: The war on infectious diseases and invasive species', *Science Communication*, vol. 26, no. 243, pp. 243–268.

Merton, R. (1968) 'The Matthew effect in science: The rewards and communication system in science are considered', *Science*, vol. 159, pp. 1–8.

Morley, L. (2003) *Quality and Power in Higher Education*. The Open University Press, Maidenhead.

Munro, D., Steer, J. and Linklater, W. (2019) 'On allegations of invasive species denialism', *Conservation Biology*, vol. 33, no. 4, pp. 797–802.

Nackley, L.L., West, A.G., Skowno, A.L. and Bond, W.J. (2017) 'The nebulous ecology of native invasions', *Trends in Ecology & Evolution*, vol. 32, pp. 814–824.

O'Brien, W. (2006) 'Exotic invasions, nativism and ecological restoration: On the persistence of a contentious debate', *Ethics, Place and Environment*, vol. 9, pp. 63–77.

Olwig, K.R. (2003) 'Natives and Aliens in the national landscape', *Landscape Research*, vol. 28, p. 61.

Oreskes, N. (1996) 'Objectivity or heroism? On the invisibility of women in science', *History of Science Society*, pp. 87–113.

Osborne, M.A. (2000) 'Acclimatizing the world: A history of the paradigmatic colonial science', *Osiris*, 2nd series, vol. 15, pp. 135–151.

Pereyra, P.J. (2016) 'Revisiting the use of the invasive species concept: An empirical approach', *Austral Ecology*, vol. 41, pp. 519–528.

Preston, C.D. (2009) 'The terms "native" and "alien": A biogeographical perspective', *Progress in Human Geography*, vol. 33, pp. 702–711.

Rémy, E. and Beck, C. (2008) 'Allochtone, autochtone, invasif: Catégorisations animales et perception d'autrui', *Politix*, vol. 21, no. 82, pp. 193–209.

Ricciardi, A. and Macisaac, H.J. (2008) 'The book that began invasion ecology: Charles Elton's 50-year-old text founded a field and is now cited more than ever', *Nature*, vol. 452, p. 34.

Ricciardi, A. and Ryan, R. (2018) 'The exponential growth of invasive species denialism', *Biological Invasions*, vol. 20, pp. 549–553.

Richardson, D.M., Pyšek, P. and Carlton, J.T. (2011) 'A compendium of essential concepts and terminology in invasion ecology', in D.M. Richardson (ed.) *Fifty Years of Invasion Ecology: The Legacy of Charles Elton*. Wiley-Blackwell, Chichester.

Ritvo, H. (1997) *The Platypus and the Mermaid and Other Figments of the Classifying Imagination*. Harvard University Press, Cambridge.

Robbins, P. (2004) 'Comparing invasive networks: Cultural and political biographies of invasive species', *Geographical Review*, vol. 94, pp. 139–156.

Roth, W.-M. (2005) 'Making classifications (at) work: Ordering practices in science', *Social Studies of Science*, vol. 35, pp. 581–621.

Rotherham, I.D. (2017) *Recombinant Ecology: A Hybrid Future?* Springer, Cham.

Russell, J.C. and Blackburn, T.M. (2017) 'Invasive Alien species: Denialism, disagreement, definitions, and dialogue', *Trends in Ecology & Evolution*, vol. 32, no. 5, pp. 312–314.

Sagoff, M. (2018) 'What is invasion biology?', *Ecological Economics*, vol. 154, pp. 22–30.

Saul, W.-C., Shackleton, R.T. and Yannelli, F.A. (2017) 'Ecologists winning arguments: Ends don't justify the means: A response to Begon', *Trends in Ecology & Evolution*, vol. 32, pp. 722–723.

Schlaepfer, M.A. (2018) 'Do non-native species contribute to biodiversity?', *PLoS Biology*, vol. 16, no. 4, p. e2005568.

Schlaepfer, M.A., Sax, D.F. and Olden, J.D. (2011) 'The potential conservation value of non-native species', *Conservation Biology*, vol. 25, pp. 428–437.

Shackleton, R.T., Richardson, D.M., Shackleton, C.M., Bennett, B., Crowley, S.L., Dehnen-Schmutz, K., Estevez, R.A., Fisher, A., Kueffer, C., Kull, C.A., Marchante, E., Novoa, A., Potgieter, L.J., Vaas, J., Vaz, A.S. and Larson, B.M.H. (2019) 'Explaining people's perceptions of invasive alien species: A conceptual framework', *Journal of Environmental Management*, vol. 229, pp. 10–26.

Simberloff, D. (2011) 'Non-natives: 141 scientists object', *Nature*, vol. 475, p. 36.

Simberloff, D. and Vitule, J.R.S. (2013) 'A call for an end to calls for the end of invasion biology', *Oikos*, vol. 123, pp. 408–413.

Smout, T.C. (2003) 'The Alien species in 20th-century Britain: Constructing a new vermin', *Landscape Research*, vol. 28, p. 11.

Smyth, J. (2017) *The Toxic University: Zombie Leadership, Academic Rock Stars and Neoliberal Ideology*. Palgrave Macmillan, London.

Subramaniam, B. (2001) 'The aliens have landed! Reflections on the rhetoric of biological invasions', *Meridians: Feminism, Race, Transnationalism*, vol. 2, pp. 26–40.

Takacs, D. (1996) *The Idea of Biodiversity: Philosophies of Paradise*. Johns Hopkins Press, Philadelphia.

Thompson, K. (2014) *Where Do Camels Belong? The Story and Science of Invasive Species*. Profile Books, London.

Tiles, M. (1996, original 1987) 'A science of mars or of venus?', in E. Fox Keller and H. Longino (eds.) *Feminism & Science*. Oxford University Press, Oxford and New York.

Uekötter, F. (2007) 'Native plants: A Nazi obsession?', *Landscape Research*, vol. 32, no. 3, pp. 379–383.

Valéry, L., Fritz, H. and Lefeuvre, J.-C. (2013) 'Another call for the end of invasion biology', *Oikos*, vol. 122, pp. 1143–1146.

Warren, C.R. (2007) 'Perspectives on the "alien" versus "native" species debate: A critique of concepts, language and practice', *Progress in Human Geography*, vol. 31, pp. 427–446.

Warren, C.R. (2008) 'Alien concepts: A response to Richardson et al.', *Progress in Human Geography*, vol. 32, pp. 299–300.

Waterton, C. (2002) 'From field to fantasy: Classifying nature, constructing Europe', *Social Studies of Science*, vol. 32, pp. 177–204.

Watson, H.C. (1835) *Remarks on the Geographical Distribution British Plants Chiefly in Connection to Latitude, Elevation and Climate*. Longman Rees Orme Brown Green and Longman, London.

Willis, K.J. and Birks, H.J.B. (2006) 'What is natural? The need for a long-term perspective in biodiversity conservation', *Science*, vol. 314, pp. 1261–1265.

Wilson, J.R.U., Dormontt, E., Prentis, P.J., Lowe, A. and Richardson, D.M. (2009) 'Something in the way you move: Dispersal pathways affect invasion success', *Trends in Ecology & Evolution*, vol. 24, pp. 136–144.

Wilson, J.R.U., Garcia-Diaz, P., Cassey, P., Richardson, D.M., Pyšek, P. and Blackburn, T.M. (2016) 'Biological invasions and natural colonisations are different: The need for invasion science', *NeoBiota*, vol. 31, p. 87.

3

INDIGENOUS BIOSECURITY

Past, present and future

Simon J. Lambert and Melanie Mark-Shadbolt

Introduction

Environmental management has always been culturally and economically significant to Indigenous communities, and these communities increasingly recognise biosecurity as vital to the maintenance of their traditional ecosystems and therefore fundamental to the continuation of their cultures. The arrival of myrtle rust (*Austropuccinia psidii*) in New Zealand in 2017 serves as an example of how Indigenous collectives, in this case Māori, can work towards a better biosecurity strategy. The incursion of myrtle rust was foreseen as early as 2010 (Ramsfield et al., 2010) and threatens several culturally and economically important species of the Myrtaceae family. These include Mānuka (*Leptospermum scoparium*), whose honey possesses medicinal properties and is a valuable export commodity, and Pōhutukawa (*Metrosideros excels*), a coastal tree whose bright red flowers emerge in December and so is known as the New Zealand Christmas tree. In 2005 the disease had become established in Hawaii, where it inflicted considerable damage to the Ohia (*Metrosideros polymorpha*), an endemic forest species that holds an important place in native Hawaiian culture (Loope, 2010). Appearing in 2010 in 'upwind' Australia (Teulon et al., 2015), it was a matter of when, not if, this pathogen would arrive in Aotearoa New Zealand.

From initial discovery of the rust on Raoul Island (over 1,000 kilometres to the north of New Zealand) in April 2017, the disease was identified in a commercial nursery on mainland New Zealand in May. Within a year it had spread to over 500 locations, with some commentators criticising the government's response (Morton, 2018). Despite the risks of myrtle rust being well known among scientific experts, New Zealand authorities seemed poorly prepared to deal with this biosecurity threat (Morton, 2018). It was in the context of rapid spread, an impromptu official response and unknown effects on native biodiversity that Māori self-organised and acted within their cultural networks (Mark-Shadbolt, 2017). A Māori Biosecurity Network was established – named Te Tira Whakamātaki (TTW), 'The Watchful Ones' in Māori – who lobbied government, organised community meetings to inform people and initiated basic training for *kaitiaki* (environmental guardians) to help manage what was clearly a rapidly spreading disease (Lambert et al., 2018).

Another significant plant pathogen for Maori is a Phytophthera that threatens the giant Kauri tree (*Agathis australis*) (Beever et al., 2007). While some biological tools are being explored, including projects led by Māori, the most fundamental response was a call by local Māori for a *rāhui*, a traditional ban or quarantine used by tribal leaders to modify human behaviour in protecting people and ecosystems (Mark-Shadbolt et al., 2018). Such an approach would accept Indigenous rights – as articulated in the Treaty of Waitangi, signed between Māori and the British Crown in 1840 – to manage their resources, albeit in collaboration with other communities. Although a ban was later formalised by local government authorities (Tokalau, 2019), breaches continued to occur, and fences put in place to prevent access were vandalised.

The concept of Indigenous biosecurity exceeds general and expert understanding and use of the term. Scientific terminology is often unhelpful in convincing Indigenous people of the possibilities of engagement in biosecurity research. The terms that are often employed by scientists – 'alien', 'colonising', 'native' – are emotionally loaded when used in the context of Indigenous communities, and the disciplinary label 'invasion science' is a tragically ironic term. For Indigenous Peoples the modern history of biothreats began with the spread of virulent infectious diseases by Europeans as an outcome of contact. Dunbar-Ortiz (2014) argues against an ahistorical interpretation of these epidemics, pointing out that war, oppression, loss of land and access to resources and the breakdown of Indigenous trading networks ultimately framed how infections impacted on Indigenous populations. We take Dunbar-Ortiz's critical stance to draw attention to narratives of risk and security concerning Indigenous communities: colonisation positioned Indigenous Peoples as savage, uncivilised – indeed, perhaps un-civilisable, even subhuman. However, this chapter will not unpack Eurocentric fallacies, as the racism at their heart is addressed through a growing body of easily accessed literature (Baber, 1996; Fanon, 1967; Smith, 1999). Our point is that imperial expansion was a *biological* phenomenon that effected, inter alia, social and cultural changes, as well as ecological disruption (Crosby, 1986). The marginalisation of Indigenous communities negatively impacted, in multiple and compounding ways, their ability to secure their biotic environments at all scales. While the legacy of settler colonial resource extraction still centres these supposedly peripheral lands in production, tourism, recreation and national identity, Indigenous Peoples are politically sidelined and pushed or constrained in often isolated and fragmented territories to maintain settler colonial political economic power (Stewart-Harawira, 2005). Indigenous Peoples address biosecurity from the margins; inserting Indigenous voices and centring Indigenous worldviews is central to rethinking more effective biosecurity approaches.

Indigenous development goals, including secure and healthy biological resources, are inherently political. Goals are built collectively through ongoing Indigenous resistance; they involve socio-cultural movements inspired by Indigenous renaissance and are economically significant at the scale of Indigenous life experiences.

The literature related to biosecurity is diverse; not only are there many areas of fundamental science relevant to biosecurity – soil science, biochemistry and molecular genetics, to name three – but integrative disciplines such as environmental history, ecology and political ecology also have important insights. In addition, Indigenous perspectives will transcend academic disciplinary structures. Oral narratives of Indigenous spirituality are as likely to feature as the latest biocontrol innovation, an approach best described as trans-systemic. Climate change, environmental degradation from toxic mal-development and the marginal status of Indigenous groups in many states overloads Indigenous communities and those members tasked with implementing Indigenous ways into practical and sustainable outcomes.

We argue that all Indigenous Peoples have the right to inform and manage biosecurity in the protection of their economic and cultural wealth. TTW and other groups (for example, Marsh et al., 2018) provide a logistical efficiency (through a culturally framed community volunteerism) that may be necessary (if not sufficient) for an effective response to ongoing biosecurity alerts. But the empowerment of Indigenous participants also goes some way to affirming a social

Table 3.1 United Nations Declaration on the Rights of Indigenous Peoples, 2007

Introduction	'Recogniz[es] that respect for indigenous knowledge, cultures and traditional practices contributes to sustainable and equitable development and proper management of the environment'.
Article 29, Section 1	'Indigenous peoples have the right to the conservation and protection of the environment and the productive capacity of their lands or territories and resources. States shall establish and implement assistance programmes for indigenous peoples for such conservation and protection, without discrimination'.
Article 32, Section 3	'Indigenous peoples have the right to the conservation and protection of the environment and the productive capacity of their lands or territories and resources. States shall establish and implement assistance programmes for indigenous peoples for such conservation and protection, without discrimination'.

license for the biosecurity sector, a sector that can and does impose onerous duties on citizens as well as regularly deploying toxic chemicals. This empowerment would be an extension of previous calls for the right of Indigenous Peoples to determine their own development, rights now enshrined in the United Nations Declaration on the Rights of Indigenous Peoples (UNDRIP; United Nations, 2007). UNDRIP acknowledges the rights of Indigenous Peoples and their communities, with several sections directly or indirectly relevant to biosecurity (see Table 3.1).

In addition to UNDRIP, some guidance has been produced in the extensive multilateral organisations that have developed since the end of World War II, notably on biological diversity (UN Convention on Biological Diversity, 2018), with Indigenous Knowledges (IK) now accepted as having a key role in future sustainable development (Rahman, 2016; Segger and Phillips, 2015). Lambert and Scott (2019), for example, detail how IK can contribute to disaster risk reduction discourse, and environmental degradation has undermined Indigenous strengths (including cultural well-being) in coping with hazards and underlined biosecurity as a feature of ecosystem maintenance. Such cross-fertilisation of ideas through the synthesis of IK and mainstream science is necessary for ethical and efficacious biosecurity.

In this chapter we review literature on traditional Indigenous biosecurity practices. This literature demonstrates the considerable insight contained within IK in protecting key species and ecosystems on which Indigenous communities relied. We discuss how colonisation undermined and dismantled Indigenous environmental practices. The definition and treatment of so-called 'feral' species provides a useful window on Indigenous worldviews and illustrates how these worldviews are marginalised. We then briefly outline the outcome of a long history of Indigenous resistance that has led to the reassertion of IK through an acknowledgement of the rights of Indigenous Peoples. These rights are evident in the growing demand for Indigenous data sovereignty as the cultural renaissance of Indigenous Peoples supports Indigenous engagement in data collection on, among (many) other things, biosecurity. Indigenous participants are increasingly well informed as, with the support of their communities, they venture beyond any territorial boundaries to take part in national and international debates on the sustainability of their space and place that includes, of course, the future of the planet.

Traditional biosecurity approaches

The development of subsistence agriculture and horticulture by Indigenous Peoples, often over millennia, offers the clearest evidence of IK informing biosecurity. At their most simple, these methods involve careful plot location, systematic crop rotation and intercropping with specific actions to reduce pests, such as removing insects, disposing of infected plants, timing of weeding, placement of traps and the use of scarecrows and smoke, methods that are recognisable in some forms of modern agroecological expressions but often absent from large-scale industrial agriculture (Abate et al., 2000; Chandola et al., 2011; Grzywacz et al., 2014; see Wallace et al., this collection, on biosecurity and industrial agriculture). Indigenous rice farmers in Sri Lanka provided perches in their paddy fields for birds that preyed on insect pests and protected their breeding habitats near their fields. They also encouraged a non-venomous snake that ate rats, a major pest of the paddy farmers, by leaving woodland areas near the paddy fields for shelter and breeding (Ulluwishewa, 1993). Other methods involved the use of light traps to attract and kill insect pests, a technique often accompanied by religious iconography and ceremony. Biochemical approaches utilised natural plant toxins to repel or control pests. These include the crushing of the leaves and seeds of certain plant species or hanging the leaves and flowers around fields, essentially exposing pests to naturally occurring compounds that act as toxins for specific cases. Māori burned the pungent leaves of the kawakawa (*Piper excelsum*) near their valuable kumara crops (*Ipomoa batatas*). Mukandiwa et al. (2016) reviewed the various uses of *Clausena anisate*, a member of the citrus family, to control pests and parasites in Africa. Nearly 50 compounds have been identified within the plant that can control various pests, including mosquitoes (*Anopheles* spp.), red flour weevil (*Tribolium castaneum*), cowpea weevil (*Callosobruchus maculatus*) and cotton bollworm (*Helicoverpa armigera*). Such methods were refined over time, transmitted inter-generationally and effective in the contexts of community practice. At the larger scale of landscape, fire management has also been practised by many Indigenous Peoples to manage forest pests including by the Yucatec Mayan of Mexico (Humphries, 1993), the Bambara and Malinké of Southern Mali (Laris, 2002) and Native Americans of California (Lewis, 1993 [1973]).

Not only have these older approaches been sidelined and forgotten, but new agricultural as well as pest management techniques have led to an increasing reliance on external inputs, the indebtedness of farmers and toxic residues accumulating in the environment (Chamala, 1990; Yapa, 1993). Yucatec Mayan farmers have sought to participate in the large Mexican agricultural sector, but transitioning into more sedentary agricultural approaches saw a growing reliance on pesticides and a loss of security for those farmers with smaller holdings and less resources. The IK held by elderly Sri Lankan paddy growers was undermined by the spread of new technologies associated with the so-called 'Green Revolution' that also saw the embedding of externally sourced technologies and often crippling debt, the rationalisation of small land holdings into larger units and reduced sustainability (Yapa, 1993). Concerns over the sustainability of modern production have not abated as varieties remain susceptible to pests and disease and the ongoing use of synthetic pesticides and herbicides continues to poison the environment.

The commercial potential of IK in biosecurity is barely appreciated and offers alternatives to the expensive and problematic synthetic chemicals becoming the primary option promoted by state and agricultural industry discourse (Ulluwishewa, 1993). However, Indigenous conceptions of intellectual property are often at odds with Western-defined 'IP' and its ownership and commercialisation. For modern biosecurity to be robust from an Indigenous perspective, there must be an appropriately framed approach and respectful partnering not in isolated silos but across the entire expanse of Indigenous biological heritage.

Colonisation as new invasive species, pest and diseases

Colonisation saw a massive change to political-economic and socio-ecological systems; these two aspects are inextricably linked, as Indigenous cultures have always acknowledged. To illustrate the political-economic rupture with pre-contact existence that Indigenous Peoples underwent, one example will serve our purpose: sovereign Indigenous Peoples became research objects. Their artworks, sacred objects, clothes and stories were 'collected' (gifted, bought, stolen), measured, stored (often in museum basements) and displayed around the imperial worlds. Indigenous lives, skulls, sexuality, foods – *everything they valued* – were cursorily examined, superficially analysed and then cast aside when new ('innovative') objects appeared. Indigeneity became so much data in a world where data was increasingly interpreted as a resource from which wealth could be extracted.

European settlers also deliberately introduced species for food, sport and aesthetic reassurance, simultaneously culling those species deemed 'pests'. While not all settler introductions were successful, the expansion of the post-contact European ecological portmanteau was inexorable (Crosby, 1986). European travellers to Australia in the early 19th century found about 20 weeds, most from England but also from South Africa and South America. One commentator in the mid-1900s complained the Tasmanian vegetation was boring because it was too English (Low, 1999). This colonial policy of biotic replacement was exemplified by the eradication of the bison, a key food species for Indigenous Peoples of the North American prairies, the removal of which caused famine, disease and the loss of a major cultural icon (Daschuk, 2013). Other key cultural species have been undermined by the modern spread of pests and disease. Several tribes in North America, including Wabanakis and Mohawk, value the ash tree (*Fraxinus* spp.) for its use in basketry. Emerald ash borer (*Agrilus planipennis* Fairmaire), an invasive insect native to Asia that attacks ash, was found in southeastern Michigan and southern Ontario in 2002. The disease spread rapidly to 27 US states and two Canadian provinces and has killed millions of North American ash trees (Costanza et al., 2017). While there are many factors determining the arrival of a new pest, the imposition of colonial development has led to a catastrophic loss of biological heritage, and Indigenous Peoples have been forced to adapt to new bioeconomies.

Exotic species: feral to whom?

The worldwide movement of people, and the purposeful and accidental transfer of other biota – fellow travellers of imperial expansion and modern globalisation – continues apace with globalisation and climate change. The impacts on Indigenous Peoples are generally seen as negative, not least by Indigenous Peoples themselves. However, as Pfeiffer and Voeks (2008) observe, invasive species will affect Indigenous communities in many often unpredictable and even contradictory ways. Although culturally significant species may be displaced, augmentation of the variety of potentially useful resources can also occur, and some culturally valued species may themselves 'invade' or be brought with migrating Indigenous groups, 'providing biocultural continuity to diaspora communities' (p. 282). The Anishnaabe of North America interpret plants, as they interpret other beings, as persons that are assembled into nations and not a disjointed 'species'. Therefore, the appearance of new plant nations is seen as 'a natural form of migration' (Reo and Ogden, 2018). There is no clear-cut categorisation of 'pest' or 'invasive' species for Indigenous communities in biosecurity debates.

Again drawing on Māori experiences, Haami (1992) records a number of cultural traditions that centre on kiore, the Polynesian rat (*Rattus exulans*) introduced by Māori many hundreds of years ago. Oral and recorded histories include genealogical links, prayers and proverbs and traditional harvesting practices. Haami makes a compelling argument that Māori did not interpret the kiore as 'a nuisance or worse but as something of considerable value' (p. 75). Disagreement between a local tribe and government representatives over the status of kiore (that can predate on endangered native birds and lizards) opens up an Indigenous worldview into biosecurity. A tribal spokesperson said the so-called 'pest' was valued by the tribe and was considered significant as they were deliberately brought to Aotearoa New Zealand by their ancestors, meaning descendants inherited guardianship duties (Tahana, 2010). It was also argued that the kiore's arrival paralleled the story of how Māori journeyed across the Pacific, a heritage they wished to acknowledge and protect. Eventually an agreement between the tribe and the Department of Conservation saw the kiore protected on a group of islands, and the protection and cultural harvest of this species continues (Waitangi Tribunal, 2019).

Similar cultural-ecological relationships are found in other Indigenous Peoples. In Australia, Aborigines acknowledge the camel (*Camelus dromedaries*) as introduced (the animal was brought by European settlers in the mid-1800s) but accept this feral species as 'belonging to country' as the animals they observed had grown up there (Vaarzon-Morel and Edwards, 2012). Due to evolving historical relationships with the animals – as food, fibre, transport and pets – and the inclusive character of Aboriginal society, many Aborigine believe the camel has 'earned its place' in Australia.

Indigenous concerns include not just the targeting of a species constructed by Western science and biosecurity regimes as unwanted but also the management approach adopted. This is exemplified in conflicts over the use of toxins to 'manage' a biosecurity issue: the '1080' debate in New Zealand. Many Māori are angered by the method of aerial drops of bait poisoned with the pesticide sodium fluoroacetate (Compound 1080) to control introduced Brush-tailed possums (*Trichosurus vulpecula*) and other vertebrates (Ogilvie et al., 2010). Although research strongly points to the success of 1080 in allowing native vegetation to regenerate and endangered birds to recover, the debate has seen impassioned, even violent, opposition (Northcott and Persico, 2018). Māori researchers and communities sought acceptable solutions, and, through integrating science and mātauranga approaches, a naturally occurring, native plant-sourced toxin (tutin, from *Coriaria arborea*) was identified and found to be effective (Ogilvie et al., 2019). Further research and testing are taking place.

These brief examples show that Indigenous participants enable the reinterpretation of events and processes, possibly challenging fundamental concepts and approaches of invasion biology and biosecurity (see also Himsworth et al., 2010; Robinson and Whitehead, 2003). The value of any species will be culturally framed but is also not static as Indigenous Peoples adapt and demand a voice in pest management programmes. In fighting back from the margins, Māori and other Indigenous Peoples contest approaches, processes and funding within political, economic, epistemological and social spheres.

The reassertion of Indigenous biosecurity approaches

Acknowledging Indigenous rights and supporting Indigenous participation in wider environmental management, including biosecurity, remains a significant challenge for many state and private actors. Mescalero Apache synthesise youth training, community development and tourism through fish farms that engage with federal science programmes to maintain disease-free stock for repopulating local waterways (Southwestern Native Aquatic Resources and Recovery Centre, 2014). In Canada, First Nations seeking to re-establish bison on the prairies are engaged in debates on the ownership of land, control of weeds, sourcing of genetic stock that is 'pure' (bison were interbred with cattle from the early 1900s; see Boyd, 1914) and the risk of bovine diseases such as tuberculosis and brucellosis (Nishi et al., 2002). The mobilisation by Māori to combat myrtle rust followed a series of biosecurity incursions that undermined the credibility and capacity of state institutions. The Māori biosecurity approach of TTW was to argue that local Māori communities were ideally placed to monitor myrtle rust as community members were present on the ground and had cultural roles as environmental guardians. Alongside the clear value of IK, a key role for Indigenous communities therefore comes from their location and connection to land as first responders to an incursion of a pest or disease. All three Indigenous ventures are inherently political through Indigenous resistance, involve socio-cultural movement inspired by Indigenous renaissance and are economically significant at the scale of Indigenous lived experience. Each demonstrates an Indigenous community aspiring and acting to move from the margins of biosecurity debates to a central position of influence and benefit.

The use of traditional pest control methods has received considerable interest in the agriculture sector as opposition to synthetic control methods has grown (Cherry and Gwynn, 2007). In Africa, the pesticidal plant *Tephrosia vogelii*, and the baculovirus *Spodoptera exempta* nucleopolyhedrovirus, have shown potential as locally sourced, inexpensive pest control tools (Grzywacz et al., 2014), among other examples. While ethical guidelines exist and are increasingly made compulsory in research proposals (for example the Canadian OCAP framework governing Ownership, Control, Access and Possession of Aboriginal knowledge; see First Nations Information Governance Centre, 2014), these are neither universal nor strictly adhered to. Where IK are drawn on for insight, understanding and the development and commercialisation of natural pesticides and herbicides, the timely (as in an initial) invitation to Indigenous knowledge holders and elders is essential for any ethical process (Coombes et al., 2014).

Biosecurity incursions are often subject to extreme state control, with community concerns – Indigenous or non-Indigenous – sidelined or ignored (Marzano et al., 2017). While the ratification of the United Nations Declaration on Indigenous Peoples in 2007 (United Nations, 2007) has articulated the growing demands for Indigenous rights, Indigenous communities too often remain on the periphery of debate and policy formation. Yet as Indigenous Peoples assert their rights and responsibilities in maintaining their biological heritage, spokespeople are threatened and even at risk of assassination by state or private militia (The Guardian/Global Witness, 2018). The murder of Honduran Berta Cáceres, a female environmental activist and Indigenous leader, in 2016 reverberated beyond her networks to reveal the risks faced by Indigenous spokespeople (Ardon and Flores, 2017). Attempts by communities to engage with research and policy institutions are fraught as non-Indigenous participants typically lack awareness of the expectations and accountability inherent in Indigenous worlds. Engagement can also be challenging for Indigenous participants as urbanisation, ongoing poverty, alienated youth and structural racism undermine Indigenous self-determination.

We finish this chapter with a brief description of the latest Indigenous attempts to build capacity and capability so communities can drive relevant biosecurity strategies in protecting their biological heritage.

An international Indigenous biosecurity gathering

The Māori biosecurity network noted previously has argued that biosecurity is the foundation of environmental management because a single significant biological invasion could undermine all of a nation's conservation work and primary sectors. A strong biosecurity system, with knowledge of risk pathways, is a first step in protecting and managing Indigenous environmental resources.

With this in mind, New Zealand researchers, supported by TTW, bought together Indigenous representatives from mainland US and the US territories Hawaii and Guam, Fiji, Samoa, two Australian Aboriginal nations and three nations based in Canada for the inaugural International Indigenous Knowledge and Values *Hui Taumata* (conference) on Biosecurity in Tauranga in September 2019. Attendees were greeted at the event through cultural ceremonies of welcome and protection, with the first day dedicated to what Māori call *whakawhanaungatanga*, the forging of family-like relationships among participants (an interaction widespread among Indigenous communities).

The event was notable to many international Indigenous visitors as it was a Māori-centric event over an extended time, with opportunity for cultural exchanges. People shared stories about environmental issues threatening Indigenous ways of being. A range of topics were discussed, including better frameworks for decision-making, early sharing of successes and failures to support better environmental outcomes and the engagement of Indigenous youth. A constant refrain was 'Nothing about us without us' and ensuring that engagement models promote inclusivity and collectivity.

One key theme was that communities from around the globe needed opportunities to find solutions to biosecurity issues in their respective lands and waters. Australian Aboriginal attendees presented a multi-year project, 'Building Resilience in Indigenous Communities through Engagement: A Focus on Biosecurity Threats'. The project focused on 'developing and implementing scientific knowledge, tools, resources and capacity to safeguard the plant industries and regional communities of Australia, New Zealand and Indonesia from the economic, environmental and social consequences of plant pests and diseases' (Marsh et al., 2018). Supported by the mainstream Australian Plant Biosecurity Cooperative Research Centre, the research led to the development of the Indigenous Engagement Model (IEM), a new mechanism for engaging with Indigenous communities for more effective surveillance of pest and disease incursion (Baxter and Hamilton, 2016). The project argues that 'success' for a biosecurity response depends on all stakeholders understanding the impact of a biosecurity risk to native ecosystems. Aboriginal participants emphasised the importance of industry maintaining connections with Indigenous communities. Beyond guiding all Australian biosecurity stakeholders and developing more understanding of what is required in a peacetime and/or rapid biosecurity response, the IEM approach also enables the benchmarking of Indigenous engagement. The model provides universal guidelines applicable for all end users, but the authors note a risk is that the IEM model is used in an ad hoc manner. Overall, they argue that the adoption of this model by communities, industry and stakeholders will lead to more responsiveness and preparedness to future incursion.

The space forced open by the response to the threat of myrtle rust by Te Tira Whakamātaki led to a coordinated Māori response in policy, research and community contexts. The network also collaborated on the development of a phone app with several agency partners to aid community members in identifying and managing the disease (SCION, 2017). Participating community-led culturally appropriate seed-banking was identified by Māori community members as important in securing the long-term future of their valued plant species; equipment, training and collecting were successfully coordinated, with the support of international collaborators. And while the disease seems established in New Zealand and unlikely to be eradicated, a Māori Biosecurity Network is likewise well established and has a growing presence in New Zealand's biosecurity strategies.

At the end of the summit, all participants joined with local Māori in declaring the need for an international Indigenous biosecurity network to raise awareness and encourage nation states to open space for Indigenous representatives and institutions in their biosecurity and environmental decision-making. Effective biosecurity strategies will need to engage with Indigenous communities and institutions as a matter of course. Planning is underway for funding proposals and exchange visits (an expression of the culturally framed commitment to reciprocity) with pledges to work collaboratively.

This one event enables some forecasting into the future of biosecurity in many regions of the world: Indigenous communities will continue to claim rights; seek empowerment in legal, political and economic contexts; participate and inform technological development; assist in access to and accuracy of relevant data; and fulfil surveillance duties that can add value to state strategies and operations. Increasingly, non-Indigenous allies are appearing in key supporting roles as events such as the Australian wildfires of 2019–20 see Indigenous practices receiving serious and more respectful political and scientific attention (Kuz, 2020).

Conclusions

Protecting biotic resources is both a concept and practice that is common to Indigenous experiences and ways of being. While 'biosecurity' may not be a familiar term to many Indigenous communities, traditional practices included pest and disease management methods that were sufficient for their needs, grounded in Indigenous philosophies and place-based local knowledges and sensitive to community understanding and aspirations. Knowledge of these traditional approaches remains in many Indigenous communities but is, like so much other IK, vulnerable to a decline in knowledge holders, dilution or dismissal by settler colonial forces and urbanisation and changes in Indigenous lifestyles.

However, as Indigenous Peoples (re)assert themselves in new research and policy spaces such as biosecurity, they engage in multiple ways through collaborative programmes, governance processes such as councils, political debates through advocacy, activism and protest. Such actions are a modern expression of their traditional rights. As the experiences of Māori show, integrating IK into modern biosecurity networks will be contested politically, epistemologically, methodologically, socially and culturally; all Indigenous Peoples know how brutal these contests can be. Highlighting Indigenous governance and cultural institutions is important as Indigenous approaches have been much maligned and attacked as a deliberate strategy of settler colonisation. Resistance and empowered participation by Indigenous communities is both an expression of their culture and an opening for important insights and logistical efficiencies for biosecurity events within their territories.

Indigenous philosophies also challenge the designation of value at multiple scales, from the molecular and individual species to whole ecosystems, landscapes and waterways. Indigenous worldviews in the biosecurity space challenge interpretations of 'introduced', 'pest', 'control' and 'management' that can only strengthen collective understanding and effectiveness. These adaptive and innovative Indigenous frameworks embody the expression of Indigenous rights in self-determined development strategies and demonstrate the unlimited potential of Indigenous Peoples as they step in from the margins and quite possibly take central roles in modern biosecurity.

References

Abate, T., van Huis, A. and Ampofo, J.K.O. (2000) 'Pest management strategies in traditional agriculture: An African perspective', *Annual Review of Entomology*, vol. 45, no. 1, pp. 631–659.

Ardon, P. and Flores, D. (2017) 'Berta lives: COPINH continues', *International Journal on Human Rights*, vol. 25, pp. 109–118.

Baber, Z. (1996) *The Science of Empire: Scientific Knowledge, Civilisation and Colonial Rule in India*. State University of New York Press, Albany.

Baxter, P. and Hamilton, G. (2016) *An Analysis of Future Spatiotemporal Surveillance for Biosecurity*. Plant Biosecurity CRC/School of Earth, Environmental and Biological Sciences, Queensland University of Technology, Brisbane.

Beever, R.E., Harman, H. Waipara, N., Paynter, Q., Barker, G. and Burns, B. (2007) *Native Flora Biosecurity Impact Assessment*. Landcare Research Ltd, Lincoln.

Boyd, M.M. (1914) 'Crossing bison and cattle: First cross dangerous but results are better in each succeeding generation: Hope of taking fur and hump of bison and placing them upon back of Domestic Ox', *Journal of Heredity*, vol. 5, no. 5, pp. 189–197.

Chamala, S. (1990) 'Social and Environmental Impacts of Modernization of Agriculture in Developing Countries', *Environmental Impact Assessment Review*, vol. 10, no. 1–2, pp. 219–231.

Chandola, M., Rathore, S. and Kumar, B. (2011) 'Indigenous Pest Management Practices Prevalent among Hill Farmers of Uttarakhand', *Indian Journal of Traditional Knowledge*, vol. 10, no. 2, pp. 311–315.

Cherry, A.J. and Gwynn, R.L. (2007) 'Perspectives on the development of biological control agents in Africa', *Biocontrol Science and Technology*, vol. 17, no. 7, pp. 665–676.

Coombes, B., Johnson, J.T. and Howitt, R. (2014) 'Indigenous geographies III: Methodological innovation and the unsettling of participatory research', *Progress in Human Geography*, vol. 38, no. 6, pp. 845–854.

Costanza, K.K.L., Livingston, W.H., Kashian, D.M., Slesak, R.A., Tardif, J.C., Dech, J.P., Diamond, A.K., Daigle, J.J., Ranco, D.J., Neptune, J.S., Benedict, L., Fraver, S.R., Reinikainen, M. and Siegert, N.W. (2017) 'The precarious state of a cultural keystone species: Tribal and biological assessments of the role and future of black ash', *Journal of Forestry*, vol. 115, no. 5, pp. 435–446.

Crosby, A.W. (1986) *Ecological Imperialism: The Biological Expansion of Europe, 900–1900*. Cambridge University Press, Cambridge.

Daschuk, J. (2013) *Clearing the Plains: Disease, Politics of Starvation, and the Loss of Aboriginal Life*. University of Regina Press, Regina.

Dunbar-Ortiz, R. (2014) *An Indigenous Peoples' History of the United States*. Beacon Press, Boston.

Fanon, F. (1967) *The Wretched of the Earth*. Penguin, London.

First Nations Information Governance Centre (2014) *The Path to First Nations Information Governance: The First Nations Information Governance Centre: Ownership, Control, Access and Possession (OCAP™): The Path to First Nations Information Governance*. The First Nations Information Governance Centre, Ottawa.

Grzywacz, D., Stevenson, P.C., Mushobozi, W.L., Belmain, S. and Wilson, K. (2014) 'The use of indigenous ecological resources for pest control in Africa', *Food Security*, vol. 6, no. 1, pp. 71–86.

The Guardian/Global Witness (2018) *Almost Four Environmental Defenders a Week Killed in 2017*. Guardian/Global Witness, London.

Haami, B.J.T.M. (1992) 'Cultural knowledge and traditions relating to the Kiore Rat in Aotearoa: Part 1: A Maori perspective', pp. 5–22, in R.J. Morrison, P.A. Geraghty and L. Crowl (eds.) *Science of Pacific Island Peoples: Fauna, Flora, Food and Medicine*. Institute of Pacific Studies, Suva, pp. 65–76.

Himsworth, C.G., Elkin, B.T., Nishi, J.S., Neimanis, A.S., Wobeser, G.A., Turcotte, C. and Leighton, F.A. (2010) 'An outbreak of bovine tuberculosis in an intensively managed conservation herd of wild bison in the Northwest territories', *The Canadian Veterinary Journal*, vol. 51, no. 6, pp. 593–597.

Humphries, S. (1993) 'The intensification of traditional agriculture among Yucatec Maya Farmers: Facing up to the dilemma of livelihood sustainability', *Human Ecology*, vol. 21, no. 2, pp. 87–102.

Kuz, M. (2020) 'Fighting "bad" fire with good: Australia revisits an Aboriginal tradition', in *Christian Science Monitor*. Church of Christ, Boston.

Lambert, S. and Scott, J.C. (2019) 'International disaster risk reduction strategies and indigenous peoples', *International Indigenous Policy Journal*, vol. 10, no. 2. doi: 10.18584/iipj.2019.10.2.2

Lambert, S., Waipara, N., Black, A., Mark-Shadbolt, M. and Wood, W. (2018) 'Indigenous biosecurity: Māori responses to Kauri Dieback and Myrtle Rust in Aotearoa New Zealand', in J. Urquhart, M. Marzano and C. Potter (eds.) *The Human Dimensions of Forest and Tree Health: Global Perspectives*. Springer International Publishing, Cham, pp. 109–137.

Laris, P. (2002) 'Burning the seasonal mosaic: Preventative burning strategies in the wooded savanna of Southern Mali', *Human Ecology*, vol. 30, no. 2, pp. 155–186.

Lewis, H.T. (1993 [1973]) 'Patterns of Indian burning in California: Ecology and ethnohistory', in T.C. Blackburn and K. Anderson (eds.) *Before the Wildreness: Environmental Management by Native Indian Californians*. Ballena Press, Menlo Park, pp. 55–116.

Loope, L. (2010) *A Summary of Information on the Rust Puccinia psidii Winter (Guava Rust) with Emphasis on Means to Prevent Introduction of Additional Strains to Hawaii*. US Geological Survey, Reeston.

Low, T. (1999) *Feral Future: The Untold Story of Australia's Exotic Invaders*. Viking, Ringwood, Australia.

Mark-Shadbolt, M. (2017) *Biosecurity (Report to Natural Resource Iwi Leaders Group)*. Te Tira Whakamataki, Lincoln.

Mark-Shadbolt, M., Wood, W. and Ataria, J. (2018) *Why Aren't People Listening? Māori Scientists on Why rāhui Are Important*. The Spinoff, Auckland.

Marsh, A., Wallace, R., Ford, L., Guthadjaka, K., Yuhun, P., Ford, C. and Funk, J. (2018) *Building Resilience in Indigenous Communities through Engagement: A Focus on Biosecurity Threats: Final Report for PBCRC4041*. Plant Biosecurity Cooperative Research Centre, Bruce, ACT.

Marzano, M., Allen, W., Haight, R.G., Holmes, T.P., Keskitalo, E.C.H., Lisa Langer, E.R., Shadbolt, M., Urquhart, J. and Dandy, N. (2017) 'The role of the social sciences and economics in understanding and informing tree biosecurity policy and planning: A global summary and synthesis', *Biological Invasions*, vol. 19, no. 11, pp. 3317–3332.

Morton, J. (2018) '"A losing battle": Myrtle rust spreads south', *New Zealand Herald*, https://www.nzherald.co.nz/the-country/news/a-losing-battle--rust-spreads-south/JPTADDXOQQMEXG47LQYMZ7HG3M/, accessed 4 October 2019, NZME Publishing Ltd, Auckland.

Mukandiwa, L., Naidoo, V. and Katerere, D.R. (2016) 'The use of Clausena anisata in insect pest control in Africa: A review', *Journal of Ethnopharmacology*, vol. 194, pp. 1103–1111.

Nishi, J.S., Stephen, C. and Elkin, B.T. (2002) 'Implications of agricultural and wildlife policy on management and eradication of bovine tuberculosis and brucellosis in free-ranging wood bison of Northern Canada', *Annals of the New York Academy of Sciences*, vol. 969, no. 1, pp. 236–244.

Northcott, M. and Persico, C. (2018) 'Protesters make a nationwide stand against 1080', in *Stuff*. Stuff Ltd, Wellington.

Ogilvie, S.C., Miller, A. and Ataria, J. (2010) 'There's a rumble in the jungle: 1080: Poisoning our forests or a necessary tool', in R. Selby, P. Moore and M. Mulholand (eds.) *Māori and the Environment: Kaitiaki*. Huia Publishers, Wellington, pp. 251–261.

Ogilvie, S.C., Sam, S., Barun, A., Van Schravendijk-Goodman, C., Doherty, J., Waiwai, J., Pauling, C.A., Selwood, A.I., Ross, J.G., Bothwell, J.C., Murphy, E.C. and Eason, C.T. (2019) 'Investigation of tutin, a naturally-occurring plant toxin, as a novel, culturally-acceptable rodenticide in New Zealand', *New Zealand Journal of Ecology*, vol. 43, no. 3, pp. 1–8.

Pfeiffer, J.M. and Voeks, R.A. (2008) 'Biological invasions and biocultural diversity: Linking ecological and cultural systems', *Environ Conserv*, vol. 35.

Rahman, K. (2016) 'Realizing Sustainable Development Goals (SDGs): Need for an indigenous approach', *Policy Perspectives*, vol. 13, no. 2, pp. 3–28.

Ramsfield, T., Dick, M., Bulman, L. and Ganley, R. (2010) *Briefing Document on the Myrtle Rust, a Member of the Guava Rust Complex, and the Risk to New Zealand*. SCION, Rotorua.

Reo, N.J. and Ogden, L.A. (2018) 'Anishnaabe Aki: An indigenous perspective on the global threat of invasive species', *Sustainability Science*, vol. 13, no. 5, pp. 1443–1452.

Robinson, C.J. and Whitehead, P. (2003) 'Cross-cultural management of pest animal damage: A case study of feral buffalo control in Australia's Kakadu National Park', *Environmental Management*, vol. 32, no. 4, pp. 445–458.

SCION (2017) *Myrtle Rust Reporter App Now Available*. SCION, Rotorua.

Segger, M.C.C. and Phillips, F-K. (2015) 'Indigenous traditional knowledge for sustainable development: The biodiversity convention and plant treaty regimes', *Journal of Forest Research*, vol. 20, no. 5, pp. 430–437.

Smith, L.T. (1999) *Decolonizing Methodologies: Research and Indigenous Peoples*. University of Otago Press, Dunedin.

Southwestern Native Aquatic Resources and Recovery Centre (2014) *Southwestern Native Aquatic Resources and Recovery Centre*. US Fish and Wildlife Service, Dexter, NM.

Stewart-Harawira, M. (2005) *The New Imperial Order: Indigenous Responses to Globalisation*. Zed Books, London.

Tahana, Y. (2010) 'Rare rats off the hook as DoC gives them island sanctuary', *New Zealand Herald*, https://www.nzherald.co.nz/environment/news/article.cfm?c_id=39&objectid=10649358, accessed 17 October 2018, NZME Publishers, Auckland.

Teulon, D.A.J., Alipia, T.T., Ropata, H.T., Green, J.M., Viljanen- Rollinson, S.L.H., Cromey, M.G., Arthur, K., MacDiarmid, R.M., Waipara, N.W. and Marsh, A.T. (2015) 'The threat of myrtle rust to Māori taonga plant species in New Zealand', *New Zealand Plant Protection*, vol. 68, pp. 66–75.

Tokalau, T. (2019) *West Aucklanders Face $20,000 Fine for Walking Dogs in Bush Closed over Kauri Dieback*. Stuff, Wellington.

Ulluwishewa, R. (1993) 'Indigenous knowledge systems for sustainable development', *Journal of Sustainable Agriculture*, vol. 3, no. 1, pp. 51–63.

UN Convention on Biological Diversity (2018) *Convention on Biological Diversity*. Secretariat of the Convention on Biological Diversity/UN, Montreal.

United Nations (2007) *United Nations Declaration on the Rights of Indigenous Peoples*. United Nations, New York.

Vaarzon-Morel, P. and Edwards, G. (2012) 'Incorporating Aboriginal people's perceptions of introduced animals in resource management: Insights from the feral camel project', *Ecological Management & Restoration*, vol. 13, no. 1, pp. 65–71.

Waitangi Tribunal (2019) *Brief Evidence of Kristan John MacDonald*. Waitangi Tribunal (ed.). Waitangi Tribunal, Whangarei.

Yapa, L. (1993) 'What are improved seeds? An epistemology of the green revolution (Theme Issue: Environment and Development, part 1)', *Economic Geography*, vol. 69, no. 3, p. 254(20).

4

GEOGRAPHIES OF VETERINARY EXPERTS AND EXPERTISE

Gareth Enticott

Veterinary worlds of work

If disease knows no boundary, is the same true of its experts and expertise – the people and their practices that seek to protect us from the effects of exotic and endemic disease? Historically, the veterinary profession has never been too far from the global spread of infectious animal diseases. As colonial expansion took agricultural techniques to distant lands, experts in animal health were required to make sense of these new environments, adapt and develop new practices to ensure their productivity (Brooking and Pawson, 2011). For the veterinary field, colonialism marked its beginning as a global profession in which a topology of veterinary professionalism was able to draw distant places together through a shared veterinary vision, establishing common practices in different places and facilitating movement between them (Mishra, 2011). The mobility and translation of veterinary practices was not always smooth: place-specific social (e.g. forms of colonial rule), biological (e.g. disease severity and environmental factors) and techno-logical actors and relations meant that veterinary practices were not immutable and translated in different ways (Gilfoyle, 2003; Mishra, 2011; Davis, 2008).

These days, the veterinary profession is no less global – it is central to international flows of agricultural produce through its role in classifying diseases and ratifying biosecurity protocols. These material veterinary practices seek to control the global mobility of agriculture (Busch, 2010). At the same time, just as in other areas of policy (Peck and Theodore, 2010), biosecurity expertise, practices and policies are rapidly translated around the world to bring new policy solutions to animal health challenges while also reflecting a shift to neoliberal values (Enticott et al., 2011).

Increasingly, too, the veterinary profession has come to reflect the characteristics of other global professions – such as international finance and accounting – referred to as 'global work' (Jones, 2008). Central to global work is the routine flow of labour between world cities (Smith, 2003; Faulconbridge et al., 2009). For the veterinary profession, these labour flows are not exclusively urban but from and to rural sites of veterinary practice. For some countries, such as the UK, these labour flows have become essential for the delivery of animal health services. Making global work normal and not a hardship, however, involves the creation and valorisa-tion of mobile subjectivities by discursive and institutional constructions of the ideal profes-sional (Ackers, 2005; Findlay et al., 2013; Cranston, 2016; Cranston, 2017). Contributing to the circulation of professional practice is therefore a sign of appropriate conduct (Jöns, 2015; Mahroum, 2000).

If biosecurity belongs to this world of 'global work', then two things become important. The first is recognising that animal disease is emergent from and produced by a range of human actors, animals, pathogens, materials, technologies and institutions that are held together in a 'veterinary world of work' (cf. Becker, 1982). This thinking reflects how assemblage approaches have been used to understand the way different agricultural goods are configured (Jones et al., 2019) and in doing so draws attention to the range of different actors that participate in 'bringing them into being' (Pawson and Perkins, 2013). Understanding biosecurity and animal disease therefore requires attention to what we might call 'disease ecologies' in which materials, animals, practices and people come together to make and settle on a specific version of disease (Lavau and Bingham, 2017). Disease ecologies weave together natures (such as pathogens, wildlife and farm animals), materialities (such as protocols and farming infrastructures) and social relations (such as institutions and subjectivities). Like any network, however, they may be vulnerable to disruptions (such as disease outbreaks), prompting *circulations* (of disease, people and practices) resulting in a subsequent re-ordering of disease ecologies. In this sense, biosecurity itself may be seen as 'global work', placing geography at the heart of the veterinary profession and understandings of disease. These geographies may take different forms: in the understanding of the migratory flows between different countries; the implicit spatialities within forms of veterinary knowledge, their circulations and translations in marginal places (Enticott, 2017); or the institutional infrastructure that creates and perpetuates vets as global and mobile actors in combatting animal disease.

This chapter therefore seeks to analyse the geography of veterinary experts and expertise. It does so by analysing three interrelated aspects of the disease ecology of one disease – bovine tuberculosis (bTB), a zoonosis affecting farmed cattle and wildlife – and tracing its geography across three different countries (the UK, Australia and New Zealand) affected by the disease. First, the chapter analyses the *rationalities* of disease management and their *mobilities*, examining not just how disease has moved around the world but biosecurity management practices, too. Here the focus is on neoliberal approaches to animal disease management, specifically ideas of partnership and cost-sharing in which governments seek to involve the agricultural industry in the governance of animal disease. These approaches are shown to have geographically unique origins and distinct reasons for moving from country to country to counter the threat of disease. Second, the chapter analyses the *materialities* of veterinary knowledges in order to reveal their implicit *spatialities* that conflict and frustrate disease management programmes. Here the focus is on diagnostic techniques and practices and how disciplinary boundaries to veterinary science can be transgressed to create more acceptable disease control policies. Finally, the chapter analyses the *subjectivities* of veterinarians, how attempts to define master narratives of professional subjectivity contribute to the *global circulation* of vets and the consequences for disease control and biosecurity.

Global veterinary rationalities

'Partnership' and 'cost-sharing' have become leitmotifs in the management of animal disease in the UK since 2001 and emblematic of an era of international policy mobility, 'fast policy' and local translation (Peck and Theodore, 2014; Clarke et al., 2015). While not strictly reflecting a clean break in the history of animal disease policy (see Woods, 2009; Woods, 2013), in the space of ten years, the UK government's approach to animal policy has nevertheless radically transformed from one of paternalism to one reflecting the key facets of a neoliberal agenda, which has drawn on techniques and rationalities first developed on the other side of the world.

Generally referring to attempts to make the agricultural industry take a greater financial ownership of the consequences of animal disease, cost-sharing became increasingly prominent in the wake of the costly outbreak of foot and mouth disease (FMD) in 2001. In subsequent plans and strategies, the governments of the UK[1] resolved to work in partnership together to make a 'lasting and continuous improvement in the health and welfare of kept animals' (Defra, 2004, 11). A new Animal Health and Welfare Strategy produced in the aftermath of inquiries into the FMD outbreak was clear in identifying 'prevention rather than cure' as the rationality through which disease should be managed. In doing so it shifted the burden of responsibility away from the government to all animal owners, challenging them to 'practise biosecurity standards and monitor the health of their animals to ensure that the spread of exotic diseases is contained as well as protecting against endemic disease' (Defra, 2004, 21).

In seeking to create a new culture of individual biosecurity responsibility (cf. Barker, 2010), the government explicitly sought to pass the costs of disease control on to those most affected by them, arguing that farming should be no different than other industries in terms of the support it receives from the taxpayer: 'Animal owners should individually and collectively take responsibility for managing animal health and welfare risks. . . . Taxpayers cannot be expected to pay for the animal health and welfare costs and risks to farmers which affect their own businesses' (Defra, 2004, 32). These arrangements, noted Defra, were already present in some other European countries. Indeed, the European Union's own animal health and welfare strategy echoed the calls for individual responsibility, cost-sharing and working in partnership (European Commission, 2007).

The cost-sharing agenda was concomitant with discourses of partnership working. By 2010, a report from Defra's cost and responsibility advisory group recommended a new partnership board be established to manage cost-sharing arrangements (Radcliffe, 2010). The following year, the desire for partnership and greater ownership of animal disease had led to what the UK government described as 'a completely new way of working' (Paice in Defra, 2011) through the creation of the Animal Health and Welfare Board for England (AHWBE). The new board was to 'put [biosecurity] decisions in the hands of those doing the work on the ground' through the appointment of farmers and veterinarians to the board and relying on their 'wealth of experience and knowledge' (Defra, 2011). Similar partnerships were created elsewhere in the UK's devolved administrations to tackle specific animal health challenges (see Enticott and Franklin, 2009).

The mobility of the rationality of cost-sharing is perhaps most evident in relation to the management of bovine tuberculosis. In 2008, the UK government established the TB Eradication Group (TBEG), which, following a recommendation by the AHWBE in 2012, later become the TB Eradication Advisory Group for England (TBEAG). Like the AHWBE, the TBEAG encouraged vets, farmers and policymakers to work together to tackle bTB. However, the genesis of this approach lay in countries that had eradicated bTB (Australia) or were on the cusp of doing so (New Zealand).

In the management of bTB, the origins of cost-sharing and partnership can be traced to the Australian approach to disease eradication. As described by Lehane (1996; see also Cousins and Roberts, 2001; Radunz, 2006; Tweddle and Livingstone, 1994), cost-sharing and partnership arrangements in the management of bTB began in the 1970s on the back of previous successful disease eradication programmes. However, it was not until 1984 that the role of the farming industry in managing the eradication programme became significant. Following opposition to disease eradication policies (More et al., 2015), members of the farming industry came to occupy key decision-making roles on national and regional animal health committees, working closely with government. Moreover, while a levy had existed since 1973 to help fund disease control, these management changes led to a formal agreement that the agricultural industry would contribute 50% of disease control costs. The impact of this approach was seen not just in the eradication of bTB by 1997 but in a long-term change in the management of disease in Australia. As More et al. (2015) noted, the model of industry and government partnership was instrumental in creating trust between the two sectors, leading to the creation of Animal Health Australia, a not-for-profit company responsible for the development and coordination of the delivery of national animal health programmes today.

Once Australia had eradicated bTB, farmers in New Zealand were desperate to do the same, believing their ability to export would otherwise suffer. Compounding this was a rise in incidence of bTB during the 1980s and 1990s as the New Zealand government reduced and withdrew from funding disease control operations. The origins of New Zealand's disease management were paternalistic but had always sought to incorporate farmers' views. In the 1960s, farmers were involved in deciding how compensation should be distributed, and these local animal health committees became formalised into Regional Animal Health Advisory Committees (RAHACs) in the 1970s. New Zealand's economic crisis in the 1980s, however, fundamentally changed how the state supported agriculture. If the removal of farming subsidies impacted on the nature of farming (Haggerty et al., 2009), so did the decision to cease funding bTB disease control operations. However, in the vacuum that followed, agricultural and veterinary leaders drove the creation of a new way of governing and managing bTB (Warburton and Livingstone, 2015; Livingstone et al., 2015; Hutchings et al., 2013). In the place of a top-down government scheme, a new Biosecurity Act allowed the creation of National Pest Management Agencies, the first of which – the Animal Health Board (AHB) – was responsible for managing bTB. The AHB was an industry-led partnership, majority funded by farmers with the state a minority partner. Farmers came to lead the organisation, have a direct say in policy, develop their own innovative solutions to influence the behaviour of other farmers and determine levels of financial support they received following a bTB incident.

The sharing of knowledge and best practice on bTB was not strictly a new phenomenon. First, the original 1958 New Zealand bTB eradication plan was based on British expertise and its model of stamping out disease. These plans were developed by expatriate vets, such as the Scottish vet Sam Jamieson, who had moved to New Zealand after working in the Scottish meat inspection system. While Jamieson shaped New Zealand's approach in a distinctly British manner (see the section Prescribed and negotiated material spatialities), he also visited other countries such as Canada, Australia and Europe to evaluate best veterinary and disease management practice. Second, as both New Zealand and the UK realised that wildlife were involved in the spread of bTB (possums in New Zealand; badgers in the UK), veterinarians paid attention to each other's wildlife management practices.

The mobility of New Zealand's neoliberal management of bTB, however, was facilitated by an increasing 'social infrastructure' of policy exchange. One site in which this has occurred is at the international *M. bovis* conference. Held approximately every five years, the conference began in 1990 as a forum for academics and policymakers to share their progress and strategies for combatting bTB. Gradually, conference presentations have featured subjects on the margins of veterinary science, usually referred to as 'non-technical' aspects of disease eradication – i.e. the social context and institutional and governance arrangements in which disease eradication takes place. Second, the conferences have offered the chance to demonstrate the manner and effectiveness of the host country's bTB governance system by offering extended field trips and discussions following the conference. In doing so, key members of staff have become 'policy entrepreneurs', advocating for certain solutions both scientific but also in terms of governance. Members of the AHB's management board have been invited to other countries to tell New Zealand's story of disease eradication while UK politicians and veterinary policymakers have visited New Zealand to see their approach first-hand. Such visits are framed as fact-finding missions, but they also represent opportunities to justify specific policy choices. Thus, by the time Defra published their new bTB eradication strategy (Defra, 2014), the mythologising of New Zealand's bTB programme was complete. In a score, appropriately of rugby sized proportions, the New Zealand approach to managing bTB was mentioned 40 times: more than the sum of other countries combined (see also the latest strategy review: Godfray et al., 2018).

What these policy mobilities show is how context-dependent biosecurity solutions in distant places become influential in shaping how disease is understood and managed on a global stage. Biosecurity policy mobilities are shaped by key individuals responsible for creating and distributing specific narratives of disease control, but they are also emergent from historical socio-economic contexts and specific disease environments. Moreover, these global flows are not necessarily dependent upon the success of their disease management practices: the narratives surrounding them are inevitably partial, misunderstood and mistranslated. For example, Australian officials have stated their concern around the interpretation of their eradication programme. By contrast, in New Zealand some aspects of disease management do not feature in their 'official' narrative (see the section Mobile veterinary subjectivities), while for others – such as the use of risk management techniques – evidence that they work is limited (Enticott, 2016). Thus, despite these global flows, neoliberal management practices of animal health cannot be considered to be mirror images of those practices in the countries they have travelled to and from.

Prescribed and negotiated material spatialities

While the global mobility of veterinary expertise may be responsible for standardising approaches to governing disease, their local translation may result in a different set of geographies. This becomes particularly apparent when analysing the use of specific veterinary materialities, such as the instruments and instructions used to diagnose disease. As Bickerstaff and Simmons (2004) showed, different approaches to disease management have their own politics as well as their own spatialities. While these spatial variations point to the multiplicity of disease (Law and Mol, 2011), there may also be ways in which competing spatialities find methods of accommodating each other (Mol and Law, 1994). A number of examples of the spatial patterning of veterinary materials exist (see for example Enticott, 2012; Hinchliffe et al., 2013), but for the purposes of continuity, the management of bTB in New Zealand is the focus of this section.

The first spatiality summoned by veterinary materials is that of a *standardised and universal space*. Diagnosing bTB relies on a set of materialities: reagents, needles and measuring devices, not to mention fences, gates and sticks to corral animals into restraints to test them for disease. Protocols – a set of instructions – order these materials into a procedure to be followed on each and every farm, for each and every test (cf Chapter 19). In this way, protocols represent an 'inscription device' (Latour, 1987) that relationally establishes order across and within space by standardising practices and behaviours therein (Murdoch, 1998). For this to occur, there can be no distractions, no dissenting to the ordering of events: replication is vital (Shapin and Schaffer, 1985). This demands a uniform spatial plane with no opportunity for variation or resistance. In doing so, the collection of animal health data establishes government and veterinary authorities as a 'centre of calculation', in which they can speak for disease and control animal health from a distance (Latour, 1988).

The challenge, of course, is to ensure these networks remain immutable. Where there have been challenges to the veracity of bTB tests, governments have usually resorted to science to preserve their eradication plans, seeking to add another layer of spatial uniformity to the protocol. So it proved to be in New Zealand during the 1960s. After losing too many cattle to the results of the bTB test, farmers grew frustrated and militant. Believing that the environment in which their cattle grazed was unique and contributed to 'non-specific' reactions to the bTB test, they resisted demands to test their cattle (for complete details, see Enticott, 2017). In response, the government resorted to a scientific experiment: they railroaded 550 cattle to their experimental agricultural training centre at Flock House, subjecting them to a range of diagnostic tests before slaughtering them to compare the results with post-mortems.

The experiment at Flock House concluded in 1964, and the results were made public. The results confirmed the government's view that bTB could be universally identified – the power of the laboratory established the immutability of disease diagnosis and the government to act from a distance. This belief in a uniform spatiality was in no small part down to Dr Sam Jamieson – the Scottish vet who had migrated to New Zealand in the 1950s. As suggested in the previous section, he had brought his training and veterinary approach. In Jamieson's view, the uniformity of science was sacrosanct: clinical interpretation and flexibility were not part of his way of doing things. Indeed, his laboratory training was central to his dismissal of clinical interpretations of bTB test results. This meant that the universal immutable standards of the laboratory applied everywhere and claims of non-specific reactions were dismissed as a farmers' myth.

A second spatiality, that of *negotiated veterinary space*, sought to challenge the fixed view of the government. As it sounds, negotiated space was the diametrical opposite. Rather than fixed relations, negotiated space reflects local context, contingency and the mutability of relations. This veterinary spatiality was set firmly against the government's by one vet, Peter Malone, who had seen his farmers lose too many animals to the bTB test and was convinced there was an alternative explanation. Rebuffed by government vets, Malone adopted his own approach. Noting that it was only younger animals that reacted to the bTB test but were later found to be clear of bTB at post-mortem, Malone began to 'read light': adopt a flexible approach to interpreting the results of the bTB test based on the situation at hand. Not surprisingly, farmers were grateful to Malone, but the government was less so. When they found out, Malone was indefinitely barred from bTB testing.

Perhaps Malone's greatest contribution, though, was in preparing the ground for a third spatiality that somehow combines both fixed and flexible perspectives. *Fluid spaces* are marked by 'variation without boundaries and transformation without discontinuity' (Mol and Law, 1994, 658) in which 'it is not possible to determine identities nice and neatly, once and for all' (Mol and Law, 1994, 660). Boundaries between diseased and healthy, or between diagnosis and treatment, are never fixed, as the movement between them is unpredictable. As a result, animal disease control will be marked by gradients, not boundaries. Distinguishing between different disease practices is difficult as different disease practices can comfortably sit alongside each other (Mol, 2002). The implication for veterinary materialities like protocols is the need to transform between contexts to ensure the diagnosis of disease. Timmermans and Berg (1997, 275) call this quality 'local universality' to suggest that protocols can only function through the 'ongoing subordination and (re)articulation of the protocol to meet the primary goals of the actors involved' (ibid., 291). For protocols to 'work', there must be opportunities for leeway and discretion to ensure actors' cooperation. The protocol's dream of standardisation therefore occurs not from irreversible relations but is only achieved where they are loose fitting (Murdoch, 1998).

By the mid-1970s, this kind of loose-fitting arrangement had found its way into the disease management plan in New Zealand. Rather than a universal geography of animal disease, a more nuanced approach had evolved. Taking seriously the idea that farmers had their own stocks of knowledge that could be incorporated into disease management, new government veterinarians started to experiment with the materialities that had patrolled the fixed boundaries of veterinary practice. In working with farmers, vets came up with a new approach to be used for some animals. The materialities of disease management were not removed or fully broken but were applied variably in accordance with different contexts. In doing so, disease materialities came to invoke a fluid spatiality of disease control that transgressed traditional boundaries, such as disciplinary boundaries and styles of animal disease management. These fluid spatialities have their own geography, too, emerging from a disciplinary 'borderland' in which borrowing, blending and adapting are essential skills (Kohler, 2002; Hinchliffe, 2015). For farmers, however, the impact of this new spatiality was a better way of living with ongoing attempts to deal with animal disease.

Mobile veterinary subjectivities

The clash between Jamieson and Malone was not simply related to the appropriate use of biosecurity materialities. Rather, it also reflected contrasting views of what constituted appropriate veterinary conduct: a fight to define and proclaim appropriate veterinary subjectivities. Understanding how veterinary subjectivities are created can also help chart the global circulations of veterinary experts and expertise and the local translation of veterinary practices. Equally, veterinary subjectivities may have implications for attempts to prevent the spread of disease. Where disease control lies on the margins of dominant veterinary subjectivities, the consequences may be 'disease fatigue', alienation and exit from the profession and labour shortages that limit the effectiveness of disease prevention policies.

Professionalism can be understood as a mode of conduct rather than simply a matter of technical expertise (Grey, 1998). Professional institutions contribute to a definition of these ways of being through various rules and regulations but also more subtle discursive regimes (Gill, 2015) and 'master narratives' (Nelson, 2001) which prescribe the limits of professional identity, regulate professional behaviour and determine professional status. These master narratives serve to reinforce dominant professional identities: those on the margins are not recognised as legitimate or valuable, are less able to intervene or voice concerns and may ultimately be forced to accept the dominant view of their roles (Thomson and Jones, 2017).

In one way, these master narratives can create a form of professional ontological security by providing a clear narrative of professional expectations. On the other hand, pressure to continually conform and worries over losing identity can lead to 'status anxiety' (Burke, 1991) and the development of counter-subjectivities as a survival strategy. Another response is to leave the organisation or profession altogether or move somewhere in which those marginal subjectivities are dominant. By contrast, migration may also contribute to the globalisation of professions, spreading master narratives that seek to establish universal versions of professional conduct (Spence et al., 2015).

In the UK, the master narrative of veterinary subjectivity continues to revolve around the 'James Herriot'[2] mixed practice ideal in which vets are respected fixtures of their local community, going the extra mile to save animals while dealing with awkward clients (Herriot, 1972; Wedderburn, 2016). Despite 80% of new vets in the UK being female, this heroic masculine subjectivity continues to dominate. Other forms of veterinary conduct, such as public health, are set to one side, reserved for overseas vets and not seen as 'real' veterinary work (Enticott, 2018). In other European countries, such dominant master narratives are not present: government veterinary work is regarded as more secure and better paid but hard to achieve without the right social connections (Enticott, 2019). In each case, however, these master narratives are linked to the global mobility of vets.

In the UK, the pressure to conform to these idealised subjectivities can trigger resentment and disillusionment with the profession. The challenge of *being* a vet – i.e. the reality of the day-to-day work of vets – as opposed to the formal training involved in *becoming* a vet can act as a trigger to leave the profession or seek employment elsewhere. Long hours, poor pay and over-work in a vet's first job are key factors in this search for an alternative. The realities of the first job are also present for European vets in Spain and Italy: faced with little work on short-term low paid contracts, seeking veterinary work in other countries is a route to the veterinary career they imagine. These mobilities are assisted by international regulations and neoliberal styles of disease management. The European Union's Mutual Recognition of Professional Qualifications directive allows European vets free movement throughout the EU. At the same time, the drive for competition in the veterinary public health sector in the UK has created a market for low paid, temporary veterinary labour which UK graduates refuse to fill because of their adherence to the James Herriot master narrative. UK veterinary employment companies recruit vets from across the EU in order to complete food safety checks, although the UK's decision to leave the EU may affect this global movement of veterinary labour (Enticott, 2019; BVA, 2017).

The circulation of disease and its effect on agriculture may also contribute to veterinary mobility. Disease outbreaks such as foot and mouth disease may create a demand for vets, prompting international mobility. For farmers and vets in the UK, however, continued disease outbreaks of bTB can lead to compassion fatigue, distancing themselves from the idealised pro-fessional master narrative. This is exacerbated when disease affects farmers' abilities to invest in new and progressive farming techniques. Rather than developing new veterinary skills to match, vets can find themselves stuck in routine jobs, questioning where their future lies. Countries with low disease incidence and progressive farming attitudes can prove attractive to those vets seeking to conform to their idealised master narrative of veterinary professionalism.

Closing circulations

This chapter has examined the geography of biosecurity through a focus on veterinary experts and expertise. In doing so, it has shown how the geographies of animal disease are central to the ways in which it is relationally configured. In this view, animal disease is situated within and emergent from sets of heterogeneous relations and actors – or what we can call a disease ecology. As the chapter has shown, the disease ecology is composed of natures (such as patho-gens, wildlife and farm animals), materialities (such as protocols and farming infrastructures) and social relations (such as institutions and subjectivities). These dimensions of the disease ecology are woven together to configure animal disease and appropriate ways of dealing with it. Like any network, these ecologies are vulnerable to mutations and disruptions that threaten to alter established understandings or ways of dealing with disease. These vulnerabilities also prompt *cir-culations* that are central to the evolving geographies of biosecurity. Thus, biological disruptions (e.g. disease outbreaks) lead to the circulation of veterinary expertise and social disruptions (e.g. changes in the nature of veterinary subjectivities) lead to the circulation of veterinary experts while material disruptions (e.g. breakdowns in veterinary procedures) are also associated with circulations and evolutions of veterinary practices.

Disease ecologies vary between places, but they can also become connected thanks to these circulations, highlighting the global geography of biosecurity. The global flow of veterinary experts and expertise might be the most obvious of all biosecurity geographies. Following these global flows reveals the contextual nature of apparently global solutions to animal disease. Not only does this question the extent to which policy mobilities deliver solutions to animal disease problems, but it also suggests that these neoliberal solutions are likely to vary wherever they are enacted. Similarly, the mutability of veterinary expertise highlights other conflicting spatialities – such as between flexible and fixed – that are attached to different uses of veterinary materials. While these conflicts may create further disruptions and circulations within disease ecologies – such as dissatisfaction with the veterinary subjectivities they configure – they can also inspire a more nuanced understanding of the spatialities of veterinary expertise in which a fluid spatiality offers a more acceptable prospect of dealing with animal disease.

Finally, the significance of the local translation of global policy and the fluid geographies of disease detection also directs our attention towards the borders of the geographies of biosecurity. While veterinary experts may think of geographical boundaries as being no obstacle to the spread of disease, the focus on the disease ecologies of bTB shows that it is the disciplinary boundaries of veterinary expertise that are in flux. Indeed, it is within the disciplinary borderlands of veterinary practice where we find how biosecurity and animal disease are brought into being, how these boundaries are constantly adapting to local contexts and situations and how veterinary expertise and experts evolve and circulate accordingly. Thus, it is not so much that 'disease knows no boundary', as veterinary experts like to suggest, but that these boundaries are constantly evolving, adapting and disrupting local disease ecologies.

Notes

1 In the UK, animal disease management is devolved to the governments of Wales, Scotland and Northern Ireland. The UK government's Department of Environment, Food and Rural Affairs (Defra) sets strategy for England and coordinates joint working with the devolved administrations.
2 James Herriot was a semi-fictional character in a popular series of books and a TV series about a rural vet and his colleagues set in the English countryside.

References

Ackers, L. (2005) 'Moving people and knowledge: Scientific mobility in the European Union', *International Migration*, vol. 43, pp. 99–131.
Barker, K. (2010) 'Biosecure citizenship: Politicising symbiotic associations and the construction of biological threat', *Transactions of the Institute of British Geographers*, vol. 35, pp. 350–363.
Becker, H.S. (1982) *Art Worlds*. University of California Press, Berkeley.
Bickerstaff, K. and Simmons, P. (2004) 'The right tool for the job? Modeling, spatial relationships, and styles of scientific practice in the UK foot and mouth crisis', *Environment and Planning D: Society and Space*, vol. 22, pp. 393–412.
Brooking, T. and Pawson, E. (2011) *Seeds of Empire: The Environmental Transformation of New Zealand*. IB Tauris, New York.
Burke, P.J. (1991) 'Identity processes and social stress', *American Sociological Review*, vol. 56, pp. 836–849.
Busch, L. (2010) 'Can fairy tales come true? The surprising story of neoliberalism and world agriculture', *Sociologia Ruralis*, vol. 50, pp. 331–351.
BVA (2017) *Brexit and the Veterinary Profession*. British Veterinary Association, London.
Clarke, J., Bainton, D., Lendvai, N. et al. (2015) *Making Policy Move: Towards a Politics of Translation and Assemblage*. Policy Press, Bristol.
Cousins, D.V. and Roberts, J.L. (2001) 'Australia's campaign to eradicate bovine tuberculosis: The battle for freedom and beyond', *Tuberculosis*, vol. 81, pp. 5–15.
Cranston, S. (2016) 'Imagining global work: Producing understandings of difference in "easy Asia"', *Geoforum*, vol. 70, pp. 60–68.
Cranston, S. (2017) 'Calculating the migration industries: Knowing the successful expatriate in the global mobility industry', *Journal of Ethnic and Migration Studies*, pp. 1–18.

Davis, D.K. (2008) 'Brutes, beasts and empire: Veterinary medicine and environmental policy in French North Africa and British India', *Journal of Historical Geography*, vol. 34, pp. 242–267.

Defra (2004) *Animal Health and Welfare Strategy for Great Britain*. Defra, London.

Defra (2011) *Sharing Responsibility for Animal Health and Welfare Policy*. Defra, London.

Defra (2014) *The Strategy for Achieving Officially Bovine Tuberculosis Free Status for England*. Defra, London.

Enticott, G. (2012) 'The local universality of veterinary expertise and the geography of animal disease', *Transactions of the Institute of British Geographers*, vol. 37, pp. 75–88.

Enticott, G. (2016) 'Market instruments, biosecurity and place-based understandings of animal disease', *Journal of Rural Studies*, vol. 45, pp. 312–319.

Enticott, G. (2017) 'Navigating veterinary borderlands: "Heiferlumps", epidemiological boundaries and the control of animal disease in New Zealand', *Transactions of the Institute of British Geographers*, vol. 42, pp. 153–165.

Enticott, G. (2018) 'International migration by rural professionals: Professional subjectivity, disease ecology and veterinary migration from the United Kingdom to New Zealand', *Journal of Rural Studies*, vol. 59, pp. 118–126.

Enticott, G. (2019) 'Mobile work, veterinary subjectivity and Brexit: Veterinary surgeons' migration to the UK', *Sociologia Ruralis*, vol. 59, no. 4, pp. 718–738.

Enticott, G., Donaldson, A., Lowe, P. et al. (2011) 'The changing role of veterinary expertise in the food chain', *Philosophical Transactions of the Royal Society B: Biological Sciences*, vol. 366, pp. 1955–1965.

Enticott, G. and Franklin, A. (2009) 'Biosecurity, expertise and the institutional void: The case of bovine tuberculosis', *Sociologia Ruralis*, vol. 49, pp. 375–393.

European Commission (2007) *A New Animal Health Strategy for the European Union (2007–13) Where 'Prevention Is Better Than Cure'*. Office for Official Publications of the European Communities, Luxembourg.

Faulconbridge, J.R., Beaverstock, J.V., Hall, S. et al. (2009) 'The "war for talent": The gatekeeper role of executive search firms in elite labour markets', *Geoforum*, vol. 40, pp. 800–808.

Findlay, A., McCollum, D., Shubin, S. et al. (2013) 'The role of recruitment agencies in imagining and producing the "good" migrant', *Social and Cultural Geography*, vol. 14, pp. 145–167.

Gilfoyle, D. (2003) 'Veterinary research and the African rinderpest epizootic: The cape colony, 1896–1898', *Journal of Southern African Studies*, vol. 29, pp. 133–154.

Gill, M.J. (2015) 'Elite identity and status anxiety: An interpretative phenomenological analysis of management consultants', *Organization*, vol. 22, pp. 306–325.

Godfray, C., Donnelly, C.A., Hewinson, G. et al. (2018) *Bovine TB Strategy Review*. Defra, London.

Grey, C. (1998) 'On being a professional in a "Big Six" firm', *Accounting, Organizations and Society*, vol. 23, pp. 569–587.

Haggerty, J., Campbell, H. and Morris, C. (2009) 'Keeping the stress off the sheep? Agricultural intensification, neoliberalism, and "good" farming in New Zealand', *Geoforum*, vol. 40, pp. 767–777.

Herriot, J. (1972) *All Creatures Great and Small*. Bantam Books, London.

Hinchliffe, S. (2015) 'More than one world, more than one health: Re-configuring interspecies health', *Social Science & Medicine*, vol. 129, pp. 28–35.

Hinchliffe, S., Allen, J., Lavau, S. et al. (2013) 'Biosecurity and the topologies of infected life: From borderlines to borderlands', *Transactions of the Institute of British Geographers*, vol. 38, pp. 531–543.

Hutchings, S.A., Hancox, N. and Livingstone, P.G. (2013) 'A strategic approach to eradication of bovine TB from wildlife in New Zealand', *Transboundary and Emerging Diseases*, vol. 60, pp. 85–91.

Jones, A. (2008) 'The rise of global work', *Transactions of the Institute of British Geographers*, vol. 33, pp. 12–26.

Jones, L., Heley, J. and Woods, M. (2019) 'Unravelling the global wool assemblage: Researching place and production networks in the global countryside', *Sociologia Ruralis*, vol. 59, pp. 137–158.

Jöns, H. (2015) 'Talent mobility and the shifting geographies of latourian knowledge hubs', *Population, Space and Place*, vol. 21, pp. 372–389.

Kohler, R.E. (2002) *Landscapes and Labscapes: Exploring the Lab-Field Border in Biology*. Chicago University Press, London.

Latour, B. (1987) *Science in Action*. Harvard University Press, Cambridge, MA.

Latour, B. (1988) *The Pasteurization of France*. Harvard University Press, Cambridge, MA.

Lavau, S. and Bingham, N. (2017) 'Practices of attention, possibilities for care: Making situations matter in food safety inspection', *The Sociological Review*, vol. 65, pp. 20–35.

Law, J. and Mol, A. (2011) 'Veterinary realities: What is foot and mouth disease?', *Sociologia Ruralis*, vol. 51, pp. 1–16.

Lehane, R. (1996) *Beating the Odds in a Big Country: The Eradication of Bovine Brucellosis and Tuberculosis in Australia*. CSIRO, Collingwood, Australia.

Livingstone, P.G., Hancox, N., Nugent, G. et al. (2015) 'Development of the New Zealand strategy for local eradication of tuberculosis from wildlife and livestock', *New Zealand Veterinary Journal*, vol. 63, pp. 98–107.

Mahroum, S. (2000) 'Scientific mobility: An agent of scientific expansion and institutional empowerment', *Science Communication*, vol. 21, pp. 367–378.

Mishra, S. (2011) 'Beasts, Murrains, and the British Raj: Reassessing colonial medicine in India from the veterinary perspective, 1860–1900', *Bulletin of the History of Medicine*, pp. 587–619.

Mol, A. (2002) *The Body Multiple: Ontology in Medical Practice*. Duke University Press, London.

Mol, A. and Law, J. (1994) 'Regions, networks and fluids: Anaemia and social topology', *Social Studies of Science*, vol. 24, pp. 641–671.

More, S.J., Radunz, B. and Glanville, R.J. (2015) 'Lessons learned during the successful eradication of bovine tuberculosis from Australia', *Veterinary Record*, vol. 177, pp. 224–232.

Murdoch, J. (1998) 'The spaces of actor-network theory', *Geoforum*, vol. 29, pp. 357–374.

Nelson, H.L. (2001) *Damaged Identities, Narrative Repair*. Cornell University Press, Ithaca, NY.

Pawson, E. and Perkins, H. (2013) 'Worlds of wool: Recreating value off the sheep's back', *New Zealand Geographer*, vol. 69, pp. 208–220.

Peck, J. and Theodore, N. (2010) 'Mobilizing policy: Models, methods, and mutations', *Geoforum*, vol. 41, pp. 169–174.

Peck, J. and Theodore, N. (2014) *Fast Policy: Experimental Statecraft at the Thresholds of Neoliberalism*. University of Minnesota Press, Minneapolis.

Radcliffe, R. (2010) *Responsibility and Cost Sharing for Animal Health and Welfare: Final Report*. Defra, London.

Radunz, B. (2006) 'Surveillance and risk management during the latter stages of eradication: Experiences from Australia', *Veterinary Microbiology*, vol. 12, pp. 283–290.

Shapin, S. and Schaffer, S. (1985) *Leviathan and the Air-Pump: Hobbes, Boyle, and the Experimental Life*. Princeton University Press, Princeton, NJ.

Smith, R.G. (2003) 'World city actor-networks', *Progress in Human Geography*, vol. 27, pp. 25–44.

Spence, C., Carter, C., Belal, A. et al. (2015) 'Tracking habitus across a transnational professional field', *Work, Employment and Society*, vol. 30, pp. 3–20.

Thomson, K. and Jones, J. (2017) 'Precarious professionals: (in)secure identities and moral agency in neocolonial context', *Journal of Business Ethics*, vol. 146, pp. 747–770.

Timmermans, S. and Berg, M. (1997) 'Standardization in action: Achieving local universality through medical protocols', *Social Studies of Science*, vol. 27, pp. 273–305.

Tweddle, N.E. and Livingstone, P. (1994) 'Bovine tuberculosis control and eradication programs in Australia and New Zealand', *Veterinary Microbiology*, vol. 40, pp. 23–39.

Warburton, B. and Livingstone, P. (2015) 'Managing and eradicating wildlife tuberculosis in New Zealand', *New Zealand Veterinary Journal*, vol. 63, pp. 77–88.

Wedderburn, P. (2016) 'The James Herriot centenary: A vet who changed his profession', *Daily Telegraph*, https://www.telegraph.co.uk/pets/news-features/the-james-herriot-centenary-a-vet-who-changed-his-profession/, London.

Woods, A. (2009) '"Partnership" in action: Contagious abortion and the governance of livestock disease in Britain, 1885–1921', *Minerva*, vol. 47, pp. 195–216.

Woods, A. (2013) 'Is prevention better than cure? The rise and fall of veterinary preventive medicine, c.1950–1980', *Social History of Medicine*, vol. 26, pp. 113–131.

5

WATCHING THE GRASS GROW

How landholders learn to live with an invasive plant in conditions of uncertainty

Shaun McKiernan, Nicholas Gill and Jennifer Atchison

Introduction

Invasive plants are adversely affecting both environmental and production values in diverse ways around the world (Pyšek and Richardson, 2010). In Australia, more than 2,770 exotic plant species are naturalised (Coleman et al., 2015). The costs of weeds to agriculture in Australia are estimated to be within $4.8 billion with a further $300 million estimated for public weed control expenditure (McLeod, 2018). Despite the need to control invasive species, management is not straightforward. Lack of finances, time constraints and impediments to collective action can prevent effective control for landholders and natural resource management (NRM) groups (Graham, 2013; Klepeis et al., 2009). Further, in some regions the sheer scale of infestation is itself transforming ecosystems and can make eradication impossible (Head et al., 2015a; Hobbs et al., 2013). Government agencies and landholders alike are faced with landscapes of uncertainty and no clear paths of action as invasive plants, often facilitated by climate change (Pyšek and Richardson, 2010), continue to push ecosystems beyond past benchmarks, where spatial and temporal analogues for management no longer exist and where priorities and scales of management are changing (Head et al., 2015a). Consequently, while for many invasive plants there are known management actions of at least some efficacy, for others there is a lack of effective knowledge regarding how to manage plants and landscapes in order to maintain production or protect environmental values. These challenges indicate the need for further research into how land managers are developing the capacity to manage and live with changing and unpredictable conditions.

In this chapter we examine how rural landholders running grazing enterprises learn to manage and cope with a specific invasive plant, African lovegrass (*Eragrostis curvula*), in conditions of uncertainty. This uncertainty arises from the dense growth and extensive distribution of an invasive plant where there is also an absence of effective management advice from formal institutions and a dearth of local knowledge as to how to manage this plant. Invasive plants not only create uncertainty in regard to their control but can also create unfamiliar environments for land managers (Head et al., 2015a). In a case where the capacities of the plant exceed knowledge and experience, and when expert advice is missing, our focus turns to how people can and do learn amidst such conditions.

Social science research has begun to document the experiences of people living with invasive species that can no longer be eradicated (Head et al., 2015a; Fischer and Charnley, 2012; Udo et al., 2019). Such a research has aimed to provide a more nuanced understanding of the different ways people interact and live with invasive species, particularly invasive plants. For example, research has drawn attention to the ability of landholders to adapt existing land uses to changing ecological conditions that are being shaped by invasive plants (Bach et al., 2019; Brenner, 2011; Cooke and Lane, 2018; Shackleton and Shackleton, 2018). Other studies have highlighted the pragmatism of land managers as they attempt to rethink invasive plant management strategies with limited resources (Barker, 2008; Head et al., 2015a). There is also a growing body of research documenting how people come to value invasive species that can no longer be eradicated (Evans et al., 2008; Shackleton et al., 2019; Shrestha et al., 2019). However, despite the increasing attention on the social dimensions of living with invasive species, research is yet to focus on how people learn within these environments.

We contend that the future ecological and economic uncertainty presented by invasive plants (Ellis, 2015) requires further research into how people learn and develop the capacity to live with invasives. To do so, we extend from critical interpretations of the environment in learning studies that incorporate non-human agency into learning frameworks (Cooke and Lane, 2015; Fenwick, 2010; Plumb, 2008) by adopting a *relational learning* approach. We define a relational learning approach as an investigation into the human and more-than-human relationships where knowledge is formed, practised and contested. Relational learning aims to decentre human agency through apprehending the distinctive capacities and contributions non-humans make to learning processes and the co-production of new knowledge. In the context of living with invasive plants, in which management knowledge is often incomplete, we argue for greater attention to how people learn in order to capture the novel ways living with invasive plants emerges and is practised.

Our focus is the invasive plant African lovegrass in Bega Valley, New South Wales, Australia. In Bega Valley, African lovegrass presents a current threat to agricultural production as well as Lowland Grassy Woodlands, recognised as a threatened ecological community in New South Wales (Schedule 2, Biodiversity Conservation Act, 2016) and nationally (Section 181, Environment Protection and Biodiversity Conservation Act, 1999). However, although African lovegrass was introduced by the New South Wales government for pasture production and improvement (Firn, 2009), how it should be managed now that it is invasive, and its pasture quality is less than anticipated, remains an open question.

In light of these challenges, this chapter has two key aims. First, we aim to improve understandings of the learning process by illuminating the role of African lovegrass – with other non-human plants and animals – in guiding management responses. In doing so, we move beyond treating the environment as a homogenous whole to unpack the distinctive capacities of an invasive plant and how it affects the learning process in specific ways. Second, in forwarding a relational learning approach, we aim to broaden current research on living with invasive species by focusing on how rural graziers learn to manage their land with an invasive plant that can no longer be eradicated and which poses a significant threat to economic and natural assets. Such an approach unsettles the emphasis on social relationships – particularly between expert and stakeholder groups – that dominates invasive science studies by positioning non-humans as stakeholders in the development of individual and collective management responses.

We begin by outlining how expert and stakeholder knowledge is positioned in invasive species management research before detailing the significance of a relational learning approach to address management uncertainty and improve social learning outcomes. After detailing the context and methodology of the study, we then present the original empirical results by detailing participants' 'learning journeys' as they discover, respond, learn about and then learn to live with African lovegrass.

Living with invasive plants

The current challenges and future uncertainty presented by invasive plants raise important ecological and social questions concerning how people learn to live with invasive plants and the outcomes for natural resource management (Head et al., 2015a; Larson et al., 2013). Uncovering the social dimensions of invasive plant management has been influential in assessing the broader social and cultural processes that influence the spread of invasive plants (Robbins, 2010), as well as the diverse perceptions, knowledge and skills influencing how people manage these species (Head, 2017). The social complexity of invasive plant management has been documented extensively, with research emphasising how certain epistemic communities (i.e. the public, managers, policymakers, scientists) perceive invasive plants differently (Head, 2017) and the consequences for generating collaborative management programmes (Shackleton et al., 2019). For example, Crowley et al. (2017) identified that top-down approaches to directing management responses across the public can be inadequate for gaining acceptance and support due to the public feeling disempowered or losing trust in institutions. The risks of public dissatisfaction can also be exacerbated when invasive species management relies on the use of hyperbole, emotive language and selective evidence to guide decision-making (Crowley et al., 2017). Such research is critical in identifying the barriers to collectively managing invasive plants, as well as elucidating the ways in which certain forms of knowledge (e.g. scientific) are given legitimacy over others (e.g. the public) (Evans et al., 2008; Shackleton et al., 2019).

More recently, research has begun to focus on cases where invasive plants can no longer be removed. Broadly captured under the idea of 'living with', social science research has documented how landholders and land managers are challenged to protect natural and economic assets from invasive plants when more familiar weed management practices are no longer effective or practical (Barker, 2008; Head et al., 2015a; Ngorima and Shackleton, 2019). Research on the social dimensions of living with invasive plants has focused on both the practices of managing species that can no longer be removed and the variable and shifting perceptions of landholders and land managers towards these species (Brenner, 2011; Kull et al., 2019; Shackleton et al., 2015; Shrestha et al., 2019). At times, living with invasive plants is understood to be an inevitable facet of rural landownership (Cooke and Lane, 2015; Klepeis et al., 2009). Alternatively, living with invasive plants has also been identified as unsettling people's perception of place and the desired ways of living in and using the land (Førde and Magnussen, 2015). The spatial and temporal variability of invasive species management is also noted to create disparities between legislation that calls for species eradication and the experiences of those tasked with implementing and enforcing invasive species control (Head et al., 2015a). Overall, research on living with invasives is shifting emphasis away from the disputes between competing knowledge claims in invasive species management towards considering how knowledge is *generated*, and management actions delivered, in spaces of uncertainty (Head et al., 2015a).

Learning in uncertainty

Living with invasive plants is often a process of managing in uncertainty. This uncertainty refers to the absence of resources and information to control the species but also uncertainty on how to maintain current land uses or protect environments that are being transformed. Learning in positions of uncertainty is hence critical to living with invasive plants. As Head et al.

(2015a, 313) argued, 'if it is not possible to live without invasive plants, it is important to document – in order to improve, through social learning – the diverse ways in which we are living with them'. However, despite the growing emphasis on how people learn, research on living with invasives seldom details the learning process. Further, research on living with invasive species often focused on social dimensions (e.g. social learning outcomes), with a tendency to obscure the other-than-human agencies affecting how people learn in these environments.

Learning is central to the ability of individuals and groups of people to respond to changing environments (Tarnoczi, 2011). In NRM, learning research examines how and why individuals learn (Fazey et al., 2005), the inter-linkage of the learning process across multiple scales (Leys and Vanclay, 2011) and the outcomes of learning that are transformative (Muro and Jeffrey, 2008). Experiential, transformative and social learning are three frameworks commonly applied in NRM contexts that reflect upon the human experience of learning, both at an individual and a group level (Armitage et al., 2008; Lankester, 2013; Keen and Mahanty, 2006). While these frameworks have been influential in understanding how individuals learn through direct experiences with the environment (Fazey et al., 2006), and how knowledge is shared and developed across different groups of people (Leys and Vanclay, 2011; Krasny and Lee, 2002), there remains scope for a deeper examination into the role of the environment in the learning process.

Multiple scholars have examined the co-production of learning in relationships between people and various non-humans, lending support to the idea that the environment is not a passive backdrop to human learning but a complicated assemblage and an active part in the co-production of knowledge (Cooke and Lane, 2015; Fenwick, 2010; Plumb, 2008). As Cooke and Lane (2015) argued in the context of peri-urban residents learning to manage their properties, experiential learning needs to be considered as a type of 'dialogue' between people and the landscape, rather than the landscape being a passive recipient of human action.

The emphasis on human stakeholders and social learning networks in invasive species management (Bart and Simon, 2013; Davis and Carter, 2014), and environmental learning more generally (Plummer and Armitage, 2007), presents an opportunity to reframe non-humans themselves as *stakeholders* within the wider social networks that shape learning and decision-making. This opportunity represents an important shift from existing environmental and weed management educational contexts where the human managers might learn *about* an invasive plant, for example through discussion with experts and other managers or in text via educational resources. If we are to consider non-humans as stakeholders in the development of management responses in spaces of uncertainty, then learning studies need to avoid reproducing the dualisms separating the 'social' from the 'natural' (Latour, 1993). Such an approach first requires recognising the capacity of non-humans to take part in and influence the learning process and, second, tracing how this knowledge is shared and influences wider social networks of people.

A relational learning approach

We extend from recent work in human geography that aims to not only draw attention to non-human agency but also the need to articulate the specificity of non-human life (Bear, 2011; Lulka, 2009). In overcoming general conclusions about what species *do*, Lulka (2009, 385), for example, argued for a 'thick hybridity' that avoids lumping together non-humans as stable constructs by attending to the diversity of non-human agency, with and without humans. To extend from Ingold (2007, 9), non-humans such as plants or animals do not present themselves as tokens of some common essence – the environment – that endows every non-human with its inherent agency.

The significance of uncovering the distinctive capacities of non-humans has been most recently emphasised in cultural geography approaches to human-plant studies. For example, in their work on rubber vine in Northern Australia, Head et al. (2015b) drew attention to plant capacities, or 'plantiness', to challenge human conceptions of agency and subjectivity. Plants have their own 'dynamic manifestations' (Head et al., 2015b, 405), which affect other humans and non-humans differently. Engaging with different categories of plants (e.g. trees, seeds, grasses and flowers) reveals different insights about the plants themselves and our relationship with them (Atchison and Head, 2013).

Drawing from this research, and more recent articulations of learning with plants, we propose the term 'relational learning' to capture the experiences of people coming to grips with managing a specific invasive plant. In particular, we borrow from Pitt's (2016, 92) work on 'planty knowledge', which refers to 'a combination of what humans learn about plantiness, and that which plants themselves understand or sense of the world', as well as Myers's (2015, 42, 60) investigation into 'phytomorphism', whereby people 'get entrained to plant behaviours, rhythms and temporalities, and learn to elicit and observe a range of phenomena that many others will never behold . . . pulling and propelling them into new modes of inquiry, and new modes of flight'. Both approaches expand the definitions of agency and subjectivity to recognise how plants guide the learning process.

When existing management knowledge is no longer applicable, research needs to pay attention to how people learn through being moved by 'plant form' (Myers, 2015, 58). In extending these recent articulations of plantiness, we not only aim to provide a more robust account of how graziers learn to manage an invasive plant but also consider how turning attention to plantiness can inform new ways of thinking and acting in spaces of management uncertainty. As Brice (2014, 946) argued, 'sensitivity to differences in others proliferates capacities to act in new and different ways, so that being affected produces not passivity but a more varied, sensitive and subtle repertoire of capacities for action'. In the context of management uncertainty, we argue that decentring human understandings of agency and subjectivity also opens new possibilities for living with invasives.

Context: African lovegrass and uncertainty

In Bega Valley, African lovegrass is slowly taking hold, outcompeting existing native and improved pasture species and forcing new management arrangements. Numerous agronomic types of African lovegrass were introduced across Australia from the early 1900s to the early 1980s for pasture improvement (Firn, 2009). The species was valued in the summer rainfall regions of southern Africa, the US and Argentina as a productive pasture species. However, in Australia, African lovegrass has failed to meet original expectations largely due to the inability of landholders to intensively manage the plant. Without intensive management African lovegrass has reduced the value of pastures − the species is often unpalatable to stock and low in crude protein − and has vigorously and extensively invaded pasture communities, woodlands, riparian areas and roadsides (Firn, 2009).

However, the damage being caused by African lovegrass is not a complete surprise. During experimental testing, researchers raised alarms about the lack of palatability of African lovegrass and also noted the species' vigour as a weed (Firn, 2009). While African lovegrass's fast growth rate, seed production and tolerance for dryer climates motivated the initial trials for pasture improvement, its lack of palatability presented a significant concern. As Leigh and Davidson (1968) noted, 'with appropriate husbandry it [African lovegrass] might be a valuable pasture; without this it might be an embarrassment' (quoted in Firn, 2009, 86). Since then, African lovegrass has become a major problem across Australia. Indeed, the properties of African lovegrass that warranted its introduction (i.e. the ability to survive harsh soil and climate conditions) have facilitated its spread. Additionally, African lovegrass's lack of palatability accelerates preferential grazing, reducing pasture competition and allowing the species to dominate.

In areas of Bega Valley, the geographic distribution and density of African lovegrass means eradication is impossible, and containing its spread is increasingly difficult. Compounding the physical inability to eradicate or contain African lovegrass, formal NRM and primary production institutions (e.g. Landcare and Local Land Service [LLS][1]) remain uncertain about the precise methods needed to manage African lovegrass, and while there is agreement on what people want to achieve, there is both indeterminacy and ignorance about how to achieve it. This institutional uncertainty stems from a lack of effective information on how to control the species but also the inapplicability of existing strategies in different geographical contexts. As Firn (2009) noted, there exists a high amount of diversity within the species *Eragrostis curvula* to the extent that it is composed of seven agronomic types: Curvula, Conferta, Short Chloromelas, Tall Chloromelas, Robusta Green, Robusta Blue and Robusta Intermediate. The variations and overlap between these agronomic types create uncertainty for land managers attempting to predict the impacts of the species and assess the suitability of existing control methods. Further, while all agronomic types of African lovegrass (with the exception of Consol) are now considered undesirable plants within Australia, declaration of the complex as noxious weeds has not occurred in all states and regions (Firn, 2009). As a result, management experiments, and the distribution of results, remain uncoordinated and are often limited to local government areas.

The research project we report on here included interviews with a total of 49 landholders and 5 NRM agency staff from April–June 2014, with follow-up interviews conducted with 10 participants in November 2014. Participants included government weeds officers, LLS, Landcare staff, research scientists and landholders. This chapter focuses on interviews with NRM agency staff, production farmers (particularly beef and sheep graziers) and lifestyle agrarians (non-commercial grazing enterprises).

Learning with African lovegrass

This empirical section is structured around the key themes of the learning process: from discovering the problem of African lovegrass to realising existing knowledge was insufficient and finally experimenting with ways of living with the plant. The learning process is not prescriptive; many landholders engaged in these scenarios simultaneously and non-linearly. In detailing landholders' learning journeys, we specifically focus on how landholders develop the capacity to manage African lovegrass in spaces of uncertainty, characterised by a lack of formal management advice.

Landholders learning about African lovegrass first involves *discovery*. Neighbours, NRM staff or weed identification booklets/online sources often assisted in the identification of African lovegrass. While this is a necessary step in confirming the species' presence, landholders learned of and about the consequences of infestation differently. The process of discovering the effects of African lovegrass is connected to the plant's behaviours in relation to other non-humans and the elements. This includes the weather (e.g. drought), livestock and the reactions of other pasture species. One grazier recounts the initial shock of discovering the explosion of grass that followed from drought-relieving rain:

> When it rained, all this tall grass came up, initially green, and half a meter high along. I thought, 'Wow, look at this place go!' And then it just realised itself as it matured, and turned into this senescing [ageing] African lovegrass that nothing would eat.
>
> *(Bruce, production farmer: beef)*

While some landholders were quick to recognise African lovegrass as a significant weed, for others, the consequences of the plant were learned more gradually. African lovegrass emerges as a weed when it disrupts land management. This is an issue facing production and non-production properties. In particular, non-farming properties discover the problem of African lovegrass when attempting to manage the species for fire prevention, in gardens or along fence lines. Its fast growth rate and seed reproduction demand immediate and sustained control efforts. Landholders hand removing African lovegrass remark on the strength of the roots and density of the tussock, which make the task overwhelming: 'If you have ever tried pulling out lovegrass when it's in a clump, it is back breaking work, [even] with the amount of acreage we've got . . . it's really quite soul destroying' (Walt, amenity migrant). For graziers, African lovegrass emerges as an issue by observing how livestock interact with the plant. As one grazier describes: 'You see it in the calves, you put good fresh cattle in there out of the good country and you move them [into African lovegrass] and you see the cows dry off, the milk drops, they lose condition' (Tony, production farmer: beef).

When African lovegrass matures, the senescing leaves turn brown, becoming rank and unpalatable. The tussock of the plant grows wide and forms a hard, sharp surface, making it difficult to move through and creating problems for animals' feet. African lovegrass's 'weediness' is partly a result of the agency of the plant: how it grows, feels and changes appearance. However, these changes only become meaningful in relationship with livestock. For example, the tall brown leaves of African lovegrass become a problem when livestock avoid the plant. Also, graziers learn through observing changes in livestock condition (losing weight, drying off) that the plant is low in protein. While wider social networks warn landholders about African lovegrass, individual experiences clarify these concerns and contextualise them personally.

The discovery of African lovegrass elicited a range of *responses*. These varied from anger over lack of transparency at sale time from real estate agents, shock about the severity or extent of the weed and naivety over the difficulty of control. Following the discovery of African lovegrass, participants often sought information from NRM agencies. LLS and Landcare are common points of contact in learning about the species, its impact and available management options. These groups serve as knowledge-brokers disseminating information in digestible formats to assist in land management. However, as one grazier describes, African lovegrass stretches and exceeds existing management knowledge and conventional responses, limiting NRM agencies capacity to assist landholders:

> It started with a district agronomist. Basically, he had no knowledge at all of managing African lovegrass other than burning it. He had no idea. I asked him how to get rid of it. He said, 'When you find out, let me know'.
>
> *(Bruce, production farmer: beef)*

In response, some graziers sought advice from federal government agency staff to remove African lovegrass through pasture improvement, which involved spraying African lovegrass with herbicide and then resowing pastures with different species. However, as one grazier describes, these strategies were unsuccessful:

> I started with pasture improvement because that was the Department of Agriculture's advice. Whether through bad luck or bad judgement on my part, I don't know. I've had three goes at pasture improvement and none of them have been successful.
>
> *(Vince, production farmer: beef and sheep)*

Adding to such troubles, attempts to improve pasture with more preferred species were costly, with one participant lamenting spending thousands of dollars on 'a fizzer', or totally unsuccessful control. These experiences challenge the relevance of existing knowledge-brokers as traditional management responses fail to meet the desired outcomes.

Landcare and LLS also admit to a lack of knowledge about managing African lovegrass; however, they are engaging in a variety of strategies to learn. One strategy involves working with informal social networks such as friends and neighbours to gather information about the different strategies to control African lovegrass (e.g. trialling different herbicides, as well as altering grazing practices through varying stocking rates and paddock rotations). The value of this informal knowledge is demonstrated by the local Landcare organisation: the Far South Coast Farmers Network. This network aims to collaborate with and between farmers in managing African lovegrass. As one farmer explains: 'I'm also a member of the Farmers Network so that's a good thing that we've been getting info from . . . what my mates are doing and how it responds and just trialing and stuff' (Tony, production farmer: beef). While organised by Landcare, this information comes from informal sources. In the absence of institutional knowledge, Landcare and the LLS collate different experiences in the hope of developing strategies and a knowledge base for coping with African lovegrass. Given that both landholders and extension services value others' experiences in learning about African lovegrass, it poses a deeper question about how this learning occurs.

The extensiveness of African lovegrass across the landscape, coupled with a lack of effective management information, has created the need for landholders to develop *personal learning* and novel ways of living with the species. Learning how to manage African lovegrass is a process of trial and error. For some, this involves personal research into the plant to develop new management techniques (e.g. reading peer reviewed texts and other publications relating to African lovegrass). Bruce details his motivations for conducting this research: 'We thought, "Well, we can't find anything out there", so I just started reading; just learning about soil. I learnt along the way that the species of plant that grows is determined by the conditions of the soil' (Bruce, production farmer: beef). Bruce's personal research aims to develop management techniques that limit African lovegrass spreading and to learn of its potential use in grazing management. Subsequently, Bruce – informed by his reading – conducts various experiments such as increasing soil fertility or increasing grazing pressure in the hope of reducing African lovegrass and promoting other grasses. These experiments then become meaningful through the responses of African lovegrass and other non-humans (particularly cattle) that validate or contradict research.

Another landholder described developing new management techniques as a response to the failures of more conventional management practices:

> I tried spraying it at first, and then I thought, 'This is ridiculous. It's all over the place. I can't spray the entire property'. And I found that slashing it in the house paddock seemed to keep it down to a reasonable extent, but really, it didn't go until a) we got the sheep on, and then b) the drought disappeared.
>
> *(Josie, lifestyle agrarian: beef and sheep)*

Informed by these new observations, graziers such as Josie then attempt to maintain the colour and length of African lovegrass – either through slashing or more intensive grazing – that corresponds with livestock eating the plant. In this context, graziers are modifying familiar management techniques as they become aware of how the materiality of African lovegrass (how it looks, feels and grows) influences livestock behaviours.

Contrasting active experimentation, landholders also observed how certain non-human relationships work to suppress African lovegrass without direct human input. For example, landholders identified an absence of African lovegrass under trees and in areas with higher fertility, such as around livestock camps. This observation is summarised by one grazier:

> African lovegrass doesn't like shading, it doesn't like high nutrient [soil] and you'll notice under some trees where stock camp, around stock troughs any area that's highly fertilised and highly utilised or where ground cover is left, African lovegrass is very sparse.
>
> *(Adam, production farmer: beef)*

Observing these changes triggers more active management to replicate these functions, such as tree planting. This observe-to-manage process reveals the influence of non-humans, which is not always directed by people. Ultimately, both active experimentation and passive observation depends on graziers' 'planty knowledge' (Pitt, 2016), as they pay careful attention to the physical properties of African lovegrass (leaf colour, length and texture, seed production) in relationship with livestock behaviour (grazing preferences).

Living with African lovegrass is a reality facing many landholders in the Bega Valley. Its invasiveness and spatial extent make eradication impossible. This is not to say landholders have a defeatist attitude but rather that they need to make immediate and pragmatic decisions with or without the help of formal institutions. The inability to remove African lovegrass presents significant challenges for landholders. For some graziers, this has resulted in them rethinking the values that underpin how weeds are perceived (particularly African lovegrass): 'I see weeds as a succession thing, so we've damaged the landscape, so they're healing, but you're not going to eradicate them. We have to learn how to live with them through land management practices' (Tabatha, lifestyle agrarian: mixed grazing). The growing acceptance of living with African lovegrass reflects similar research that documents the changing attitudes of landholders towards invasive plants that can no longer be removed (Barker, 2008; Kull et al., 2019). However, we have aimed to not only provide evidence of a change in perceptions but also trace how these changing perceptions are learned *with* the plant.

Learning to live with African lovegrass is partly achieved through treating the plant as any other pasture species. Rather than inventing completely new management techniques, some graziers rely on existing knowledge and skills of pasture management. In this context, graziers' management responses are not entirely novel but rely on careful observations and adaptation of more familiar practices (Shackleton and Shackleton, 2018).

Bruce captures how learning to live with African lovegrass involves attuning existing skills to the specificities of this plant: 'I can look at the plant [African lovegrass] to tell me what the temperature's going to be for the day. If it doesn't roll its leaf up, then I know it's not going to go 35, and I don't have to be overly concerned about shade for the cattle' (Bruce, production farmer: beef). Beef and sheep graziers depend on detailed knowledge of grass: they watch how grass responds to rain; they see what happens from overgrazing; they inspect paddocks routinely and notice new or odd things in relation to grass and weeds. They move stock to manage their grasses as much as to manage the stock. This attentiveness to grass makes them keenly aware of the way small grassy plants can change and respond. However, the fact that African lovegrass cannot be completely managed the same as other pasture species highlights the diversity of plant life that also needs careful consideration (Brice, 2014). Learning African lovegrass's plantiness – the length at which it becomes unpalatable, how seeds are spread, how it competes with other grasses how it reacts to certain herbicides – is necessary in order to live with the plant. While existing knowledge of pasture grasses is useful, differentiating African lovegrass from other more 'manageable' plants provides the basis for learning how to live with this species, which is not completely novel but nevertheless depends on learning its distinctive capacities (Head et al., 2015b).

In many ways, living with African lovegrass has seen a return to Leigh and Davidson's (1968) initial observation that maintaining the value of African lovegrass in grazing systems, and avoiding the degradation of pastures and natural habitats, requires 'intensive management and appropriate husbandry'. Moreover, learning to live with African lovegrass has required graziers to learn how to cultivate the distinctive capacities of the species that justified its initial introduction. Attending to both the plant itself, and the wider relationships where its effects are co-produced, provides opportunities for learning how to cultivate African lovegrass while at the same time avoiding the properties of the plant that are least valued in grazing systems. In this context, African lovegrass becomes a stakeholder within the learning process, as graziers not only attempt to maintain desired land uses practices but also learn the conditions and management practices that best support the plant, along with other non-humans. Such attentiveness to 'plant form' (Myers, 2015) is critical in developing the capacity to live with the species and maintain land uses.

Conclusion

By articulating a relational learning approach, we have simultaneously widened and focused the scope of research that invokes some level of agency to the non-human actors involved in invasive species management. First, and most generally, we have sought to decentre the role of human-to-human networks in environmental learning research to explore the utility of extending agency to non-humans. Second, and to advance this point, we have considered how a specific invasive plant, in relation with human and other non-humans, facilitates learning and guides management practices. In detailing how African lovegrass's plantiness comes to not only trouble but inform graziers' management practices, we have broadened the conceptions of agency and subjectivity commonly applied in environmental learning. When existing management strategies became ineffective, graziers attuned themselves to the properties and capacities of African lovegrass. In acquiring African lovegrass's plantiness – rapid growth, difficulty in control, unpalatability and lack of crude protein – graziers were not only able to better understand how the plant became a significant weed but, over time, learned to develop management practices to live with the species.

By proposing a relational learning approach, we have also aimed to contribute to research on living with invasive species, and invasive sciences more generally, by shifting focus towards the distinctive capacities of non-humans. The current emphasis in invasive species management on social learning and knowledge controversies between experts and stakeholder groups, while essential, limits analysis on how learning is achieved in spaces of uncertainty. In response, we have positioned non-humans as stakeholders in the learning process to draw attention to the other-than-human factors that influence invasive plant management. Focusing on one non-human stakeholder, African lovegrass, we have traced how the species both affects, and is affected by, the development of management responses in Bega Valley. Treating non-humans as stakeholders within this learning approach opens space to not only consider how non-humans can unsettle existing management knowledge and practices but also challenges the ability of land managers to identify and create conditions that will *benefit* non-humans, particularly if effective ways of living with invasive plants are to be developed.

Note

1 Local Land Services is a state-funded NRM and primary production agency; there are 11 Local Land Services bodies across New South Wales. Landcare is a federally funded volunteer NRM program operating throughout Australia; there are up to 3,000 Landcare groups in New South Wales.

References

Armitage, D.R., Marschke, M. and Plummer, R. (2008) 'Adaptive co-management and the paradox of learning', *Global Environmental Change*, vol. 18, no. 1, pp. 86–98.

Atchison, J. and Head, L. (2013) 'Eradicating bodies in invasive plant management', *Environment and Planning D: Society and Space*, vol. 31, pp. 951–968.

Bach, T.M., Kull, C.A. and Rangan, H. (2019) 'From killing lists to healthy country: Aboriginal approaches to weed control in the Kimberley, Western Australia', *Journal of Environmental Management*, vol. 229, pp. 182–192.

Barker, K. (2008) 'Flexible boundaries in biosecurity: Accommodating gorse in Aotearoa New Zealand', *Environment and Planning A*, vol. 40, no. 7, pp. 1598–1614.

Bart, D. and Simon, M. (2013) 'Evaluating local knowledge to develop integrative invasive-species control strategies', *Human Ecology*, vol. 41, no. 5, pp. 779–788.

Bear, C. (2011) 'Being Angelica? Exploring individual animal geographies', *Area*, vol. 43, no. 3, pp. 297–304.

Biodiversity Conservation Act 2016 (NSW), NSW government legislation (Act), refers to Harvard referencing guide.

Brenner, J.C. (2011) 'Pasture conversion, private ranchers, and the invasive exotic buffelgrass (Pennisetum ciliare) in Mexico's Sonoran Desert', *Annals of the Association of American Geographers*, vol. 101, no. 1, pp. 84–106.

Brice, J. (2014) 'Attending to grape vines: Perceptual practices, planty agencies and multiple temporalities in Australian viticulture', *Social & Cultural Geography*, vol. 15, pp. 942–965.

Coleman, M.J., Sindel, B.M., Reeve, I.J. and Thompson, L.J. (2015) 'Best practice weed detection on Australian farms', *Land Use Policy*, vol. 48, pp. 567–574.

Cooke, B. and Lane, R. (2015) 'How do amenity migrants learn to be environmental stewards of rural landscapes?', *Landscape and Urban Planning*, vol. 134, pp. 43–52.

Cooke, B. and Lane, R. (2018) 'Plant-human commoning: Navigating enclosure, neoliberal conservation, and plant mobility in exurban landscapes', *Annals of the American Association of Geographers*, vol. 108, no. 6, pp. 1715–1731.

Crowley, S.L., Hinchliffe, S. and Mcdonald, R.A. (2017) 'Conflict in invasive species management', *Frontiers in Ecology and the Environment*, vol. 15, no. 3, pp. 133–141.

Davis, D. and Carter, J. (2014) 'Finding common ground in weed management: Peri-urban farming, environmental and lifestyle values and practices in southeast Queensland, Australia', *Geographical Journal*, vol. 180, no. 4, pp. 342–352.

Ellis, E.C. (2015) 'Ecology in an anthropogenic biosphere', *Ecological Monographs*, vol. 85, no. 3, pp. 287–331.

Environment Protection and Biodiversity Conservation Act 1999 (Cwlth), Australian Government legislation (Act).

Evans, J.M., Wilkie, A.C. and Burkhardt, J. (2008) 'Adaptive management of nonnative species: Moving beyond the "either-or" through experimental pluralism', *Journal of Agricultural and Environmental Ethics*, vol. 21, no. 6, pp. 521–539.

Fazey, I., Fazey, J.A. and Fazey, D.M.A. (2005) 'Learning more effectively from experience', *Ecology and Society*, vol. 10, no. 2, p. 4.

Fazey, I., Fazey, J.A., Salisbury, J.G., Lindenmayer, D.B. and Dovers, S. (2006) 'The nature and role of experiential knowledge for environmental conservation', *Environmental Conservation*, vol. 33, no. 1, pp. 1–10.

Fenwick, T. (2010) 'Re-thinking the "thing"', *Journal of Workplace Learning*, vol. 22, no. 1/2, pp. 104–116.

Firn, J. (2009) 'African lovegrass in Australia: A valuable pasture species or embarrassing invader?', *Tropical Grasslands*, vol. 43, pp. 86–97.

Fischer, A.P. and Charnley, S. (2012) 'Private forest owners and invasive plants: Risk perception and management', *Invasive Plant Science and Management*, vol. 5, no. 3, pp. 375–389.

Førde, A. and Magnussen, T. (2015) 'Invaded by weeds: Contested landscape stories', *Geografiska Annaler: Series B, Human Geography*, vol. 97, no. 2, pp. 183–193.

Graham, S. (2013) 'Three cooperative pathways to solving a collective weed management problem', *Australasian Journal of Environmental Management*, vol. 20, no. 2, pp. 116–129.

Head, L. (2017) 'The social dimensions of invasive plants', *Nature Plants*, vol. 3, p. 17075.

Head, L., Atchison, J. and Phillips, C. (2015b) 'The distinctive capacities of plants: Re-thinking difference via invasive species', *Transactions of the Institute of British Geographers*, vol. 40, pp. 399–413.

Head, L., Larson, B., Hobbs, R., Atchison, J., Gill, N., Kull, C. and Rangan, H. (2015a) 'Living with invasive plants in the anthropocene: The importance of understanding practice and experience', *Conservation and Society*, vol. 13, no. 3, pp. 1–21.

Hobbs, R., Higgs, E.S. and Hall, C.A. (eds.) (2013) *Novel Ecosystems: Intervening in the New Ecological World Order*. Wiley-Blackwell, Oxford.

Ingold, T. (2007) 'Materials against materiality', *Archeological Dialogues*, vol. 14, no. 1, pp. 1–16.

Keen, M. and Mahanty, S. (2006) 'Learning in sustainable natural resource management: Challenges and opportunities in the pacific', *Society and Natural Resources*, vol. 19, no. 6, pp. 497–513.

Klepeis, P., Gill, N. and Chisholm, L. (2009) 'Emerging amenity landscapes: Invasive weeds and land subdivision in rural Australia', *Land Use Policy*, vol. 26, no. 2, pp. 380–392.

Krasny, M.E. and Lee, S.K. (2002) 'Social learning as an approach to environmental education: Lessons from a program focusing on non-indigenous, invasive species', *Environmental Education Research*, vol. 8, no. 2, pp. 101–119.

Kull, C.A., Harimanana, S.L., Andrianoro, A.R. and Rajoelison, L.G. (2019) 'Divergent perceptions of the "neo-Australian" forests of lowland eastern Madagascar: Invasions, transitions, and livelihoods', *Journal of Environmental Management*, vol. 229, pp. 48–56.

Lankester, A.J. (2013) 'Conceptual and operational understanding of learning for sustainability: A case study of the beef industry in North-Eastern Australia', *Journal of Environmental Management*, vol. 119, pp. 182–193.

Larson, B., Kueffer, C. and ZiF Working Group on Ecological Novelty (2013) 'Managing invasive species amidst high uncertainty and novelty', *Trends in Ecology and Evolution*, vol. 28, no. 5, pp. 255–256.

Latour, B. (1993) *We Have Never Been Modern*. Harvester Wheatsheaf, London.

Leigh, J.H. and Davidson, R.L. (1968) '*Eragrostis curvula* (Schrad.) Nees and some other African Lovegrasses', *Plant Introduction Review*, vol. 5, pp. 21–46.

Leys, A.J. and Vanclay, J.K. (2011) 'Social learning: A knowledge and capacity building approach for adaptive co-management of contested landscapes', *Land Use Policy*, vol. 28, no. 3, pp. 574–584.

Lulka, D. (2009) 'The residual humanism of hybridity: Retaining a sense of the earth', *Transactions of the Institute of British Geographers*, vol. 34, no. 3, pp. 378–393.

McLeod, R. (2018) *Annual Costs of Weeds in Australia*. Centre for Invasive Species Solutions, Canberra, Australia.

Muro, M. and Jeffrey, P. (2008) 'A critical review of the theory and application of social learning in participatory natural resource management processes', *Journal of Environmental Planning and Management*, vol. 51, no. 3, pp. 325–344.

Myers, N. (2015) 'Conversations on plant sensing', *NatureCulture*, vol. 3, pp. 35–66.

Ngorima, A. and Shackleton, C.M. (2019) 'Livelihood benefits and costs from an invasive alien tree (Acacia dealbata) to rural communities in the Eastern Cape, South Africa', *Journal of Environmental Management*, vol. 229, pp. 158–165.

Pitt, H. (2016) 'Towards a more than human participatory research', in M. Bastian, O. Jones, N. Moore and E. Roe (eds.) *Participatory Research in More-Than-Human Worlds*. Routledge, London, pp. 92–106.

Plumb, D. (2008) 'Learning as dwelling', *Studies in the Education of Adults*, vol. 40, no. 1, pp. 62–79.

Plummer, R. and Armitage, D. (2007) 'Crossing boundaries, crossing scales: The evolution of environment and resource co-management', *Geography Compass*, vol. 1, no. 4, pp. 834–849.

Pyšek, P. and Richardson, D.M. (2010) 'Invasive species, environmental change and management, and health', *Annual Review of Environment and Resources*, vol. 35, no. 1, pp. 25–55.

Robbins, P. (2010) 'Comparing invasive networks: Cultural and political biographies of invasive species', *Geographical Review*, vol. 94, no. 2, pp. 139–156.

Shackleton, R.T., Le Maitre, D.C. and Richardson, D.M. (2015) 'Stakeholder perceptions and practices regarding Prosopis (mesquite) invasions and management in South Africa', *Ambio*, vol. 44, no. 6, pp. 569–581.

Shackleton, R.T., Richardson, D.M., Shackleton, C.M. et al. (2019) 'Explaining people's perceptions of invasive alien species: A conceptual framework', *Journal of Environmental Management*, vol. 229, pp. 10–26.

Shackleton, S.E. and Shackleton, R.T. (2018) 'Local knowledge regarding ecosystem services and disservices from invasive alien plants in the arid Kalahari, South Africa', *Journal of Arid Environments*, vol. 159, pp. 22–33.

Shrestha, B.B., Shrestha, U.B., Sharma, K.P., Thapa-Parajuli, R.B., Devkota, A. and Siwakoti, M. (2019) 'Community perception and prioritization of invasive alien plants in Chitwan-Annapurna Landscape, Nepal', *Journal of Environmental Management*, vol. 229, pp. 38–47.

Tarnoczi, T. (2011) 'Transformative learning and adaptation to climate change in the Canadian Prairie agro-ecosystem', *Mitigation Adaptation Strategies Global Change*, vol. 16, pp. 387–406.

Udo, N., Darrot, C. and Atlan, A. (2019) 'From useful to invasive, the status of gorse on Reunion Island', *Journal of Environmental Management*, vol. 229, pp. 166–173.

6

NOVEL SCIENTIFIC APPROACHES TO UNDERSTANDING EMERGING INFECTIOUS DISEASES

Kim B. Stevens

Introduction: the what, where, when and why of EIDs

Infectious diseases – in particular, emerging infectious diseases (EIDs) and their rapid spread – are being increasingly recognised as a global threat as a result of the considerable burden they place on global economies and public health (Morens and Fauci, 2012; Nava et al., 2017). Although a very modern problem, EIDs and knowledge of the threat they pose dates back millennia and was highlighted in 1685 by the scientist Robert Boyle, who observed that 'there are ever new forms of epidemic diseases appearing . . . among [them] the emergent variety of exotick and hurtful' (Morens and Fauci, 2012). This chapter looks at what EIDs are and why they occur, followed by a brief overview of a range of novel scientific techniques for expanding our knowledge of the ecology and evolution of such diseases, which, in turn, allows for more effective control and prevention measures.

What are EIDs?

Morens et al. (2004, 242) defined EIDs as 'infections that have newly appeared in a population or have existed previously but are rapidly increasing in incidence or geographic range'. They further subdivide EIDs into those that are 'newly emerging' (i.e. not previously recognised) and those that are 're-emerging/resurging' (i.e. diseases that were a problem before declining dramatically and then increasing again) (Morens et al., 2004). More than 60% of EIDs are zoonotic (i.e. they originated in animals), and over 70% of these originated in wild animals (Jones et al., 2008). While some of these transmit directly to humans, the transmission of emerging infections is often complex and can involve several host species (Wendelboe et al., 2010). In fact, emerging pathogens frequently exhibit a broad host range (Woolhouse and Gowtage-Sequeria, 2005), yet how animal-adapted pathogens make the jump and adapt to new species such as humans still remains largely unknown. With a view to improving surveillance and outbreak preparedness, various studies have developed geographical risk assessment frameworks in an attempt to identify areas of highest risk for spillover of key pathogens, such as avian and swine influenza, into the human population (Hill et al., 2015; Berger et al., 2018). However, others claim that it is impossible to predict where or when the next pandemic will occur, and teasing apart the details of such mechanisms remains an important challenge in understanding EIDs (Morens and Fauci, 2012).

Although EIDs are caused by a range of pathogens, Jones et al. (2008) showed that the majority (54%) are caused by bacteria (including rickettsia). Taylor et al. (2001), on the other hand, found viruses to be the primary cause of EIDs (44%), with only 30% caused by bacteria and rickettsia. However, the two studies differed in their classification of pathogens, as Taylor et al. (2001) categorised pathogens by species only while Jones et al. (2008) classified individual drug-resistant microbial strains as separate pathogens, reasoning that different strains can cause discrete outbreaks (Lederberg et al., 1992). This suggests that correct taxonomic classification is a fundamental first step to improved understanding of EIDs.

When and where do EIDs occur?

In a study of the global spatio-temporal distribution of EIDs, Jones et al. (2008) found that 335 emerging-disease 'events' were reported globally between 1940 and 2004 and that even after controlling for reporting efforts, EID events showed a significantly (p < 0.001) increasing trend between 1940 and 1980.

While studies have found that EID events are more likely to be *reported* (as opposed to occur) in *developed* countries – a likely effect of the increased and improved surveillance efforts in such regions (Jones et al., 2008; Allen et al., 2017) – risk of *emergence* is highest in *developing* countries. However, global distribution of EIDs differed depending on whether the pathogens arose as a result of drug resistance; were vector-borne, zoonotic pathogens originating from wildlife; or were zoonotic pathogens originating from non-wildlife, although risk was generally greatest in tropical regions (Jones et al., 2008). A more recent study which focused only on zoonoses originating from wildlife showed a similar EID hotspot distribution centred on the tropics, in particular parts of Southeast Asia and the east and west coasts of central Africa (Allen et al., 2017). More specifically, predictive modelling characterised areas with the highest risk for disease emergence as being regions of tropical forest, high in mammalian biodiversity and undergoing land-use changes related to agricultural practices (Allen et al., 2017). The results are concerning as they suggest that global resources may be incorrectly allocated, with the majority of the scientific and surveillance efforts focused on countries from where the next important EID is least likely to originate (Jones et al., 2008). Allen et al. (2017) therefore advocate that resources be re-allocated to focus on high-risk regions such as tropical Africa, Latin America and Asia.

Why EIDs occur: drivers of emergence

While our understanding of the drivers behind EIDs remains incomplete (Allen et al., 2017), Cohen (2000) cited 'change' as the main cause of disease emergence: climate change, changing ecosystems due to loss of biodiversity, changing land use, microbial adaption, societal changes such as changing human demographics and behaviour, changing human susceptibility to infection, improved technology and industry and increased international travel (Woolhouse and Gowtage-Sequeria, 2005; Wendelboe et al., 2010). Ecological change has been identified as the greatest driver of EIDs in tropical areas with high biodiversity (Wilcox and Gubler, 2005; Jones et al., 2008) while other studies suggest that anthropogenic drivers such as land-use change, food production changes and global trade and travel are having the greatest influence on emergence (Nava et al., 2017). However, rather than a single cause, EIDs often stem from complex interactions among several microbial, host and environmental factors together creating opportunities for pathogens to evolve into new ecological niches (Morens et al., 2004). In addition, pathogen emergence might differ between developed and developing countries (Cohen, 2000) or by geographical region (Lederberg et al., 1992), and it is therefore important to consider each EID event on its own merits as a comprehensive understanding of the often complex relationships between multi-host systems, environmental change and human populations is essential in order to implement effective control measures (Cohen, 2000).

Surveillance of emerging infections

Identification of EIDs at their earliest stages through effective surveillance strategies – followed by a rapid response – is crucial to mitigate the spread and effects of EID outbreaks. Despite its acknowledged importance as the first line of defence against EID outbreaks, a study found considerable variation in global surveillance efforts for avian and swine influenza (Berger et al., 2018). Although most countries implemented surveillance at the national level, the authors found that this varied over time and that surveillance at a subnational level (state/province) was patchy, with almost half (49%) of states/provinces with a predicted high risk for avian-to-human transmission of influenza viruses lacking any kind of surveillance for the previous 15 years (Berger et al., 2018).

Surveillance can be either active or passive, and both approaches have their advantages and limitations. Active surveillance requires the accurate pinpointing of at-risk areas and populations through the use of predictive models and spatial risk frameworks in order to identify priority areas for targeted surveillance with a view to making best use of limited resources. For example, the spatial framework presented by Berger et al. (2018) highlights the spatial variation in outbreak emergence potential for animal-to-human and secondary human-to-human transmission of avian and swine influenza viruses. However, despite the undeniable value of active surveillance, identification of EIDs generally relies largely on passive surveillance through reporting by healthcare providers and laboratories (Reingold and Lewnard, 2019). While such systems are fundamental for the detection of EIDs, they are not without their limitations, such as delayed detection of cases and lags in official reporting. While lag times for such reports as the Centers for Disease Control and Prevention (CDC) US Influenza Sentinel Provider Surveillance reports are in the order of one to two weeks (Ginsberg et al., 2009), in developing countries reporting of animal disease events can be delayed by months (Karimuribo et al., 2012), while Arsevska et al. (2018) found that official World Organisation for Animal Health (OIE) notification of an outbreak occurred a week after onset for 43% of the outbreaks studied and 30 days after onset for 27% of outbreaks.

In addition to national and subnational surveillance, the increasing number of transboundary epidemics has highlighted the need for establishing such systems at broader scales. As a result, data warehouses and disease reporting systems such as the World Animal Health Information Database (WAHID) (Jebara et al., 2012) and EMPRES Global Animal Disease Information System (EMPRES-i) (Farnsworth et al., 2010; FAO, 2011) were launched to encourage and facilitate data collection and sharing of animal disease data at a global scale. Stevens and Pfeiffer (2015) provide an overview of these data warehouses and their data sources, and while such systems are a step in the right direction, the absence of a single organisation responsible for surveillance inhibits effective EID detection. The recent Ebola outbreaks highlighted similar shortcomings, resulting in a call for global public health agencies to coordinate an improved epidemiological data management system for disease surveillance (Owada et al., 2016; Holmes et al., 2018).

Internet-based surveillance

The internet is revolutionising how epidemic intelligence is gathered, particularly in developed countries, allowing for earlier detection of disease outbreaks than when using traditional surveillance approaches, with the added bonuses of reduced costs and increased reporting transparency. Text-based information sources on the web, such as publicly available news articles, official disease reports and newsletters, have been found to be informative for early detection of EID outbreaks (Arsevska et al., 2018), and over the years, several web-focused event-based biosurveillance systems that mine secondary data sources such as internet-based media sites have been created for the timely detection of EID events. Human biosurveillance internet-based systems include BioCaster (Collier et al., 2008), HealthMap (Brownstein et al., 2008; Freifeld et al., 2008; Keller et al., 2009; Wilson and Brownstein, 2009; Brownstein et al., 2010), ProMED-mail (Cowen et al., 2006; Zeldenrust et al., 2008) (Tolentino et al., 2007), EpiSPIDER (Tolentino et al., 2007; Keller et al., 2009) and Canada's Global Public Health Intelligence Network (GPHIN) (Mykhalovskiy and Weir, 2006). Similar initiatives in developing countries include India's Media Scanning and Verification Cell (MSVC), which scans global and national media sources and flags unusual health events and successfully flagged a number of outbreaks before they were identified by traditional surveillance systems (Sharma et al., 2012). Animal biosurveillance systems include the Platform for Automated extraction of Disease Information from the web (PADI-web) (Arsevska et al., 2018).

The primary value of these internet surveillance systems currently lies in their ability to act as early warning systems, thereby lessening the consequences of an outbreak (Wilson and Brownstein, 2009; Hartley et al., 2013). For example, a local online media article, identified and posted by the ProMED-mail system (ProMED-mail, 2014) on 13 January 2014 which mentioned complaints by hunters of increased mortality in wild boars on the Lithuania-Belorussia border, was likely one of the first signs of the spread of African swine fever (ASF) to a new territory well before official government reports were issued on 24 January 2014 (OIE, 2014). In another example, GPHIN identified the 2002 severe acute respiratory syndrome (SARS) outbreak in Guangdong Province, China, more than two months before the World Health Organization's (WHO) official announcement (Mykhalovskiy and Weir, 2006). Similarly, HealthMap identified news stories reporting a strange fever in Guinea nine days before official notification of the 2014 West African Ebola outbreak (Milinovich et al., 2015).

Internet-based data sources exist outside traditional reporting channels and, as such, are invaluable to public-health agencies that rely on official surveillance activities or the timely flow of information across administrative borders. While such novel surveillance systems are still a long way from replacing traditional surveillance methods, they have been shown to usefully complement conventional approaches as syndromic surveillance (Milinovich et al., 2014) to the extent that they have become an important component of the influenza surveillance scene. For example, the World Health Organization's Global Outbreak Alert and Response Network uses such data as part of its day-to-day surveillance activities (Grein et al., 2000; Heymann and Rodier, 2001) and is authorised to act on this information (Wilson et al., 2008).

However, such automated systems are not without problems. A comparison of BioCaster, EpiSPIDER and HealthMap identified significant differences in their ability to obtain relevant disease information owing mainly to differences in sources searched, languages read, regions of occurrence and types of cases (Lyon et al., 2012; Barboza et al., 2014), although running the three surveillance systems in parallel was shown to enhance early detection of disease anomalies over traditional surveillance approaches (Barboza et al., 2014). In addition, the location detection tool of all three systems assumed the number of articles plotted for a country reflected the number of articles found about that country, which was not necessarily true (Lyon et al., 2012). However, the sheer amount of information available on the internet makes timely detection of relevant disease events challenging, and therefore the development of internet-based surveillance systems that use a sufficiently broad range of intelligence sources, while limiting the misleading effect of 'noise', is crucial in the development of effective systems that might one day supplant traditional surveillance efforts.

Spatial analysis

Visualising disease distribution

Mapping of EID events is fundamental for delineating the extent of the emergence, especially when emergence occurs in a new geographical location, and the advent of interactive maps and virtual globes such as Google Maps and Google Earth allows for easy visualisation of disease data. Two examples serve to highlight the value of Google Earth technology in creating effective information resources that increase understanding of EID events. First, *Nature* used the platform to track the global spatio-temporal spread of avian influenza (H5N1) (Butler, 2006) – a project that won the Association of Online Publishers (AOP) Use of a New Digital Platform Award in 2006. Second, when visualising unconventional forms of georeferenced data, Google Earth is a useful alternative to geographic information systems (GIS) software. In a modern-day reprise of John Snow's 1856 cholera investigation, Google Earth enabled Baker et al. (2011) to map the spread of a typhoid outbreak in Kathmandu – where street names were not used – and trace the cause of the epidemic to low-lying public water resources.

Identification of significant clusters

Identification of significant disease clusters can advance EID understanding in several ways, including suggesting potential risk factors for further investigation or indicating likely disease transmission routes. For example, clustering of disease events over short distances can suggest local transmission such as direct contact or vector- or airborne dispersal of infected droplets (Picado et al., 2007) while clustering over distances of hundreds of kilometres can suggest long-distance spread (Ahmed et al., 2010; Picado et al., 2011) as a result of movement of infected animals or windborne dispersal of vectors (up to 100 km for some *Aedes* and *Culex* species, Service, 1997) or infectious isolates (up to 300 km for the *C Noville* foot and mouth disease [FMD] isolate; (Alexandersen et al., 2003). Furthermore, transmission mechanisms for a specific disease can differ between outbreaks. In their study of the spatio-temporal pattern of Rift Valley fever (RVF) in South Africa, Métras et al. (2012) suggest that while mosquito bites appeared to be the main mechanism of RVF virus infection, the differing extents and intensities of the space-time interactions exhibited by the five epidemics occurring between 2008 and 2011 suggest that other, concomitant transmission mechanisms may have been involved in long-distance spread of the disease. In addition, cluster detection can also be used to highlight possible regional differences in disease transmission (Vander Kelen et al., 2012); identify areas where vectors and hosts coincide, resulting in potentially increased risk of disease transmission (Hennebelle et al., 2013; Swirski et al., 2013); or track the direction and geographical extent of disease spread (Wilesmith et al., 2003; Denzin et al., 2013), all of which helps increase understanding of EID transmission.

Predictive spatial modelling

Spatial variation in disease risk is determined by interactions among pathogens, vectors and hosts and among these agents and their environment. As a result, spatially explicit models of disease occurrence are valuable tools for furthering understanding of disease dynamics while the associated risk maps provide a geographical representation of disease risk for informing risk-based disease control and surveillance strategies. The inclusion of uncertainty in model outputs (Clements et al., 2006; Stevens et al., 2013) using, for example, Dempster–Shafer theory (Dempster, 1966, 1967; Beynon et al., 2000) allows for identification of areas for which the collection of additional information might be beneficial in increasing understanding of disease events.

The majority of spatial modelling studies focus on the identification of factors associated with disease introduction, spread, persistence or occurrence using presence and absence data in a regression model. However, a common problem with regression modelling of disease is that, while the.outcome variable may consist of fairly reliable disease presence data, absence data may not be available (e.g. when using surveillance data). This has led to the development of different sampling approaches for generating pseudoabsence data (Phillips et al., 2009) that can be used with modelling methods that require both presence and absence data such as regression or novel machine learning methods (e.g. boosted regression trees, Elith et al., 2008). In addition, ecological niche modelling methods such as Maxent (Phillips et al., 2006; Phillips and Dudík, 2008; Phillips, 2012) are also useful in the predictive spatial modelling of EIDs for which there are only disease presence data available. Maxent has been used successfully with very small sample sizes – less than 100 observations (Phillips et al., 2004) or even as few as 10 presence records (Wisz et al., 2008) – a characteristic that is very useful in attempts to model the ecological niche of EIDs in the early stages of an outbreak.

Novel spatial modelling methods, such as Maxent and boosted regression trees (Elith et al., 2008), can also provide additional information over that of the traditional odds ratios and p-values of logistic regression models, such as percentage contribution of variables together with profiles of the effect of each individual predictor on the outcome over the range of its values (i.e. partial dependence plots). Such information can be useful for characterising the ecological niche of an EID or designing surveillance strategies that target regions with these characteristics. For example, although studies using traditional regression methods have frequently identified an increasing density of domestic waterfowl (Gilbert et al., 2006; Pfeiffer et al., 2007; Gilbert et al., 2008; Minh et al., 2009; Martin et al., 2011) and rice growing (Pfeiffer et al., 2007; Gilbert et al., 2008) as important risk factors for occurrence of highly pathogenic avian influenza (HPAI) H5N1 in Asia, Stevens and Pfeiffer (2014) used boosted regression tree and Maxent models to show that roughly three-quarters of the spatial variation in distribution of HPAI H5N1 in domestic poultry in Asia could be attributed to these two variables.

This increasingly wide range of flexible spatial modelling approaches has been used to model the spatial distribution of an array of EID pathogens and disease vectors, including HPAI H5N1 (Williams et al., 2008; Williams and Peterson, 2009; Hogerwerf et al., 2010), *Bacillus anthracis* – the causative agent of anthrax (Joyner et al., 2010; Mullins et al., 2011; Chikerema et al., 2013), Ebola (Funk and Piot, 2014; Pigott et al., 2014; Pigott et al., 2016) and Marburg virus (Pigott et al., 2015; Nyakarahuka et al., 2017), to name a few. In addition, studies have defined the climate niches of tick species involved in disease transmission in the Mediterranean region and forecast changes in their habitat suitability as a result of the effects of climate change (Estrada-Pena and Venzal, 2007) and have predicted the global risk of spread of the mosquito *Aedes albopictus* (Benedict et al., 2007).

However, risk maps designed to identify suitable habitats beyond those currently occupied by a pathogen might seem unreliable if apparently false-positive predictions show up on new continents unless previous long-term events are taken into account. For example, 30 years ago, few would have believed that bluetongue would spread to Europe, yet the virus found not only suitable habitats but also competent indigenous vectors. Bluetongue has shown two distinct patterns of geographical emergence over the past decade – a steady northwards spread of the virus, associated with range expansion of the midge *Culicoides imicola* vector, together with long-distance jumps into new territories. The risk of the former was correctly predicted using statistical risk-mapping methods (Tatem et al., 2003), although the latter was unforeseen as the areas of emergence were beyond the range of *C. imicola*. However, northern expansion of *C. imicola* led to range overlap with indigenous northern *Culicoides* species which proved to be competent vectors for the disease (Randolph and Rogers, 2010).

Molecular genome sequencing

Genome sequencing is a relatively novel technique in the arsenal of tools available for understanding the evolution and biology of EIDs and can provide a range of information including identifying common ancestors and the source of disease outbreaks (Dudas et al., 2017) and routes of transmission (Arias et al., 2016); reconstructing the geographic spread of pathogens (Reingold and Lewnard, 2019); determining whether spread is local or further reaching, which provides some indication of the mechanisms involved in transmission; determining whether an outbreak is the result of constant seeding from numerous individual introductions or maintained through local transmission (Dudas et al., 2017); and revealing the size and duration of case clusters resulting from each introduction. For example, frequent cluster extinction within a large outbreak can suggest that individual outbreaks are constrained by the degree of connectedness among contact networks (Dudas et al., 2017).

Retrospective virus genome sequencing of the 2013–2016 Ebola epidemic provided substantial insight into the evolution and spread of the outbreak (Dudas et al., 2017). For example, molecular clock dating, together with phylogeographic estimation, identified the likely temporal (between December 2013 and February 2014) and spatial (the Guéckédou Prefecture, Nzérékoré Region, Guinea) origin of the epidemic's common ancestor. The authors were also able to determine that rather than being the result of independent zoonotic introduction, disease in Sierra Leone was caused by the virus crossing the border from Guinea, and that cases were the result of sustained transmission rather than sporadic independent outbreaks. In addition, they identified risk factors associated with geographic spread of the virus (tended to disperse between geographically close regions and positively correlated with population size of both origin and destination) and that the epidemic's dispersal followed a classic gravity-model dynamic. The authors also identified factors that predicted cumulative case count (population size, a negative correlation with travel time to large urban areas) and highlighted the areas in which cross-border transmission mainly occurred and the non-infected areas that were at risk of becoming part of the epidemic if the virus had spread to them. They also determined that in Sierra Leone, the bulk of transmission was the result of a single early introduction, while Guinea experienced repeated reintroductions of viral lineages from disparate transmission chains from both Sierra Leone and Liberia. From this single example, it is apparent that genome sequencing of viral pathogens can provide a wealth of information for advancing understanding of EIDs, and, while not offering the complete picture, such studies provide a framework for predicting the potential behaviour of future Ebola outbreaks and suggesting successful control strategies.

Conclusion

Despite the medical and scientific advances of the 20th and 21st centuries, EIDs remain a constant threat to public and animal health worldwide. At the same time, the world is experiencing considerable changes – in climate, land use, biodiversity and microbial resistance – which is particularly disturbing in light of the fact that 'change' has been cited as the main driver of infectious disease emergence. Conversely, though, we never had such a varied range of tools at our disposal with which to develop our understanding of the complex mechanisms leading to emergence and thereby reduce the risks associated with EIDs.

References

Ahmed, S.S.U., Ersboll, A.K., Biswas, P.K. and Christensen, J.P. (2010) 'The space-time clustering of highly pathogenic avian influenza (HPAI) H5N1 outbreaks in Bangladesh', *Epidemiology and Infection*, vol. 138, pp. 843–852.

Alexandersen, S., Zhang, Z., Donaldson, A.I. and Garland, A.J.M. (2003) 'The pathogenesis and diagnosis of foot-and-mouth disease', *Journal of Comparative Pathology*, vol. 129, pp. 1–36.

Allen, T., Murray, K.A., Zambrana-Torrelio, C., Morse, S.S., Rondinini, C., Di Marco, M., Breit, N., Olival, K.J. and Daszak, P. (2017) 'Global hotspots and correlates of emerging zoonotic diseases', *Nature Communications*, vol. 8, pp. 1124–1124.

Arias, A., Watson, S.J., Asogun, D., Tobin, E.A., Lu, J., Phan, M.V.T., Jah, U., Wadoum, R.E.G., Meredith, L., Thorne, L., Caddy, S., Tarawalie, A., Langat, P., Dudas, G., Faria, N.R., Dellicour, S., Kamara, A., Kargbo, B., Kamara, B.O., Gevao, S., Cooper, D., Newport, M., Horby, P., Dunning, J., Sahr, F., Brooks, T., Simpson, A.J.H., Groppelli, E., Liu, G., Mulakken, N., Rhodes, K., Akpablie, J., Yoti, Z., Lamunu, M., Vitto, E., Otim, P., Owilli, C., Boateng, I., Okoror, L., Omomoh, E., Oyakhilome, J., Omiunu, R., Yemisis, I., Adomeh, D., Ehikhiametalor, S., Akhilomen, P., Aire, C., Kurth, A., Cook, N., Baumann, J., Gabriel, M., Wölfel, R., Di Caro, A., Carroll, M.W., Günther, S., Redd, J., Naidoo, D., Pybus, O.G., Rambaut, A., Kellam, P., Goodfellow, I. and Cotten, M. (2016) 'Rapid outbreak sequencing of Ebola virus in Sierra Leone identifies transmission chains linked to sporadic cases', *Virus Evolution*, vol. 2, no. 1, p. vew016.

Arsevska, E., Valentin, S., Rabatel, J., de Goër de Hervé, J., Falala, S., Lancelot, R. and Roche, M. (2018) 'Web monitoring of emerging animal infectious diseases integrated in the French Animal Health Epidemic Intelligence System', *PLoS One*, vol. 13, p. e0199960.

Baker, S., Holt, K.E., Clements, A.C.A., Karkey, A., Arjyal, A., Boni, M.F., Dongol, S., Hammond, N., Koirala, S., Duy, P.T., Nga, T.V.T., Campbell, J.I., Dolecek, C., Basnyat, B., Dougan, G. and Farrar, J.J. (2011) 'Combined high-resolution genotyping and geospatial analysis reveals modes of endemic urban typhoid fever transmission', *Open Biology*, vol. 1, no. 2, p. 10008.

Barboza, P., Vaillant, L., Le Strat, Y., Hartley, D.M., Nelson, N.P., Mawudeku, A., Madoff, L.C., Linge, J.P., Collier, N., Brownstein, J.S. and Astagneau, P. (2014) 'Factors influencing performance of internet-based biosurveillance systems used in epidemic intelligence for early detection of infectious diseases outbreaks', *PLoS One*, vol. 9, p. e90536.

Benedict, M., Levine, R., Hawley, W. and Lounibos, L. (2007) 'Spread of the tiger: Global risk of invasion by the mosquito Aedes albopictus', *Vector-Borne and Zoonotic Diseases*, vol. 7, pp. 76–85.

Berger, K.A., Pigott, D.M., Tomlinson, F., Godding, D., Maurer-Stroh, S., Taye, B., Sirota, F.L., Han, A., Lee, R.T.C., Gunalan, V., Eisenhaber, F., Hay, S.I. and Russell, C.A. (2018) 'The geographic variation of surveillance and zoonotic spillover potential of influenza viruses in domestic poultry and swine', *Open Forum Infectious Diseases*, vol. 5, no. 12, p. ofy318.

Beynon, M., Curry, B. and Morgan, P. (2000) 'The Dempster-Shafer theory of evidence: An alternative approach to multicriteria decision modelling', *Omega*, vol. 28, pp. 37–50.

Brownstein, J.S., Freifeld, C.C., Chan, E.H., Keller, M., Sonricker, A.L., Mekaru, S.R. and Buckeridge, D.L. (2010) 'Information technology and global surveillance of cases of 2009 H1N1 influenza', *New England Journal of Medicine*, vol. 362, pp. 1731–1735.

Brownstein, J.S., Freifeld, C.C., Reis, B.Y. and Mandl, K.D. (2008) 'Surveillance sans frontieres: Internet-based emerging infectious disease intelligence and the HealthMap project', *PLoS Medicine*, vol. 5, p. e151.

Butler, D. (2006) 'Mashups mix data into global service', *Nature*, vol. 439, pp. 6–7.

Chikerema, S.M., Murwira, A., Matope, G. and Pfukenyi, D.M. (2013) 'Spatial modelling of Bacillus anthracis ecological niche in Zimbabwe', *Preventive Veterinary Medicine*, vol. 111, pp. 25–30.

Clements, A., Pfeiffer, D. and Martin, V. (2006) 'Application of knowledge-driven spatial modelling approaches and uncertainty management to a study of Rift Valley fever in Africa', *International Journal of Health Geographics*, vol. 5, p. 57.

Cohen, M.L. (2000) 'Changing patterns of infectious disease', *Nature*, vol. 406, pp. 762–767.

Collier, N., Doan, S., Kawazoe, A., Goodwin, R.M., Conway, M., Tateno, Y., Ngo, Q.-H., Dien, D., Kawtrakul, A., Takeuchi, K., Shigematsu, M. and Taniguchi, K. (2008) 'BioCaster: Detecting public health rumors with a web-based text mining system', *Bioinformatics*, vol. 24, pp. 2940–2941.

Cowen, P., Garland, T., Hugh-Jones, M.E., Shimshony, A., Handysides, S., Kaye, D., Madoff, L.C., Pollack, M.P. and Woodall, J. (2006) 'Evaluation of ProMED-mail as an electronic early warning system for emerging animal diseases: 1996 to 2004', *Journal of the American Veterinary Medical Association*, vol. 229, pp. 1090–1099.

Dempster, A.P. (1966) 'New methods for reasoning towards posterior distributions based on sample data', *The Annals of Mathematical Statistics*, vol. 37, pp. 355–374.

Dempster, A.P. (1967) 'Upper and lower probabilities induced by a multivalued mapping', *The Annals of Mathematical Statistics*, vol. 38, pp. 325–339.

Denzin, N., Borgwardt, J., Freuling, C. and Müller, T. (2013) 'Spatio-temporal analysis of the progression of Aujeszky's disease virus infection in wild boar of Saxony-Anhalt, Germany', *Geospatial Health*, vol. 8, pp. 2013–2213.

Dudas, G., Carvalho, L.M., Bedford, T., Tatem, A.J., Baele, G., Faria, N.R., Park, D.J., Ladner, J.T., Arias, A., Asogun, D., Bielejec, F., Caddy, S.L., Cotten, M., D'Ambrozio, J., Dellicour, S., Di Caro, A., Diclaro, J.W., Duraffour, S., Elmore, M.J., Fakoli, L.S., Faye, O., Gilbert, M.L., Gevao, S.M., Gire, S., Gladden-Young, A., Gnirke, A., Goba, A., Grant, D.S., Haagmans, B.L., Hiscox, J.A., Jah, U., Kugelman, J.R., Liu, D., Lu, J., Malboeuf, C.M., Mate, S., Matthews, D.A., Matranga, C.B., Meredith, L.W., Qu, J., Quick, J., Pas, S.D., Phan, M.V.T., Pollakis, G., Reusken, C.B., Sanchez-Lockhart, M., Schaffner, S.F., Schieffelin, J.S., Sealfon, R.S., Simon-Loriere, E., Smits, S.L., Stoecker, K., Thorne, L., Tobin, E.A., Vandi, M.A., Watson, S.J., West, K., Whitmer, S., Wiley, M.R., Winnicki, S.M., Wohl, S., Wölfel, R., Yozwiak, N.L., Andersen, K.G., Blyden, S.O., Bolay, F., Carroll, M.W., Dahn, B., Diallo, B., Formenty, P., Fraser, C., Gao, G.F., Garry, R.F., Goodfellow, I., Günther, S., Happi, C.T., Holmes, E.C., Kargbo, B., Keïta, S., Kellam, P., Koopmans, M.P.G., Kuhn, J.H., Loman, N.J., Magassouba, N.F., Naidoo, D., Nichol, S.T., Nyenswah, T., Palacios, G., Pybus, O.G., Sabeti, P.C., Sall, A., Ströher, U., Wurie, I., Suchard, M.A., Lemey, P. and Rambaut, A. (2017) 'Virus genomes reveal factors that spread and sustained the Ebola epidemic', *Nature*, vol. 544, pp. 309–315.

Elith, J., Leathwick, J. and Hastie, T. (2008) 'A working guide to boosted regression trees', *Journal of Animal Ecology*, vol. 77, pp. 802–813.

Estrada-Pena, A. and Venzal, J.M. (2007) 'Climate niches of tick species in the Mediterranean region: Modeling of occurrence data, distributional constraints and impact of climate change', *Journal of Medical Entomology*, vol. 44, pp. 1130–1138.

FAO (2011) 'EMPRES transboundary animal disease bulletin', pp. 7–8, www.fao.org/docrep/014/i2249e/i2249e2200.pdf, accessed 11 June 2011.

Farnsworth, M.L., Hamilton-West, C., Fitchett, S., Newman, S.H., de La Rocque, S., De Simone, L., Lubroth, J. and Pinto, J. (2010) 'Comparing national and global data collection systems for reporting, outbreaks of H5N1 HPAI', *Preventive Veterinary Medicine*, vol. 95, pp. 175–185.

Freifeld, C.C., Mandl, K.D., Reis, B.Y. and Brownstein, J.S. (2008) 'HealthMap: Global infectious disease monitoring through automated classification and visualization of internet media reports', *Journal of the American Medical Informatics Association*, vol. 15, pp. 150–157.

Funk, S. and Piot, P. (2014) 'Mapping Ebola in wild animals for better disease control', *eLife*, vol. 3, pp. e04565–e04565.

Gilbert, M., Chaitaweesub, P., Parakamawongsa, T., Premashthira, S., Tiensin, T., Kalpravidh, W., Wagner, H. and Slingenbergh, J. (2006) 'Free-grazing ducks and highly pathogenic avian influenza, Thailand', *Emerging Infectious Diseases*, vol. 12, pp. 227–234.

Gilbert, M., Xiao, X., Pfeiffer, D., Epprecht, M., Boles, S., Czarnecki, C., Chaitaweesub, P., Kalpravidh, W., Minh, P. and Otte, M. (2008) 'Mapping H5N1 highly pathogenic avian influenza risk in Southeast Asia', *Proceedings of the National Academy of Sciences, USA*, vol. 105, pp. 4769–4774.

Ginsberg, J., Mohebbi, M.H., Patel, R.S., Brammer, L., Smolinski, M.S. and Brilliant, L. (2009) 'Detecting influenza epidemics using search engine query data', *Nature*, vol. 457, pp. 1012–1014.

Grein, T.W., Kamara, K.B., Rodier, G., Plant, A.J., Bovier, P., Ryan, M.J., Ohyama, T. and Heymann, D.L. (2000) 'Rumors of disease in the global village: Outbreak verification', *Emerging Infectious Diseases*, vol. 6, pp. 97–102.

Hartley, D.M., Nelson, N.P., Arthur, R.R., Barboza, P., Collier, N., Lightfoot, N., Linge, J.P., van der Goot, E., Mawudeku, A., Madoff, L.C., Vaillant, L., Walters, R., Yangarber, R., Mantero, J., Corley, C.D. and Brownstein, J.S. (2013) 'An overview of internet biosurveillance', *Clinical Microbiology and Infection*, vol. 19, pp. 1006–1013.

Hennebelle, J.H., Sykes, J.E., Carpenter, T.E. and Foley, J. (2013) 'Spatial and temporal patterns of Leptospira infection in dogs from northern California: 67 cases (2001–2010)', *Journal of the American Veterinary Medical Association*, vol. 242, pp. 941–947.

Heymann, D.L. and Rodier, G.R. (2001) 'Hot spots in a wired world: WHO surveillance of emerging and re-emerging infectious diseases', *The Lancet Infectious Diseases*, vol. 1, pp. 345–353.

Hill, A.A., Dewé, T., Kosmider, R., Von Dobschuetz, S., Munoz, O., Hanna, A., Fusaro, A., De Nardi, M., Howard, W., Stevens, K., Kelly, L., Havelaar, A. and Stärk, K. (2015) 'Modelling the species jump: Towards assessing the risk of human infection from novel avian influenzas', *Royal Society Open Science*, vol. 2, no. 9, p. 150173.

Hogerwerf, L., Wallace, R., Ottaviani, D., Slingenbergh, J., Prosser, D., Bergmann, L. and Gilbert, M. (2010) 'Persistence of highly pathogenic avian influenza H5N1 virus defined by agro-ecological niche', *EcoHealth*, vol. 7, no. 2, pp. 213–225.

Holmes, E., Rambaut, A. and Andersen, K. (2018) 'Pandemics: Spend on surveillance, not prediction', *Nature*, vol. 558, pp. 180–182.

Jebara, K.B., Cáceres, P., Berlingieri, F. and Weber-Vintzel, L. (2012) 'Ten years' work on the World Organisation for Animal Health (OIE) worldwide animal disease notification system', *Preventive Veterinary Medicine*, vol. 107, pp. 149–159.

Jones, K., Patel, N., Levy, M., Storeygard, A., Balk, D., Gittleman, J. and Daszak, P. (2008) 'Global trends in emerging infectious diseases', *Nature*, vol. 451, pp. 990–993.

Joyner, T.A., Lukhnova, L., Pazilov, Y., Temiralyeva, G., Hugh-Jones, M.E., Aikimbayev, A. and Blackburn, J.K. (2010) 'Modeling the potential distribution of Bacillus anthracis under multiple climate change scenarios for Kazakhstan', *PLoS One*, vol. 5, p. e9596.

Karimuribo, E., Sayalel, K., Beda, E., Short, N., Wambura, P., Mboera, L., Kusiluka, L. and Rweyemamu, M. (2012) 'Towards one health disease surveillance: The Southern African Centre for infectious disease surveillance approach', *Onderstepoort Journal of Veterinary Research*, vol. 79, no. 2, pp. 31–37.

Keller, M., Blench, M., Tolentino, H., Freifeld, C., Mandl, K., Mawudeku, A., Eysenbach, G. and Brownstein, J. (2009) 'Use of unstructured event-based reports for global infectious disease surveillance', *Emerging Infectious Diseases*, vol. 15, pp. 689–695.

Lederberg, J., Shope, R.E. and Oaks, S.C. Jr. (1992) *Emerging Infections: Microbial Threats to Health in the United States.* National Academy Press, Washington, DC. Institute of Medicine (US) Committee on Emerging Microbial Threats to Health.

Lyon, A., Nunn, M., Grossel, G. and Burgman, M. (2012) 'Comparison of web-based biosecurity intelligence systems: BioCaster, EpiSPIDER and HealthMap', *Transboundary and Emerging Diseases*, vol. 59, pp. 223–232.

Martin, V., Pfeiffer, D., Zhou, X., Xiao, X., Prosser, D., Guo, F. and Gilbert, M. (2011) 'Spatial distribution and risk factors of highly pathogenic avian influenza (HPAI) H5N1 in China', *PLoS Pathogens*, vol. 7, p. e1001308.

Métras, R., Porphyre, T., Pfeiffer, D.U., Kemp, A., Thompson, P.N., Collins, L.M. and White, R.G. (2012) 'Exploratory space-time analyses of Rift Valley fever in South Africa in 2008–2011', *PLoS Neglected Tropical Diseases*, vol. 6, p. e1808.

Milinovich, G.J., Magalhães, R.J.S. and Hu, W. (2015) 'Role of big data in the early detection of Ebola and other emerging infectious diseases', *The Lancet Global Health*, vol. 3, pp. e20–e21.

Milinovich, G.J., Williams, G.M., Clements, A.C.A. and Hu, W. (2014) 'Internet-based surveillance systems for monitoring emerging infectious diseases', *The Lancet Infectious Diseases*, vol. 14, pp. 160–168.

Minh, P.Q., Morris, R.S., Schauer, B., Stevenson, M., Benschop, J., Nam, H.V. and Jackson, R. (2009) 'Spatio-temporal epidemiology of highly pathogenic avian influenza outbreaks in the two deltas of Vietnam during 2003–2007', *Preventive Veterinary Medicine*, vol. 89, pp. 16–24.

Morens, D.M. and Fauci, A.S. (2012) 'Emerging infectious diseases in 2012: 20 years after the institute of medicine report', *mBio*, vol. 3, pp. e00494–e00412.

Morens, D.M., Folkers, G.K. and Fauci, A.S. (2004) 'The challenge of emerging and re-emerging infectious diseases', *Nature*, vol. 430, pp. 242–249.

Mullins, J., Lukhnova, L., Aikimbayev, A., Pazilov, Y., Van Ert, M. and Blackburn, J. (2011) 'Ecological niche modelling of the Bacillus anthracis A1.a sub-lineage in Kazakhstan', *BMC Ecology*, vol. 11, p. 32.

Mykhalovskiy, E. and Weir, L. (2006) 'The global public health intelligence network and early warning outbreak detection: A Canadian contribution to global public health', *Canadian Journal of Public Health*, vol. 97, pp. 42–44.

Nava, A., Shimabukuro, J.S., Chmura, A.A. and Luz, S.L.B. (2017) 'The impact of global environmental changes on infectious disease emergence with a focus on risks for Brazil', *ILAR Journal*, vol. 58, pp. 393–400.

Nyakarahuka, L., Ayebare, S., Mosomtai, G., Kankya, C., Lutwama, J., Mwiine, F.N. and Skjerve, E. (2017) 'Ecological niche modeling for Filoviruses: A risk map for Ebola and Marburg virus disease outbreaks in Uganda', *PLoS Currents*, vol. 9, ecurrents.outbreaks.07992a87522e07991f07229c07997cb023270a0 23272af023271

OIE (2014) 'African swine fever, Lithuania. Immediate notification', www.oie.int/wahis_2/public/wahid. php/Reviewreport/Review/viewsummary?reportid=14690, accessed 10 May 2015.

Owada, K., Eckmanns, T., Kamara, K.-B.O.B. and Olu, O.O. (2016) 'Epidemiological data management during an outbreak of Ebola virus disease: Key issues and observations from Sierra Leone', *Frontiers in Public Health*, vol. 4, p. 163.

Pfeiffer, D., Minh, P., Martin, V., Epprecht, M. and Otte, M. (2007) 'An analysis of the spatial and temporal patterns of highly pathogenic avian influenza occurrence in Vietnam using national surveillance data', *The Veterinary Journal*, vol. 174, pp. 302–309.

Phillips, S. (2012) 'A brief tutorial on Maxent', *Lessons in Conservation*, vol. 3, pp. 107–135.

Phillips, S., Anderson, R. and Schapire, R. (2006) 'Maximum entropy modeling of species geographic distributions', *Ecological Modelling*, vol. 190, pp. 231–259.

Phillips, S.J. and Dudík, M. (2008) 'Modeling of species distributions with Maxent: New extensions and a comprehensive evaluation', *Ecography*, vol. 31, pp. 161–175.

Phillips, S.J., Dudik, M., Elith, J., Graham, C.H., Lehmann, A., Leathwick, J. and Ferrier, S. (2009) 'Sample selection bias and presence-only distribution models: Implications for background and pseudo-absence data', *Ecological Applications*, vol. 19, pp. 181–197.

Phillips, S.J., Dudik, M. and Schapire, R.E. (2004) 'A maximum entropy approach to species distribution modeling', *Proceedings of the 21st International Conference on Machine Learning*, New York, pp. 665–662.

Picado, A., Guitian, F. and Pfeiffer, D. (2007) 'Space-time interaction as an indicator of local spread during the 2001 FMD outbreak in the UK', *Preventative Veterinary Medicine*, vol. 79, pp. 3–19.

Picado, A., Speybroeck, N., Kivaria, F., Mosha, R.M., Sumaye, R.D., Casal, J. and Berkvens, D. (2011) 'Foot-and-mouth disease in Tanzania from 2001 to 2006', *Transboundary and Emerging Diseases*, vol. 58, pp. 44–52.

Pigott, D.M., Golding, N., Mylne, A., Huang, Z., Henry, A.J., Weiss, D.J., Brady, O.J., Kraemer, M.U., Smith, D.L., Moyes, C.L., Bhatt, S., Gething, P.W., Horby, P.W., Bogoch, I.I., Brownstein, J.S., Mekaru, S.R., Tatem, A.J., Khan, K. and Hay, S.I. (2014) 'Mapping the zoonotic niche of Ebola virus disease in Africa', *eLife*, vol. 3, p. e04395.

Pigott, D.M., Golding, N., Mylne, A., Huang, Z., Weiss, D.J., Brady, O.J., Kraemer, M.U.G. and Hay, S.I. (2015) 'Mapping the zoonotic niche of Marburg virus disease in Africa', *Transactions of the Royal Society of Tropical Medicine and Hygiene*, vol. 109, no. 6, pp. 366–378.

Pigott, D.M., Millear, A.I., Earl, L., Morozoff, C., Han, B.A., Shearer, F.M., Weiss, D.J., Brady, O.J., Kraemer, M.U., Moyes, C.L., Bhatt, S., Gething, P.W., Golding, N. and Hay, S.I. (2016) 'Updates to the zoonotic niche map of Ebola virus disease in Africa', *eLife*, vol. 5, p. e16412.

ProMED-mail (2014) 'Undiagnosed deaths, swine–Lithuania: Wild boar', www.promedmail.org/post/2175896, accessed 25 May 2019.

Randolph, S.E. and Rogers, D.J. (2010) 'The arrival, establishment and spread of exotic diseases: Patterns and predictions', *Nature Reviews Microbiology*, vol. 8, pp. 361–371.

Reingold, A.L. and Lewnard, J.A. (2019) 'Emerging challenges and opportunities in infectious disease epidemiology', *American Journal of Epidemiology*, vol. 188, no. 5, pp. 873–882.

Service, M. (1997) 'Mosquito (Diptera: Culicidae) dispersal: The long and the short of it', *Journal of Medical Entomology*, vol. 34, pp. 579–588.

Sharma, R., Karad, A.B., Dash, B., Dhariwal, A.C., Chauhan, L.S. and Lal, S. (2012) 'Media scanning and verification system as a supplemental tool to disease outbreak detection & reporting at National Centre for Disease Control, Delhi', *Journal of Communicable Diseases*, vol. 44, pp. 9–14.

Stevens, K.B., Gilbert, M. and Pfeiffer, D.U. (2013) 'Modeling habitat suitability for occurrence of highly pathogenic avian influenza virus H5N1 in domestic poultry in Asia: A spatial multicriteria decision analysis approach', *Spatio and Spatio-Temporal Epidemiology*, vol. 4, pp. 1–14.

Stevens, K.B. and Pfeiffer, D.U. (2014) 'Enhanced decision-making of HPAI H5N1 in domestic poultry in Asia: A comparison of spatial-modelling methods', in *Society for Veterinary Epidemiology and Preventive Medicine (SVEPM)*. Ghent, Belgium.

Stevens, K.B. and Pfeiffer, D.U. (2015) 'Sources of spatial animal and human health data: Casting the net wide to deal more effectively with increasingly complex disease problems', *Spatial and Spatio-Temporal Epidemiology*, vol. 13, pp. 15–29.

Swirski, A.L., Pearl, D.L., Williams, M.L., Homan, H.J., Linz, G.M., Cernicchiaro, N. and LeJeune, J.T. (2013) 'Spatial epidemiology of Escherichia coli O157:H7 in dairy cattle in relation to night roosts of Sturnus vulgaris (European starling) in Ohio, USA (2007–2009)', *Zoonoses and Public Health*, vol. 61, no. 6, pp. 427–435.

Tatem, A.J., Baylis, M., Mellor, P.S., Purse, B.V., Capela, R., Pena, I. and Rogers, D.J. (2003) 'Prediction of bluetongue vector distribution in Europe and north Africa using satellite imagery', *Veterinary Microbiology*, vol. 97, pp. 13–29.

Taylor, L.H., Latham, S.M. and Woolhouse, M.E. (2001) 'Risk factors for human disease emergence', *Philosophical Transactions of the Royal Society of London: Series B, Biological Sciences*, vol. 356, pp. 983–989.

Tolentino, H., Kamadjeu, R., Fontelo, P., Liu, F., Matters, M., Pollack, M. and Madoff, L. (2007) 'Scanning the emerging infectious diseases horizon: Visualizing ProMED emails using EpiSPIDER', *Advances in Disease Surveillance*, vol. 2, p. 169.

Vander Kelen, P., Downs, J., Stark, L., Loraamm, R., Anderson, J. and Unnasch, T. (2012) 'Spatial epidemiology of eastern equine encephalitis in Florida', *International Journal of Health Geographics*, vol. 11, p. 47.

Wendelboe, A.M., Grafe, C. and Carabin, H. (2010) 'The benefits of transmission dynamics models in understanding emerging infectious diseases', *The American Journal of the Medical Sciences*, vol. 340, pp. 181–186.

Wilcox, B. and Gubler, D. (2005) 'Disease ecology and the global emergence of zoonotic pathogens', *Environmental Health and Preventive Medicine*, vol. 10, pp. 263–272.

Wilesmith, J.W., Stevenson, M.A., King, C.B. and Morris, R.S. (2003) 'Spatio-temporal epidemiology of foot-and-mouth disease in two counties of Great Britain in 2001', *Preventive Veterinary Medicine*, vol. 61, pp. 157–170.

Williams, R., Fasina, F. and Peterson, A. (2008) 'Predictable ecology and geography of avian influenza (H5N1) transmission in Nigeria and West Africa', *Transactions of the Royal Society of Tropical Medicine and Hygiene*, vol. 102, no. 5, pp. 471–479.

Williams, R. and Peterson, A. (2009) 'Ecology and geography of avian influenza (HPAI H5N1) transmission in the Middle East and northeastern Africa', *International Journal of Health Geographics*, vol. 8, p. 47.

Wilson, K. and Brownstein, J.S. (2009) 'Early detection of disease outbreaks using the Internet', *Canadian Medical Association Journal*, vol. 180, pp. 829–831.

Wilson, K., von Tigerstrom, B. and McDougall, C. (2008) 'Protecting global health security through the international health regulations: Requirements and challenges', *Canadian Medical Association Journal*, vol. 179, pp. 44–48.

Wisz, M.S., Hijmans, R.J., Li, J., Peterson, A.T., Graham, C.H., Guisan, A. and Group, N.P.S.D.W. (2008) 'Effects of sample size on the performance of species distribution models', *Diversity and Distributions*, vol. 14, no. 5, pp. 763–773.

Woolhouse, M.E.J. and Gowtage-Sequeria, S. (2005) 'Host range and emerging and reemerging pathogens', *Emerging Infectious Diseases*, vol. 11, pp. 1842–1847.

Zeldenrust, M., Rahamat-Langendoen, J., Postma, M. and van Vliet, J. (2008) 'The value of ProMED-mail for the early warning committee in the Netherlands: More specific approach recommended', *Eurosurveillance*, vol. 13, p. 8033.

PART 2

Terrains

7

FOREST ECOSYSTEMS

Tommaso Sitzia, Thomas Campagnaro, Giuseppe Brundu,
Massimo Faccoli, Alberto Santini and Bruce L. Webber

Introduction

There are few better study ecosystems than forests to reveal global insights into the patterns and processes associated with biological invasions. Forests range from single species stands to some of the most biodiverse terrestrial ecosystems and are managed for a wide variety of environmental, social and economic values in myriad ways with equally contrasting intensity of effort. There are no forests left on the planet that are untouched by global environmental change. The impacts of non-native species invasions and anthropogenic climate change act individually and synergistically to drive considerable change in the direct and indirect interactions between species in these ecosystems. If we are to improve our management of non-native species, as well as conserve forest habitats worldwide, there is merit in bringing together contrasting examples to improve our understanding. Here we provide an overview of non-native species invasions in forest ecosystems using broad to fine scale examples to illustrate key patterns and highlight priority areas for future research. Our aim is to represent complex human interactions with the invasion phenomenon through the use of emblematic species groups (plants, pathogens and insects) and forest ecosystems (temperate and tropical). Specific attention is also given to the active management of the tree component because such management could be either the major factor of disturbance favouring non-native species spread or, conversely, the only way to prevent further invasions, which could happen without proactive mitigation.

Invasions in forest ecosystems: context and process

Forests still cover around 4,000 million hectares (FAO, 2020), and forested biomes represent 41.5% of the earth's terrestrial area (Dinerstein et al., 2017). However, the ongoing clearing of forests across the planet and conversion into different land-use types represents one of the greatest threats to native biodiversity. Together with climate change and the introduction of non-native invasive species, these drivers of change represent the three biggest challenges for future conservation management (Sala et al., 2000; Ferrier et al., 2016). Forest ecosystems have been invaded by a wide range of organisms, including plants, insects and mammals fulfilling novel roles in herbivory, decomposition and predation interactions (Rejmánek, 2015; Liebhold et al., 2017). Such interactions often involve synergies between invaders and with climate change and land clearing and can act both directly and indirectly to threaten biodiversity and its related services (Didham et al., 2007).

Forests are usually viewed, compared to other ecosystems, as resistant to invasion of non-native plant species because of increased competition for finite resources, such as light availability under continuous canopy cover (Rejmánek, 2015). Nevertheless, there is strong evidence that even forests with dense, intact canopy cover can be invaded (Martin et al., 2009). The type of forest ecosystem is also one factor usually underlined as important when looking to forest invasibility. The threat of invasion in tropical forests is predicted to increase (Fine, 2002), as confirmed by recent increasing evidence of non-native occurrences (Bellingham et al., 2018; Florens et al., 2016). Moreover, temperate riparian forests seem to be more susceptible to plant invasions than other woodland habitats (Wagner et al., 2017; Campagnaro et al., 2018a).

Indeed, anthropogenic disturbances are one major cause for invasions in forests. Human actions, primarily forestry-related activities, have triggered and accelerated invasion by non-native species (Martin et al., 2009; Wagner et al., 2017). In addition, proximity to stocks of non-native species increases the risk of forest invasions – for example, with insect species (Piel et al., 2008; Skarpaas and Økland, 2009).

Multiple factors can play a role in making non-native species effective invaders of forest ecosystems. Trees are the fundamental element of forest ecosystems, and therefore non-native species of this life form can be significant drivers of change in composition and conservation trajectories in natural forests. A recent global assessment reported a total of 434 non-native invasive tree species, with Asia, South America, Europe and Australia as major sources (Rejmánek and Richardson, 2013). Non-native invasive tree species can have detrimental effects on ecosystem properties and functions in natural and semi-natural woodlands (Richardson, 1998; Richardson and Rejmánek, 2011). Furthermore, a number of these tree species are expanding and forming new forests in landscapes with different anthropic pressures and, in particular, by reforesting abandoned lands (Hobbs et al., 2006). At the same time, invasion of non-native species may also prevent forest succession processes. For example, Japanese stilt grass (*Microstegium vimineum* (Trin.) A. Camus) can prevent the establishment of native tree seedlings, as observed in a three-year field experiment in the US, with a possible reduction in the rate of reforestation on grasslands (Flory and Clay, 2010). In other cases, the negative effect of invasive weeds, like Himalayan balsam (*Impatiens glandulifera* Royle; Figure 7.1), on survival and growth of tree seedlings is controversial because it is mediated by management or confounded with the effect of native competitive species, like bramble (*Rubus* spp.; Ammer et al., 2011).

Species life traits can be used to explain the success of non-native species establishment and spread in forest ecosystems. Different ecological groups may share or have different suites of factors related to their potential to invade after introduction (Novoa et al., 2020). The ability to germinate and establish rapidly, or to maintain fast growth into maturity, is one trait that could be generally important for different taxa. A number of studies have investigated these characteristics for specific groups. For example, conifer species with small seed mass (< 50 mg), a short juvenile period (< 10 years) and short intervals between masting seem to have a higher invasive potential (Richardson and Rejmánek, 2004).

Without overlooking economic and social impacts, invasion of non-native species may have a wide array of ecological impacts on forests. Impacts with negative outcomes on biodiversity and ecosystem functioning are those that have received the most attention in forest ecosystems. Different ecological groups have been shown to impact the biodiversity of different forest ecosystems – for example, by favouring other non-native species, a process called invasional meltdown (Simberloff and Von Holle, 1999; Essl et al., 2011), or with negative cascading effects on native species (Stinson et al., 2006). The invasion of forests by non-native species can represent

Figure 7.1   Himalayan balsam is native to the Himalayas and naturalised and widespread across riverbanks in Europe, North America and New Zealand. Railway management measures, like topsoil disturbance and application of herbicides, are important human drivers of further expansion. Its negative effects on the survival and growth of tree seedlings are controversial and require further research.

Source: Photo by T. Sitzia, Valsugana, Northern Italy.

different ecological processes according to the 'passengers' and 'drivers' model (MacDougall and Turkington, 2005), which can help identify the mechanisms behind community composition and function in invasions (O'Leary et al., 2018). Non-native species can be considered passengers when modified disturbance regimes or other changes in ecosystem properties are the cause of their invasion (Chabrerie et al., 2008) while drivers are those species identified as causes of environmental change (Jäger et al., 2007). Furthermore, a number of non-native species have been shown to cause pronounced effects on the functioning of forests (Ehrenfeld et al., 2001) and as such have been defined as 'transformers' (*sensu* Richardson et al., 2000) as they alter ecosystem dynamics beyond local thresholds (O'Leary et al., 2018).

Case studies: the ecology of forest invasions

Depending on which ecological group is considered, as well as the context of global environmental change impacts, forests may be considered more or less susceptible to harmful invasions. Rather than seeking to draw out patterns in forest invasions, we use case studies to highlight the diversity of factors that mediate and facilitate forest invasions. We begin with an example of a tree species that behaves both as a driver and a passenger of ecosystem change, the black locust (*Robinia pseudoacacia* L.), before exploring invasions by non-native decomposers and pathogens.

Tommaso Sitzia et al.

Controversial and widely spread invasive trees: the example of **Robinia pseudoacacia**

Black locust is a fast-growing nitrogen-fixing tree species native to the Appalachian and Ozark Mountains in North America. It is one of the most widespread non-native tree species in the world and now has naturalised in all continents (Richardson and Rejmánek, 2011; Rejmánek and Richardson, 2013). It was introduced into different countries initially as an ornamental species and for forestry and reclamation purposes (Cierjacks et al., 2013; Rejmánek and Richardson, 2013). Currently, it is highly valued by land managers for timber, firewood and honey production, even in its non-native range (Nicolescu et al., 2020). For example, significant positive economic outcomes at the local and national levels derived from these products were recently reported in Romania and Hungary (Nicolescu et al., 2018).

This species has one of the broadest habitat ranges among non-native invasive trees and can tolerate a wide range of physical-chemical soil conditions (Vítková et al., 2015). Climatic tolerance is also quite broad, although frost is an important limitation factor in its non-native range (Vítková et al., 2017). It has been observed invading a variety of natural and semi-natural ecosystems where it can cause severe impacts on biodiversity (Cierjacks et al., 2013; Vítková et al., 2017; Campagnaro et al., 2018a). Moreover, black locust is expected to expand its distribution with future climate change projections, potentially increasing its pressure on habitats of conservation importance (Kleinbauer et al., 2010; Camenen et al., 2016; Puchałka et al., 2021). Perversely from a biodiversity threat perspective, this increasing potential range is likely to expand forestry-related industries using black locust.

Studies have shown that black locust can be considered a passenger of change. This species can invade open grassland habitats in both rural and urban contexts (Sitzia et al., 2012; Sitzia et al., 2016b), leading to the formation of forest in previously cleared and abandoned land (Campagnaro et al., 2018b). Other changes in forest management regimes, including coppicing, favour the spread of this species, reinforcing its role as a passenger in invading existing native forests (Radtke et al., 2013).

The dominance of black locust stands is related to specific soil characteristics leading to differences in understory plant flowering periods, reproduction types and life forms, when compared with native forests (Figure 7.2) derived from the same reforestation processes (Sitzia et al., 2018). Authors associated the presence of this non-native species to higher soil available P and C/N ratio; however, soil legacies may have been an important factor mediating the observed impacts. For these reasons, and as highlighted for floodplain forests (Terwei et al., 2016), it was not clear whether the species is a driver or passenger of change. Nevertheless, a general driver of change in this context can be traced back to the abandonment of agriculture activities and the development of stands on non-forested areas. But, according to the highlighted impacts, the presence of black locust can also be considered a driver of a certain degree of change compared to the conditions observed under the canopy of native species.

Therefore, this species might be defined as both a passenger and a driver of change not only based on changing contexts but also in the same invaded region. This species may be more properly called a 'back-seat driver' that benefits from changes to ecosystem processes or properties and causes additional changes to ecosystem properties and influences native species communities, in addition to those driven by initial changes of the ecosystem (Bauer, 2012). Furthermore, this species was described as a transformer (*sensu* Richardson et al., 2000) because it provides limiting resources – e.g. nitrogen (Rejmánek et al., 2005). Indeed, in addition to the variety of changes occurring to soil characteristics (Medina-Villar et al., 2016; Lazzaro et al., 2018), black locust invasion can trigger cascading effects on different trophic levels, causing strong biodiversity changes (Hejda et al., 2017).

Finally, while categorisation of non-native invasive tree species using such models can be useful for management and prioritisation (Sádlo et al., 2017), case-to-case considerations for their

108

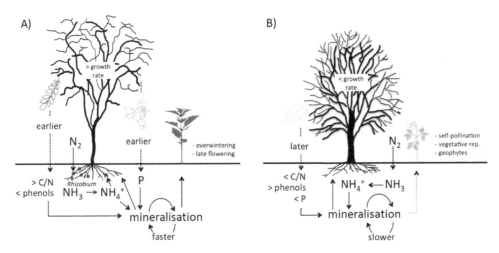

Figure 7.2 Significant differences are possible between soil properties and plant trait composition under non-native black locust (a) and native tree (b) canopy, making the former both a passenger and a driver of ecosystem change.

Source: From Sitzia et al., 2018.

identification should be made by taking into account recipient forest ecosystems and possible trends over time (Didham et al., 2005; Chabrerie et al., 2008). Nevertheless, black locust has been shown to take advantage of ecosystem changes due to modifications of management regimes for its invasion and subsequently can cause important ecological changes to native ecosystems.

Forest pathogens: invasiveness, invasibility and risk mitigation in Europe

Invasion by non–native forest pathogens is one of the main causes of decline or extirpations in tree populations, and even when there is not a risk of species extinction, the strong and sudden reduction in tree numbers produces significant changes in the impacted ecosystems (Lovett et al., 2016).

One of the main questions in invasion science is how organisms originating in a different habitat could spread in a new environment. This issue implies not just the invasiveness of the species but also the invasibility of the habitat. Why some organisms have a better chance to become invasive and why some regions of the world are more prone to invasions are two questions worth answering if an effective risk mitigation strategy is to be established. Social and economic factors are crucial for species introduction (Sakai et al., 2001; Guo et al., 2012), whereas biogeographic and ecological factors are important for establishment, with phenotypic plasticity and rapid evolution being key mediators of invasiveness (Sax, 2001; Davidson et al., 2011).

The global trade of live plants is widely acknowledged as the main entrance pathway of tree pathogens (Santini et al., 2013). The trade of ornamental plants is one of the most important industries in Europe and North America, and, thanks to advances in transport technologies, it is possible to produce plants where climates are more favourable and labour costs are much lower. For example, black locust clones developed in Hungary (Keresztesi, 1983), where the species is non-native (see previous case study), are currently exported to the US. The result is the development of an intricate network of a wide commerce. One peculiarity of this industry is the great flexibility in terms of number of species and varieties (Bradley et al., 2012). The large volumes of plants traded and the rapid turnover of commercial varieties and origins hinders efforts to control the establishment of new forest pathogens, despite the intensification of phytosanitary measures worldwide (Aukema et al., 2010; Eschen et al., 2015).

Invasiveness (i.e. the ability to naturalise and expand rapidly in range) is generally associated with biological traits of the invasive organism: in the case of plant pathogens, these include the virulence of the strain, host specificity and mode of action, as well as the host's abundance, demography, phytosociology and variation in susceptibility (Lovett et al., 2006; Schulze-Lefert and Panstruga, 2011). Residence time (time since introduction or first record), lifestyle (generalists vs. specialists), phylogenic order, mode of reproduction and dispersal, spore shape and size, optimal temperature for growth, parasitic specialisation and adaptive rapid evolution of phenotypic traits seem to play a major role in invasiveness of forest pathogens (Desprez-Loustau et al., 2010; Philibert et al., 2011; Santini et al., 2013; Garbelotto et al., 2015). The ability of an introduced non-native pathogen to exploit well-developed interactions between native non-aggressive organisms and insects associated with trees, determines the success of the invader and the establishment of an epidemic (Santini and Battisti, 2019). In addition, pathogens able to attack both ornamental (i.e. in parks and gardens within a urban setting) and forest trees generally show higher spread rates (Santini et al., 2013).

Physical factors such as climate or topography have been suggested as useful predictors of a country's invasibility (Dukes et al., 2009; Huang et al., 2011). The environmental and biological diversity of a country is also strictly and directly related to the establishment of new invasive pathogens (Santini et al., 2013). At the same time, human activity expressed as volume of imports or human population density (Desprez-Loustau et al., 2009; 2010) is an important factor for determining the invasibility of a country. Even a country's history matters: in Europe, for example, countries previously included in the Soviet bloc experienced a limited number of pathogen invasions, probably owing to their commercial isolation. On the other hand, Great Britain, which used to pursuit species exchange from around the world, taking advantage of ruling and adminis-trating a huge territory, has experienced several impressive invasions (Santini et al., 2013).

It is evident that globalisation, technological developments and societal changes achieved in recent decades have facilitated the introduction and establishment of new forest pathogens that have adversely affected society and shaped social development. However, since the very beginning, human civilisation played a role in spreading pathogens that become invasive, with catastrophic ecological and even societal effects (Santini et al., 2018). Ultimately, the solution to these issues does not lie in trying to stop globalisation, which is neither desirable nor pos-sible, if not unhistorical, but should instead rely on scientific knowledge informing improved biosecurity and control programmes. For example, in the European Union, the risks inherent in the import of live plants in pots should be addressed by phytosanitary regulations, as the soil contained in pots is a primary introduction pathway for micro- and macro-organisms (Ghelar-dini et al., 2016). In addition, as phytosanitary inspections mostly rely on visual assessment of the aerial parts, the use of innovative molecular techniques would improve interception efficiency, especially in the case of latent fungi in asymptomatic plants (Migliorini et al., 2015).

Non-native bark and ambrosia beetles: a key taxon of invasiveness, invasibility and ecological risk in forest ecosystems

Bark and ambrosia beetles (Coleoptera: Curculionidae, Scolytinae) are small beetles with larvae developing in the phloem (bark beetles) or within the sapwood (ambrosia beetles) of living trees growing in forest ecosystems. Infesting and killing thousands of trees, these beetles are known as one of the most important threats to both managed (including plantations) and natural forests worldwide (Brockerhoff et al., 2006), with damage of the same threat magnitude as climate change and clear felling of tropical forests (Grégoire et al., 2015). Total bark beetle damage in

Europe from 1958 to 2001 was estimated at about 124 million square metres (Seidl et al., 2011). At the same time, they are considered one of the most successful groups of invasive species as they can be transported in almost all kinds of wood products (Brockerhoff et al., 2006). Given the increase of international trade, the rate of introduction of non-native scolytins – mainly native from Asia – has sharply increased in recent decades, and new species are intercepted or established almost every year in Europe (Kirkendall and Faccoli, 2010; Rassati et al., 2015). Southern Europe and the Mediterranean have been identified as especially suitable for the arrival and establishment of non-native bark and ambrosia beetles (Kirkendall, 1993; Kirkendall and Faccoli, 2010). The habitat diversity – offering a large number of potential host trees – and the dry and warm climate with mild winters occurring in the Mediterranean regions have been identified as the main features favouring the establishment of non-native species native to subtropical regions (Marini et al., 2011). The Mediterranean region is, in fact, disproportionately rich in non-native bark and ambrosia beetles compared to temperate areas of Europe (Kirkendall and Faccoli, 2010).

As the introduction of non-native scolytins has always been unintentional, it is difficult to determine precisely when a certain species arrived (Kirkendall and Faccoli, 2010). In fact, non-native species can be detected years or even decades after their arrival (Kenis et al., 2007; Mattson et al., 2007; Roques et al., 2009). Ambrosia beetles develop in the sapwood but with larvae feeding on symbiotic fungi carried on and farmed by the adults inside the host trees. These species are thus usually polyphagous (Beaver, 1979; Kirkendall, 1983), and the lack of host specificity is considered to be a major reason for their success as invaders (Kirkendall and Faccoli, 2010). Bark beetles, however, develop in a more restricted number of hosts (i.e., monophagous or oligophagous), which makes their invasiveness related to the presence of specific potential hosts.

Although it is difficult to determine when bark and ambrosia beetle invasions began, the link between this phenomenon and globalisation is well recognised worldwide (Perrings et al., 2005; Meyerson and Mooney, 2007; Hulme, 2009). Wood-packaging materials, such as crating, dunnage and pallets, and importation of ornamental plants are recognised as the main pathway of introduction of non-native wood-boring beetles (Haack, 2001; 2006; Zahid et al., 2008; Colunga-Garcia et al., 2009; Haack and Rabaglia, 2013; Rassati et al., 2014a, 2014b). Given that such materials are always associated with international trade and that during transport insects can find protection within the wood and escape detection at points of entry (Haack, 2001; McCullough et al., 2006), they represent an optimal vector for non-native species introduction. This trend will probably continue or even increase in the future, with the rising speed and frequency of movement of goods around the globe alongside climate change (Levine and D'Antonio, 2003).

Many non-native species may significantly impact on the invaded environment both economically and ecologically (Kenis et al., 2009). From an ecological perspective, species that are able to kill living trees can threaten indigenous species by decreasing or increasing the distribution and abundance of certain plants, altering habitat and food supply (Gandhi and Herms, 2010). From an economic perspective, the costs required for the eradication of non-native species may be extremely high in the few cases where eradication is even possible. Moreover, non-native species can cause serious damage to native forests, induce severe alterations of the landscape in many natural areas and produce economic losses in tree plantations and managed forests, killing thousands of trees and altering the timber quality.

Case studies: the management of the tree component

All major forest ecosystems of the world are significantly impacted, both directly and indirectly, by global environmental change. Therefore, dealing with non-native invasive species is critical across a wide range of forest communities and management systems. Invasibility applies to both tropical and temperate forests, but there are important differences between these regions that can be harnessed to mitigate the negative impacts of non-native species.

Managing plant invasions in tropical forests: context, challenges and priorities

Invasive non-native plants represent a multi-billion-dollar problem for the world annually. Their control should be a major consideration in efforts to enhance global food security, maintain biodiversity and reduce environmental degradation. Managing these factors links directly to alleviating poverty, and the brunt of their impact is borne across diverse environmental, social and economic sectors, in agreement with the sustainable development goals set by United Nations General Assembly (United Nations, 2015). For example, threats from invasive plants are both driven by and impact on the forestry and agricultural sectors, as well as having knock-on effects on native ecosystem values that can underpin significant tourism-related income. Our knowledge on the impacts of non-native invasive plants on forests is dominated by studies focusing on temperate environments (Hulme et al., 2013). In contrast, documented information on the introduction, distribution and impacts of non-native invasive plants in tropical forests is particularly limited (Kueffer et al., 2013). It is highly likely that this paucity of information from tropical forests is merely due to a lack of information on invasions that have already occurred, rather than genuinely low levels of introduced plants (Fine, 2002).

Tropical forests have a number of high-risk invasion pathways that make them equally if not more likely to be exposed to plant invasion threats. First, there has been a long history of plant movement between tropical regions, driven in part by the exchange of plants among colonial botanical gardens in the late 18th and early 19th centuries (Crosby, 2004; Alpern, 2008). Considerable efforts were made to give due consideration to plants with potential economic value that would thrive in commercial production in new areas. The very traits that confer commercial appeal are those traits that increase the risk of invasion threats, and unsurprisingly many of these early introductions into botanical gardens have subsequently become threatening invaders (Dawson et al., 2008; Hulme, 2011).

Second, logging and the conversion of tropical forests into plantation forestry and agriculture have been closely aligned with the increase in globalisation and movement of plants to new areas, increasing the risk of introductions into those forests that remain (Putz et al., 2012). Logging activities can severely impact forest canopy, understory plant communities and substrate, and, as a consequence, they may facilitate the spread of non-native plant species (Vilà and Ibáñez, 2011). Forestry activities can also target non-native species that 'escape' and become invasion threats (Richardson, 1998) for harvest, as well as introducing an associated suite of invasive plants via contaminated equipment, novel dispersal methods for invasive plant seeds and increased landscape fragmentation and disturbance providing suitable sites for establishment.

Of concern for tropical forests in relation to the management and mitigation of non-native invasive plant impacts is that many tropical forests occur in less developed nations. In contrast to temperate forests, which are concentrated in countries that generally place a higher priority on managing non-native invasive species or have greater capacity to do so, in less developed nations there is often little awareness of the potential threats from such invasions (Early et al., 2016) and a lack of substantive strategies for their control (Pyšek et al., 2008). As non-native invasive plants often represent a future threat with impacts that are often unclear or poorly known, the management of invasions becomes a low priority against more pressing issues such as health and education. Compounding the problem, many plant invasions are not recognised or even perceived as having negative impacts by the general public, and often very little is known about the nature and dynamics of these invasions.

While some are starting to address this considerable challenge for conserving tropical forest biodiversity (Zenni et al., 2017), there remains a genuine risk that non-native invasive plants and their impacts on tropical forests are often going unnoticed and unmanaged. Where there has been research into the impacts of invasive plants on tropical forests, it has been shown that these invaders cause fundamental impacts on their recipient native ecosystems (Asner et al., 2008; Murphy et al., 2008). These changes include a decrease in native plant diversity (Lugo and Helmer, 2004), the complete replacement of native vegetation (Meyer and Lavergne, 2004) and changes in forest structure and composition (Asner et al., 2008). In regions where access to health services is relatively more difficult, the expansion of non-native invasive plants may also represent a direct threat to human security. For example, the tickberry (*Lantana camara* L.) is a vigorous light demanding pioneer shrub with multicoloured flowers that has been introduced to many tropical regions. This shrub forms dense tickets in vacant lots and in heterogeneous land mosaics (Figure 7.3), increasing the availability of breeding habitats for the tsetse fly (*Glossina* spp.) (Syed and Guerin, 2004) and slowing the establishment of native tree species.

Most of these better-studied tropical regions, however, are in developed nations that only account for a relatively small proportion of tropical forests. Recent studies have shown that it is more likely that the lack of documented non-native invasive species impacts on tropical forests is merely because of a lack of knowledge rather than a fundamental difference from temperate forests. On the island of Borneo, Padmanaba and Sheil (2014) found that the spread of the introduced spiked pepper (*Piper aduncum* L.) tracked the installation of logging roads through the rainforest while Döbert et al. (2018) found low overall levels of invasion but a strong positive relationship between logging and non-native plant biomass and leaf area index. It can therefore be expected that the current increase in logging and plantation activities in tropical forests will significantly increase the risk of plant invasions, with potential long-term implications for their successional trajectory (Padmanaba and Corlett, 2014; Brown and Gurevitch, 2004; Friday et al., 2008).

There is a growing call for improving invasive species management strategies in ways that are relevant to and realistic for less developed countries (Corlett, 2009; Shaw et al., 2010). There is also increasing awareness of and action to manage the issue of invasive plants in tropical forests (Zenni, 2015; Zenni et al., 2017). A priority for tropical forests is to improve the outcomes of these efforts by leveraging our existing understanding of forest invasions and their management solutions that have been conceived in developed nations, in temperate climates and under differing systems of governance, infrastructure and land tenure (Nuñez and Pauchard, 2010; Webber et al., 2011). It is imperative that such actions also take into account the interactions between non-native species and other aspects of global environmental change (Lee-Yaw et al., 2019), if we are to achieve optimal environmental, social and economic outcomes for forests worldwide.

The role of forestry in reducing the risk of non-native species invasion

The worldwide share of forest area that is currently managed or has been managed is clearly beyond that of primary forests, and a significant proportion of that is under management certification (Kraxner et al., 2017; Potapov et al., 2017). Therefore, forest management is a key tool if the aim is to counter invasion of non-native species in forest ecosystems. Consequently, a number of stakeholders can play an important role in identifying and applying appropriate management steps towards achieving such an aim.

Non-native species represent an enormous challenge but also an important opportunity for foresters (Sitzia et al., 2016a; Muzika, 2017; Pötzelsberger et al., 2020a). Indeed, due to the multifaceted nature of the non-native invasive problem, forest management must be strategic and adaptive (van Wilgen and Richardson, 2014; Sitzia et al., 2016a; Muzika, 2017). For non-native

(a)

(b)

Figure 7.3   The tickberry forms dense thickets in vacant lots and road banks in tropical regions (a) and in
mosaics of agricultural, forest, grassland and urban land uses (b), increasing the availability of
breeding habitats for the tsetse fly.

Source: Photo by T. Sitzia, Kasese district, Uganda, not far from the gate of the Rwenzori national park.

invasive tree species in particular, novel approaches are required to deal with conflicts that arise between key actors driven by contrasting interests relating to forestry and conservation motivations (Nuñez et al., 2017; Brundu et al., 2020). Furthermore, prioritisation should properly take into account existing legal instruments, specifically in terms of non-native species that are more of a concern (e.g. see list of the European Union; European Commission, 2019) and of limitations at the administrative and geographical level (e.g. protected areas, habitat types and species; Sádlo et al., 2017; Campagnaro et al., 2018a).

Forest planning acts at the landscape scale and is an important tool to tackle problems related to biodiversity conservation (Trentanovi et al., 2018). This characteristic enables planners to consider non-forest habitats as well and take a broader approach towards non-native invasive species (Sitzia et al., 2016a).

A key part of forest management is the application of appropriate actions, mainly silvicultural activities that shape forest features. A wealth of experience and tradition characterise current silvicultural practices, representing an important base on which to build actions to control non-native invasive species. A number of silvicultural interventions can be used to modify forest structure and processes in order to support specific features, like species composition, that can counter invasions (Sitzia et al., 2016a). Nevertheless, specific conditions due to non-native invasions – for example those associated with novel forest pests – call for innovative or unusual approaches going beyond the standard silvicultural actions (Muzika, 2017).

Different species require different actions. Increasing attention is given to management protocols and guidelines for specific species; recent examples include the already mentioned black locust in central Europe (Sádlo et al., 2017) and mesquite (*Prosopis* spp.) in South Africa (Shackleton et al., 2017). However, a number of non-native species combinations form forests that were not observed until recent years (i.e. novel ecosystems; Hobbs et al., 2006; Kowarik, 2011; Sitzia et al., 2018). This specificity calls for their full consideration in modern silviculture. Furthermore, actions must take into account invasion stages (Blackburn et al., 2011), which can require different silvicultural solutions (Waring and O'Hara, 2005). As already mentioned, disturbance regimes are fundamental for the invasion of many non-native species in forest ecosystems, particularly from those that can be defined as 'passengers' of change. Therefore, for these species, knowing the effect derived by applying different silvicultural systems can help control their spread. In the case of 'drivers' of change, knowledge of outcomes derived by silvicultural applications is needed in relation to their efficacy for reducing negative impacts. Furthermore, restoration of forest ecosystems invaded by non-native organisms can be achieved through silvicultural activities and prescriptions with the objective of maintaining and forming resilient forest communities (Sitzia et al., 2016a; Muzika, 2017). Indeed, careful decision as to which silvicultural actions to implement and their thorough planning must consider current presence of non-native species together with the future threat of invasions.

For example, the fungus *Hymenoscyphus fraxineus* (T. Kowalski) Baral, Queloz, Hosoya, which has been introduced from Asia, is currently causing a wide dieback of common ash (*Fraxinus excelsior* L.) in Europe. This pathogen causes more severe damage on smaller than on larger trees; hence it might be paradoxically favoured by the uncontrolled encroachment of abandoned agricultural lands by young trees. Alternative silvicultural strategies depending on stand age and the severity of dieback are needed (Cleary et al., 2017). Some actions may trigger further invasions – such as, for example, the intentional accumulation of deadwood, which is usually intended to conserve biodiversity and increase forest ecosystem resilience but might be a driver of invasion by non-native scolytins (see the previous discussion). It may happen that the mechanism of impact by non-native species is mainly biological, with indirect habitat-related effects difficult to demonstrate. For instance, the cane toad (*Rhinella marina* [Linnaeus, 1758]), a large highly toxic anuran native to South and Central America, was introduced to many other regions in attempts to control agricultural pests (Figure 7.4). The mechanism of impact for larger animals is mostly due to a single pathway – lethal ingestion of toads by frog-eating predators (Shine, 2010). This amphibian concentrates in the disturbed habitats created by human activity, but it may penetrate pristine ecosystems as well, particularly where there is easy access to water. Its habitat overlaps with native species in a broad range of tropical environments, but it seems that the invader occurs at far lower densities in primary intact forest (Thomas et al., 2011; Markula et al., 2016). Therefore, managers of forest ecosystems have a responsibility to engage in the control of non-native species that have their persistence, and hence impacts, facilitated or enhanced by such landscapes (NSW, 2012).

Can a voluntary code for existing and future non-native tree plantations mitigate invasion risk?

Planted forests make significant contributions to regional and national economies and provide multiple products and ecosystem services that support livelihoods and biodiversity conservation (Brockerhoff et al., 2008; FAO, 2020). However, many widely used forestry trees are invasive, spreading from planting sites into adjoining areas, where they may establish and cause substantial damage to biodiversity and ecosystem services. The challenge is to manage existing and future plantation forests of non-native trees and the introduction, breeding and use of new species/ provenances to maximise current benefits while minimising present and future risks, reducing negative impacts and not compromising future benefits and land uses (Brundu and Richardson, 2016; Brundu et al., 2020). This process cannot be solely supported by black letter legislation but needs to be assisted by a plethora of actions in the framework of regional and national strategies on biodiversity and invasive species and dedicated action plans.

The European Union, in particular, has a long history of introduction of non-native trees (Pavari and De Philippis, 1941; Streets, 1962): some parts of Europe, particularly in the south, lack highly productive native tree species with timber or growth characteristics suited to plantation forestry, and foresters rely largely upon non-native tree species (Brundu and Richardson, 2016). This long tradition has produced a very important and widespread legacy in the environment and landscape – as well as in the region's economy, history and culture – and influenced many social aspects, including legislation and land uses, so that nowadays a significant number of controversial tree forest species (*sensu* van Wilgen and Richardson, 2014), plantation types and land uses are present, and new ones may emerge in the near future.

Although forests have been and are affected by a broad array of community policies, in the European Union, the formulation of forest policies is in the power of the member states within a clearly defined framework of established ownership rights and with a long history of national and regional laws and regulations based on long-term planning, including the possibility to use non-native tree species (Pötzelsberger et al., 2020b). In fact, there is a long list of EU measures supporting certain forest-related activities, coordinated with member states mainly through the Standing Forestry Committee (EU Commission, 2016), or even agroforestry and environmental measures, and these have, in some cases, promoted the use of non-native trees, as in the case of the Council Regulation (EEC) No. 2080/92 of 30 June 1992 instituting a community aid scheme for forestry measures in agriculture or, on the other hand, funded the local control or eradication of non-native trees (as in the case of many LIFE projects, e.g. Scalera et al., 2017, and the CAP High Nature Value farming).

Importantly, nearly half of European forests are privately owned (Schmithüsen and Hirsch, 2010; Nichiforel et al., 2018), and contemporary policy on private forest management is considered guided by sustainable forest management concepts (Nichiforel et al., 2018 and references

(a)

(b)

Figure 7.4   The cane toad is an anuran that had been introduced to several regions of the world. In its non-native range it can penetrate pristine forests (a) and threaten native species via predation and because of its toxicity. Agamids and other frog-eating lizards that share their habitats, like the Boyd's forest dragon (*Lophosaurus boydii* [MacLeay, 1884]) are particularly threatened (Smith and Phillips, 2006) (b).

Source: Photo by T. Sitzia, Daintree forest, Queensland, Australia.

cited therein). Depending on the region and forest type, these emphasise different aspects of sustainability, such as 'sustainable yield', which focuses on sustained timber production; 'multi-purpose forestry', which highlights multiple goods and services; and 'ecosystem management', which stresses the status and evolution of forest ecosystems (Nichiforel et al., 2018 and references cited therein), but also provide guidelines or restrictions for the use of non-native trees (Brundu and Richardson, 2016). This aspect clearly stresses the importance of involving all the stake-holders in the mitigation of the risk posed by invasive or potentially non-native invasive trees. Stakeholders are expected to seek to influence the outcomes of policy development in order to secure their own interests and positions as described by Huttunen (2014) in the case of forest bioenergy legislation in Finland.

In the European Union, non-native invasive species and (invasive) pests are regulated under two different sections of the European Commission. Non-native invasive species fall under the responsibility of Environment Directorate-General. This directorate has passed legislation that entered into force 1 January 2015 (Regulation EU n. 1143/2014). On the other hand, the leg-islation within the EU regarding (invasive) pests falls under the responsibility of the Directorate-General Health and Food Safety, which formulates the regulations regarding plant health and biosecurity. These two legislations are kept separate in their respective aims (environment, trade, see Klapwijk et al., 2016). To date, only a limited number of non-native trees have been included in the list of non-native invasive species of EU concern: *Acacia saligna* (Labill.) H.L. Wendl., *Ailanthus altissima* (Mill.) Swingle, *Prosopis juliflora* (Sw.) DC and *Triadica sebifera* (L.) Small. On the other hand, it encourages the use of codes of good practice to address the priority pathways and prevent the unintentional introduction and spread of non-native invasive species into or within the EU.

To encourage national authorities to implement general principles of prevention and miti-gation of the risks posed by invasive non-native tree species into their national environmental policies, the Council of Europe has promoted the preparation of a Code of Conduct on Invasive Alien Trees (Brundu and Richardson, 2017). The hope is that this code that provides guidelines focusing on key pathways and core groups will be taken up by relevant sectors of society and eventually be included in national legislation. The code itself is voluntary and does not replace any statutory requirements under international or national legislation. The code was endorsed by the Standing Committee of the Convention on the Conservation of European Wildlife and Natural Habitats in December 2017 with the Recommendation no. 193 (2017) calling the con-tracting countries to take the European Code of Conduct mentioned prior into account while drawing up other relevant codes or, where appropriate, to draw up national codes of conduct on non-native invasive trees and to collaborate as appropriate with the actors involved in forestry activities in implementing and helping disseminate good practices and codes of conduct aimed at preventing and managing of introduction, release and spread of non-native invasive trees. The endorsement of this code is therefore a promising first step to mitigate the risk posed by unregu-lated introductions of non-native trees in forestry.

An international workshop was held in 2019 to develop global guidelines for the sustain-able use of non-native trees, using the Bern Convention Code of Conduct on Invasive Alien Trees as a starting point. These global guidelines consist of eight recommendations: 1) use native trees, or non-invasive non-native trees, in preference to invasive non-native trees; 2) be aware of and comply with international, national, and regional regulations concerning non-native trees; 3) be aware of the risk of invasion and consider global change trends; 4) design and adopt tailored practices for plantation site selection and silvicultural management; 5) pro-mote and implement early detection and rapid response programmes; 6) design and adopt tailored practices for invasive non-native tree control, habitat restoration, and for dealing with highly modified ecosystems; 7) engage with stakeholders on the risks posed by invasive non-native trees, the impacts caused, and the options for management; and 8) develop and support global networks, collaborative research, and information sharing on native and non-native trees (Brundu et al., 2020).

References

Alpern, S.B. (2008) 'Exotic plants of Western Africa: Where they came from and when', *History in Africa*, vol. 35, pp. 63–102.

Ammer, C., Schall, P., Wördehoff, R., Lamatsch, K. and Bachmann, M. (2011) 'Does tree seedling growth and survival require weeding of Himalayan balsam (*Impatiens glandulifera*)?', *European Journal of Forest Research*, vol. 130, pp. 107–116.

Asner, G.P., Knapp, D.E., Kennedy-Bowdoin, T., Jones, M.O., Martin, R.E., Boardman, J. and Hughes, R.F. (2008) 'Invasive species detection in Hawaiian rainforests using airborne imaging spectroscopy and LiDAR', *Remote Sensing of Environment*, vol. 112, pp. 1942–1955.

Aukema, J.E., McCullough, D.G., Von Holle, B., Liebhold, A.M., Britton, K. and Frankel, S.J. (2010) 'Historical accumulation of nonindigenous forest pests in the continental United States', *Bioscience*, vol. 60, pp. 886–897.

Bauer, J.T. (2012) 'Invasive species: "Back-seat drivers" of ecosystem change?', *Biological Invasions*, vol. 14, pp. 1295–1304.

Beaver, R.A. (1979) 'Host specificity of temperate and tropical animals', *Nature*, vol. 281, pp. 139–141.

Bellingham, P.J., Tanner, E.V., Martin, P.H., Healey, J.R. and Burge, O.R. (2018) 'Endemic trees in a tropical biodiversity hotspot imperiled by an invasive tree', *Biological Conservation*, vol. 217, pp. 47–53.

Blackburn, T.M., Pyšek, P., Bacher, S., Carlton, J.T., Duncan, R.P., Jarošík, V., Wilson, J.R.U and Richardson, D.M. (2011) 'A proposed unified framework for biological invasions', *Trends in Ecology & Evolution*, vol. 26, pp. 333–339.

Bradley, B.A., Blumenthal, D.M., Early, R., Grosholz, E.D., Lawler, J.J., Miller, L.P., Sorte, J.B., D'Antonio, C.M., Diez, J.M., Dukes, J.S., Ibáñez, I. and Olden, J.D. (2012) 'Global change, global trade, and the next wave of plant invasions', *Frontiers in Ecology and the Environment*, vol. 10, pp. 20–28.

Brockerhoff, E.G., Bain, J., Kimberley, M.O. and Knížek, M. (2006) 'Interception frequency of exotic bark and ambrosia beetles (Coleoptera: Scolytinae) and relationship with establishment in New Zealand and worldwide', *Canadian Journal of Forest Research*, vol. 36, pp. 289–298.

Brockerhoff, E.G., Jactel, H., Parrotta, J.A., Quine, C.P. and Sayer, J. (2008) 'Plantation forests and biodiversity: Oxymoron or opportunity?', *Biodiversity and Conservation*, vol. 17, pp. 925–951.

Brown, K.A. and Gurevitch, J. (2004) 'Long-term impacts of logging on forest diversity in Madagascar', *Proceedings of the National Academy of Sciences of the United States of America*, vol. 101, pp. 6045–6049.

Brundu, G., Pauchard, A., Pyšek, P., Pergl, J., Bindewald, A.M., Brunori, A., Canavan, S., Campagnaro, T., Celesti-Grapow, L., Dechoum, M. de S., Dufour-Dror, J.-M., Essl, F., Flory, S.L., Genovesi, P., Guarino, F., Guangzhe, L., Hulme, P.E., Jäger, H., Kettle, C.J., Krumm, F., Langdon, B., Lapin, K., Lozano, V., Le Roux, J.J., Novoa, A., Nuñez, M.A., Porté, A.J., Silva, J.S., Schaffner, U., Sitzia, T., Tanner, R., Tshidada, N., Vítková, M., Westergren, M., Wilson, J.R.U. and Richardson, D.M. (2020) 'Global guidelines for the sustainable use of non-native trees to prevent tree invasions and mitigate their negative impacts', *NeoBiota*, vol. 61, pp. 65–116.

Brundu, G. and Richardson, D.M. (2016) 'Planted forests and invasive alien trees in Europe: A Code for managing existing and future plantings to mitigate the risk of negative impacts from invasions', *NeoBiota*, vol. 30, pp. 5–47.

Brundu, G. and Richardson, D.M. (2017) Code of conduct for invasive Alien trees: T-PVS/Inf (2017) 8, Council of Europe, Strasbourg, France.

Camenen, E., Porté, A.J. and Benito Garzón, M. (2016) 'American trees shift their niches when invading Western Europe: Evaluating invasion risks in a changing climate', *Ecology and Evolution*, vol. 6, pp. 7263–7275.

Campagnaro, T., Brundu, G. and Sitzia, T. (2018a) 'Five major invasive alien tree species in European Union forest habitat types of the Alpine and Continental biogeographical regions', *Journal for Nature Conservation*, vol. 43, pp. 227–238.

Campagnaro, T., Nascimbene, J., Tasinazzo, S., Trentanovi, G. and Sitzia, T. (2018b) 'Exploring patterns, drivers and structure of plant community composition in alien *Robinia pseudoacacia* secondary woodlands', *iForest*, vol. 11, pp. 586–593.

Chabrerie, O., Verheyen, K., Saguez, R. and Decocq, G. (2008) 'Disentangling relationships between habitat conditions, disturbance history, plant diversity, and American black cherry (*Prunus serotina* Ehrh.) invasion in a European temperate forest', *Diversity and Distributions*, vol. 14, pp. 204–212.

Cierjacks, A., Kowarik, I., Joshi, J., Hempel, S., Ristow, M., Lippe, M. and Weber, E. (2013) 'Biological flora of the British Isles: *Robinia pseudoacacia*', *Journal of Ecology*, vol. 101, pp. 1623–1640.

Cleary, M., Nguyen, D., Stener, L.G., Stenlid, J. and Skovsgaard, J.P. (2017) 'Ash and ash dieback in Sweden: A review of disease history, current status, pathogen and host dynamics, host tolerance and management options in forests and landscapes', in R. Vasaitis and R. Enderle (eds.) *Dieback of European Ash (Fraxinus spp.): Consequences and Guidelines for Sustainable Management*. Swedish University of Agricultural Sciences, Uppsala, Sweden.

Colunga-Garcia, M., Haack, R.A. and Adesoji, O.A. (2009) 'Freight transportation and the potential for invasions of exotic insects in urban and periurban forests of the United States', *Journal of Economic Entomology*, vol. 102, pp. 237–246.

Corlett, R.T. (2009) 'Invasive aliens on tropical East Asian islands', *Biodiversity and Conservation*, vol. 19, pp. 411–423.

Crosby, A.W. (2004) *Ecological Imperialism: The Biological Expansion of Europe, 900–1900*. Cambridge University Press, Cambridge, UK.

Davidson, A.M., Jennions, M. and Nicotra, A.B. (2011) 'Do invasive species show higher phenotypic plasticity than native species and, if so, is it adaptive? A meta-analysis', *Ecology Letters*, vol. 14, pp. 419–431.

Dawson, W., Mndolwa, A.S., Burslem, D.F.R.P. and Hulme, P.E. (2008) 'Assessing the risks of plant invasions arising from collections in tropical botanical gardens', *Biodiversity and Conservation*, vol. 17, pp. 1979–1995.

Desprez-Loustau, M.-L. (2009) 'The alien fungi of Europe', in DAISIE (ed.) *Handbook of Alien Species in Europe*. Springer, Dordrecht, Netherlands.

Desprez-Loustau, M.-L., Courtecuisse, R., Robin, C., Husson, C., Moreau, P.A., Blancard, D., Selosse, M.A., Lung-Escarmant, B., Piou, D. and Sache, I. (2010) 'Species diversity and drivers of spread of alien fungi (sensu lato) in Europe with a particular focus on France', *Biological Invasions*, vol. 12, pp. 157–172.

Didham, R.K., Tylianakis, J.M., Gemmell, N.J., Rand, T.A. and Ewers, R.M. (2007) 'Interactive effects of habitat modification and species invasion on native species decline', *Trends in Ecology & Evolution*, vol. 22, pp. 489–496.

Didham, R.K., Tylianakis, J.M., Hutchison, M.A., Ewers, R.M. and Gemmell, N.J. (2005) 'Are invasive species the drivers of ecological change?', *Trends in Ecology & Evolution*, vol. 20, pp. 470–474.

Dinerstein, E., Olson, D., Joshi, A., Vynne, C., Burgess, N.D., Wikramanayake, E., Hahn, N., Palminteri, S., Hedao, P., Noss, R., Hansen, M., Locke, H., Ellis, E.C., Jones, B., Barber, C.V., Hayes, R., Kormos, C., Martin, V., Crist, E., Sechrest, W., Price, L., Baillie, J.E.M., Weeden, D., Suckling, K., Davis, C., Sizer, N., Moore, R., Thau, D., Birch, T., Potapov, P., Turubanova, S., Tyukavina, A., de Souza, N., Pintea, L., Brito, J.C., Llewellyn, O.A., Miller, A.G., Patzelt, A., Ghazanfar, S.A., Timberlake, J., Klöser, H., Shennan-Farpón, Y., Kindt, R., Lillesø, J.-P.B., van Breugel, P., Graudal, L., Voge, M., Al-Shammari, K.F. and Saleem, M. (2017) 'An ecoregion-based approach to protecting half the terrestrial realm', *Bioscience*, vol. 67, pp. 534–545.

Döbert, T.F., Webber, B.L., Sugau, J.B., Dickinson, K.J. and Didham, R.K. (2018) 'Logging, exotic plant invasions, and native plant reassembly in a lowland tropical rain forest', *Biotropica*, vol. 50, pp. 254–265.

Dukes, J.S., Pontius, J., Orwig, D., Garnas, J.R., Rodgers, V.L., Brazee, N., Cooke, B., Theoharides, K.A., Stange, E.E., Harrington, R., Ehrenfeld, J., Gurevitch, J., Lerdau, M., Stinson, K., Wick, R. and Ayres, M. (2009) 'Responses of insect pests, pathogens, and invasive plant species to climate change in the forests of northeastern North America: What can we predict?', *Canadian Journal of Forest Research*, vol. 39, pp. 231–248.

Early, R., Bradley, B., Dukes, J., Lawler, J.J., Olden, J.D., Blumenthal, D.M., Gonzalez, P., Grosholz, E.D., Ibáñez, I., Miller, L.P., Sorte, C.J.B. and Tatem, A.J. (2016) 'Global threats from invasive alien species in the twenty-first century and national response capacities', *Nature Communications*, vol. 7, article ID 12485.

Ehrenfeld, J.G., Kourtev, P. and Huang, W. (2001) 'Changes in soil functions following invasions of exotic understory plants in deciduous forests', *Ecological Applications*, vol. 11, pp. 1287–1300.

Eschen, R., Britton, K., Brockerhoff, E., Burgess, T., Dalley, V., Epanchin-Niell, R.S., Gupta, K., Hardy, G., Huang, Y., Kenis, M., Kimani, E., Li, H.-M., Olsen, S., Ormrod, R., Otieno, W., Sadof, C., Tadeu, E. and Theyse, M. (2015) 'International variation in phytosanitary legislation and regulations governing importation of plants for planting', *Environmental Science & Policy*, vol. 51, pp. 228–237.

Essl, F., Milasowszky, N. and Dirnböck, T. (2011) 'Plant invasions in temperate forests: Resistance or ephemeral phenomenon?', *Basic and Applied Ecology*, vol. 12, pp. 1–9.

European Commission (2016) *EU Forest Policies*, http://ec.europa.eu/environment/forests/fpolicies.htm, accessed 10 July 2018.

European Commission (2019) 'Commission implementing regulation (EU) 2019/1262 of 25 July 2019 amending implementing regulation (EU) 2016/1141 to update the list of invasive alien species of Union concern', *Official Journal of the European Union*, vol. 62, pp. 1–4.

FAO (2020) *Global Forest Resources Assessment 2020 – Key Findings*. Food and Agriculture Organization of the United Nations, Rome, Italy.

Ferrier, S., Ninan, K.N., Leadley, P., Alkemade, R., Acosta, L.A., Akçakaya, H.R., Brotons, L., Cheung, W.W.L., Christensen, V., Harhash, K.A., Kabubo-Mariara, J., Lundquist, C., Obersteiner, M., Pereira, H.M., Peterson, G., Pichs-Madruga, R., Ravindranath, N., Rondinini, C. and Wintle, B.A. (eds.) (2016) *The Methodological Assessment Report on Scenarios and Models of Biodiversity and Ecosystem Services*. Secretariat of the Intergovernmental Science-Policy Platform on Biodiversity and Ecosystem Services, Bonn, Germany.

Fine, P.V.A. (2002) 'The invasibility of tropical forests by exotic plants', *Journal of Tropical Ecology*, vol. 18, pp. 687–705.

Florens, F.V., Baider, C., Martin, G.M., Seegoolam, N.B., Zmanay, Z. and Strasberg, D. (2016) 'Invasive alien plants progress to dominate protected and best-preserved wet forests of an oceanic island', *Journal for Nature Conservation*, vol. 34, pp. 93–100.

Flory, S.L. and Clay, K. (2010) 'Non-native grass invasion suppresses forest succession', *Oecologia*, vol. 164, pp. 1029–1038.

Friday, J.B., Scowcroft, P.G. and Ares, A. (2008) 'Responses of native and invasive plant species to selective logging in an *Acacia koa-Metrosideros polymorpha* forest in Hawai'i', *Applied Vegetation Science*, vol. 11, pp. 471–482.

Gandhi, K.J.K. and Herms, D.A. (2010) 'Direct and indirect effects of alien insect herbivores on ecological processes and interactions in forests of eastern North America', *Biological Invasions*, vol. 12, pp. 389–405.

Garbelotto, M., Della Rocca, G., Osmundson, T., di Lonardo, V. and Danti, R. (2015) 'An increase in transmission-related traits and in phenotypic plasticity is documented during a fungal invasion', *Ecosphere*, vol. 6, pp. 1–16.

Ghelardini, L., Pepori, A.L., Luchi, N., Capretti, P. and Santini, A. (2016) 'Drivers of emerging fungal diseases of forest trees', *Forest Ecology and Management*, vol. 381, pp. 235–246.

Grégoire, J.-C., Raffa, K.F. and Lindgren, B.S. (2015) 'Economics and politics of bark beetles', in F.E. Vega and R.W. Hofstetter (eds.) *Bark Beetles: Biology and Ecology of Native and Invasive Species*. Academic Press, London, UK.

Guo, Q., Rejmánek, M. and Wen, J. (2012) 'Geographical, socioeconomic, and ecological determinants of exotic plant naturalization in the United States: Insights and updates from improved data', *NeoBiota*, vol. 12, pp. 41–55.

Haack, R.A. (2001) 'Intercepted Scolytidae (Coleoptera) at US ports of entry: 1985–2000', *Integrated Pest Management Reviews*, vol. 6, pp. 253–282.

Haack, R.A. (2006) 'Exotic bark- and wood-boring Coleoptera in the United States: Recent establishments and interceptions', *Canadian Journal of Forest Research*, vol. 36, pp. 269–288.

Haack, R.A. and Rabaglia, R.J. (2013) 'Exotic bark and ambrosia beetles in the USA: Potential and current invaders', in J.E. Peña (ed.) *Potential Invasive Pests of Agricultural Crops*. CABI, Wallingford, UK.

Hejda, M., Hanzelka, J., Kadlec, T., Štrobl, M., Pyšek, P. and Reif, J. (2017) 'Impacts of an invasive tree across trophic levels: Species richness, community composition and resident species' traits', *Diversity and Distributions*, vol. 23, pp. 997–1007.

Hobbs, R.J., Arico, S., Aronson, J., Baron, J.S., Bridgewater, P., Cramer, V.A., Epstein, P.R., Ewel, J.J., Klink, C.A., Lugo, A.E., Norton, D., Ojima, D., Richardson, D.M., Sanderson, E.W., Valladares, F., Vilà, M., Zamora, R. and Zobel, M. (2006) 'Novel ecosystems: Theoretical and management aspects of the new ecological world order', *Global Ecology and Biogeography*, vol. 15, pp. 1–7.

Huang, D., Haack, R.A. and Zhang, R. (2011) 'Does global warming increase establishment rates of invasive alien species? A centurial time series analysis', *PLoS One*, vol. 6, p. e24733.

Hulme, P.E. (2009) 'Trade, transport and trouble: Managing invasive species in an era of globalization', *Journal of Applied Ecology*, vol. 46, pp. 10–18.

Hulme, P.E. (2011) 'Addressing the threat to biodiversity from botanic gardens', *Trends in Ecology & Evolution*, vol. 26, pp. 168–174.

Hulme, P.E., Pyšek, P., Jarošík, V., Pergl, J., Schaffner, U. and Vilà, M. (2013) 'Bias and error in understanding plant invasion impacts', *Trends in Ecology & Evolution*, vol. 28, pp. 212–218.

Huttunen, S. (2014) 'Stakeholder frames in the making of forest bioenergy legislation in Finland', *Geoforum*, vol. 53, pp. 63–73.

Jäger, H., Tye, A. and Kowarik, I. (2007) 'Tree invasion in naturally treeless environments: Impacts of quinine (*Cinchona pubescens*) trees on native vegetation in Galápagos', *Biological Conservation*, vol. 140, pp. 297–307.

Kenis, M., Auger-Rozenberg, M.A., Roques, A., Timms, L., Péré, C., Cock, M.J., Settele, J., Augustin, S. and Lopez-Vaamonde, C. (2009) 'Ecological effects of invasive alien insects', *Biological Invasions*, vol. 11, pp. 21–45.

Kenis, M., Rabitsch, W., Auger-Rozenberg, M.A. and Roques, A. (2007) 'How can alien species inventories and interception data help us prevent insect invasions?', *Bulletin of Entomological Research*, vol. 97, pp. 489–502.

Keresztesi, B. (1983) 'Breeding and cultivation of black locust, *Robinia pseudoacacia*, in Hungary', *Forest Ecology and Management*, vol. 6, pp. 217–244.

Kirkendall, L.R. (1983) 'The evolution of mating systems in bark and ambrosia beetles (Coleoptera: Scolytidae and Platypodidae)', *Zoological Journal of the Linnean Society*, vol. 77, pp. 293–352.

Kirkendall, L.R. (1993) 'Ecology and evolution of biased sex ratios in bark and ambrosia beetles (Scolytidae)', in D.L. Wrensch and M.A. Ebbert (eds.) *Evolution and Diversity of Sex Ratio: Insects and Mites.* Chapman and Hall, New York.

Kirkendall, L.R. and Faccoli, M. (2010) 'Bark beetles and pinhole borers (Curculionidae, Scolytinae, Platypodinae) alien to Europe', *ZooKeys*, vol. 56, pp. 227–251.

Klapwijk, M.J., Hopkins, A.J.M., Eriksson, L., Pettersson, M., Schroeder, M., Lindelöw, Å., Rönnberg, J., Keskitalo, E.C.H. and Kenis, M. (2016) 'Reducing the risk of invasive forest pests and pathogens: Combining legislation, targeted management and public awareness', *Ambio*, vol. 45, no. 2, pp. 223–234.

Kleinbauer, I., Dullinger, S., Peterseil, J. and Essl, F. (2010) 'Climate change might drive the invasive tree *Robinia pseudacacia* into nature reserves and endangered habitats', *Biological Conservation*, vol. 143, pp. 382–390.

Kowarik, I. (2011) 'Novel urban ecosystems, biodiversity, and conservation', *Environmental Pollution*, vol. 159, pp. 1974–1983.

Kraxner, F., Schepaschenko, D., Fuss, S., Lunnan, A., Kindermann, G., Aoki, K., Dürauer, M., Shvidenko, A. and See, L. (2017) 'Mapping certified forests for sustainable management: A global tool for information improvement through participatory and collaborative mapping', *Forest Policy and Economics*, vol. 83, pp. 10–18.

Kueffer, C., Pyšek, P. and Richardson, D.M. (2013) 'Integrative invasion science: Model systems, multi-site studies, focused meta-analysis and invasion syndromes', *New Phytologist*, vol. 200, pp. 615–633.

Lazzaro, L., Mazza, G., d'Errico, G., Fabiani, A., Giuliani, C., Inghilesi, A.F., Lagomarsino, A., Landi, S., Lastrucci, L., Pastorelli, R., Roversi, P.F., Torrini, G., Tricarico, E. and Foggi, B. (2018) 'How ecosystems change following invasion by *Robinia pseudoacacia*: Insights from soil chemical properties and soil microbial, nematode, microarthropod and plant communities', *Science of the Total Environment*, vol. 622–623, pp. 1509–1518.

Lee-Yaw, J.A., Zenni, R.D., Hodgins, K.A., Larson, B.M.H., Cousens, R. and Webber, B.L. (2019) 'Range shifts and local adaptation: Integrating data and theory towards a new understanding of species' distributions in the Anthropocene', *New Phytologist*, vol. 221, pp. 644–647.

Levine, J.M. and D'Antonio, C.M. (2003) 'Forecasting biological invasions with increasing international trade', *Conservation Biology*, vol. 17, pp. 322–326.

Liebhold, A.M., Brockerhoff, E.G., Kalisz, S., Nuñez, M.A., Wardle, D.A. and Wingfield, M.J. (2017) 'Biological invasions in forest ecosystems', *Biological Invasions*, vol. 19, pp. 3437–3458.

Lovett, G.M., Canham, C.D., Arthur, M.A., Weathers, K.C. and Fitzhugh, R.D. (2006) 'Forest ecosystem responses to exotic pests and pathogens in Eastern North America', *BioScience*, vol. 56, pp. 395–405.

Lovett, G.M., Weiss, M., Liebhold, A.M., Holmes, T.P., Leung, B., Lambert, K.F., Orwig, D.A., Campbell, F.T., Rosenthal, J., McCullough, D.G., Wildova, R., Ayres, M.P., Canham, C.D., Foster, D.R., LaDeau, S.L. and Weldy, T. (2016) 'Nonnative forest insects and pathogens in the United States: Impacts and policy options', *Ecological Applications*, vol. 26, pp. 1437–1455.

Lugo, A.E. and Helmer, E. (2004) 'Emerging forests on abandoned land: Puerto Rico's new forests', *Forest Ecology and Management*, vol. 190, pp. 145–161.

MacDougall, A.S. and Turkington, R. (2005) 'Are invasive species the drivers or passengers of change in degraded ecosystems?', *Ecology*, vol. 86, pp. 42–55.

Marini, L., Haack, R.A., Rabaglia, R.J., Petrucco Toffolo, E., Battisti, A. and Faccoli, M. (2011) 'Exploring associations between international trade and environmental factors with establishment patterns of alien Scolytinae', *Biological Invasions*, vol. 13, pp. 2275–2288.

Markula, A., Csurhes, S. and Hannan-Jones, M. (2016) *Invasive Animal Risk Assessment. Cane toad. Bufo marinus*. Queensland Government, Department of Agriculture and Fisheries, Biosecurity Queensland, Brisbane, Australia.

Martin, P.H., Canham, C.D. and Marks, P.L. (2009) 'Why forests appear resistant to exotic plant invasions: Intentional introductions, stand dynamics, and the role of shade tolerance', *Frontiers in Ecology and the Environment*, vol. 7, pp. 142–149.

Mattson, W., Vanhanen, H., Veteli, T., Sivonen, S. and Niemelä, P. (2007) 'Few immigrant phytophagous insects on woody plants in Europe: Legacy of the European crucible?', *Biological Invasions*, vol. 9, pp. 957–974.

McCullough, D.G., Work, T.T., Cavey, J.F., Liebhold, A.M. and Marshall, D. (2006) 'Interceptions of nonindigenous plant pests at US ports of entry and border crossings over a 17-year period', *Biological Invasions*, vol. 8, pp. 611–630.

Medina-Villar, S., Rodríguez-Echeverría, S., Lorenzo, P., Alonso, A., Pérez-Corona, E. and Castro-Díez, P. (2016) 'Impacts of the alien trees *Ailanthus altissima* (Mill.) Swingle and *Robinia pseudoacacia* L. on soil nutrients and microbial communities', *Soil Biology and Biochemistry*, vol. 96, pp. 65–73.

Meyer, J.Y. and Lavergne, C. (2004) 'Beautés fatales: Acanthaceae species as invasive alien plants on tropical Indo-Pacific Islands', *Diversity and Distributions*, vol. 10, pp. 333–347.

Meyerson, L.A. and Mooney, H.A. (2007) 'Invasive alien species in an era of globalization', *Frontiers in Ecology and the Environment*, vol. 5, pp. 199–208.

Migliorini, D., Ghelardini, L., Tondini, E., Luchi, N. and Santini, A. (2015) 'The potential of symptomless potted plants for carrying invasive soil-borne plant pathogens', *Diversity and Distributions*, vol. 21, pp. 1218–1229.

Murphy, H.T., Hardesty, B.D., Fletcher, C.S., Metcalfe, D.J., Westcott, D.A. and Brooks, S.J. (2008) 'Predicting dispersal and recruitment of *Miconia calvescens* (Melastomataceae) in Australian tropical rainforests', *Biological Invasions*, vol. 10, pp. 925–936.

Muzika, R.M. (2017) 'Opportunities for silviculture in management and restoration of forests affected by invasive species', *Biological Invasions*, vol. 19, pp. 3419–3435.

Nichiforel, L., Keary, K., Deuffic, P., Weiss, G., Thorsen, B.J., Winkel, G., Avdibegović, M., Dobšinská, Z., Feliciano, D., Gatto, P., Górriz-Mifsud, E., Hoogstra-Klein, M., Hrib, M., Hujala, T., Jager, L., Jarský, V., Jodłowski, K., Lawrence, A., Lukmine, D., Malovrh, Š.P., Nedeljković, J., Nonić, D., Ostoić, S.K., Pukall, K., Rondeux, J., Samara, T., Sarvašová, Z., Scriban, R.E., Šilingienė, R., Sinko, M., Stojanovska, M., Stojanovski, V., Stoyanov, N., Teder, M., Vennesland, B., Vilkriste, L., Wihelmsson, E., Wilkes-Allemann, J. and Bouriaud, L. (2018) 'How private are Europe's private forests? A comparative property rights analysis', *Land Use Policy*, vol. 76, pp. 535–552.

Nicolescu, V.-N., Rédei, K., Mason, W.L., Vor, T., Pöetzelsberger, E., Bastien, J.C., Brus, R., Benčaťa, T., Đodan, M., Cvjetkovic, B., Andrašev, S., La Porta, N., Lavnyy, V., Mandžukovski, D., Petkova, K., Roženbergar, D., Wąsik, R., Mohren, G.M.J., Monteverdi, M.C., Musch, B., Klisz, M., Perić, S., Keça, L., Bartlett, D., Hernea, C. and Pástor, M. (2020) 'Ecology, growth and management of black locust (*Robinia pseudoacacia* L.), a non-native species integrated into European forests', *Journal of Forestry Research*, vol. 31, pp. 1081–1101.

Nicolescu, V.-N., Hernea, C., Bakti, B., Keserű, Z., Antal, B. and Rédei, K. (2018) 'Black locust (*Robinia pseudoacacia* L.) as a multi-purpose tree species in Hungary and Romania: A review', *Journal of Forestry Research*, vol. 29, no. 6, pp. 1449–1463.

Novoa, A., Richardson, D.M., Pyšek, P., Meyerson, L.A., Bacher, S., Canavan, S., Catford, J.A., Čuda, J., Essl, F., Foxcroft, L.C., Genovesi, P., Hirsch, H., Hui, C., Jackson, M.C., Kueffer, C., Le Roux, J.J., Measey, J., Mohanty, N.P., Moodley, D., Müller-Schärer, H., G. Packer, J.G., Pergl, J., Robinson, T.B., Saul, W.-C., Shackleton, R.T., Visser, V., Weyl, O.L.F., Yannelli, F.A. and Wilson, J.R.U. (2020) 'Invasion syndromes: a systematic approach for predicting biological invasions and facilitating effective management', *Biological Invasions*, vol. 22, pp. 1801–1820.

NSW (2012) *Management plan for cane toads in national parks and reserves 2012*. State of New South Wales and Office of Environment and Heritage, Sydney, Australia.

Nuñez, M.A., Chiuffo, M.C., Torres, A., Paul, T., Dimarco, R.D., Raal, P., Policelli, N., Moyano, J., García, R.A., van Wilgen, B.W., Pauchard, A. and Richardson, D.M. (2017) 'Ecology and management of invasive Pinaceae around the world: Progress and challenges', *Biological Invasions*, vol. 19, pp. 3099–3120.

Nuñez, M.A. and Pauchard, A. (2010) 'Biological invasions in developing and developed countries: Does one model fit all?', *Biological Invasions*, vol. 12, pp. 707–714.

O'Leary, B., Burd, M., Venn, S.E. and Gleadow, R. (2018) 'Integrating the passenger-driver hypothesis and plant community functional traits to the restoration of lands degraded by invasive trees', *Forest Ecology and Management*, vol. 408, pp. 112–120.

Padmanaba, M. and Corlett, R.T. (2014) 'Minimizing risks of invasive alien plant species in tropical production forest management', *Forests*, vol. 5, pp. 1982–1998.

Padmanaba, M. and Sheil, D. (2014) 'Spread of the invasive alien species *Piper aduncum* via logging roads in Borneo', *Tropical Conservation Science*, vol. 7, pp. 35–44.

Pavari, A. and De Philippis, A. (1941) 'La sperimentazione di specie forestali esotiche in Italia. Risultati del primo ventennio', *Annali della Sperimentazione Agraria*, vol. 38, pp. 1–646.

Perrings, C., Dehnen-Schmutz, K., Touza, J. and Williamson, M. (2005) 'How to manage biological invasions under globalization', *Trends in Ecology & Evolution*, vol. 20, pp. 212–215.

Philibert, A., Desprez-Loustau, M.L., Fabre, B., Frey, P., Halkett, F., Husson, C., Lung-Escarmant, B., Marçais, B., Robin, C., Vacher, C. and Makowski, D. (2011) 'Predicting invasion success of forest pathogenic fungi from species traits', *Journal of Applied Ecology*, vol. 48, pp. 1381–1390.

Piel, F., Gilbert, M., De Cannière, C. and Grégoire, J.-C. (2008) 'Coniferous round wood imports from Russia and Baltic countries to Belgium: A pathway analysis for assessing risks of exotic pest insect introductions', *Diversity and Distributions*, vol. 14, pp. 318–328.

Potapov, P., Hansen, M.C., Laestadius, L., Turubanova, S., Yaroshenko, A., Thies, C., Smith, W., Zhuravleva, I., Komarova, A., Minnemeyer, S. and Esipova, E. (2017) 'The last frontiers of wilderness: Tracking loss of intact forest landscapes from 2000 to 2013', *Science Advances*, vol. 3, article ID e1600821.

Pötzelsberger, E., Lapin, K., Brundu, G., Adriaens, T., Andonovski, V., Andrašev, S., Bastien, J.C., Brus, R., Čurović, M., Čurović, Ž. and Cvjetković, B. (2020b) 'Mapping the patchy legislative landscape of non-native tree species in Europe', *Forestry*, vol. 93, pp. 567–586.

Pötzelsberger, E., Spiecker, H., Neophytou, C., Mohren, F., Gazda, A. and Hasenauer, H. (2020a) 'Growing non-native trees in European forests brings benefits and opportunities but also has its risks and limits', *Current Forestry Reports*, vol. 6, pp. 339–353.

Puchałka, R., Dyderski, M.K., Vítková, M., Sádlo, J., Klisz, M., Netsvetov, M., Prokopuk, Y., Matisons, R., Mionskowski, M., Wojda, T., Koprowski, M. and Jagodziński, A.M. (2021) 'Black locust (*Robinia pseudoacacia* L.) range contraction and expansion in Europe under changing climate', *Global Change Biology*, doi: 10.1111/gcb.15486

Putz, F.E., Zuidema, P.A., Synnott, T., Peña-Claros, M., Pinard, M.A., Sheil, D., Vanclay, J.K., Sist, P., Gourlet-Fleury, S., Griscom, B., Palmer, J. and Zagt, R. (2012) 'Sustaining conservation values in selectively logged tropical forests: The attained and the attainable', *Conservation Letters*, vol. 5, pp. 296–303.

Pyšek, P., Richardson, D.M., Pergl, J., Jarošík, V., Sixtová, Z. and Weber, E. (2008) 'Geographical and taxonomic biases in invasion ecology', *Trends in Ecology & Evolution*, vol. 23, pp. 237–244.

Radtke, A., Ambraß, S., Zerbe, S., Tonon, G., Fontana, V. and Ammer, C. (2013) 'Traditional coppice forest management drives the invasion of *Ailanthus altissima* and *Robinia pseudoacacia* into deciduous forests', *Forest Ecology and Management*, vol. 291, pp. 308–317.

Rassati, D., Faccoli, M., Marini, L., Haack, R.A., Battisti, A. and Petrucco Toffolo, E. (2014a) 'Exploring the role of wood waste landfills in early detection of non-native wood boring beetles', *Journal of Pest Science*, vol. 88, pp. 563–572.

Rassati, D., Faccoli, M., Petrucco Toffolo, E., Battisti, A. and Marini, L. (2015) 'Improving the early detection of alien wood-boring beetles in ports and surrounding forests', *Journal of Applied Ecology*, vol. 52, pp. 50–58.

Rassati, D., Petrucco Toffolo, E., Roques, A., Battisti, A. and Faccoli, M. (2014b) 'Trapping wood-boring beetles in Italian ports: A pilot study', *Journal of Pest Science*, vol. 87, pp. 61–69.

Rejmánek, M. (2015) 'Biological invasions in forests and forest plantations', in K.S.-H. Peh, R.T. Corlett and Y. Bergeron (eds.) *Routledge Handbook of Forest Ecology*. Routledge, Abingdon, UK.

Rejmánek, M. and Richardson, D.M. (2013) 'Trees and shrubs as invasive alien species 2013 update of the global database', *Diversity and Distributions*, vol. 19, pp. 1093–1094.

Rejmánek, M., Richardson, D.M. and Pyšek, P. (2005) 'Plant invasions and invasibility of plant communities', in E. van der Maarel (ed.) *Vegetation Ecology*. Blackwell Publishing, Chichester, UK.

Richardson, D.M. (1998) 'Forestry trees as invasive aliens', *Conservation Biology*, vol. 12, pp. 18–26.

Richardson, D.M., Pyšek, P., Rejmánek, M., Barbour, M.G., Panetta, F.D. and West, C.J. (2000) 'Naturalization and invasion of alien plants: Concepts and definitions', *Diversity and Distributions*, vol. 6, pp. 93–107.

Richardson, D.M. and Rejmánek, M. (2004) 'Conifers as invasive aliens: A global survey and predictive framework', *Diversity and Distributions*, vol. 10, pp. 321–331.

Richardson, D.M. and Rejmánek, M. (2011) 'Trees and shrubs as invasive alien species: A global review', *Diversity and Distributions*, vol. 17, pp. 788–809.

Roques, A., Rabitsch, W., Rasplus, J.Y., Lopez-Vaamode, C., Nentwig, W. and Kenis, M. (2009) 'Alien terrestrial invertebrates of Europe', in DAISIE (ed.) *Handbook of Alien Species in Europe*. Springer, Dordrecht, Netherlands.

Sádlo, J., Vítková, M., Pergl, J. and Pyšek, P. (2017) 'Towards site-specific management of invasive alien trees based on the assessment of their impacts: The case of *Robinia pseudoacacia*', *NeoBiota*, vol. 35, pp. 1–34.

Sakai, A.K., Allendorf, F.W., Holt, J.S., Lodge, D.M., Molofsky, J., With, K.A., Baughman, S., Robert, J.C., Cohen, J.E., Ellstrand, N.C., McCualey, D.E., O'Neil, P., Parker, I.M., Thompson, J.N. and Weller, S.G. (2001) 'The population biology of invasive species', *Annual Review of Ecology and Systematics*, vol. 32, pp. 305–332.

Sala, O.E., Chapin III, F.S., Armesto, J.J., Berlow, E., Bloomfield, J., Dirzo, R., Huber-Sandwald, E., Huenneke, L.F., Jackson, R.B., Kinzig, A., Leemans, R., Lodge, D.M., Mooney, H.A., Oesterheld, M., LeRoy Poff, N., Sykes, M.T., Walker, B.H., Walker, M. and Wall, D.H. (2000) 'Global biodiversity scenarios for the year 2100', *Science*, vol. 287, pp. 1770–1774.

Santini, A. and Battisti, A. (2019) 'Complex insect–pathogen interactions in tree pandemics', *Frontiers in Physiology*, vol. 10, p. 550.

Santini, A., Ghelardini, L., De Pace, C., Desprez-Loustau, M.L., Capretti, P., Chandelier, A., Cech, T., Chira, D., Diamandis, S., Gaitniekis, T., Hantula, J., Holdenrieder, O., Jankovsky, L., Jung, T., Jurc, D., Kirisits, T., Kunca, A., Lygis, V., Malecka, M., Marcais, B., Schmitz, S., Schumacher, J., Solheim, H., Solla, A., Szabò, I., Tsopelas, P., Vannini, A., Vettraino, A.M., Webber, J., Woodward, S. and Stenlid, J. (2013) 'Biogeographical patterns and determinants of invasion by forest pathogens in Europe', *New Phytologist*, vol. 197, pp. 238–250.

Santini, A., Liebhold, A., Migliorini, D. and Woodward, S. (2018) 'Tracing the role of human civilization in the globalization of plant pathogens', *The ISME Journal*, vol. 12, pp. 647–652.

Sax, D.F. (2001) 'Latitudinal gradients and geographic ranges of exotic species: Implications for biogeography', *Journal of Biogeography*, vol. 28, pp. 139–150.

Scalera, R., Cozzi, A., Caccamo, C. and Rossi, I. (2017) *A Catalogue of LIFE Projects Contributing to the Management of Alien Species in the European Union: Platform Meeting on Invasive Alien Species (IAS) 29–30 November 2017, Milan (Italy)*, Regione Lombardia, Milan, Italy.

Schmithüsen, F. and Hirsch, F. (2010) *Private Forest Ownership in Europe*. United Nations Economic Commission for Europe & Food and Agriculture Organization of the United Nations, Geneva, Switzerland.

Schulze-Lefert, P. and Panstruga, R. (2011) 'A molecular evolutionary concept connecting nonhost resistance, pathogen host range, and pathogen speciation', *Trends in Plant Science*, vol. 16, pp. 117–125.

Seidl, R., Schelhaas, M. and Lexer, M.J. (2011) 'Unraveling the drivers of intensifying forest disturbance regimes in Europe', *Global Change Biology*, vol. 17, pp. 2842–2852.

Shackleton, R.T., Le Maitre, D.C., van Wilgen, B.W. and Richardson, D.M. (2017) 'Towards a national strategy to optimise the management of a widespread invasive tree (*Prosopis* species; mesquite) in South Africa', *Ecosystem Services*, vol. 27, pp. 242–252.

Shaw, J.D., Wilson, J.R.U. and Richardson, D.M. (2010) 'Initiating dialogue between scientists and managers of biological invasions', *Biological Invasions*, vol. 12, pp. 4077–4083.

Shine, R. (2010) 'The ecological impact of invasive cane toads (*Bufo marinis*) in Australia', *Quarterly Review of Biology*, vol. 85, pp. 253–291.

Simberloff, D. and Von Holle, B. (1999) 'Positive interactions of non indigenous species: Invasional meltdown?', *Biological Invasions*, vol. 1, pp. 21–32.

Sitzia, T., Campagnaro, T., Dainese, M. and Cierjacks, A. (2012) 'Plant species diversity in alien black locust stands: A paired comparison with native stands across a north-Mediterranean range expansion', *Forest Ecology and Management*, vol. 285, pp. 85–91.

Sitzia, T., Campagnaro, T., Kotze, D.J., Nardi, S. and Ertani, A. (2018) 'The invasion of abandoned fields by a major alien tree filters understory plant traits in novel forest ecosystems', *Scientific Reports*, vol. 8, p. 8410.

Sitzia, T., Campagnaro, T., Kowarik, I. and Trentanovi, G. (2016a) 'Using forest management to control invasive alien species: Helping implement the new European regulation on invasive alien species', *Biological Invasions*, vol. 18, pp. 1–7.

Sitzia, T., Campagnaro, T. and Weir, R.G. (2016b) 'Novel woodland patches in a small historical Mediterranean city: Padova, Northern Italy', *Urban Ecosystems*, vol. 19, pp. 475–487.

Skarpaas, O. and Økland, B. (2009) 'Timber import and the risk of forest pest introductions', *Journal of Applied Ecology*, vol. 46, pp. 55–63.

Smith, J.G. and Phillips, B.L. (2006) 'Toxic tucker: The potential impact of Cane Toads on Australian reptiles', *Pacific Conservation Biology*, vol. 12, pp. 40–49.

Stinson, K.A., Campbell, S.A., Powell, J.R., Wolfe, B.E., Callaway, R.M., Thelen, G.C., Hallett, S.G., Prati, D. and Klironomos, J.N. (2006) 'Invasive plant suppresses the growth of native tree seedlings by disrupting belowground mutualisms', *PLoS Biology*, vol. 4, p. e140.

Streets, R.J. (1962) *Exotic Forest Trees in the British Commonwealth*. Clarendon Press, Oxford, UK.

Syed, Z. and Guerin, P.M. (2004) 'Tsetse flies are attracted to the invasive plant Lantana camara', *Journal of Insect Physiology*, vol. 50, pp. 43–50.

Terwei, A., Zerbe, S., Mölder, I., Annighöfer, P., Kawaletz, H. and Ammer, C. (2016) 'Response of floodplain understorey species to environmental gradients and tree invasion: A functional trait perspective', *Biological Invasions*, vol. 18, pp. 2951–2973.

Thomas, N., Morrison, C., Winder, L. and Morley, C. (2011) 'Spatial distribution of co-occurring vertebrate species: Case study of an endangered frog and an introduced toad in Fiji', *Pacific Conservation Biology*, vol. 17, pp. 68–77.

Trentanovi, G., Campagnaro, T., Rizzi, A. and Sitzia, T. (2018) 'Synergies of planning for forests and planning for Natura 2000: Evidences and prospects from northern Italy', *Journal for Nature Conservation*, vol. 43, pp. 239–249.

United Nations (2015) Transforming our world: The 2030 agenda for sustainable development. A/RES/70/1. United Nations, New York.

van Wilgen, B.W. and Richardson, D.M. (2014) 'Challenges and trade-offs in the management of invasive alien trees', *Biological Invasions*, vol. 16, pp. 721–734.

Vilà, M. and Ibáñez, I. (2011) 'Plant invasions in the landscape', *Landscape Ecology*, vol. 26, pp. 461–472.

Vítková, M., Müllerová, J., Sádlo, J., Pergl, J. and Pyšek, P. (2017) 'Black locust (*Robinia pseudoacacia*) beloved and despised: A story of an invasive tree in Central Europe', *Forest Ecology and Management*, vol. 384, pp. 287–302.

Vítková, M., Tonika, J. and Müllerová, J. (2015) 'Black locust: Successful invader of a wide range of soil conditions', *Science of the Total Environment*, vol. 505, pp. 315–328.

Wagner, V., Chytrý, M., Jiménez-Alfaro, B., Pergl, J., Hennekens, S., Biurrun, I., Knollová, I., Berg, C., Vassilev, K., Rodwell, J.S., Škvorc, Ž., Jandt, U., Ewald, J., Jansen, F., Tsiripidis, I., Botta-Dukát, Z., Casella, L., Attorre, F., Rašomavičius, V., Ćušterevska, R., Schaminée, J.H.J., Brunet, J., Lenoir, J., Svenning, J.-C., Kącki, Z., Petrášová-Šibíková, M., Šilc, U., García-Mijangos, I., Campos, J.A., Fernández-González, F., Wohlgemuth, T., Onyshchenko, V. and Pyšek, P. (2017) 'Alien plant invasions in European woodlands', *Diversity and Distributions*, vol. 23, pp. 969–981.

Waring, K.M. and O'Hara, K.L. (2005) 'Silvicultural strategies in forest ecosystems affected by introduced pests', *Forest Ecology and Management*, vol. 209, pp. 27–41.

Webber, B.L., Born, C., Conn, B.J., Hadiah, J.T. and Zalamea, P.C. (2011) 'What is in a name? That which we call *Cecropia peltata* by any other name would be as invasive?', *Plant Ecology & Diversity*, vol. 4, pp. 289–293.

Zahid, M.I., Grgurinovic, C.A. and Walsh, D.J. (2008) 'Quarantine risks associated with solid wood pack-aging materials receiving ISPM-15 treatments', *Australian Forestry*, vol. 71, pp. 287–293.

Zenni, R.D. (2015) 'The naturalized flora of Brazil: A step towards identifying future non-native invasive species', *Rodriguésia*, vol. 66, pp. 1137–1144.

Zenni, R.D., Ziller, S.R., Pauchard, A., Rodriguez-Cabal, M. and Nuñez, M.A. (2017) 'Invasion science in the developing world: A response to Ricciardi et al.', *Trends in Ecology & Evolution*, vol. 32, pp. 807–808.

8

ISLAND ECOSYSTEMS

Qinfeng Guo

Background

Islands have long attracted special attention from both the public and research communities mostly due to curiosities related to their smaller area, isolation and uniqueness (Wu and Vankat, 1995). Islands also have a long reputation of being regarded as excellent 'natural laboratories' for field and experimental studies in population biology, ecology and evolution (MacArthur and Wilson, 1967).

Nevertheless, the unique features of islands also make island ecosystems particularly sensitive to natural and human disturbances. Indeed, while biological invasions of non-native (i.e. exotic, alien, introduced) species pose significant threats to native ecosystems and biomes across the globe, island ecosystems now increasingly suffer greater impacts from ongoing global change than mainland ecosystems (Bellard et al., 2017). For example, for a long time, islands have been perceived to be substantially more vulnerable to species invasions (but see Simberloff, 1995; Sol, 2000; Gimeno et al., 2006).

Relative to studies on continental or mainland areas, a great advantage of studying islands is that they have clear natural boundaries. This is especially important for research in species invasions because the definitions of native, non-native and endemic species are boundary dependent (Guo, 2011). Regardless of species origins, islands offer convenient settings for population-level studies, which usually involve meta-population analysis and distribution-abundance relationships (Morrison, 2002). The former is closely linked to habitat fragmentation (patch occupancy and patch analysis), especially when caused by human activities. The latter has been used to generate species-area relationships, mostly for basic ecology research and conservation purposes.

In invasion-related research, whether islands are more invasible than mainlands remains a central topic and is still heavily debated (Simberloff, 1995; Sol, 2000; Gimeno et al., 2006). However, many invasion and other ecological or evolutionary studies often produce inconclusive and even inconsistent results. The main cause could be due to the fact that most studies to date have only used certain (or certain sets of) islands (not the majority of islands around the globe) and different taxonomic groups. Major changes associated with biotic invasions observed on islands so far include: (1) reduced native species diversity in some extreme cases; (2) biotic homogenisation across multiple scales (Drake et al., 1989; McKinney, 1998); (3) different or novel interspecific interactions such as hybridisation between native and non-natives and also among non-native species; and (4) altered ecological states due to invasion-caused landscape metamorphosis (Fei et al., 2014). Based on the definitions of native vs. non-native species, as the regional species pool considered becomes larger, the proportion of native species in a specific flora will always increase while correspondingly that of non-natives will always decrease (as fewer species are classified as non-native).

In invasion biology, a frequently used method for examining invasion patterns, processes and associated mechanisms is comparing natives and non-natives in the same taxonomic group or habitat. Alongside this, comparing different taxonomic groups in different geographic settings could offer new insights into island invasions. Even within one major taxonomic type such as plants, invasion biologists often examine species-area-isolation relationships and then make meaningful comparisons. To date, researchers have shown that the positive relationship between area and species richness is mostly stronger for natives than for non-natives while the opposite is true for species-isolation relationships (e.g., Blackburn et al., 2008). Also, as expected, the proportion of endemic species that usually suffer greater impacts from invasive species almost always increases with isolation (Moody, 2000).

With accelerated human traffic linking mainlands and islands, many remote islands across the globe are no longer truly isolated but increasingly connected (Gillespie et al., 2008), resulting in many (and some unexpected or disruptive) impacts on island biotas and ecosystems. While islands are under many similar threats to mainland ecosystems, such as pollution and overexploitation (of land and other resources), the focus here is on species invasions.

Islands and island ecosystems

Islands are usually isolated landmasses surrounded by water. The true number of islands across the globe can only be estimated when a specific area is used as a threshold for defining islands (Weigelt et al., 2014). As there is no standard definition for islands in terms of size (area), all landmasses (regardless their origin) across the globe could considered within the 'island-mainland continuum' (Rosenzweig, 1995; Guo et al., 2017). Depending on the contexts and objectives of scrutiny, islands can be classified into different categories (e.g., Irl et al., 2017). In general, there are two types of islands surrounded by water – that is, oceanic islands (formed by volcanic eruption) and continental islands (including those once connected to a continent, i.e., the so-called 'land-bridge islands'). Alternatively, one may classify islands into continental, tidal, barrier, oceanic, coral and artificial islands, among others, including land-bridge islands (Hu et al., 2011).

Globally, although islands (n > 100,000) occupy only 3% – 3.6% of global landmass (Whittaker and Fernández-Palacios, 2007; Kier et al., 2009), about 25% (70,000) of the world's vascular plant species are endemic to islands (Kreft et al., 2008). Partly for this reason, 9 out of the world's 25 most significant/largest biodiversity hotspots include islands or archipelagos (Kreft et al., 2008). Currently, islands around the globe also support more than 600 million people, or about 10% of the world's population, a higher density (thus impact per unit area) than on mainlands (Wong et al., 2005).

In the much broader sense of ecology and evolutionary biology, however, many insular areas or entities on mainlands such as mountain tops, lakes, springs and fragmented habitats (e.g. forest patches) may also be termed as 'islands' as they are usually small, isolated from each other and support unique species assemblages (e.g. Guo, 2015). Although comprehensive studies on islands are still lacking, research to date shows some interesting patterns. For example, islands are small and thus support fewer species but higher endemism (Kier et al., 2009). Competition and predation tend to have a stronger effect on species on islands (Bowen and Vuren, 1997).

Relatively and proportionally, islands tend to have more reptiles and birds and fewer mammals (Kier et al., 2009; Yu et al., 2012).

Because of their smaller area, many species that need larger areas of habitat to survive may not be present on islands; thus the diversity is generally lower relative to mainland habitats with the same area. And because of their isolation (thus lack of or reduced dispersal and gene flow with other biotas), island habitats often develop distinct biotas.

Due to the intertwined effects of area and isolation, plus the interactive effects of other factors such as climate, latitude, island type and human disturbances, making generalisations on ecological and evolutionary patterns across islands remains a significant challenge.

Island vs. mainland habitats

Mainly due to varying island sizes and levels of isolation, not only are ecological communities on oceanic islands different from each other, but they are also very different from those on mainlands. In many aspects, the patterns found on islands and their implications can only be more meaningful and useful when compared with those found on mainland habitats (see next section). Some major features unique to islands include:

(1) It has been found that species-area relationships are different between island and mainland habitats. For example, the slope in the species-area relationships for natives are steeper across islands than on mainlands (Rosenzweig, 1995). Given the same area, islands usually have fewer native species than mainland sites. Islands tend to have higher z in species-area curves ($S = cAz$, where S is the number of species, c is a constant, A is area and z describes the slope in the relationship) than mainland sites (Rosenzweig, 1995).

(2) Because of their long-time isolation relative to mainlands, islands often support more distinctive species assemblages with more rare and endemic species.

(3) Island ecosystems suffer more species extinctions than those on mainlands due to smaller population sizes and reduced opportunities for gene flow with mainlands and among islands. Indeed, an assessment by the World Conservation Monitoring Centre (1992) shows that, before the massive human population growth in the last few decades, almost all reptile, roughly 90% of bird, 79% of mollusk and more than half (59%) mammal extinctions occurred on islands, and many of these extinctions were due to species invasions caused by humans.

(4) Life history strategies often differ between island and mainland species. For example, species with large body sizes on mainlands often have reduced body size on islands (the so-called 'insular dwarfism') while some small-sized species tend to have enlarged body sizes (the so-called 'gigantism'), and both have been commonly described as the 'island rule' (Lomolino et al., 2013).

(5) Island species also tend to exhibit reduced dispersal (Cody and Overton, 1996), increased population density, higher frequency of self-compatibility (Baker's law) and 'slow' life histories (the so-called 'island syndrome'; Novosolov and Meiri, 2013). Among bryophytes, the proportion of bisexual species is higher on oceanic islands than on mainlands (Patino et al., 2013).

In general, taking 'marine islands' as a group, they are cooler (due to many being in Arctic areas), wetter and less seasonal than mainlands mainly because they are surrounded by water (oceans). Past climate change velocity is higher on islands. Also, elevation differences are smaller among islands, and there is a negative relationship between island age and elevation that may reflect erosion over time (Weigelt et al., 2014).

Island biogeography

A truly comprehensive theory of biogeography will treat islands and continents together.
– MacArthur and Wilson (1967), 182

Island biogeography has been a key topic in research related to species diversity and distribution and the underlying ecological processes and mechanisms. A central theme of island biogeography is the theory of island biogeography developed by Robert H. MacArthur and E. O. Wilson (1967) and its implications for application. This was partly based upon the species-area curves reported much earlier in accumulated field observations of many naturalists and ecologists during their explorations around the world (e.g., Preston, 1962; Wu and Vankat, 1995) but jointly considered area with distance (or isolation; Figure 8.1). Their theory effectively integrates ecology, evolution and population biology of island species with broad implications for conservation biology. The theory predicts that the number of species (richness) usually increases with island area but decreases with isolation (i.e., distance from outside species pools; Wu and Vankat, 1991b, 1995). In addition to area and isolation, some recent studies have also adopted other measures to examine the effects of specific island features – such as fragmentation, time, the perimeter to area ratio of the islands and economics – on species diversity (e.g., Wu and Vankat, 1991a; Hu et al., 2011; Helmus et al., 2014).

As an extension of the original theme, the equilibrium island biogeography theory takes both immigration (*I*) and extinction rates (*E*) into account. When *I* and *E* are balanced (*I* = *E*), the species richness on the islands reaches an equilibrium point. According to this theory, *I* decreases while *E* increases with isolation. However, field studies show that an 'equilibrium' point, although theoretically valid, rarely exists and even when achieved would be transient and may not last for long because speciation, extinction, immigration and emigration ('species turnover') take place almost constantly, although not necessary concurrently. At any particular time, it is possible that the diversity on many islands may be in a state 'near equilibrium'. However,

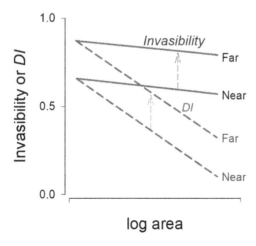

Figure 8.1   Invasibility (solid lines) and degree of invasion (*DI* – dashed lines) by non-native plants are assumed to be higher on remote islands due to the lower native diversity based on the diversity-invasibility theory (Elton, 1958). However, non-native birds often show the opposite patterns, with more isolated islands having more non-native birds due to more intentional introductions.

Source: e.g. Blackburn et al., 2008.

over the long-term, for many reasons such as disturbance and even taxon cycles, diversity on islands varies (often quite significantly), and new (transient) equilibria are being formed and broken. Now with species invasions due to human assistance, the chances for such equilibria in diversity on islands would be even smaller (see the following).

In general, the island biogeography theory concerns (1) the effects of area and isolation and (2) the 'equilibrium' between *I* and *E*. With species introductions to islands by humans, *I* will be higher than *E*, leading to overall diversity increase (Rosenzweig, 1995), at least initially. With so many non-native species on islands, ecologists and biogeographers ask whether the island biogeography theory still applies. More important, the original theory designed to describe native species may be now tested for three kinds of species separately: i.e., natives, non-natives and all species combined.

The isolation of islands from mainlands and each other also provides ideal opportunities and settings for global interactive network analyses based on shared species threatened by invasive species (Bellard et al., 2017). The island biogeography theory makes connections between meta-population dynamics and design of nature reserves for conservation. Initial results show that, although island biogeography theory still applies to native species in most cases, it is possible that island biogeography theory may be less applicable to native species if the current rate of invasion persists or increases in the future (Guo, 2015). This would be more likely the case when species invasions cause extinction of native species.

Small vs. large islands

In addition to the major differences between islands and mainlands, some major differences exist between small and large islands. This information is useful when we make paired comparisons between two islands in similar geographical settings (e.g. latitude, climate) and age. In general, species turnover rate would be higher on small islands mostly because of low diversity. Given the same level of disturbance, small islands would have greater extinction rates partly because the population size of component species is usually smaller.

Similar analogies may be made to both near and remote islands where endemism on remote islands may be higher due to isolation (thus lack of gene flow) from mainland species pools and populations. More isolated islands may be more easily invaded by non-native species due to their lower native species diversity that may lack key functional groups (Daehler, 2006) and low similarity in phylogenetic relatedness between native and non-native species (as shown by higher endemism; Figure 8.1). Also, similar to the island-mainland comparisons, smaller islands usually show greater impacts from human activities than larger islands (Rykken and Farrell, 2018; Figure 8.2).

Biological invasions on global islands

At noted earlier, species invasion is increasingly becoming a major issue for island ecosystems and species conservation. Partly because the island systems could serve as natural laboratories and also partly because non-native species can be better defined on islands due to their clear boundaries, islands have been useful subjects for invasion studies. Relative to mainland regions, a major factor in this process is that isolation which hinders or delays invasions is now no longer an issue due to intentional species introductions by humans. On the other hand, isolation continues to be a major barrier that reduces the chance to rescue a species when it is endangered (Irl et al., 2017).

Accumulated findings in island invasions provide an opportunity to confirm some general patterns. For example, in current conditions, despite the effects of isolation and slopes of the

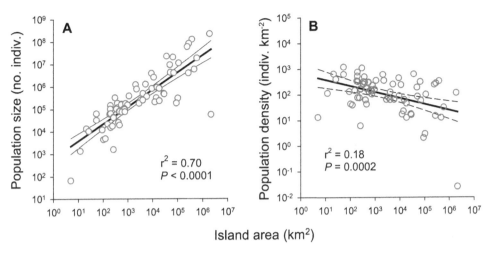

Figure 8.2 Although in general larger islands (*n* = 77) have larger human populations, smaller islands have higher population densities. Similar patterns also exist on mainlands when geopolitical units are used.

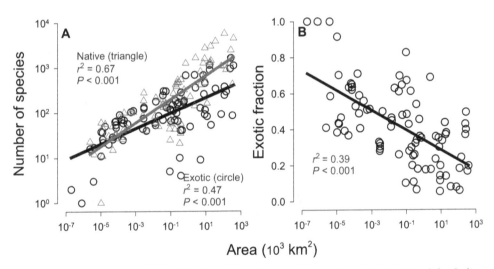

Figure 8.3 The species–area relationships for native and non–native species on islands (A) and the decline in non–native fraction or degree of invasion (*DI*) with island area (B; n = 89).

Source: Modified from Guo et al., 2017.

species–area relationships for natives still being steeper across islands than on mainlands, biotas across global islands, similar to the mainland trends, are increasingly homogenised (Castro et al., 2010). At the present level of invasion, greater than 50% of the total variability in the degree of invasion (*DI*) across the globe can be explained by landmass area alone (Guo et al., 2017).

Human population density tends to be higher on islands, especially smaller islands (Figure 8.2), thus increasing the propagule pressure per unit of area. Given the same area, islands usually have fewer native species than mainland sites; as a result, even if the same number of non–native species have invaded, *DI* on islands tends to be higher. Among islands, those with smaller areas tend to have higher *DI* (Figure 8.3). Because of isolation and small size, island ecosystems are likely to experience greater impacts than those of mainland areas even if *DI* remains the same (Sol, 2000). Thus, islands seem more important for species conservation and are extremely attractive for basic research and conservation.

A few recent studies found no statistically significant difference in the relationships between area and non-native fraction (a measure of *DI*) between islands and mainland areas, supporting the 'island-mainland continuum concept' in terms of species invasions or invasibility (e.g. Guo et al., 2017). In other words, islands may not be more susceptible to invaders than continents (e.g. Simberloff, 1995; Sol, 2000), but rather, because of their smaller size (low native richness), the non-native species pool is larger (Lonsdale, 1999; Guo, 2014), and *DI* (not invasibility) appears to be larger than on mainlands. This is also enhanced by the fact that non-native species are defined more accurately using the clear island boundaries (and are often underestimated on mainland sites based on larger geopolitical [e.g. national] boundaries). In addition, people have a preference for travel to islands for tourism, which leads to higher propagule pressure and so *DI* (Guo, 2014).

Sax and Gaines (2008) reported that richness of non-native plants increased the most on oceanic islands (basically doubled) and the least on mainlands, with land-bridge islands in between. Mammals exhibited similar trends but to a lesser degree. However, the overall richness of birds on islands has not changed much, although species composition has changed due to invasions. Despite the drastic increase of plant species on islands, few extinctions due to non-native plant invasion have been confirmed. In contrast, extinction of many birds from islands have been observed. Such significant inconsistencies and even contradictory patterns in invasions and extinction among taxonomic groups have been frequently reported (Kueffer et al., 2010). Thus, identifying mechanisms for both general patterns and exceptions are urgently needed.

Well-documented examples of heavily invaded islands include Hawaii, New Zealand, Guam and the Galapagos Islands. Frequently mentioned species include: (1) rodents (*Rattus exulands*, *R. norvegicus* and *R. rattus*), which have eliminated more than 50 bird species on about 40 islands; (2) the brown tree snake (*Boiga irregularis*), which led to the extinction of 12 bird species on the Pacific Islands of Guam; (3) pig (*Sus scrofa*) on the California Channel Islands; and (4) introduced cats (*Felis catus*), which have caused the extinction of about 30 island bird species worldwide.

Bowen and Vuren (1997) pointed out that native island plants often are defenceless against introduced herbivores. Because competition and predation (including herbivory) tend to have a stronger effect on island species, island communities suffer greater impacts even if *DI* stays the same as that in similar mainland habitats. However, other factors such as hybridisation and genetic swamping (genetic pollution) are also among the leading causes of native species extinction on islands. The 'island rule' (i.e., dwarfism vs. gigantism; Lomolino et al., 2013) offers baseline information for predicting what kind of species (using traits like body size) may be more likely to successfully invade islands and how their traits might change over time.

Another factor is time, which is critically related to ongoing species invasions. Many islands are still experiencing high rates of species invasion, and thus any observations made today might change in the near future. With the same level of human assistance (i.e., trade, travel), species are more likely to migrate from species-rich mainland regions to species-poor islands.

Previous studies have found that the areas exporting most non-native species are also likely to receive many non-native species (Guo et al., 2006; Turbelin et al., 2017), partly due to the two-way traffic in trade and travel. This is probably the case among islands as well (Turbelin et al., 2017). Similar to natural species migrations that occurred without human intervention, modern species invasions due to human assistance are also most likely to be asymmetrical. This could be the consequence of a greater release from competition experienced in species-rich areas (Grace, 2012; Iannone et al., 2016). Until effective prevention and management are in place, *DI* is likely to increase at multiple scales.

Eradication, management and conservation

As stated previously, due to their isolation and small size, islands are often used as natural laboratories for ecological studies (Whittaker and Fernández-Palacios, 2007). For example, the many examples of the impacts of invasive species on islands provided by Sax and Gaines (2008) offer excellent opportunities for research and management related to eradication of invasive species (Veitch et al., 2011; Glen et al., 2013).

Indeed, many attempts/efforts have been made in this regard, and there have been some successful invasive species eradications from islands (Howald et al., 2007; Glen et al., 2013; Bellard et al., 2017). In some cases, endemic flora and fauna that are on the verge of extinction have immediately started to recover within a short period of time after eradication of invasive species (Nogales et al., 2004; Jones et al., 2016). However, the 'sudden' removal of dominant invasive species can lead to surprising consequences (Bergstrom et al., 2009; Caut et al., 2009), especially once the post-invasion ecosystems have evolved or become stabilised. Clearly, the entire network of species interactions, especially those that involve multi-trophic levels and may therefore lead to cascading effects once the most dominant invasive species are removed, need to be taken into account before eradication is attempted (Caut et al., 2009). Carefully designed restoration, close monitoring and adaptive management are thus needed after target invasive species have been removed (Glen et al., 2013).

Related comparative studies

New insights and improved understanding of species invasions on islands are increasing worldwide. Accumulated knowledge in island studies naturally leads to increased comparative studies among: (1) islands or island groups (archipelagoes), (2) island types (e.g. continental vs. oceanic), (3) organism or taxonomic groups (e.g. plants vs. animals or birds vs. mammals) and (4) latitudes or climate zones (tropical vs. temperate), among others. Temporally, one could also compare between young and old islands or across other time scales. Often, such data are already available, and new analysis can produce explanations or clarification for inconsistent results from individual original work (Daehler, 2006).

Comparative studies between native vs. non-native species offer excellent opportunities for better understanding of invasion history, patterns and underlying mechanisms (Leihy et al., 2018). For example, recent studies have found both positive (e.g. Blackburn et al., 2008) and negative isolation-non-native richness correlations (Long et al., 2009). These studies also found inconsistent areal effects on non-native richness and *DI*. Such patterns are likely due to the fact that, relative to natural dispersal/migration of native species, the history of species introductions by humans is too short and not balanced across islands (also among world regions) and taxonomic groups (Yu et al., 2012).

Recent studies show that island biogeography theory can still better describe the patterns (species-area and species-isolation relationships) for native species and overall diversity (natives plus non-natives) than for non-natives alone (e.g. Guo, 2014). Patterns for non-native species are highly variable among islands or island groups and depend on the level of human disturbance. Native and non-native species may show different patterns across the same (sets of) islands. Native species are more likely to follow the general theory of island biogeography (MacArthur and Wilson, 1967; Whittaker and Fernández-Palacios, 2007) because they have had enough time to colonise the islands and form more or less 'equilibrium' points with island conditions. For example, Blackburn et al. (2008) and Long et al. (2009) reported positive species-area relationships for both native and non-native species. When all species are combined in such considerations, the species-area relationships are often stronger (Figure 8.2).

Reasons for the differences among taxonomic groups are multifold. First, different taxonomic groups may have different pathways of introductions. For example, humans often intentionally introduce non-native birds to remote oceanic islands where native bird richness is low (Blackburn et al., 2008). This is usually not the case for plants, which can also experience natural dispersal from nearby regions/islands (Long et al., 2009). Second, the different dispersibility among taxonomic groups is also responsible for the major differences in invasion patterns. For example, although no dispersal was found for non-native birds among the remotely isolated islands examined in Blackburn et al. (2008), significant movement of viable seeds among islands in Boston Harbor was found after initial introduction to the New England region (i.e., 'secondary dispersal'; Long et al., 2009).

Different sets of islands chosen for study would also be likely to exhibit different patterns depending on their particular contexts such as location, geographic arrangement, number, size and isolation. This is partly because the level of human influences and spatial autocorrelation may be significant factors that determine *DI* or non-native richness. In such cases, natives, non-natives and their combinations should be considered separately and also jointly when applying island biogeography theory.

Based on many studies that have compared *DI*s and invasibility across islands and/or island groups (Daehler, 2006), recent attention has been increasingly extended to include many other aspects related to invasions. Some studies have closely examined the composition of the non-native species, especially what specific damage and function a particular species group may have in the introduced regions and habitats. For example, among the numerous introduced plant species in Hawaii, many are found to be fire-promoting grasses that can increase fire hazards. Also, colonisation by humans causes many disturbances that native species cannot adapt to (e.g. changes in land use; introduced pests/diseases; and animals such as pigs, cats, rats and dogs, among others; Cox, 2004; Bellard et al., 2017).

Recent comparative studies on island invasions have important implications. In general, species invasions on islands lead to higher overall diversity, although more time is needed to see whether the increased diversity may be maintained or if it is only transitional (i.e. exists in the near term; Sax and Gaines, 2008). It would also be interesting to see whether the non-native species could increase overall ecosystem functionality on islands or lead to the extinction of native species.

Currently, due to logistical convenience and costs related to distance, most island studies have been on those islands that are close to continents and over a relatively short period of time. Most studies have compared different sets of islands, leading to biased observations as many remote and/or large islands remain underexplored, thus hindering comparative island studies globally.

Long-term trends towards global homogenisation

At the present level of invasion, and on some highly visited islands such as Hawaii and New Zealand, the number of non-native species have either approached or already exceeded that of native species (Guo et al., 2018). On the global level, including most islands, non-native species are still mostly distributed at lower elevations (Figure 8.4; see also Guo et al., 2018). However, similar to the situation on mainland sites, due to increasing human population and associated expansion along elevation gradients, human activities are shifting upward from lower elevations (Small and Cohen, 2004) onto high elevations (island tops) on islands due to tourism and activities such as mining. The spread of invasive species is likely to follow this trend (Pauchard et al., 2009). As high elevations usually mean smaller areas, with climate warming

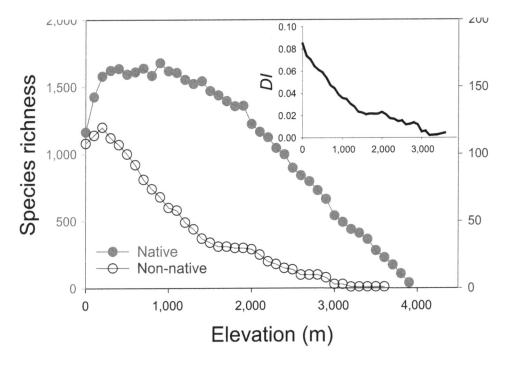

Figure 8.4 Invasions of non-native plants on the island of Taiwan as an example of elevational invasion patterns. Note the different scales used for native (solid circles) vs. non-native (open circles) plant species (*DI* = degree of invasion).

Source: Data from Shen et al. (2017), available in the Purdue University Research Repository data archive (https://purr.purdue.edu/).

(McDougall et al., 2011), the upward shifts of invasive species may further reduce the space/resources for native species.

Compared to most ecological studies in which the observed patterns are likely to change over time and time lags are a major causal factor, the importance of time in species invasions is more important and the patterns are much more sensitive. This is because species invasions directly or indirectly caused by humans are occurring at much faster rates than natural species dispersal or migration. Thus, the presently observed patterns only constitute a snapshot of ongoing processes and are likely to be fairly different in the near future. Long-term repeated observations and experimental studies are needed to monitor invasions across island ecosystems. In general, prevention, close monitoring and early detection/eradication would still be the most efficient management options to reduce the impacts of species invasions on islands.

Challenges and perspectives

Many challenging and important questions remain to be answered given ongoing climate change and species invasions on islands around the world. Examples of research questions and gaps that need to be addressed to better understand island invasions include:

(1) Which islands might show species saturation due to species invasions? Unlike large continental areas, species saturation could actually occur on islands, especially the small ones.

(2) How speciation/extinction rates might be related to island age, area and isolation in both direct and indirect ways such as through edge and microclimatic effects.

(3) Are species-rich islands (e.g., tropical) more difficult to invade?

(4) With similar levels of propagule pressure, what species (traits) are likely to successfully invade islands (and what kind of islands)?

(5) How do characteristics such as island age, latitude, elevation, type (continental vs. oceanic vs. volcanic), isolation (absolute and relative: to mainlands only vs. to other islands) and context (e.g. geographic locations; oceanic vs. inland islands such as those in large lakes) affect observational results of studies?

(6) How patterns change with time (e.g. succession, taxon cycle), species invasions (especially when native richness is affected) and human activities.

(7) How global change affects islands: for example, climate warming and especially sea-level rise reduces island size (area) and so may push resident species upward and may exacerbate the 'small island effect'.

(8) How to take early actions to effectively reduce further introduction of invasive species onto islands and how to control and even eradicate established invasive species on islands? Any related action to be taken should take the advantage of the natural boundary and isolation of islands.

At present, island biogeography theory continues to play a key role in both basic research and native species conservation. It is clear that, at current level of invasions, the theory of island biogeography does not apply to non-native species in many cases due to human intervention. However, with continuing species invasions, it is possible that at small scales, such as in smaller islands, species saturation could occur. It would be useful to test when and where the theory of island biogeography may be applied to non-native species and whether species invasions have affected the applicability of the theory for native species. To do this, long-term studies and monitoring across a large sample of islands would be needed (Guo, 2014).

The observed differences in life history, ecology and evolution between island and mainland species and between populations of the same species on islands vs. mainlands are critically important for predicting which species may successfully invade islands and what may happen over time after introduction. Such information can help guide prevention and management efforts.

Given that many kinds of organisms are still being introduced to islands from mainlands, two pressing questions remain for future studies: (1) while the mechanisms of invasion on islands appear to be similar to those on mainlands, what particular species or functional groups (e.g. large: plants vs. mammals; small: herbs vs. forbs) are likely to succeed on islands (and which species may become extinct because of this; Sax and Gaines, 2008) and (2) which groups of species are likely to go extinct under species invasions of either the same (competition) or different kinds of organisms (e.g. predation). Some of the issues could be resolved by re-examining and comparing the species-area curves for different groups of species and their combinations over time. Furthermore, possible asymmetrical invasions between islands need to be monitored and the underlying forces need to be explored.

Conclusions

Human assistance, either intentional or unintentional, causes many non-native species to overcome natural barriers to colonisation and migration. One of the consequences is more homogenised biotas across the globe. There is no sign that the rate of species introduction across the globe is declining. In some heavily invaded habitats, native species extinctions are evident, leading to a loss of global biodiversity. Observational results across islands (or groups of islands) at roughly the same time and on the same islands across years offer new insights into invasion patterns, processes and mechanisms. Because of the dynamic nature of ongoing species invasions, searching for general patterns and broad applications of island biogeography theory in island invasions is difficult. Previous studies show that patterns of invasion on islands depend on the geographic context, human influences, history and pathways of species introductions. For example, the types of islands (e.g. oceanic vs. continental, large vs. small) as well as the number and locations of selected islands for study may affect the results (Blackburn et al., 2008; Bellard et al., 2017). For this reason, however, comparisons among case studies (also between islands and mainlands and between island types) can offer new insights into the relative roles of geographic and human contexts in island invasions. To take advantage of the isolation features of islands, both research and management should make more efforts to remove invasive species, probably starting with the most isolated islands (as they have the lowest possible reinvasion due to natural dispersal).

Acknowledgements

I thank M. Scobie and D. Sovilla of the University of North Carolina at Asheville and several others who assisted with data collection on island invasions. Robert Francis and Jianguo Wu provided constructive comments and suggestions. This study was partly supported by NSF Macrosystems Biology grants (DEB-1241932 and DEB-1638702).

References

Bellard, C., Rysman, J.F., Leroy, B., Claud, C. and Mace, G.M. (2017) 'A global picture of biological invasion threat on islands', *Nature Ecology and Evolution*, vol. 1, pp. 1862–1869.

Bergstrom, D.M., Lucieer, A., Kiefer, K., Wasley, J., Belbin, L., Pedersen, T.K. and Chown, S.L. (2009) 'Indirect effects of invasive species removal devastate World Heritage Island', *Journal of Applied Ecology*, vol. 46, pp. 73–81.

Blackburn, T.M., Cassey, P. and Lockwood, J. (2008) 'The island biogeography of exotic bird species', *Global Ecology and Biogeography*, vol. 17, pp. 246–251.

Bowen, L. and Vuren, D.V. (1997) 'Insular endemic plants lack defenses against herbivores', *Conservation Biology*, vol. 11, pp. 1249–1254.

Castro, S.A., Daehler, C.C., Silva, L., Torres-Santana, C.W., Reyes-Betancort, J.A., Atkinson, R., Jaramillo, P., Guezou, A. and Jaksic, F.M. (2010) 'Floristic homogenization as a teleconnected trend in oceanic islands', *Diversity and Distributions*, vol. 16, pp. 902–910.

Caut, S., Angulo, E. and Courchamp, F. (2009) 'Avoiding surprise effects on Surprise Island: Alien species control in a multitrophic level perspective', *Biological Invasions*, vol. 11, pp. 1689–1703.

Cody, M.L. and Overton, J.M.C. (1996) 'Short-term evolution of reduced dispersal in island plant populations', *Journal of Ecology*, vol. 84, pp. 53–61.

Cox, G.W. (2004) *Alien Species and Evolution: The Evolutionary Ecology of Exotic Plants, Animals, Microbes, and Interacting Native Species*. Island Press, Washington, DC.

Daehler, C.C. (2006) 'Invasibility of tropical islands by introduced plants: Partitioning the influence of isolation and propagule pressure', *Preslia*, vol. 78, pp. 389–404.

Drake, J.A., Mooney, H.A., Di Castri, F., Groves, R.H., Kruger, F.J., Rejmánek, M. and Williamson, M. (1989) *Biological Invasions: A Global Perspective*. J. Wiley, New York.

Elton, C.S. (1958) *The Ecology of Invasions by Plants and Animals*. Methuen, London.

Fei, S., Phillips, J. and Shouse, M. (2014) 'Biogeomorphic impacts of invasive species', *Annual Review of Ecology, Evolution, and Systematics*, vol. 45, pp. 69–87.

Gillespie, R.G., Claridge, E.M. and Roderick, G.K. (2008) 'Biodiversity dynamics in isolated island communities: Interaction between natural and human-mediated processes', *Molecular Ecology*, vol. 17, pp. 45–57.

Gimeno, I., Vila, M. and Hulme, P.E. (2006) 'Are islands more susceptible to plant invasion than continents? A test using Oxalis pes-caprae L. in the western Mediterranean', *Journal of Biogeography*, vol. 33, pp. 1559–1565.

Glen, A.S., Atkinson, R., Campbell, K.J., Hagen, E., Holmes, N.D., Keitt, B.S., Parkes, J.P., Saunders, A., Sawyer, J. and Torres, H. (2013) 'Eradicating multiple invasive species on inhabited islands: The next big step in island restoration?', *Biological Invasions*, vol. 15, pp. 2589–2603.

Grace, J. (2012) *Perspectives on Plant Competition*. Elsevier, Amsterdam.

Guo, Q. (2011) 'Counting "exotics"', *NeoBiota*, vol. 9, p. 71.

Guo, Q. (2014) 'Species invasions on islands: Searching for general patterns and principles', *Landscape Ecology*, vol. 29, pp. 1123–1131.

Guo, Q.F. (2015) 'Island biogeography theory: Emerging patterns and human effects', *Earth Systems and Environmental Sciences*, vol. 32, pp. 1–5.

Guo, Q.F., Fei, S., Shen, Z., Iannone, B.V. III, Knott, J. and Chown, S.L. (2018) 'A global analysis of elevational distribution of non-native vs. native plants', *Journal of Biogeography*, vol. 45, pp. 793–803.

Guo, Q.F., Iannone, B.V. III, Nunez-Mir, G.C., Potter, K.M., Oswalt, C.M. and Fei, S. (2017) 'Species pool, human population, and global versus regional invasion patterns', *Landscape Ecology*, vol. 32, pp. 229–238.

Guo, Q.F., Qian, H., Ricklefs, R.E. and Xi, W.M. (2006) 'Distributions of exotic plants in eastern Asia and North America', *Ecology Letters*, vol. 9, pp. 827–834.

Helmus, M.R., Mahler, D.L. and Losos, J.B. (2014) 'Island biogeography of the Anthropocene', *Nature*, vol. 513, pp. 543–546.

Howald, G., Donlan, C.J., Galvan, J.P., Russell, J.C., Parkes, J., Samaniego, A., Wang, Y., Veitch, D., Genovesi, P. and Pascal, M. (2007) 'Invasive rodent eradication on islands', *Conservation Biology*, vol. 21, pp. 1258–1268.

Hu, G., Feeley, K.J., Wu, J., Xu, G. and Yu, M. (2011) 'Determinants of plant species richness and patterns of nestedness in fragmented landscapes: Evidence from land-bridge islands', *Landscape Ecology*, vol. 26, pp. 1405–1417.

Iannone, B.V., Potter, K.M., Hamil, K.A.D., Huang, W., Zhang, H., Guo, Q.F., Oswalt, C.M., Woodall, C.W. and Fei, S.L. (2016) 'Evidence of biotic resistance to invasions in forests of the Eastern USA', *Landscape Ecology*, vol. 31, pp. 85–99.

Irl, S.D., Schweiger, A.H., Medina, F.M., Fernández-Palacios, J.M., Harter, D.E., Jentsch, A., Provenzale, A., Steinbauer, M.J. and Beierkuhnlein, C. (2017) 'An island view of endemic rarity: Environmental drivers and consequences for nature conservation', *Diversity and Distributions*, vol. 23, pp. 1132–1142.

Jones, H.P., Holmes, N.D., Butchart, S.H., Tershy, B.R., Kappes, P.J., Corkery, I., Aguirre-Muñoz, A., Armstrong, D.P., Bonnaud, E. and Burbidge, A.A. (2016) 'Invasive mammal eradication on islands results in substantial conservation gains', *Proceedings of the National Academy of Sciences*, vol. 113, pp. 4033–4038.

Kier, G., Kreft, H., Lee, T.M., Jetz, W., Ibisch, P.L., Nowicki, C., Mutke, J. and Barthlott, W. (2009) 'A global assessment of endemism and species richness across island and mainland regions', *Proceedings of the National Academy of Sciences*, vol. 106, pp. 9322–9327.

Kreft, H., Jetz, W., Mutke, J., Kier, G. and Barthlott, W. (2008) 'Global diversity of island floras from a macroecological perspective', *Ecology Letters*, vol. 11, pp. 116–127.

Kueffer, C., Daehler, C.C., Torres-Santana, C.W., Lavergne, C., Meyer, J.-Y., Otto, R. and Silva, L. (2010) 'A global comparison of plant invasions on oceanic islands', *Perspectives in Plant Ecology, Evolution and Systematics*, vol. 12, pp. 145–161.

Leihy, R.I., Duffy, G.A. and Chown, S.L. (2018) 'Species richness and turnover among indigenous and introduced plants and insects of the Southern Ocean Islands', *Ecosphere*, vol. 9, p. e02358.

Lomolino, M. V., van der Geer, A.A., Lyras, G.A., Palombo, M.R., Sax, D.F. and Rozzi, R. (2013) 'Of mice and mammoths: Generality and antiquity of the island rule', *Journal of Biogeography*, vol. 40, pp. 1427–1439.

Long, J.D., Trussell, G.C. and Elliman, T. (2009) 'Linking invasions and biogeography: Isolation differentially affects exotic and native plant diversity', *Ecology*, vol. 90, pp. 863–868.

Lonsdale, W.M. (1999) 'Global patterns of plant invasions and the concept of invasibility', *Ecology*, vol. 80, pp. 1522–1536.

MacArthur, R.H. and Wilson, E.O. (1967) *The Theory of Island Biogeography*. Princeton University Press, Princeton, NJ.

McDougall, K.L., Khuroo, A.A., Loope, L.L., Parks, C.G., Pauchard, A., Reshi, Z.A., Rushworth, I. and Kueffer, C. (2011) 'Plant invasions in mountains: Global lessons for better management', *Mountain Research and Development*, vol. 31, pp. 380–387.

McKinney, M.L. (1998) 'On predicting biotic homogenization: Species-area patterns in marine biota', *Global Ecology and Biogeography Letters*, vol. 7, pp. 297–301.

Moody, A. (2000) 'Analysis of plant species diversity with respect to island characteristics on the Channel Islands, California', *Journal of Biogeography*, vol. 27, pp. 711–723.

Morrison, L.W. (2002) 'Island biogeography and metapopulation dynamics of Bahamian ants', *Journal of Biogeography*, vol. 29, pp. 387–394.

Nogales, M., Martín, A., Tershy, B.R., Donlan, C.J., Veitch, D., Puerta, N., Wood, B. and Alonso, J. (2004) 'A review of feral cat eradication on islands', *Conservation Biology*, vol. 18, pp. 310–319.

Novosolov, M. and Meiri, S. (2013) 'The effect of island type on lizard reproductive traits', *Journal of Biogeography*, vol. 40, pp. 2385–2395.

Patino, J., Bisang, I., Hedenäs, L., Dirkse, G., Bjarnason, A.H., Ah-Peng, C. and Vanderpoorten, A. (2013) 'Baker's law and the island syndromes in bryophytes', *Journal of Ecology*, vol. 101, pp. 1245–1255.

Pauchard, A., Kueffer, C., Dietz, H., Daehler, C.C., Alexander, J., Edwards, P.J., Arevalo, J.R., Cavieres, L.A., Guisan, A., Haider, S., Jakobs, G., McDougall, K., Millar, C.I., Naylor, B.J., Parks, C.G., Rew, L.J. and Seipel, T. (2009) 'Ain't no mountain high enough: Plant invasions reaching new elevations', *Frontiers in Ecology and the Environment*, vol. 7, pp. 479–486.

Preston, F.W. (1962) 'The canonical distribution of commonness and rarity: Part I', *Ecology*, vol. 43, pp. 185–215.

Rosenzweig, M.L. (1995) *Species Diversity in Space and Time*. Cambridge University Press, Cambridge.

Rykken, J.J. and Farrell, B.D. (2018) 'Six-legged colonists: The establishment and distribution of non-native beetles in Boston Harbor Islands NRA', *Northeastern Naturalist*, vol. 25, pp. 1–22.

Sax, D.F. and Gaines, S.D. (2008) 'Species invasions and extinction: The future of native biodiversity on islands', *Proceedings of the National Academy of Sciences of the United States of America*, vol. 105, pp. 11490–11497.

Shen, Z., Guo, Q. and Fei, S. (2017) 'A database of elevational distribution of non-native plants across 11 mountains in China', *Purdue University Research Repository*. doi: 10.4231/R7610XHR.

Simberloff, D. (1995) 'Why do introduced species appear to devastate islands more than mainland areas?', *Pacific Science*, vol. 49, pp. 87–97.

Small, C. and Cohen, J.E. (2004) 'Continental physiography, climate, and the global distribution of human population', *Current Anthropology*, vol. 45, pp. 269–277.

Sol, D. (2000) 'Are islands more susceptible to be invaded than continents? Birds say no', *Ecography*, vol. 23, pp. 687–692.

Turbelin, A.J., Malamud, B.D. and Francis, R.A. (2017) 'Mapping the global state of invasive alien species: Patterns of invasion and policy responses', *Global Ecology and Biogeography*, vol. 26, pp. 78–92.

Veitch, C., Clout, M. and Towns, D. (2011) *Island Invasives: Eradication and Management*. IUCN, Gland, Switzerland.

Weigelt, P., Jetz, W. and Kreft, H. (2014) 'Bioclimatic and physical characterization of the world's islands', *Proceedings of the National Academy of Sciences*, vol. 111, pp. 18400–18400.

Whittaker, R.J. and Fernández-Palacios, J.M. (2007) *Island Biogeography: Ecology, Evolution, and Conservation*. Oxford University Press, Oxford.

Wong, P.P., Marone, E., Lana, P., Fortes, M., Moro, D., Agard, J., Vicente, L., Thonell, J., Deda, P. and Mulongoy, J. (2005) 'Island systems', in R. Hassan, R. Scholes and N. Ash (eds.) *Millennium Ecosystem Assessment*. Island Press, Washington, DC, pp. 663–680.

World Conservation Monitoring Centre (1992) *Global Biodiversity: Status of the Earth's Living Resources*. Chapman and Hall, London, England.

Wu, J. and Vankat, J.L. (1991a) 'An area-based model of species richness dynamics of forest islands', *Ecological Modelling*, vol. 58, pp. 249–271.

Wu, J. and Vankat, J.L. (1991b) 'A system dynamics model of island biogeography', *Bulletin of Mathematical Biology*, vol. 53, pp. 911–940.

Wu, J. and Vankat, J.L. (1995) 'Island biogeography: Theory and applications', *Encyclopedia of Environmental Biology*, vol. 2, pp. 371–379.

Yu, M., Hu, G., Feeley, K.J., Wu, J. and Ding, P. (2012) 'Richness and composition of plants and birds on land-bridge islands: Effects of island attributes and differential responses of species groups', *Journal of Biogeography*, vol. 39, pp. 1124–1133.

9

MARINE AND COASTAL ECOSYSTEMS

Rebecca J. Giesler and Elizabeth J. Cottier-Cook

Introduction

The introduction of novel organisms is contributing to global change in marine systems (Carlton, 1996; Occhipinti-Ambrogi, 2007). The study of biological invasions has been primarily concentrated in terrestrial and freshwater systems, although rapid expansion in the study of marine species introductions has occurred in the last decade (Chan and Briski, 2017). Marine invasive species have been documented across all continents, including in the polar regions (Chan et al., 2018). Species introduced range from fish to seaweeds, sedentary invertebrates to microorganisms, and can have a wide range of impacts in their introduced regions. While the majority of introduced non-native species (NNS) cause little impact to their recipient marine environment, some go on to cause significant negative impacts to eco-systems and the services and industries they support (Molnar et al., 2008; Roy et al., 2014; Cook et al., 2016).

Globalisation, and the resulting growth of transoceanic trade and maritime activities, has resulted in increasing numbers of species introductions into the marine environment (Streftaris et al., 2005; Meyerson and Mooney, 2007; Seebens et al., 2013; Seebens et al., 2017). Advances in international shipping have reduced transport times, increased ballast water movement and escalated the potential for NNS introduction (Levine and D'Antonio, 2003; Hulme, 2009; Clarke Murray et al., 2012). Accompanying this growth in transportation activities is a rising demand for the oceans to provide solutions to food and energy supply shortages and to sup-port growing coastal populations (Bulleri and Chapman, 2010). The development of maritime industries to meet these demands has resulted in the construction of a wide range of support-ing infrastructure, from shoreline modifications to offshore installations. This infrastructure includes harbours, ports, coastal defence structures, aquaculture farms, artificial reefs, renew-able energy devices and oil and gas platforms (Sheehy and Vik, 2010; Mineur et al., 2012; Miller et al., 2013). These artificial structures have been shown to facilitate the establishment and spread of NNS by providing hard substrate for colonisation, increasing connectivity and creating corridors for the spread of species (Bulleri and Chapman, 2010; Mineur et al., 2012; Johnston et al., 2017). This chapter covers the ways in which human activity can intentionally or unintentionally move marine organisms around the globe. It goes on to discuss the potential impact of these species in their recipient environments and the challenges of management and control in marine systems.

Spread of marine non-native species

Mobile vectors

In the marine environment, the main pathways of species transfer to new regions are via mobile vectors, primarily commercial shipping and aquaculture (Gollasch, 2006; Minchin et al., 2013; Seebens et al., 2013). Vectors associated with shipping include the release of organisms contained within ballast water, the entrapment of species within sea chests and the attachment of fouling species on ship hulls (Gollasch, 2002; Seebens et al., 2013; Figure 9.1). The discharge of ballast water at different ports has been implicated in the global transfer of a large number of aquatic species ranging from unicellular organisms to fish (David and Gollasch, 2018). Ballast water and hull fouling transference of organisms via shipping have been identified as being responsible for over half of the nearly 1,400 marine and brackish NNS introductions into European waters (Gollasch et al., 2007; David and Gollasch, 2018). Ballast water transfer is not limited to solely marine species, as many global ports are located in estuaries; thus, euryhaline species are frequently introduced through ballast water exchange (Leppäkoski and Olenin, 2000; Paavola et al., 2005). Estuaries containing large ports are among the most invaded global systems, with the concentration of large volumes of shipping traffic contributing to high rates of species introductions.

The role of small private and commercial vessels in transporting species, primarily via hull fouling, has become increasingly recognised in the last decade (Minchin et al., 2006; Clarke Murray et al., 2011; Ashton et al., 2014). Studies of the hulls on recreational yachts have shown that these vessels may have high levels of biofouling, including NNS, especially if anti-fouling

Figure 9.1 Biofouling growing attached to a vessel hull in Scotland.

Source: Photograph by C. Beveridge.

paint is not maintained or the vessel remains stationary for prolonged periods of time (Ashton et al., 2006). Niche areas, where application of anti-fouling paint is difficult, such as propeller shafts or keels, may be especially susceptible to colonisation by NNS (Ashton et al., 2006; Clarke Murray et al., 2011; Peters et al., 2017). The frequent movement of smaller vessels between larger ports and marinas in more remote areas may be an important vector of intraregional spread, especially for organisms with restricted natural dispersal ability (Acosta and Forrest, 2009; Davidson et al., 2010; Zabin et al., 2014). Marinas may thus act as entry points of NNS secondary introduction to a region, supporting the establishment of NNS, which may then spread into natural systems (Floerl and Inglis, 2005; López-Legentil et al., 2015; Ferrario et al., 2017). Less well studied are the fouling communities on slower moving barges and supply vessels, although their role in connecting offshore infrastructure with heavily invaded port areas could also be an important pathway for the secondary spread of NNS (Godwin, 2003; Mineur et al., 2012).

Aquaculture is another important driver of NSS introduction. Aquaculture production around the world has grown significantly in recent decades. The cultivation of fish, shellfish and seaweed has involved the use of non-native species for over 50 years (Minchin, 2007). The movement of species for aquaculture has resulted in non-native introductions both as a result of wild escapes of stock species and the unintentional release of species used as packing material or as epibionts or parasites of aquaculture species (Minchin, 2007; Williams and Smith, 2007). For example, the cultivation of the Pacific oyster, *Magallana gigas*, has not only resulted in the establishment of wild populations in some regions but also the introduction of species, such as the invasive wireweed, *Sargassum muticum* (Critchley, 1983; Figure 9.2). *Sargassum muticum* is

Figure 9.2 A tidal pool in the Isle of Man dominated by *Sargassum muticum*. The long strands of *S. muticum* can be seen floating on the surface of the pool.

Source: Photo by J. Giesler.

thought to have been transported first to the West Coast of North America and then to western Europe in this manner (Critchley, 1983).

There are a wide variety of other examples of the active movement of species, which may result in the introduction of novel species via accidental or intentional releases. This includes the transport of species for use in aquariums or as part of the live food or bait trade (Padilla and Williams, 2004; Passarelli and Pernet, 2019). For example, in the US, marine bloodworms are used as bait by recreational fishermen. The worms are collected from intertidal mudflats in Maine before being packed in 'wormweed' (*Ascophyllum nodosum*) and transported around the country. Numerous epibiotic organisms have been found to be transported with the wormweed and bait, which are often discarded after use, resulting in the potential introduction of a wide range of species (Fowler et al., 2016).

Static structures

The role of vessels and aquaculture as pathways of active species transfer is well recognised, but more recent work has identified that static infrastructure also contributes to facilitating species dispersal (reviewed in Mineur et al., 2012). Once an introduced species is established in an area, it may spread via the movement of reproductive individuals, propagules or vegetative fragments by the wind, ocean currents or other species (Harries et al., 2007). However, species will be limited in their ability to naturally disperse either by environmental barriers (e.g. land-masses, current patterns, temperature, habitat boundaries) or by their dispersal potential (duration of planktonic phase). The creation of artificial structures in the sea can alter these dispersal limitations – for instance by providing a corridor for species movement or stepping stones to increase the dispersal range of a species, thus artificially influencing levels of ecological connectivity (Gollasch et al., 2006; Strain et al., 2018). The installation of a range of artificial structures in the marine environment may therefore facilitate the spread of NNS to new regions or within their invaded range (Bishop et al., 2017; Miller et al., 2013).

The creation of corridors connecting two previously isolated areas of suitable habitat can occur on a variety of scales (Bulleri and Airoldi, 2005). The largest examples are the Suez and Panama Canals, which connect separate seas and oceans (Gollasch et al., 2006; Ruiz et al., 2009b). The Suez Canal was first constructed in 1869 to connect the Mediterranean and Red Seas and provides passageway for over 17,000 ships a year (Galil et al., 2014a). As a result, the Mediterranean Sea is the most heavily invaded marine ecosystem in the world (Edelist et al., 2013). Of the 751 multicellular NNS recorded in the Mediterranean in 2012, approximately two-thirds are thought to have arrived through the Suez Canal (Galil et al., 2014a; Galil et al., 2014b). The effect of the Panama Canal on patterns of marine invasion is less well known, although it is thought that the presence of a freshwater lake along the route may limit transfer of some hull fouling organisms (Ruiz et al., 2009b; Ros et al., 2014). Artificial structures may also facilitate movement of species across areas of unsuitable habitat (Bulleri and Airoldi, 2005). For example, the presence of breakwalls along the Adriatic coast has been shown to enable colonisation and dispersal of the seaweed *Codium fragile* ssp. *fragile* (Bulleri and Airoldi, 2005; Bulleri et al., 2006). Thus, breakwalls and other coastal defences can also provide corridors of hard substrate across areas of otherwise unsuitable habitat, enabling movement of species with limited dispersal distances.

Artificial structures can provide 'stepping stones' for species dispersal (Apte et al., 2000). This term was originally used to refer to the role artificial habitats can play in providing suitable habitat for enough time to enable onward dispersal, even if the species does not ultimately establish in this environment (Apte et al., 2000). NNS propagules released in ports and harbours can inoculate visiting vessels which go on to spread organisms further afield (Floerl and Inglis, 2005; Floerl et al., 2009). Environmental conditions in ports and harbours could trigger spawning of species arriving on vessels, enabling transfer of species between vessels (Apte et al., 2000;

Minchin and Gollasch, 2003). Ports and harbours also have high resident populations of NNS, so these areas can act as hubs or refuges for the onward dispersal of species (Floerl et al., 2009; Ashton et al., 2010).

In offshore areas, NNS have been shown to colonise oil and gas platforms and renewable energy devices (Page et al., 2006; Sheehy and Vik, 2010; De Mesel et al., 2015; Miller et al., 2013). Species may be spread by the movement of the devices themselves, through transfer on construction and maintenance vessels or by facilitating the dispersal of pelagic larvae (Mineur et al., 2012). Modelling studies have shown the installation of wind farms could enable species with pelagic larvae to spread across the Irish Sea (Adams et al., 2014). NNS have also been recorded on floating offshore monitoring buoys and floating wave devices in Scotland, suggesting novel renewable energy infrastructure such as floating wind turbines might also be colonised by NNS (Nall, 2015; Macleod et al., 2016).

Increasing levels of plastic pollution in the oceans present another pathway by which species can be moved around – by rafting on marine litter (Gregory, 2009; Minchin et al., 2013; Miralles et al., 2018). This represents a global threat, including the transportation of NNS to more remote locations with less human activity and fewer traditional pathways of species introduction (Barnes, 2002; Barnes and Milner, 2005; Campbell et al., 2017). Further effects may be seen by the movement of damaged infrastructure. The 2011 earthquake and resulting tsunami in Japan enabled large amounts of debris to travel across the Pacific Ocean. This debris later arrived on the West Coast of the US and Hawaii with a number of living eastern Pacific species still attached (Calder et al., 2014; Carlton et al., 2017). Storm damage can also disrupt species control efforts. A storm in the UK resulted in fragments of damaged pontoons from a marina with the highly invasive tunicate *Didemnum vexillum* attached drifting into open waters, potentially disrupting efforts to contain this species within the marina (Scottish Government, 2018). As climate change is expected to increase the intensity of storms, the frequency of this sort of event is likely to increase (Patricola and Wehner, 2018). Most of the plastic pollution in the ocean comes from land-based sources, but debris from aquaculture and coastal infrastructure may be particularly concerning for the spread of NNS, as well as the colonisation of plastic debris by NNS once it reaches coastal environments (Campbell et al., 2017).

NNS establishment in urbanised marine and coastal environments

The introduction of hard structures to the marine environment provides large amounts of uncolonised substrate for settlement by marine fouling organisms (Bulleri and Chapman, 2010). NNS are particularly prevalent on artificial structures in comparison to natural environments (Vaselli et al., 2008; Ruiz et al., 2009a; Dafforn et al., 2012; Airoldi et al., 2015). In one study, NNS were found to occupy 80% more space on pilings and pontoons compared with natural reefs (Dafforn et al., 2012). Ruiz et al. (2009a) found that, in a survey of NNS in North America, 71% of recorded species were found on hard substrata, with the majority colonising artificial structures in docks and marinas. As increasingly larger amounts of the coastline are becoming hardened, identification of why these habitats support colonisation of NNS over native species is important in order to minimise their role in facilitating the spread of NNS (Airoldi and Beck, 2007; Johnston et al., 2017).

Invasion success is thought to be determined by three main factors: the frequency and volume of propagules introduced (propagule pressure), the characteristics of the invading species and the susceptibility of the receiving environment to invasion (Lonsdale, 1999; Davis et al., 2000). A significant contributor to the predominance of NNS in estuaries and bays may be the concentration of vector activity in areas with high volumes of connected artificial infrastructure

(Ruiz et al., 2009a). The probability of species establishment will therefore be increased in areas where the propagule pressure is high (Britton-Simmons and Abbott, 2008; Clark and Johnston, 2009; Johnston et al., 2009). Artificial structures may also alter circulation dynamics, resulting in the retainment of propagules as a result of reduced flushing within enclosed ports and harbours (Floerl and Inglis, 2003; Rivero et al., 2013; Dafforn et al., 2015a). However, experimental studies have shown that propagule pressure alone does not maximise invasion success; instead the interaction between propagule pressure and disturbance regulates success of marine NNS establishment (Britton-Simmons and Abbott, 2008; Clark and Johnston, 2009). The concentration of large numbers of transport vectors in areas with high densities of marine infrastructure and anthropogenic disturbance may create conditions with elevated rates of NNS introduction and establishment (Cohen and Carlton, 1998; Grosholz, 2002).

Biotic interactions

Artificial structures may also alter the biotic interactions between native and non-native species. Byers (2002) proposed that artificial structures may present a novel habitat to which both native and NNS are not adapted due to the differences in substrate and orientation between natural and artificial habitats (Glasby, 2000; Tyrrell and Byers, 2007; Chase et al., 2016; Megina et al., 2016). This may reduce the competitive advantage of established native species and reduce biotic resistance to colonisation of fouling NNS (Shea and Chesson, 2002; Tyrrell and Byers, 2007). The materials, placement and design of structures have been shown to alter the composition of the communities able to colonise artificial structures (Perkol-Finkel et al., 2006; Firth et al., 2014). Airoldi et al. (2015) found that islands of hard infrastructure located in soft sediment habitat had two to three times more NNS than rocky reefs or artificial structures located near rocky habitat, demonstrating that native and NNS colonisation may be affected by the surrounding environment, as well as the inherent properties of artificial structures (Perkol-Finkel et al., 2006).

The position of artificial structures in the water column, or their distance from the shore, may also alter the biotic resistance of natural communities by preventing native species from curtailing NNS establishment in natural habitats. Where native predators are capable of controlling NNS, artificial structures may create refuges (Johnston et al., 2017). This may explain why floating structures, such as pontoons, have been observed to have greater abundances of NNS than in benthic habitats or on static columns, which are accessible to benthic predators (Connell, 2000; Boos et al., 2011; Dumont et al., 2011; Forrest et al., 2013).

NNS may also be better competitors in artificial environments than native species as a result of pre-selection of organisms capable of surviving oceanic transfer by artificial vectors (Floerl and Inglis, 2005; Clarke Murray et al., 2012). Species that survive transfer across biogeographic boundaries on vessel hulls are likely to be good colonisers of artificial substrates or may be more tolerant to certain levels of pollutants or anti-fouling paints (Floerl and Inglis, 2005). As floating docks and pontoons present a similar habitat to vessel hulls in terms of surface and orientation, NNS may have a competitive advantage over other species in these environments (Dafforn et al., 2009b). NNS have also been shown to be more tolerant of metal pollution and the presence of anti-fouling chemicals, potentially as a result of exposure during transfer by shipping vectors (Piola and Johnston, 2008; Dafforn et al., 2009a; Crooks et al., 2010; Canning-Clode et al., 2011).

Disturbance

Other characteristics NNS possess which may facilitate establishment in urbanised environments include wider environmental tolerance and greater adaption to disturbance (Lenz et al., 2011;

Clarke Murray et al., 2012). Communities associated with marine artificial environments may be subject to increased disturbance as a result of maintenance activities, frequent movement of vessels, eutrophication and chemical pollution from run-off and toxic paints and the construction work involved in building (or decommissioning) artificial structures (Bulleri and Airoldi, 2005; Piola and Johnston, 2008; Crooks et al., 2010; Airoldi and Bulleri, 2011). Higher levels of disturbance may increase availability of resources, such as space, which is a major limiting resource in intertidal epifaunal communities (Shea and Chesson, 2002; Britton-Simmons and Abbott, 2008). In space-limited systems, such as fouling communities, processes which provide new substrate for colonisation or decrease the cover of fouling species have been shown to facilitate establishment of a number of marine NNS (Stachowicz et al., 2002; Britton-Simmons and Abbott, 2008; Janiak et al., 2013).

The establishment of NNS has also been shown to be influenced by substrate material (Glasby et al., 2007; Tyrrell and Byers, 2007; Vaz-Pinto et al., 2014), position in the water column (Glasby and Connell, 2001; Dafforn et al., 2009b; Boos et al., 2011) and placement of artificial structures (Airoldi et al., 2015; Bishop et al., 2017). The changed environmental conditions created by infrastructure design (Floerl and Inglis, 2003; Rivero et al., 2013; Dafforn et al., 2015a), and higher levels of physical and chemical disturbance (Piola and Johnston, 2008; Crooks et al., 2010), also affect susceptibility of different environments to NNS establishment. However, while the role of these factors in facilitating invasions on artificial structures compared with natural habitats is clear, elucidating which of these factors might best result in reduced invasion success if altered is more challenging. For example, differences in the type of artificial substrate have been shown to affect native and NNS colonisation, with fibreglass structures more susceptible to NNS colonisation than concrete or wood (Glasby, 2000; Vaz-Pinto et al., 2014; Chase et al., 2016). Furthermore, the role of disturbance in facilitating NNS establishment has been shown to vary both temporally and spatially (Glasby et al., 2007).

Fundamental environmental parameters, such as temperature and salinity, affect the survival and distribution of all aquatic species (Smyth and Elliott, 2016). However, NNS have been shown to have broader environmental tolerances than related native species and may also be more tolerant of environmental stress and disturbance (Sorte et al., 2010; Lenz et al., 2011). In estuaries, where many of the world's largest ports can be found, the mixing of freshwater and saline waters results in a gradient of environmental conditions which may affect NNS establishment (Sorte et al., 2010; Pardo et al., 2011). Rapid changes in salinity can act as major disturbance events, flushing bays and estuaries and either removing NNS populations or opening up novel space for colonisation (Mineur et al., 2012; Jimenez et al., 2017). Variation in winter rainfall and resultant low and high salinity events have been suggested as the driving force behind turnover of fouling species communities in the San Francisco Bay and the resulting dominance of NNS (Chang et al., 2018). As anthropogenic climate change is expected to result in changes to rainfall patterns and increased storm events, marine communities may more frequently be exposed to this type of environmental disturbance (Sorte et al., 2010; Cottier-Cook et al., 2017).

Impacts of invasive NNS

Ecological impacts

The introduction of NNS is contributing to changes in species assemblages in aquatic environments (Occhipinti-Ambrogi and Savini, 2003). Of the 12,000 terrestrial and aquatic NNS in Europe, 10%–15% are estimated to be invasive (Sundseth, 2014). In a study of the aquatic NNS in Chesapeake Bay in 1999, it was determined that approximately 20% of the known 196 NNS were thought to be having an ecological impact (Ruiz et al., 1999). However, it was highlighted that quantitative data was likely only available for approximately 5% of NNS present in other bays and estuaries. NNS introductions may interact with other anthropogenic stresses on the marine environment, making quantifying the level of impact of invasive species difficult; this is further exacerbated by a lack of baseline data (Ruiz et al., 1999; Manchester and Bullock, 2000; Ojaveer et al., 2015). Despite these restrictions, invasive species are thought to be one of the main contributors to global biodiversity loss (Bax et al., 2003; Vilà et al., 2010).

Invasive NNS may negatively impact native biodiversity through a variety of mechanisms (Molnar et al., 2008). These include competition, predation, spread of parasites and disease, hybridisation with native species and habitat modification (Ruiz et al., 1999; Grosholz, 2002; Cook et al., 2016). The Indo-Pacific lionfish *Pterois volitans* (Linnaeus, 1758), thought to have been introduced to the Florida coast as an accidental aquarium release, has now spread across much of the Caribbean, as far south as Brazil and northwards on the East Coast of the US (Côté and Maljković, 2010). As a voracious predator, the lionfish has had a significant impact on coral reef fish populations, and predation of this species on herbivorous fish is thought to have resulted in a phase shift from coral to algal dominated systems along some reefs (Côté and Maljković, 2010; Lesser and Slattery, 2011). Invasive macroalgae species, such as *Caulerpa taxifolia* (M.Vahl) C.Agardh 1817 and *Undaria pinnatifida* (Harvey) Suringar, 1873, which are both included on the list of the world's 100 worst invasive species (Lowe et al., 2000), are capable of extensive spread within their introduced ranges and may compete with native macroalgae and seagrass species (Anderson, 2005; Maggi et al., 2015; Epstein and Smale, 2017). The effects of invasive macroalgae on higher trophic levels are more variable, though, due to their role in forming habitat for both epiphytic and mobile species (Maggi et al., 2015; Katsanevakis et al., 2016).

NNS which fundamentally alter the ecological and/or physical characteristics of habitats into which they are introduced are sometimes termed 'ecosystem engineers' (Ruiz et al., 1997; Guy-Haim et al., 2018). The impact of these species can be extensive as they may alter the availability of resources by providing additional hard substrate or altering light, nutrient or space availability (Crooks, 2002). For example, the dispersal of the Pacific oyster *Magallana gigas* following its introduction for shellfish aquaculture has resulted in the transformation of muddy intertidal areas to dense oyster reefs (Cook et al., 2016). Habitat formation can have positive or negative effects for native biota – for instance, by either reducing habitat suitability for native species or facilitating growth through provision of additional habitat complexity (Rodriguez, 2006; Gribben et al., 2013). The establishment of NNS has also been shown to facilitate further establishment of other NNS, termed 'invasional meltdown' (Simberloff et al., 1999).

Socio-economic impacts

The economic impacts of marine invasive NNS stem from impacts on human health, productivity of marine industries and costs associated with removal of unwanted or fouling species (Bax et al., 2003). Invasive NNS may reduce fishing catches, such as in the case of the introduction of the invasive ctenophore *Mnemiospis leidyi* Agassiz, 1865, into the Black Sea. Introduced in the 1980s to a system already suffering from the impacts of eutrophication and overfishing, huge blooms of *M. leidyi* competed with and predated on fish larvae, causing the collapse of the anchovy fishery by the early 1990s (Shiganova, 1998). Thus, systems which are already degraded as a result of anthropogenic impact may be especially vulnerable to impacts of invasive NNS.

Much of the marine aquaculture around the world is increasingly reliant on using specific NNS as stock. However, unwanted pest species can cause problems by introducing diseases or impacting the growth of cultivated stock and reducing yield (Watson et al., 2009; Cottier-Cook et al., 2016). Fouling of stock and equipment can result in increased cleaning costs and reduced opportunities for sale (Bourque et al., 2005; Watson et al., 2009). Non-native ascidian species in particular have caused problems for shellfish aquaculture companies. Species such as *Styela clava* Herdman 1881 and *Didemnum vexillum* Kott 2002 are both capable of settling in high densities on aquaculture structures and smothering mussel species (Dafforn et al., 2012; Ferguson et al., 2017; Figure 9.3). Larvae of the bay scallop *Argopectin irradians irradians* were even shown to avoid settlement on colonies of *D. vexillum*, potentially affecting commercial sea scallop fisheries in the US (Morris et al., 2009).

NNS may even detrimentally impact public health. For example, large swarms of the non-native tropical scyphozoan *Rhopilema nomadica* Galil, 1990, in the Levantine Sea have blocked intake pipes of power plant cooling systems and interfered with fishing, and their severe stings represent a health risk, driving tourists away from beaches (Yahia et al., 2013; Galil et al., 2015). The economic costs associated with NNS can thus be significant through both direct costs (e.g. removal or control), and indirect costs (e.g. losses to tourist industry). The combined costs of control and biosecurity of terrestrial and aquatic species to the UK economy was calculated at £1.7 billion in 2010 (Williams et al., 2010). In 2005, Pimentel et al. (2005) identified that the environmental damage and loss associated with invasive NNS in the US totalled around $120 billion annually.

Management of marine invasives

In recognition of the profound economic and environmental impacts invasive NNS may have in their introduced range, several policies and regulations have addressed the need to reduce the number of species introductions and control existing invasive populations (Genovesi et al., 2014; Hulme, 2015). Policy on the control of invasive NNS has been part of international agenda for over 60 years, and there are a number of international and regional agreements, as well as national policies, which support the management of NNS. For example, in acknowledgement of the issues of species transfer in ships' ballast water, in 2004 the International Maritime Organisation (IMO) adopted the International Convention for the Control and Management of Ships' Ballast Water and Sediment (BWM Convention), and this was ratified in 2018 (David and Gollasch, 2018). The IMO has also developed voluntary hull fouling guidelines, which set out measures for ports, shipping companies and member states to reduce risk of transfer on hulls (IMO, 2011). Post-arrival measures, which have been trialled to control and eradicate marine invasive NNS, include physical removal of individuals, air drying (Forrest and Blakemore, 2006; Hopkins et al., 2016), wrapping in plastic to allow application of chemicals (LeBlanc et al., 2007; Piola et al., 2009; Roche et al., 2015; Atalah et al., 2016) and immersion in freshwater (McCann et al., 2013; Moreira et al., 2014). Successful eradication examples include that of the mussel *Mytilopsis sp.* in Australia and the localised removal of *Caulerpa taxifolia* in California (Bax et al., 2002; Anderson, 2005). Where established marine NNS cannot be eradicated, the control of established populations may reduce their further spread and impact on native biodiversity (Genovesi, 2005; Epstein and Smale, 2017).

The predominance of NNS in communities on artificial structures has led to research focusing on ways to prevent establishment and further spread in these environments (Dafforn, 2017). Johnston et al. (2017) reviewed the ways in which anthropogenic activity influences the establishment of NNS and suggested management actions to minimise the role artificial structures play in facilitating NNS spread. Suggestions for altering infrastructure design to reduce the susceptibility of urban environments to NNS establishment have incorporated research from the field of ecological engineering, which has historically been focused on enhancing native biodiversity on

Figure 9.3 The carpet sea squirt *Didemnum vexillum* growing on a rope trailing from a pontoon in a Scottish marina.

Source: Photograph by C. Beveridge.

hard coastal structures (Chapman and Underwood, 2011; Dafforn et al., 2015a; Mayer-Pinto et al., 2017). The biotic resistance of a site may be increased by promoting the growth of native species or by increasing predation pressure (Bulleri and Chapman, 2010; Mayer-Pinto et al., 2017). Methods of promoting native biodiversity have included 'seeding' structures with habitat forming species such as kelp or altering the design of structures to increase colonisation by native species (Mayer-Pinto et al., 2017). The integration of microhabitats such as crevices and rock-pools into coastal protection structures has been trialled in a number of locations in the UK and further afield and has shown success in attracting key intertidal species (Firth et al., 2013; Naylor et al., 2017). Studies have also shown that certain factors such as the presence of high volumes of floating infrastructure, high levels of disturbance or restricted water circulation due to the enclosure of an area by seawalls can promote NNS establishment (Nall et al., 2015; Foster et al., 2016). Altering the siting or design of marinas and harbours to reduce the volume of floating structures, increase their proximity to freshwater or incorporate quarantine and treatment facilities might also help reduce the establishment of NNS (Bax et al., 2002; Minchin and Gollasch, 2003; Holt and Cordingley, 2011; Foster et al., 2016; Mayer-Pinto et al., 2017). Research investigating factors that affect the vulnerability of different environments to colonisation by NNS could also help identify management strategies to mitigate NNS spread and establishment (Dafforn et al., 2015a; Bishop et al., 2017; Johnston et al., 2017).

References

Acosta, H. and Forrest, B.M. (2009) 'The spread of marine non-indigenous species via recreational boating: A conceptual model for risk assessment based on fault tree analysis', *Ecological Modelling*, vol. 220, pp. 1586–1598. http://dx.doi.org/10.1016/j.ecolmodel.2009.03.026

Adams, T., Miller, R., Aleynik, D. and Burrows, M. (2014) 'Offshore marine renewable energy devices as stepping stones across biogeographical boundaries', *Journal of Applied Ecology*, vol. 51, pp. 330–338. doi: 10.1111/1365-2664.12207

Airoldi, L. and Beck, M.W. (2007) 'Loss, status and trends for coastal marine habitats of Europe', *Oceanography and Marine Biology: An Annual Review*, vol. 45, pp. 345–405. doi: 10.1201/9781420050943.ch7

Airoldi, L. and Bulleri, F. (2011) 'Anthropogenic disturbance can determine the magnitude of opportunistic species responses on marine urban infrastructures', *PLoS One*, vol. 6, p. e22985. doi: 10.1371/journal.pone.0022985

Airoldi, L., Turon, X., Perkol-Finkel, S. and Rius, M. (2015) 'Corridors for aliens but not for natives: Effects of marine urban sprawl at a regional scale', *Diversity and Distributions*, vol. 21, no. 7, pp. 755–768. doi: 10.1111/ddi.12301

Anderson, L. (2005) 'California's reaction to Caulerpa taxifolia: A model for invasive species rapid response', *Biological Invasions*, vol. 7, pp. 1003–1016. doi: 10.1007/s10530-004-3123-z

Apte, S., Holland, B.S., Godwin, L.S. and Gardner, J.P.A. (2000) 'Jumping ship: A stepping stone event mediating transfer of a non-indigenous species via a potentially unsuitable environment', *Biological Invasions*, vol. 2, pp. 75–79. doi: 10.1023/A:1010024818644

Ashton, G.V., Boos, K., Shucksmith, R. and Cook, E.J. (2006) 'Risk assessment of hull fouling as a vector for marine non-natives in Scotland', *Aquatic Invasions*, vol. 1, pp. 214–218.

Ashton, G.V., Burrows, M.T., Willis, K.J. and Cook, E.J. (2010) 'Seasonal population dynamics of the non-native Caprella mutica Crustacea, Amphipoda on the west coast of Scotland', *Marine and Freshwater Research*, vol. 61, pp. 549–559. http://dx.doi.org/10.1071/MF09162

Ashton, G.V., Davidson, I. and Ruiz, G. (2014) 'Transient small boats as a long-distance coastal vector for dispersal of biofouling organisms', *Estuaries and Coasts*, vol. 37, pp. 1572–1581. doi: 10.1007/s12237-014-9782-9

Atalah, J., Brook, R., Cahill, P., Fletcher, L.M. and Hopkins, G.A. (2016) 'It's a wrap: Encapsulation as a management tool for marine biofouling', *Biofouling*, vol. 32, pp. 277–286. doi: 10.1080/08927014.2015.1137288

Barnes, D.K.A. (2002) 'Invasions by marine life on plastic debris', *Nature*, vol. 416, p. 808. doi: 10.1038/416808a

Barnes, D.K.A. and Milner, P. (2005) 'Drifting plastic and its consequences for sessile organism dispersal in the Atlantic Ocean', *Marine Biology*, vol. 146, pp. 815–825. doi: 10.1007/s00227-004-1474-8

Bax, N., Hayes, K., Marshall, A., Parry, D. and Thresher, R. (2002) 'Man-made marinas as sheltered islands for alien marine organisms: Establishment and eradication of an alien invasive marine species', in C.R. Veitch and M.N. Clout (eds.) *Turning the Tide: The Eradication of Invasive Species*. IUCN, Gland, Switzerland and Cambridge, UK.

Bax, N., Williamson, A., Aguero, M., Gonzalez, E. and Geeves, W. (2003) 'Marine invasive alien species: A threat to global biodiversity', *Marine Policy*, vol. 27, pp. 313–323. doi: 10.1016/S0308-597X(03) 00041-1

Bishop, M.J., Mayer-Pinto, M., Airoldi, L., Firth, L.B., Morris, R.L., Loke, L.H.L., Hawkins, S.J., Naylor, L.A., Coleman, R.A., Chee, S.Y. and Dafforn, K.A. (2017) 'Effects of ocean sprawl on ecological connectivity: Impacts and solutions', *Journal of Experimental Marine Biology and Ecology*, vol. 492, pp. 7–30. http://dx.doi.org/10.1016/j.jembe.2017.01.021

Boos, K., Ashton, G. and Cook, E.J. (2011) 'The Japanese skeleton shrimp Caprella mutica (Crustacea, Amphipoda): A global invader of coastal waters', in B.S. Galil, P. Clark and J. Carlton (eds.) *In the Wrong Place: Alien Marine Crustaceans: Distribution, Biology and Impacts*. Springer, Netherlands.

Bourque, D., Le Blanc, A., Landry, T., McNair, N. and Davidson, J. (2005) 'Tunicate infested mussel aquaculture sites in Prince Edward Island, Canada', *Journal of Shellfish Research*, vol. 24, p. 1261.

Britton-Simmons, K. and Abbott, K. (2008) 'Short- and long-term effects of disturbance and propagule pressure on a biological invasion', *Journal of Ecology*, vol. 96, pp. 68–77. doi: 10.1111/j.1365-2745.2007.01319.x

Bulleri, F., Abbiati, M. and Airoldi, L. (2006) 'The colonisation of human-made structures by the invasive alga Codium fragile ssp. tomentosoides in the north Adriatic Sea (NE Mediterranean)', *Hydrobiologia*, vol. 555, pp. 263–269. doi: 10.1007/s10750-005-1122-4

Bulleri, F. and Airoldi, L. (2005) 'Artificial marine structures facilitate the spread of a non-indigenous green alga, Codium fragile ssp. tomentosoides, in the north Adriatic Sea', *Journal of Applied Ecology*, vol. 42, pp. 1063–1072. doi: 10.1111/j.1365-2664.2005.01096.x

Bulleri, F. and Chapman, M.G. (2010) 'The introduction of coastal infrastructure as a driver of change in marine environments', *Journal of Applied Ecology*, vol. 47, pp. 26–35. doi: 10.1111/j.1365-2664.2009. 01751.x

Byers, J.E. (2002) 'Impact of non-indigenous species on natives enhanced by anthropogenic alteration of selection regimes', *Oikos*, vol. 97, pp. 449–458. doi: 10.1034/j.1600-0706.2002.970316.x

Calder, D., Choong, H., Carlton, J., Chapman, J., Miller, J. and Geller, J. (2014) 'Hydroids (Cnidaria: Hydrozoa) from Japanese tsunami marine debris washing ashore in the northwestern United States', *Aquatic Invasions*, vol. 9, pp. 425–440. http://dx.doi.org/10.3391/ai.2014.9.4.02

Campbell, M.L., King, S., Heppenstall, L.D., van Gool, E., Martin, R. and Hewitt, C.L. (2017) 'Aquaculture and urban marine structures facilitate native and non-indigenous species transfer through generation and accumulation of marine debris', *Marine Pollution Bulletin*, vol. 123, pp. 304–312. https://doi. org/10.1016/j.marpolbul.2017.08.040

Canning-Clode, J., Fofonoff, P., Riedel, G.F., Torchin, M. and Ruiz, G.M. (2011) 'The effects of copper pollution on fouling assemblage diversity: A tropical-temperate comparison', *PLoS One*, vol. 6, p. e18026. doi: 10.1371/journal.pone.0018026

Carlton, J.T. (1996) 'Marine bioinvasions: The alteration of marine ecosystems by nonindigenous species', *Oceanography*, vol. 9, pp. 36–43. https://doi.org/10.5670/oceanog.1996.25

Carlton, J.T., Chapman, J.W., Geller, J.B., Miller, J.A., Carlton, D.A., McCuller, M.I., Treneman, N.C., Steves, B.P. and Ruiz, G.M. (2017) 'Tsunami-driven rafting: Transoceanic species dispersal and implications for marine biogeography', *Science*, vol. 357, pp. 1402–1406. doi: 10.1126/science.aao1498

Chan, F.T. and Briski, E. (2017) 'An overview of recent research in marine biological invasions', *Marine Biology*, vol. 164, p. 121. doi: 10.1007/s00227-017-3155-4

Chan, F.T., Stanislawczyk, K.C., Sneekes, A., Dvoretsky, A., Gollasch, S., Minchin, D., David, M., Jelmert, A., Albretsen, J. and Bailey, S. (2018) 'Climate change opens new frontiers for marine species in the Arctic: Current trends and future invasion risks', *Global Change Biology*, vol. 25, no. 1, pp. 25–38.

Chang, A.L., Brown, C.W., Crooks, J.A. and Ruiz, G.M. (2018) 'Dry and wet periods drive rapid shifts in community assembly in an estuarine ecosystem', *Global Change Biology*, vol. 24, pp. e627–e642. doi: 10.1111/gcb.13972

Chapman, M.G. and Underwood, A.J. (2011) 'Evaluation of ecological engineering of "armoured" shorelines to improve their value as habitat', *Journal of Experimental Marine Biology and Ecology*, vol. 400, pp. 302–313. doi: 10.1016/j.jembe.2011.02.025

Chase, A.L., Dijkstra, J.A. and Harris, L.G. (2016) 'The influence of substrate material on ascidian larval settlement', *Marine Pollution Bulletin*, vol. 106, pp. 35–42. http://dx.doi.org/10.1016/j.marpolbul.2016.03.049

Clark, G.F. and Johnston, E.L. (2009) 'Propagule pressure and disturbance interact to overcome biotic resistance of marine invertebrate communities', *Oikos*, vol. 118, pp. 1679–1686. doi: 10.1111/j.1600-0706.2009.17564.x

Clarke Murray, C., Pakhomov, E. and Therriault, T. (2011) 'Recreational boating: A largely unregulated vector transporting marine invasive species', *Diversity and Distributions*, vol. 17, pp. 1161–1172. doi: 10.1111/j.1472-4642.2011.00798.x

Clarke Murray, C., Therriault, T.W. and Martone, P.T. (2012) 'Adapted for invasion? Comparing attachment, drag and dislodgment of native and nonindigenous hull fouling species', *Biological Invasions*, vol. 14, pp. 1651–1663. doi: 10.1007/s10530-012-0178-0

Cohen, A.N. and Carlton, J.T. (1998) 'Accelerating invasion rate in a highly invaded estuary', *Science*, vol. 279, pp. 555–558.

Connell, S.D. (2000) 'Floating pontoons create novel habitats for subtidal epibiota', *Journal of Experimental Marine Biology and Ecology*, vol. 247, pp. 183–194. doi: 10.1016/S0022-0981(00)00147-7

Cook, E., Payne, R., Macleod, A. and Brown, S. (2016) 'Marine biosecurity: Protecting indigenous marine species', *Research and Reports in Biodiversity Studies*, vol. 5, pp. 1–14. doi: 10.2147/rrbs.s63402

Côté, I.M. and Maljković, A. (2010) 'Predation rates of Indo-Pacific lionfish on Bahamian coral reefs', *Marine Ecology Progress Series*, vol. 404, pp. 219–225.

Cottier-Cook, E.J., Beveridge, C., Bishop, J.D.D., Brodie, J., Clark, P.F., Epstein, G., Jenkins, S.R., Johns, D.G., Loxton, J., MacLeod, A., Maggs, C., Minchin, D., Mineur, F., Sewell, J. and Wood, C.A. (2017) 'Non-native species', *MCCIP Science Review 2017*, pp. 47–61. doi: 10.14465/2017.arc10.005-nns

Cottier-Cook, E.J., Nagabhatla, N., Badis, Y., Campbell, M., Chopin, T., Dai, W., Fang, J., He, P., Hewitt, C. and Kim, G. (2016) 'Safeguarding the future of the global seaweed aquaculture industry', *United Nations University (INWEH) and Scottish Association for Marine Science Policy Brief*. Hamilton, Canada, p. 12.

Critchley, A.T. (1983) 'Sargassum muticum: A taxonomic history including worldwide and western Pacific distribution', *Journal of the Marine Biological Association of the UK*, vol. 63, pp. 617–625.

Crooks, J.A. (2002) 'Characterizing ecosystem-level consequences of biological invasions: The role of ecosystem engineers', *Oikos*, vol. 97, pp. 153–166. doi: 10.1034/j.1600-0706.2002.970201.x

Crooks, J.A., Chang, A.L. and Ruiz, G.M. (2010) 'Aquatic pollution increases the relative success of invasive species', *Biological Invasions*, vol. 13, pp. 165–176. doi: 10.1007/s10530-010-9799-3

Dafforn, K.A. (2017) 'Eco-engineering and management strategies for marine infrastructure to reduce establishment and dispersal of non-indigenous species', *Management of Biological Invasions*, vol. 8, pp. 153–161.

Dafforn, K.A., Glasby, T.M., Airoldi, L., Rivero, N.K., Mayer-Pinto, M. and Johnston, E.L. (2015a) 'Marine urbanization: An ecological framework for designing multifunctional artificial structures', *Frontiers in Ecology and the Environment*, vol. 13, no. 2, pp. 82–90. doi: 10.1890/140050

Dafforn, K.A., Glasby, T.M. and Johnston, E.L. (2009a) 'Links between estuarine condition and spatial distributions of marine invaders', *Diversity and Distributions*, vol. 15, pp. 807–821. https://doi.org/10.1080/08927010802710618

Dafforn, K.A., Glasby, T.M. and Johnston, E.L. (2012) 'Comparing the invasibility of experimental "reefs" with field observations of natural reefs and artificial structures', *PLoS One*, vol. 7, p. e38124. doi: 10.1371/journal.pone.0038124

Dafforn, K.A., Johnson, E. and Glasby, T.M. (2009b) 'Shallow moving structures promote marine invader dominance', *Biofouling*, vol. 25, pp. 277–287. doi: 10.1080/08927010802710618

David, M. and Gollasch, S. (2018) 'How to approach ballast water management in European seas', *Estuarine, Coastal and Shelf Science*, vol. 201, pp. 248–255. https://doi.org/10.1016/j.ecss.2016.10.018

Davidson, I.C., Zabin, C.J., Chang, A.L., Brown, C.W., Sytsma, M.D. and Ruiz, G.M. (2010) 'Recreational boats as potential vectors of marine organisms at an invasion hotspot', *Aquatic Biology*, vol. 11, pp. 179–191. doi: 10.3354/ab00302

Davis, M.A., Grime, J.P. and Thompson, K. (2000) 'Fluctuating resources in plant communities: A general theory of invasibility', *Journal of Ecology*, vol. 88, pp. 528–534. doi: 10.1046/j.1365-2745.2000.00473.x

De Mesel, I., Kerckhof, F., Norro, A., Rumes, B. and Degraer, S. (2015) 'Succession and seasonal dynamics of the epifauna community on offshore wind farm foundations and their role as stepping stones for non-indigenous species', *Hydrobiologia*, vol. 756, no. 1, pp. 37–50. doi: 10.1007/s10750-014-2157-1

Dumont, C.P., Gaymer, C.F. and Thiel, M. (2011) 'Predation contributes to invasion resistance of benthic communities against the non-indigenous tunicate Ciona intestinalis', *Biological Invasions*, vol. 13, pp. 2023–2034. doi: 10.1007/s10530-011-0018-7

Edelist, D., Rilov, G., Golani, D., Carlton, J.T. and Spanier, E. (2013) 'Restructuring the sea: Profound shifts in the world's most invaded marine ecosystem', *Diversity and Distributions*, vol. 19, pp. 69–77. doi: 10.1111/ddi.12002

Epstein, G. and Smale, D.A. (2017) 'Undaria pinnatifida: A case study to highlight challenges in marine invasion ecology and management', *Ecology and Evolution*, vol. 7, pp. 8624–8642. doi: 10.1002/ece3.3430

Ferguson, L.F., Davidson, J.D.P., Landry, T., Clements, J.C. and Therriault, T.W. (2017) 'Didemnum vexillum: Invasion potential via harvesting and processing of the Pacific oyster (*Crassostrea gigas*) in British Colombia, Canada', *Management of Biological Invasions*, vol. 8, no. 4, pp. 553–558.

Ferrario, J., Caronni, S., Occhipinti-Ambrogi, A. and Marchini, A. (2017) 'Role of commercial harbours and recreational marinas in the spread of non-indigenous fouling species', *Biofouling*, vol. 33, no. 8, pp. 651–660 and 1–10. doi: 10.1080/08927014.2017.1351958

Firth, L.B., Thompson, R.C., Bohn, K., Abbiati, M., Airoldi, L., Bouma, T.J., Bozzeda, F., Ceccherelli, V.U., Colangelo, M.A., Evans, A., Ferrario, F., Hanley, M.E., Hinz, H., Hoggart, S.P.G., Jackson, J.E., Moore, P., Morgan, E.H., Perkol-Finkel, S., Skov, M.W., Strain, E.M., van Belzen, J. and Hawkins, S.J. (2014) 'Between a rock and a hard place: Environmental and engineering considerations when designing coastal defence structures', *Coastal Engineering*, vol. 87, pp. 122–135. doi: 10.1016/j.coastaleng.2013.10.015

Firth, L.B., Thompson, R.C., White, F.J., Schofield, M., Skov, M.W., Hoggart, S.P.G., Jackson, J., Knights, A.M. and Hawkins, S.J. (2013) 'The importance of water-retaining features for biodiversity on artificial intertidal coastal defence structures', *Diversity and Distributions*, vol. 19, pp. 1275–1283. doi: 10.1111/ddi.12079

Floerl, O. and Inglis, G.J. (2003) 'Boat harbour design can exacerbate hull fouling', *Austral Ecology*, vol. 28, pp. 116–127. doi: 10.1046/j.1442-9993.2003.01254.x

Floerl, O. and Inglis, G.J. (2005) 'Starting the invasion pathway: The interaction between source populations and human transport vectors', *Biological Invasions*, vol. 7, pp. 589–606. doi: 10.1007/s10530-004-0952-8

Floerl, O., Inglis, G.J., Dey, K. and Smith, A. (2009) 'The importance of transport hubs in stepping-stone invasions', *Journal of Applied Ecology*, vol. 46, pp. 37–45. doi: 10.1111/j.1365-2664.2008.01540.x

Forrest, B.M. and Blakemore, K.A. (2006) 'Evaluation of treatments to reduce the spread of a marine plant pest with aquaculture transfers', *Aquaculture*, vol. 257, pp. 333–345. https://doi.org/10.1016/j.aquaculture.2006.03.021

Forrest, B.M., Fletcher, L.M., Atalah, J., Piola, R.F. and Hopkins, G.A. (2013) 'Predation limits spread of Didemnum vexillum into natural habitats from refuges on anthropogenic structures', *PLoS One*, vol. 8, p. e82229. doi: 10.1371/journal.pone.0082229

Foster, V., Giesler, R.J., Wilson, A.M.W., Nall, C.R. and Cook, E.J. (2016) 'Identifying the physical features of marina infrastructure associated with the presence of non-native species in the UK', *Marine Biology*, vol. 163, pp. 1–14. doi: 10.1007/s00227-016-2941-8

Fowler, A.E., Blakeslee, A.M.H., Canning-Clode, J., Repetto, M.F., Phillip, A.M., Carlton, J.T., Moser, F.C., Ruiz, G.M. and Miller, A.W. (2016) 'Opening Pandora's bait box: A potent vector for biological invasions of live marine species', *Diversity and Distributions*, vol. 22, pp. 30–42. doi: 10.1111/ddi.12376

Galil, B.S., Boero, F., Campbell, M.L., Carlton, J.T., Cook, E., Fraschetti, S., Gollasch, S., Hewitt, C.L., Jelmert, A., Macpherson, E., Marchini, A., McKenzie, C., Minchin, D., Occhipinti-Ambrogi, A., Ojaveer, H., Olenin, S., Piraino, S. and Ruiz, G.M. (2014a) '"Double trouble": The expansion of the Suez Canal and marine bioinvasions in the Mediterranean Sea', *Biological Invasions*, vol. 17, no. 4, pp. 973–976. doi: 10.1007/s10530-014-0778-y

Galil, B.S., Boero, F., Fraschetti, S., Piraino, S., Campbell, M., Hewitt, C., Carlton, J., Cook, E., Jelmert, A. and Macpherson, E. (2015) 'The enlargement of the Suez Canal and introduction of non-indigenous species to the Mediterranean Sea', *Limnology and Oceanography Bulletin*, vol. 24, pp. 43–45.

Galil, B.S., Marchini, A., Occhipinti-Ambrogi, A., Minchin, D., Narscius, A., Ojaveer, H. and Olenin, S. (2014b) 'International arrivals: Widespread bioinvasions in European Seas', *Ethology Ecology and Evolution*, vol. 26, pp. 152–171. doi: 10.1080/03949370.2014.897651

Genovesi, P. (2005) 'Eradications of invasive alien species in Europe: A review', *Biological Invasions*, vol. 7, pp. 127–133.

Genovesi, P., Carboneras, C., Vilà, M. and Walton, P. (2014) 'EU adopts innovative legislation on invasive species: A step towards a global response to biological invasions?', *Biological Invasions*, vol. 17, no. 5, pp. 1307–1311. doi: 10.1007/s10530-014-0817-8

Glasby, T.M. (2000) 'Surface composition and orientation interact to affect subtidal epibiota', *Journal of Experimental Marine Biology and Ecology*, vol. 248, pp. 177–190. doi: 10.1016/S0022-0981(00)00169-6

Glasby, T.M. and Connell, S.D. (2001) 'Orientation and position of substrata have large effects on epibiotic assemblages', *Marine Ecology Progress Series*, vol. 214, pp. 127–135. doi: 10.3354/meps214127

Glasby, T.M., Connell, S.D., Holloway, M.G. and Hewitt, C.L. (2007) 'Nonindigenous biota on artificial structures: Could habitat creation facilitate biological invasions?', *Marine Biology*, vol. 151, pp. 887–895. doi: 10.1007/s00227-006-0552-5

Godwin, L.S. (2003) 'Hull fouling of maritime vessels as a pathway for marine species invasions to the Hawaiian Islands', *Biofouling*, vol. 19, pp. 123–131. doi: 10.1080/0892701031000061750

Gollasch, S. (2002) 'The importance of ship hull fouling as a vector of species introductions into the North Sea', *Biofouling: The Journal of Bioadhesion and Biofilm*, vol. 18, pp. 105–121. doi: 10.1080/089270 10290011361

Gollasch, S. (2006) 'Overview on introduced aquatic species in European navigational and adjacent waters', *Helgoland Marine Research*, vol. 60, pp. 84–89. doi: 10.1007/s10152-006-0022-y

Gollasch, S., David, M., Voigt, M., Dragsund, E., Hewitt, C. and Fukuyo, Y. (2007) 'Critical review of the IMO international convention on the management of ships' ballast water and sediments', *Harmful Algae*, vol. 6, pp. 585–600. doi: 10.1016/j.hal.2006.12.009

Gollasch, S., Galil, B. and Cohen, A.E. (2006) 'Bridging divides: Maritime canals as invasion corridors', *Monographiae Biologicae*, vol. 83. Springer.

Gregory, M.R. (2009) 'Environmental implications of plastic debris in marine settings: Entanglement, ingestion, smothering, hangers-on, hitch-hiking and alien invasions', *Philosophical Transactions of the Royal Society B: Biological Sciences*, vol. 364, pp. 2013–2025. doi: 10.1098/rstb.2008.0265

Gribben, P.E., Byers, J.E., Wright, J.T. and Glasby, T.M. (2013) 'Positive versus negative effects of an invasive ecosystem engineer on different components of a marine ecosystem', *Oikos*, vol. 122, pp. 816–824. doi: 10.1111/j.1600-0706.2012.20868.x

Grosholz, E. (2002) 'Ecological and evolutionary consequences of coastal invasions', *Trends in Ecology and Evolution*, vol. 17, pp. 22–27.

Guy-Haim, T., Lyons, D.A., Kotta, J., Ojaveer, H., Queirós, A.M., Chatzinikolaou, E., Arvanitidis, C., Como, S., Magni, P., Blight, A.J., Orav-Kotta, H., Somerfield, P.J., Crowe, T.P. and Rilov, G. (2018) 'Diverse effects of invasive ecosystem engineers on marine biodiversity and ecosystem functions: A global review and meta-analysis', *Global Change Biology*, vol. 24, pp. 906–924. doi: 10.1111/gcb.14007

Harries, D.B., Cook, E., Donnan, D., Mair, J., Harrow, S. and Wilson, J. (2007) 'The establishment of the invasive alga Sargassum muticum on the west coast of Scotland: Rapid northwards spread and identification of potential new areas for colonisation', *Aquatic Invasions*, vol. 2, pp. 367–377. doi: 10.3391/ai.2007.2.4.5

Holt, R. and Cordingley, A. (2011) *Eradication of the Non-Native Carpet Ascidian (Sea Squirt) Didemnum Vexillum in Holyhead Harbour: Progress, Methods and Results to Spring 2011.* Countryside Council for Wales, Cardiff, UK.

Hopkins, G.A., Prince, M., Cahill, P.L., Fletcher, L.M. and Atalah, J. (2016) 'Desiccation as a mitigation tool to manage biofouling risks: Trials on temperate taxa to elucidate factors influencing mortality rates', *Biofouling*, vol. 32, pp. 1–11. doi: 10.1080/08927014.2015.1115484

Hulme, P.E. (2009) 'Trade, transport and trouble: Managing invasive species pathways in an era of globalisation', *Journal of Applied Ecology*, vol. 46, pp. 10–18. doi: 10.1111/j.1365-2664.2008.01600.x

Hulme, P.E. (2015) 'Invasion pathways at a crossroad: Policy and research challenges for managing alien species introductions', *Journal of Applied Ecology*, vol. 52, pp. 1418–1424. doi: 10.1111/1365-2664.12470

IMO (2011) *Guidelines for the Control and Management of Ships' Biofouling to Minimize the Transfer of Invasive Aquatic Species.* Marine Environment Protection Committee, International Maritime Organization. Resolution MEPC.207(62), London, UK.

Janiak, D.S., Osman, R.W. and Whitlatch, R.B. (2013) 'The role of species richness and spatial resources in the invasion success of the colonial ascidian Didemnum vexillum Kott, 2002 in eastern Long Island Sound', *Journal of Experimental Marine Biology and Ecology*, vol. 443, pp. 12–20. doi: 10.1016/j.jembe.2013.02.030

Jimenez, H., Keppel, E., Chang, A.L. and Ruiz, G.M. (2017) 'Invasions in marine communities: Contrasting species richness and community composition across habitats and salinity', *Estuaries and Coasts*, vol. 41, pp. 484–494. doi: 10.1007/s12237-017-0292-4

Johnston, E.L., Dafforn, K.A., Clark, G.F., Rius, M. and Floerl, O. (2017) 'How anthropogenic activities affect the establishment and spread of non-indigenous species post arrival', *Oceanography and Marine Biology: An Annual Review*, vol. 55, pp. 2–33.

Johnston, E.L., Piola, R.F. and Clark, G.F. (2009) 'The role of propagule pressure in invasion success', in G. Rilov and J.A. Crooks (eds.) *Biological Invasions in Marine Ecosystems: Ecological, Management, and Geographic Perspectives.* Springer, Berlin, Germany.

Katsanevakis, S., Tempera, F. and Teixeira, H. (2016) 'Mapping the impact of alien species on marine ecosystems: The Mediterranean Sea case study', *Diversity and Distributions*, vol. 22, pp. 694–707. doi: 10.1111/ddi.12429

LeBlanc, N., Davidson, J., Tremblay, R., McNiven, M. and Landry, T. (2007) 'The effect of anti-fouling treatments for the clubbed tunicate on the blue mussel, Mytilus edulis', *Aquaculture*, vol. 264, pp. 205–213. doi: 10.1016/j.aquaculture.2006.12.027

Lenz, M., da Gama, B.A.P., Gerner, N.V., Gobin, J., Gröner, F., Harry, A., Jenkins, S.R., Kraufvelin, P., Mummelthei, C., Sareyka, J., Xavier, E.A. and Wahl, M. (2011) 'Non-native marine invertebrates are more tolerant towards environmental stress than taxonomically related native species: Results from a globally replicated study', *Environmental Research*, vol. 111, pp. 943–952. doi: 10.1016/j.envres.2011.05.001

Leppäkoski, E. and Olenin, S. (2000) 'Non-native species and rates of spread: Lessons from the brackish Baltic Sea', *Biological Invasions*, vol. 2, pp. 151–163.

Lesser, M.P. and Slattery, M. (2011) 'Phase shift to algal dominated communities at mesophotic depths associated with lionfish (*Pterois volitans*) invasion on a Bahamian coral reef', *Biological Invasions*, vol. 13, pp. 1855–1868. doi: 10.1007/s10530-011-0005-z

Levine, J.M. and D'Antonio, C.M. (2003) 'Forecasting biological invasions with increasing international trade', *Conservation Biology*, vol. 17, pp. 322–326. doi: 10.1046/j.1523-1739.2003.02038.x

Lonsdale, W.M. (1999) 'Global patterns of plant invasions and the concept of invasibility', *Ecology*, vol. 80, pp. 1522–1536. doi: 10.1890/0012-9658(1999)080

López-Legentil, S., Legentil, M.L., Erwin, P.M. and Turon, X. (2015) 'Harbor networks as introduction gateways: Contrasting distribution patterns of native and introduced ascidians', *Biological Invasions*, vol. 17, pp. 1623–1638. doi: 10.1007/s10530-014-0821-z

Lowe, S., Browne, M., Boudjekas, S. and De Poorter, M. (2000) *100 of the World's Worst Invasive Alien Species.* The Invasive Species Specialist Group (ISSG) of the Species Survival Commission (SSC) of the World Conservation Union (IUCN), Auckland, New Zealand.

Macleod, A.K., Stanley, M.S., Day, J.G. and Cook, E.J. (2016) 'Biofouling community composition across a range of environmental conditions and geographical locations suitable for floating marine renewable energy generation', *Biofouling*, vol. 32, pp. 261–276. doi: 10.1080/08927014.2015.1136822

Maggi, E., Benedetti-Cecchi, L., Castelli, A., Chatzinikolaou, E., Crowe, T.P., Ghedini, G., Kotta, J., Lyons, D.A., Ravaglioli, C., Rilov, G., Rindi, L. and Bulleri, F. (2015) 'Ecological impacts of invading seaweeds: A meta-analysis of their effects at different trophic levels', *Diversity and Distributions*, vol. 21, pp. 1–12. doi: 10.1111/ddi.12264

Manchester, S.J. and Bullock, J.M. (2000) 'The impacts of non-native species on UK biodiversity and the effectiveness of control', *Journal of Applied Ecology*, vol. 37, no. 5, pp. 845–864.

Mayer-Pinto, M., Johnston, E.L., Bugnot, A.B., Glasby, T.M., Airoldi, L., Mitchell, A. and Dafforn, K.A. (2017) 'Building "blue": An eco-engineering framework for foreshore developments', *Journal of Environmental Management*, vol. 189, pp. 109–114. https://doi.org/10.1016/j.jenvman.2016.12.039

McCann, L.D., Holzer, K.K., Davidson, I.C., Ashton, G.V., Chapman, M.D. and Ruiz, G.M. (2013) 'Promoting invasive species control and eradication in the sea: Options for managing the tunicate invader Didemnum vexillum in Sitka, Alaska', *Marine Pollution Bulletin*, vol. 77, pp. 165–171. doi: 10.1016/j.marpolbul.2013.10.011

Megina, C., González-Duarte, M.M. and López-González, P.J. (2016) 'Benthic assemblages, biodiversity and invasiveness in marinas and commercial harbours: An investigation using a bioindicator group', *Biofouling*, vol. 32, pp. 465–475. doi: 10.1080/08927014.2016.1151500

Meyerson, L.A. and Mooney, H.A. (2007) 'Invasive alien species in an era of globalization', *Frontiers in Ecology and the Environment*, vol. 5, pp. 199–208. doi: 10.1890/1540-295(2007)5[199:IASIAE]2.0.CO;2

Miller, R.G., Hutchison, Z.L., Macleod, A.K., Burrows, M.T., Cook, E.J., Last, K.S. and Wilson, B. (2013) 'Marine renewable energy development: Assessing the Benthic Footprint at multiple scales', *Frontiers in Ecology and the Environment*, vol. 11, no. 8, pp. 433–440.

Minchin, D. (2007) 'Aquaculture and transport in a changing environment: Overlap and links in the spread of alien biota', *Marine Pollution Bulletin*, vol. 55, pp. 302–313. doi: 10.1016/j.marpolbul.2006.11.017

Minchin, D., Cook, E. and Clark, P. (2013) 'Alien species in British brackish and marine waters', *Aquatic Invasions*, vol. 8, pp. 3–19. doi: 10.3391/ai.2013.8.1.02

Minchin, D., Floerl, O., Savini, D. and Occhipinti-Ambrogi, A. (2006) 'Small craft and the spread of exotic species', in J. Davenport and J.L. Davenport (eds.) *The Ecology of Transportation: Managing Mobility for the Environment*. Springer, Netherlands.

Minchin, D. and Gollasch, S. (2003) 'Fouling and ships' hulls: How changing circumstances and spawning events may result in the spread of exotic species', *Biofouling*, vol. 19, pp. 111–122. doi: 10.1080/0892701021000057891

Mineur, F., Cook, E., Minchin, D., Bohn, K., MacLeod, A. and Maggs, C. (2012) 'Changing coasts: Marine aliens and artificial structures', *Oceanography and Marine Biology: An Annual Review*, vol. 50, pp. 187–232. doi: 10.1201/b12157-5

Miralles, L., Gomez-Agenjo, M., Rayon-Viña, F., Gyraitė, G. and Garcia-Vazquez, E. (2018) 'Alert calling in port areas: Marine litter as possible secondary dispersal vector for hitchhiking invasive species', *Journal for Nature Conservation*, vol. 42, pp. 12–18. doi: 10.1016/j.jnc.2018.01.005

Molnar, J.L., Gamboa, R.L., Revenga, C., Spalding, M.D. and Molnar, J. (2008) 'Assessing the global threat of invasive species to marine biodiversity', *Frontiers in Ecology and the Environment*, vol. 6, pp. 485–492. doi: 10.1890/070064

Moreira, P.L., Ribeiro, F.V. and Creed, J.C. (2014) 'Control of invasive marine invertebrates: An experimental evaluation of the use of low salinity for managing pest corals (*Tubastraea spp.*)', *Biofouling*, vol. 30, pp. 639–650. doi: 10.1080/08927014.2014.906583

Morris, J.A., Carman, M.R., Hoagland, K.E., Green-Beach, E. and Karney, R.C. (2009) 'Impact of the invasive colonial tunicate Didemnum vexillum on the recruitment of the bay scallop (*Argopecten irradians irradians*) and implications for recruitment of the sea scallop (*Placopecten magellanicus*) on Georges Bank', *Aquatic Invasions*, vol. 4, pp. 207–211.

Nall, C.R. (2015) *Marine Non-Native Species in Northern Scotland and the Implications for the Marine Renewable Energy Industry*. PhD thesis, University of Aberdeen, UK.

Nall, C.R., Guerin, A. and Cook, E. (2015) 'Rapid assessment of marine non-native species in northern Scotland and a synthesis of existing Scottish records', *Aquatic Invasions*, vol. 10, pp. 107–121. doi: 10.3391/ai.2015.10.1.11

Naylor, L.A., MacArthur, M., Hampshire, S., Bostock, K., Coombes, M.A., Hansom, J.D., Byrne, R. and Folland, T. (2017) 'Rock armour for birds and their prey: Ecological enhancement of coastal engineering', *Proceedings of the Institution of Civil Engineers: Maritime Engineering*, vol. 170, pp. 67–82. doi: 10.1680/jmaen.2016.28

Occhipinti-Ambrogi, A. (2007) 'Global change and marine communities: Alien species and climate change', *Marine Pollution Bulletin*, vol. 55, pp. 342–352. doi: 10.1016/j.marpolbul.2006.11.014

Occhipinti-Ambrogi, A. and Savini, D. (2003) 'Biological invasions as a component of global change in stressed marine ecosystems', *Marine Pollution Bulletin*, vol. 46, pp. 542–551. doi: 10.1016/S0025-326X(02)00363-6

Ojaveer, H., Galil, B.S., Campbell, M.L., Carlton, J.T., Canning-Clode, J., Cook, E.J., Davidson, A.D., Hewitt, C.L., Jelmert, A., Marchini, A., McKenzie, C.H., Minchin, D., Occhipinti-Ambrogi, A., Olenin, S. and Ruiz, G. (2015) 'Classification of non-indigenous species based on their impacts: Considerations for application in marine management', *PLoS Biology*, vol. 13, p. e1002130. doi: 10.1371/journal.pbio.1002130

Paavola, M., Olenin, S. and Leppäkoski, E. (2005) 'Are invasive species most successful in habitats of low native species richness across European brackish water seas?', *Estuarine, Coastal and Shelf Science*, vol. 64, pp. 738–750. doi: 10.1016/j.ecss.2005.03.021

Padilla, D.K. and Williams, S.L. (2004) 'Beyond ballast water: Aquarium and ornamental trades as sources of invasive species in aquatic ecosystems', *Frontiers in Ecology and the Environment*, vol. 2, pp. 131–138.

Page, H.M., Dugan, J.E., Culver, C.S. and Hoesterey, J.C. (2006) 'Exotic invertebrate species on offshore oil platforms', *Marine Ecology Progress Series*, vol. 325, pp. 101–107. doi: 10.3354/meps325101

Pardo, L.M., González, K., Fuentes, J.P., Paschke, K. and Chaparro, O.R. (2011) 'Survival and behavioral responses of juvenile crabs of Cancer edwardsii to severe hyposalinity events triggered by increased runoff at an estuarine nursery ground', *Journal of Experimental Marine Biology and Ecology*, vol. 404, pp. 33–39. doi: 10.1016/j.jembe.2011.05.004

Passarelli, B. and Pernet, B. (2019) 'The marine live bait trade as a pathway for the introduction of non-indigenous species into California: Patterns of importation and thermal tolerances of imported specimens', *Management of Biological Invasions*, vol. 10, no. 1, pp. 80–95.

Patricola, C.M. and Wehner, M.F. (2018) 'Anthropogenic influences on major tropical cyclone events', *Nature*, vol. 563, pp. 339–346. doi: 10.1038/s41586-018-0673-2

Perkol-Finkel, S., Shashar, N. and Benayahu, Y. (2006) 'Can artificial reefs mimic natural reef communities? The roles of structural features and age', *Marine Environmental Research*, vol. 61, pp. 121–135. doi: 10.1016/j.marenvres.2005.08.001

Peters, K., Sink, K. and Robinson, T.B. (2017) 'Raising the flag on marine alien fouling species', *Management of Biological Invasions*, vol. 8, no. 1, pp. 1–11.

Pimentel, D., Zuniga, R. and Morrison, D. (2005) 'Update on the environmental and economic costs associated with alien-invasive species in the United States', *Ecological Economics*, vol. 52, pp. 273–288. doi: 10.1016/j.ecolecon.2004.10.002

Piola, R.F., Dunmore, R.A. and Forrest, B.M. (2009) 'Assessing the efficacy of spray-delivered "eco-friendly" chemicals for the control and eradication of marine fouling pests', *Biofouling*, vol. 26, pp. 187–203. doi: 10.1080/08927010903428029

Piola, R.F. and Johnston, E.L. (2008) 'Pollution reduces native diversity and increases invader dominance in marine hard-substrate communities', *Diversity and Distributions*, vol. 14, pp. 329–342. doi: 10.1111/j.1472-4642.2007.00430.x

Rivero, N.K., Dafforn, K.A., Coleman, M.A. and Johnston, E.L. (2013) 'Environmental and ecological changes associated with a marina', *Biofouling*, vol. 29, pp. 803–815. doi: 10.1080/08927014.2013.805751

Roche, R.C., Monnington, J.M., Newstead, R.G., Sambrook, K., Griffith, K., Holt, R.H.F. and Jenkins, S.R. (2015) 'Recreational vessels as a vector for marine non-natives: Developing biosecurity measures and managing risk through an in-water encapsulation system', *Hydrobiologia*, vol. 750, pp. 187–199. doi: 10.1007/s10750-014-2131-y

Rodriguez, L.F. (2006) 'Can invasive species facilitate native species? Evidence of how, when, and why these impacts occur', *Biological Invasions*, vol. 8, pp. 927–939. doi: 10.1007/s10530-005-5103-3

Ros, M., Ashton, G.V., Lacerda, M.B., Carlton, J.T., Vázquez-Luis, M., Guerra-García, J.M. and Ruiz, G.M. (2014) 'The Panama Canal and the transoceanic dispersal of marine invertebrates: Evaluation of the introduced amphipod Paracaprella pusilla Mayer, 1890 in the Pacific Ocean', *Marine Environmental Research*, vol. 99, pp. 204–211.

Roy, H.E., Peyton, J., Aldridge, D.C., Bantock, T., Blackburn, T.M., Britton, R., Clark, P., Cook, E., Dehnen-Schmutz, K., Dines, T., Dobson, M., Edwards, F., Harrower, C., Harvey, M.C., Minchin, D., Noble, D.G., Parrott, D., Pocock, M.J.O., Preston, C.D., Roy, S., Salisbury, A., Schönrogge, K., Sewell, J., Shaw, R.H., Stebbing, P., Stewart, A.J.A. and Walker, K.J. (2014) 'Horizon scanning for invasive alien species with the potential to threaten biodiversity in Great Britain', *Global Change Biology*, vol. 20, no. 12, pp. 3859–3871. doi: 10.1111/gcb.12603

Ruiz, G.M., Carlton, J.T., Grosholz, E. and Hines, A.H. (1997) 'Global invasions of marine and estuarine habitats by non-indigenous species: Mechanisms, extent and consequences', *American Zoologist*, vol. 37, pp. 621–632.

Ruiz, G.M., Fofonoff, P., Hines, A.H. and Grosholz, E.D. (1999) 'Non-indigenous species as stressors in estuarine and marine communities: Assessing invasion impacts and interactions', *Limnology and Oceanography*, vol. 44, pp. 950–972. doi: 10.4319/lo.1999.44.3_part_2.0950

Ruiz, G.M., Freestone, A.L., Fofonoff, P.W. and Simkanin, C. (2009a) 'Habitat distribution and heterogeneity in marine invasion dynamics: The importance of hard substrate and artificial structure', in M. Wahl (ed.) *Marine Hard Bottom Communities*. Springer, Berlin, Heidelberg.

Ruiz, G.M., Torchin, M.E. and Grant, K. (2009b) 'Using the Panama Canal to test predictions about tropical marine invasions', *Smithsonian Contributions to the Marine Sciences*, vol. 38, pp. 291–299.

Scottish Government (2018) *Holyhead Marina Debris: Information and Advice*. Marine invasive non-native species advice note, www2.gov.scot/Topics/marine/marine-environment/species/non-natives/Holyhead, accessed 31 October 2019.

Seebens, H., Blackburn, T.M., Dyer, E.E., Genovesi, P., Hulme, P.E., Jeschke, J.M., Pagad, S., Pyšek, P., Winter, M., Arianoutsou, M., Bacher, S., Blasius, B., Brundu, G., Capinha, C., Celesti-Grapow, L., Dawson, W., Dullinger, S., Fuentes, N., Jäger, H., Kartesz, J., Kenis, M., Kreft, H., Kühn, I., Lenzner, B., Liebhold, A., Mosena, A., Moser, D., Nishino, M., Pearman, D., Pergl, J., Rabitsch, W., Rojas-Sandoval, J., Roques, A., Rorke, S., Rossinelli, S., Roy, H.E., Scalera, R., Schindler, S., Štajerová, K., Tokarska-Guzik, B., van Kleunen, M., Walker, K., Weigelt, P., Yamanaka, T. and Essl, F. (2017) 'No saturation in the accumulation of alien species worldwide', *Nature Communications*, vol. 8, p. 14435. doi: 10.1038/ncomms14435

Seebens, H., Gastner, M.T. and Blasius, B. (2013) 'The risk of marine bioinvasion caused by global shipping', *Ecology Letters*, vol. 16, pp. 782–790. doi: 10.1111/ele.12111

Shea, K. and Chesson, P. (2002) 'Community ecology theory as a framework for biological invasions', *Trends in Ecology and Evolution*, vol. 17, pp. 170–176. doi: 10.1016/S0169-5347(02)02495-3

Sheehy, D.J. and Vik, S.F. (2010) 'The role of constructed reefs in non-indigenous species introductions and range expansions', *Ecological Engineering*, vol. 36, pp. 1–11. doi: 10.1016/j.ecoleng.2009.09.012

Shiganova, T. (1998) 'Invasion of the Black Sea by the ctenophore Mnemiopsis leidyi and recent changes in pelagic community structure', *Fisheries Oceanography*, vol. 7, pp. 305–310. doi: 10.1046/j.1365-2419.1998.00080.x

Simberloff, D., Simberloff, D., Holle, B.V. and Holle, B.V. (1999) 'Positive interactions of nonindigenous species: Invasional meltdown?', *Biological Invasions*, vol. 1, no. 1, pp. 21–32. doi: 10.1023/a:1010086329619

Smyth, K. and Elliott, M. (2016) 'Effects of changing salinity on the ecology of the marine environment', in M. Solan and N.M. Whiteley (eds.) *Stressors in the Marine Environment.* Oxford University Press, Oxford, UK.

Sorte, C.J.B., Fuller, A. and Bracken, M.E.S. (2010) 'Impacts of a simulated heat wave on composition of a marine community', *Oikos*, vol. 119, pp. 1909–1918. doi: 10.1111/j.1600-0706.2010.18663.x

Stachowicz, J.J., Fried, H., Osman, R.W. and Whitlatch, R.B. (2002) 'Biodiversity, invasion resistance, and marine ecosystem function: Reconciling pattern and process', *Ecology*, vol. 83, pp. 2575–2590.

Strain, E.M.A., Olabarria, C., Mayer-Pinto, M., Cumbo, V., Morris, R.L., Bugnot, A.B., Dafforn, K.A., Heery, E., Firth, L.B., Brooks, P.R. and Bishop, M.J. (2018) 'Eco-engineering urban infrastructure for marine and coastal biodiversity: Which interventions have the greatest ecological benefit?', *Journal of Applied Ecology*, vol. 55, pp. 426–441. doi: 10.1111/1365-2664.12961

Streftaris, N., Zenetos, A. and Papathanassiou, E. (2005) 'Globalisation in marine ecosytems: The story of non-indigenous marine species across European seas', *Oceanography and Marine Biology: An Annual Review*, vol. 43, pp. 419–453.

Sundseth, K. (2014) *Invasive Alien Species: A European Response.* European Union, https://ec.europa.eu/environment/nature/invasivealien/docs/ias-brochure-en-web.pdf, accessed 31 October 2019.

Tyrrell, M.C. and Byers, J.E. (2007) 'Do artificial substrates favor nonindigenous fouling species over native species?', *Journal of Experimental Marine Biology and Ecology*, vol. 342, pp. 54–60. doi: 10.1016/j.jembe.2006.10.014

Vaselli, S., Bulleri, F. and Benedetti-Cecchi, L. (2008) 'Hard coastal-defence structures as habitats for native and exotic rocky-bottom species', *Marine Environmental Research*, vol. 66, pp. 395–403. doi: 10.1016/j.marenvres.2008.06.002

Vaz-Pinto, F., Torrontegi, O., Prestes, A.C.L., Alvaro, N.V., Neto, A.I. and Martins, G.M. (2014) 'Invasion success and development of benthic assemblages: Effect of timing, duration of submersion and substrate type', *Marine Environmental Research*, vol. 94, pp. 72–79. doi: 10.1016/j.marenvres.2013.12.007

Vilà, M., Basnou, C., Pyšek, P., Josefsson, M., Genovesi, P., Gollasch, S., Nentwig, W., Olenin, S., Roques, A., Roy, D. and Hulme, P.E. (2010) 'How well do we understand the impacts of alien species on ecosystem services? A pan-European, cross-taxa assessment', *Frontiers in Ecology and the Environment*, vol. 8, pp. 135–144. doi: 10.1890/080083

Watson, D.I., Shumway, S.E. and Whitlatch, R.B. (2009) 'Biofouling and the shellfish industry', in S.E. Shumway and G.E. Rodrick (eds.) *Shellfish Safety and Quality.* Woodhead Publishing, Cambridge, UK.

Williams, F., Eschen, R., Harris, A., Djeddour, D., Pratt, C., Shaw, R.S., Varia, S., Lamontagne-Godwin, J., Thomas, S.E. and Murphy, S.T. (2010) *The Economic Cost of Invasive Non-Native Species on Great Britain.* CABI report CAB/001/09, Wallingford, UK.

Williams, S.L. and Smith, J.E. (2007) 'A global review of the distribution, taxonomy, and impacts of introduced seaweeds', *Annual Review of Ecology, Evolution and Systematics*, vol. 38, pp. 327–359. doi: 10.1146/annurev.ecolsys.38.091206.095543

Yahia, M.N.D., Yahia, O.K-D., Gueroun, S.K.M., Aissi, M., Deidun, A., Fuentes, V. and Piraino, S. (2013) 'The invasive tropical scyphozoan Rhopilema nomadica Galil, 1990 reaches the Tunisian coast of the Mediterranean Sea', *BioInvasions Records*, vol. 2, pp. 319–323.

Zabin, C., Ashton, G., Brown, C., Davidson, I., Sytsma, M. and Ruiz, G. (2014) 'Small boats provide connectivity for nonindigenous marine species between a highly invaded international port and nearby coastal harbors', *Management of Biological Invasions*, vol. 5, pp. 97–112. http://dx.doi.org/10.3391/mbi.2014.5.2.03

10

SPECIES INVASIONS IN FRESHWATER ECOSYSTEMS

Robert A. Francis and Michael A. Chadwick

Introduction

Freshwater ecosystems are among the most degraded in the world, with the vast majority impacted by hard engineering and flow modification, pollution, habitat loss and species invasions (Dudgeon et al., 2006; Moorhouse and Macdonald, 2015). This is of particular concern given the wide range of ecosystem services that these ecosystems provide, including food and water provision, nutrient cycling, transport, recreation and maintenance of biodiversity (Brauman, 2016). Despite occupying somewhere between 0.8%–2.5% of the earth's surface (see Costanza et al., 1997; Dudgeon et al., 2006), rivers, lakes and other inland wetlands are among the most biodiverse ecosystems in existence. However, both the legacy of past degradation and the ongoing risk of ecological harm has made freshwater ecosystems a focus of conservation efforts, including national, regional and international policy and legislation such as the Ramsar Convention, EU Water Framework Directive and North American Wetlands Conservation Act.

Freshwater ecosystem degradation has been occurring for millennia as a result of successive societies harnessing desired ecosystem services while discouraging disservices – for example, channel engineering to improve transport capabilities of rivers while reducing local flood risk. The rate and extent of degradation increased dramatically following the industrial revolution, alongside growing human populations and increases in agriculture and urbanisation (Allan, 1995). Despite progress in the setting and implementation of policy and legislation, combined with some successful restoration and conservation interventions, freshwater ecosystems remain degraded globally, and in some regions further degradation continues apace (e.g. Xing et al., 2016).

Species invasions are common in freshwater ecosystems and can be both a symptom and driver of ecological change (Francis and Chadwick, 2012; Nunes et al., 2015). Invasions have increased dramatically over recent decades, with rates of invasion still accelerating (Lenzner et al., 2019). Keller et al. (2009) documented substantial increases in freshwater invasions in Great Britain from historical records, associated with increased population growth, economic output and globalisation resulting from the industrial revolution in the mid-19th century. Many early introductions were intentional, but, particularly from the mid-20th century, unintentional introductions have become increasingly common. Jackson and Grey (2013) found that new species invasions are recorded every 50 weeks on average in the River Thames, London, with 53% of non-natives having established in the last 50 years, while Leuven et al. (2009) found that invasion rates increased from less than 1 species per decade in the 18th century to 13 per decade in more recent years.

This chapter explores species invasions in the context of freshwater ecosystems, including mechanisms of introduction and spread, types of impact and key considerations around prevention and control of species invasions. For the purposes of this chapter, 'freshwater ecosystems' includes rivers, lakes, ponds and any other form of inland wetland, including their riparian zones. 'Species invasions' refers to the introduction and spread of non-native (also 'alien' or 'exotic') species, although most examples are drawn from the subset of non-native species that cause some form of ecological, societal or economic harm and are therefore designated as 'invasive alien species', or IAS (e.g. Turbelin et al., 2017).

A background to freshwater invasions

Freshwater ecosystems, particularly rivers, have been associated with the growth (and fates) of civilisations for millennia (Allan, 1995; Macklin and Lewin, 2015). This association established longstanding conditions for species invasions, including (1) habitat modification and loss, as the disturbance of biophysical habitat creates opportunities for non-native species to colonise, and (2) increased movement of people and goods, offering opportunities for introduction as species are moved through the freshwater systems intentionally (e.g. for trade) or unintentionally, as 'passengers' on transports or people. Historic examples of species invasions include the common carp (*Cyprinus carpio*), introduced across Europe as early as 2,000 years ago, with expanses in distribution driven particularly by monastic aquaculture (Hicks et al., 2012), and more recently the introduction of mosquitofish (*Gambusia* spp.) into a wide range of temperate and tropical countries as control agents for mosquitoes from the early 1900s (Walton et al., 2012).

These two factors remain highly relevant to current species invasions (e.g. Nunes et al., 2015). Water-based transport, both within region and internationally, is a key pathway for species introduction, with, for example, ship hulls and ballast water often harbouring freshwater or euryhaline species acquired from different biogeographical regions (Keller et al., 2011; Bailey, 2015). It has been estimated that up to 7,000 species may be in the process of international transportation in ships in any given 24-hour period (Leppäkoski et al., 2002), and shipping remains a major industry, with global shipping doubling between 1990 and 2007 (World Ocean Review, 2010), although the financial crisis of 2008 led to a decline that has since experienced some recovery but remains volatile (UNCTAD, 2018). Transoceanic transport was noted as a key vector in the extensive introduction of non-natives to the Great Lakes of North America (Ricciardi and MacIsaac, 2000) while more local recreational boating has been found to be a vector for dispersal of zebra mussels (*Dreissena polymorpha*) in lakes, with larvae contained in live wells, bilges, bait buckets and engines but the species mainly being spread via macrophytes entangled on boat trailers and anchors (Johnson et al., 2001), even when transported overland between waterbodies (De Ventura et al., 2016).

The introduction of non-native species for aquaculture, the pet/aquarium trade and stocking for recreational fishing are also major pathways for introduction, as found for freshwater species invasions of Europe (Nunes et al., 2015) and elsewhere (Padilla and Williams, 2004; Grosholz et al., 2015). This is particularly the case for non-native fish and crayfish (Savini et al., 2010; Smith, 2019).

These varied factors can result in multiple spatial points of introduction along river catchments or lakes, facilitating both population growth and geographical spread (Johnson et al., 2006). While ports represent a key source of introduction and spread (Drake and Lodge, 2004), movement of transports between waterbodies (including overland) and the dissemination of species across the landscape in fish farms, ponds and so on increases the risk of unintentional entry to individual reaches across catchments (Lee and Gordon, 2006). In addition, the landscape characteristics of freshwater ecosystems themselves make them susceptible to species invasions, increasing their 'invasibility' (Moorhouse and Macdonald, 2015). Of these, landscape position and connectivity are particularly relevant to species invasions.

Landscape position

Rivers and lakes exist in areas of low elevation as gravity draws water to these locations. As runoff and throughflow gravitate to rivers and lakes, they transport along with them sediments, nutrients and, importantly, plant propagules (de Rouw et al., 2018). Freshwater ecosystems therefore represent a 'sink' (in the 'storage' sense of the term) for these materials. Seeds and other propagules of non-native species may be carried to such ecosystems from the surrounding landscape, especially where waterbodies are fed by surface runoff (Zedler and Kercher, 2004; de Rouw et al., 2018), while the nutrients contained in runoff can cause cultural eutrophication, favouring non-native species that may have a competitive advantage over natives in terms of resource acquisition or tolerance to degraded conditions (Boers et al., 2006). Landscape position and the diffuse nature of inputs can also create multiple points of introduction throughout river networks or lakes. Lower elevations also tend to be associated with more dramatic land use change, such as urbanisation, and higher populations. Landscape position and urban land cover (along with relative patch richness as a measure of land cover type variation and game fish abundance) were found to be strong predictors of freshwater invasion by Shaker et al. (2017).

Landscape connectivity and directionality of hydrological flow

Rivers, at least in their natural form, are highly connected ecosystems. Longitudinal connectivity between reaches means that an invasion in one part of the system has the potential to propagate elsewhere with relative ease, although this depends in large part on the particular organism in question, as functional connectivity will vary for different species with different movement or dispersal abilities (e.g. Kornis and Vander Zanden, 2010). Longitudinal flows are predominantly unidirectional (downstream), and so for most aquatic or riparian organisms (especially plants and invertebrates), unassisted spread downstream from an upstream invasion point is more likely (Johnson et al., 2006; Dallimer et al., 2012; Churchill and Quigley, 2018), although for easily mobile organisms (such as fish and mammals), spread upstream is also common. As in most cases where spread is facilitated by human agency, more complex patterns of spread (including 'jump dispersal') are also common (e.g. Loo et al., 2007). Several studies have demonstrated rapid rates of longitudinal spread along river catchments, including 44–112 km year^{-1} for non-native macroinvertebrates along the River Rhine (Leuven et al., 2009) and movement of 30 km per week for *Hypopthalmichthys* spp. in high flow conditions (Garvey, 2012), although rates may vary temporally, with, for example, nutria (*Myocastor coypus*) expanding at mean annual rates of between 11 km year^{-1} and 49.7 km year^{-1} along the Nakdong River catchment in South Korea between 1999 and 2013 (Hong et al., 2015).

Alongside longitudinal connectivity, rivers (and to a lesser extent lentic waterbodies) maintain lateral connectivity between channel/edge and floodplain so that high flows connect within-channel habitat with the floodplain, offering opportunities for exchange of materials and species (Ward et al., 2002). In this way, non-native species may move between river channels and adjacent waterbodies (such as ponds, ditches and backwaters) or riparian areas and vice versa (Predick and Turner, 2008).

Rivers also maintain vertical connectivity across the hyporheic zone as surface water and groundwater move into or out of the stream bed. This zone represents an ecotone that maintains its own particular communities and as such may be vulnerable to species invasions. Despite some observations of non-native species in hyporheic zones (Marmonier et al., 2012), and exploration of how some non-native species may utilise the zone (e.g. Kouba et al., 2016), there is so far relatively little research into hyporheic zones in the context of freshwater invasions, including the role of vertical connectivity.

Lentic ecosystems such as lakes and ponds are also highly connected internally, and once a species has invaded, spread may be rapid due to lake currents dispersing individuals (e.g. Beletsky et al., 2017), as well as continued introductions from, for example, navigation and recreational boating (Mandrak and Cudmore, 2010). Lakes may also be connected through natural waterway linkages, and regional clusters of lakes can experience both waterborne and overland invasions from source populations through many of the same pathways for lotic ecosystems, over both short and long distances (MacIsaac et al., 2004). Rates of spread between isolated waterbodies such as lakes may be slower than between reaches of river catchments, with, for example, Johnson et al. (2006) identifying that only a small proportion (8%) of suitable inland lakes had been invaded in the first 15 years of *D. polymorpha* invasion in the US and that rates were slowing, in contrast to spread along navigable waterways.

Anthropogenic connections created between freshwater systems, such as transportation canals or water transfer pipelines, have been particularly problematic, sometimes resulting in 'invasion highways' (Zhan et al., 2015); examples include the Chicago Area Waterway linking Lake Michigan with, ultimately, the Mississippi River (Rasmussen et al., 2011); the Main-Danube Canal in the Rhine and Danube basins (Leuven et al., 2009); and the Champlain Canal connecting Lake Champlain with the Hudson River (Marsden and Ladago, 2017).

Hindering – or rather not restoring – connectivity has been suggested as one form of management for species invasions. 'Favourable fragmentation' may be ecologically harmful in many ways as it interrupts flow and conduit functions for native species and movement of (for example) sediments and nutrients, but in some scenarios this is potentially preferable to allowing the spread of particularly damaging invasive species (Rood et al., 2010). This concern has mainly been raised when removal of impoundments has been suggested, with the observation that doing so may inadvertently lead to species invasions (as well as the release of stored sediments, pollutants and so on), either through increasing connectivity or creating a point of disturbance that could be exploited by IAS (Stanley and Doyle, 2003; Kornis and Vander-Zanden, 2010). However, reservoirs in general remain a concentrated source of non-native species which are likely to spread even where connectivity is restricted (Johnson et al., 2008).

Impacts associated with freshwater invasions

Much research has been conducted on the impacts of species invasions in freshwater ecosystems (for reviews see Francis and Chadwick, 2012; Moorhouse and Macdonald, 2015; Francis et al., 2019), and freshwater impacts are generally considered to be more severe than in terrestrial ecosystems (Moorhouse and Macdonald, 2015). Impacts are, however, highly context-specific and will vary according to the particular ecosystem invaded, the traits of the invading species (see Chapter 1) the introduction effort and population size, the time since invasion and other biotic and abiotic factors (e.g. Higgins and Vander Zanden, 2010). Broadly, freshwater impacts can be grouped into impacts on the ecology, impacts on resources/service and economics and effects on human health. The main forms of documented impact are noted in the following, although this list is far from exhaustive.

Ecological impacts

A suite of ecological impacts has been associated with freshwater invasions. These impacts tend to be complex, context-specific and interrelated, such as changes to biophysical conditions or species behaviour that then reduce native populations and can lead to species extirpations and biodiversity loss. Examples of documented impacts include:

(1) Loss of native biodiversity through reducing populations and causing local extirpations (and, in the case of some endemics, global extinctions; Mandrak and Cudmore, 2010): Common mechanisms include (i) outcompeting natives, with some non-natives showing greater capacity for resource acquisition due to particular functional traits or because of an absence of predators, herbivores or pathogens in the invaded ecosystem ('enemy release') – as observed for non-native red-eared slider turtles (*Trachemys scripta elegans*) that outcompeted native red-bellied turtles (*Pseudemys rubriventris*) for limited food sources in mesocosm experiments (Pearson et al., 2015) or reduced initial levels of helminth parasitisation on non-native Eurasian round goby (*Neogobius menalostomus*) in the Great Lakes–St. Lawrence basin compared to native fish, even if short-lived (Gendron et al., 2012); (ii) predation on natives, particularly juveniles, as observed for the fish community of the Rondegat River in South Africa following invasion by smallmouth bass (*Micropterus dolomieu*; Woodford et al., 2005); (iii) causing infection of native species, such as the spread of crayfish plague (*Aphanomyces astaci*), which has led to declines in native crayfish populations across Europe and is attributed to North American signal crayfish (*Pacifastacus leniusculus*; James et al., 2017); and (iv) changing feeding relationships and trophic dynamics (for example greater role of non-natives in diet), sometimes leading to cascading effects (e.g. Pagnucco et al., 2016; Lisi et al., 2018). In some cases, invasions may involve several, or all, of these mechanisms of impact, such as the extirpations and extinctions resulting from fish invasions around the Great Lakes of North America (Mandrak and Cudmore, 2010) and the introduction of Nile perch (*Lates niloticus*) to Lake Victoria in East Africa (Goldschmidt et al., 1993).

(2) Hybridisation and genetic pollution. Despite historic geographical separation, often non-native and native taxa are closely related and hybridisation between species can occur (see Facon et al., 2005). Hybrids may exhibit increased fitness and consequently outcompete original native populations, especially for plants but also for some animals (Ellstrand and Schierenbeck, 2000; Facon et al., 2005). In some cases, introductions can lead to trends such as introgressive hybridisation and the decline and replacement of native species or genotypes (Dufresnes et al., 2016).

(3) Changes to species behaviour, for example disruption of mating, hunting or foraging behaviour. For example, female white-clawed crayfish (*Austropotamobius pallipes*) have been observed to respond to chemical signatures from non-native male North American signal crayfish (*Pacifastacus leniusculus*), leading to reproductive interference (Dunn, 2012), while D'Amore et al. (2009) observed similar reproductive interference (interspecific amplexus, so the adoption of mating positions that are ultimately redundant) between native and non-native frogs in North America.

(4) Changes to the biophysical environment. In freshwater ecosystems, in particular, impacts on the biophysical environment and the species habitat it represents include (i) sediment dynamics, such as non-native crayfish increasing sediment suspension and water turbidity through feeding and burrowing activity (Matsuzaki et al., 2009; Faller et al., 2016) or non-native plants causing bank erosion and sediment entrainment (Arnold and Toran, 2018; Greenwood et al., 2018); (ii) nutrient loads and cycling, for example through changes to

the amount, frequency and biochemical properties of plant litter and other organics that may enter the ecosystem, as well as breakdown rates and patterns of storage/accumulation (e.g. Fargen et al., 2015; Kuglerová et al., 2017); and (iii) changes to other resources, such as light, as in cases when non-native plants like Eurasian watermilfoil (*Myriophyllum spicatum*) cover water surfaces with dense mats of foliage and reduce light availability to other plants and aquatic animals (Havel et al., 2015). Habitat changes are not always negative and may sometimes increase populations and species diversity through mechanisms such as increasing habitat complexity and food provision (Burlakova et al., 2012; Albertson and Daniels, 2016).

(5) Facilitation of further invasions and invasional meltdown. Multiple invasions of species with similar biogeographical backgrounds may create ecological conditions that are conducive to further invasions, generating positive feedback and leading to what has been termed 'invasional meltdown' or the 'invasional meltdown hypothesis' (Gallardo and Aldridge, 2015; Braga et al., 2018). Supporting evidence for processes of invasional meltdown is complex; Braga et al. (2018) found that a majority of studies examining invasional meltdown supported the hypothesis of invaders facilitating further invasions but that the evidence was limited at community and ecosystem scales and for some habitats and organism types. Nonetheless, the hypothesis may be used to indicate potential impacts or areas at risk of facilitated invasions. In a review of Ponto-Caspian invaders, Gallardo and Aldridge (2015) found that 76% of reported interactions between invaders from this region were either positive or neutral, creating risks of large-scale complimentary Ponto-Caspian invasions in, for example, the southeast of England (see also Mills et al., 2019).

Economic and ecosystem service impacts

Freshwater ecosystems provide multiple services and resources that can be compromised by species invasions. These include increased flood risk from clogged channels or waterways (Zavaleta, 2000), clogging of water pipes (Pejchar and Mooney, 2009), reduction of stock in aquaculture or naturally fished areas (Lovell et al., 2006), loss of crops (Cowie and Hayes, 2012), loss of aesthetic appeal (Pejchar and Mooney, 2009), reduction of recreation and tourism potential (Pejchar and Mooney, 2009) and so on: for greater exploration of the impacts of species invasions on ecosystem services, see Charles and Dukes (2008), Pejchar and Mooney (2009) and Vilà and Hulme (2017).

Estimates of global economic impact of IAS exist and have highlighted that invasions of aquatic ecosystems in the US incur a cost of $7.7 billion annually for damages and control (Pimental et al., 2005; Havel et al., 2015). Detailed freshwater-specific evaluations are less common but have been conducted for specific species and ecosystems (see Lovell et al., 2006; Marbuah et al., 2014 for examples). These usually relate to loss of aquaculture production, but loss of broader ecosystem service valuations have also been made; for example Walsh et al. (2016) calculated that the invasion of Lake Mendota (Wisconsin, US) by spiny water flea (*Bythotrephes longimanus*), and the consequent declines of both the grazer *Daphnia pulicaria* and lake water clarity, cost an estimated $140 million in lost services. Horsch and Lewis (2009) estimated an average 13% decrease in land values around lakes following infestation with Eurasian watermilfoil (*M. spicatum*) in forested areas of Wisconsin.

Financial impacts also include the cost of control efforts, which can be substantial. Oreska and Aldridge (2011) estimated control costs for freshwater non-natives in Britain at £26.5 million year^{-1}, with potential costs rising to £43.5 million year^{-1} if control was applied to all invaded freshwater ecosystems. These figures seem only likely to increase along with the rate and extent of invasions.

Human health and quality of life impacts

Freshwater invasions can have a suite of human health and quality of life impacts. The most obvious examples include species that may act as vectors for transmission of human pathogens, such as mosquitoes that may transmit malaria, West Nile virus and dengue fever (e.g. *Aedes albopictus*, *A. japonicus* and *A. aegypti*; Leisnham, 2012). One of the earliest examples of human health impacts of a freshwater invader is the extensive malaria outbreak caused by the introduction of *Anopheles arabiensis* (reported as *Anopheles gambiae*) to Brazil around 1929 (Elton, 1958). It is also one of the best examples of successful eradication of an invasive species, as *A. arabiensis* was removed from 54,000 km^2 of suitable habitat by the early 1940s, utilising an integrated control programme with particular emphasis on larval control (Killeen et al., 2002).

There is a suite of invasive aquatic and riparian species that can have health and lifestyle impacts, however (Levy, 2004; Mazza and Tricarico, 2018; Souty-Grosset et al., 2018). Notable examples include giant hogweed (*Heracleum mantegazzianum*), which contains a phototoxic sap in all its tissues (including the seeds) that can cause inflammation and hyperpigmentation in skin it comes into contact with (Pergl et al., 2012), and freshwater snails that can host pathogens responsible for schistosomiasis and dermatitis (Levy, 2004; Tolley-Jordan and Chadwick, 2012).

Control of freshwater invasions

Control of species invasions can be separated into pre-introduction and post-introduction efforts. Pre-introduction efforts include steps taken to stop the movement of non-natives between countries or regions and a focus on the key pathways and vectors of spread (e.g. trade, shipping). Post-introduction efforts are aimed at reducing the spread and impact of non-native species once they have been introduced and include physical, chemical and biological control. Often these control methods are combined to increase efficacy, especially chemical and physical control. These forms of control are often enacted by a range of different government and non-governmental organisations, as well as local communities that may act on a volunteer basis and also provide data to inform control efforts (Shine and Doody, 2011; Simpson et al., 2009). Freshwater ecosystems present particular challenges for control – for example, the logistical complications that arise when rivers cross multiple riparian territories and landowners or actors do not agree on control methods or expend differential effort or resources, meaning that populations in some sections of the river persist and can repopulate areas where they have been extirpated (Arango et al., 2015). As noted previously, the landscape position and connectivity of freshwater ecosystems exacerbates reintroduction potential if control is ineffective.

The most effective form of control is prevention of introduction. Many countries have put in place legislation and policy to prevent the introduction of non-native species, both generally and in specific cases of concern (Turbelin et al., 2017). Of these, several countries have legislation or policy that explicitly focuses on (or mentions) non-native species in freshwater or aquatic eco-systems, although some consider local control and prevention of spread rather than international introduction, and the vast majority focus on aquaculture and fisheries (Table 10.1).

By necessity, legislation and preventative efforts must focus on key drivers and vectors, which are often linked to trade (e.g. regulating trade and imports/exports or incorporating checks and controls into shipping activities; Lovell et al., 2006). Some preventative efforts are well established, such as the International Maritime Organisation (IMO) guidelines for prevention of biofouling and non-native spread of marine, euryhaline and freshwater species (IMO, 2019), although checking compliance and establishing efficacy of prevention can be difficult (Zabin et al., 2018).

Table 10.1 Examples of legislation and policy from various countries/regions that refer to freshwater/aquatic non-native and/or invasive species, many of which focus on aquaculture and fisheries

Country/region	Legislation/policy including reference to non-native and/or invasive species
Antigua and Barbuda	An act to provide for the development and management of fisheries and matters incidental thereto
	Fisheries Regulations, 2013 (S.I. No. 2 of 2013)
Australia	Fisheries Management Act 2007
Belarus	Decree No. 30 of the Ministry of Agriculture and Food validating the modalities of carrying out veterinary and sanitary expertise of fish and fish products
British Virgin Islands (UK)	Fisheries (Protected Species) Order, 2014 (S.I. No. 28 of 2014)
Canada	Aquatic Invasive Species Regulations (SOR/2015–121)
	Water Protection Act (C.C.S.M. c. W65)
	Lake Simcoe Protection Act, 2008 (SO 2008, c 23)
	Ontario Fishery Regulations, 2007 (SOR/2007–237)
	Commercial Aquaculture Regulation (R.Q. c. P-9.01, r.1)
Denmark	Order No. 654 on the management of ballast water and sediments from ships' ballast water tanks
European Union	Regulation (EU) No. 304/2011 of the European Parliament and of the Council amending Council Regulation (EC) No. 708/2007 concerning use of alien and locally absent species in aquaculture
	Council Regulation (EC) No. 708/2007 concerning use of alien and locally absent species in aquaculture
	Commission Regulation (EC) No. 506/2008 amending Annex IV to Council Regulation (EC) No. 708/2007 concerning use of alien and locally absent species in aquaculture

	Commission Regulation (EC) No. 535/2008 laying down detailed rules for the implementation of Council Regulation (EC) No. 708/2007 concerning use of alien and locally absent species in aquaculture
	Commission Regulation (EC) No. 710/2009 amending Regulation (EC) No. 889/2008 laying down detailed rules for the implementation of Council Regulation (EC) No. 834/2007, as regards laying down detailed rules on organic aquaculture animal and seaweed production
Gambia	Fisheries Act, 2007 (Act No. 20 of 2007)
Germany	Lower Saxony Coastal Fishery Ordinance
	Ordinance on the implementation of Council Regulation (EC) No. 708/2007 of 11 June 2007 concerning use of alien and locally absent species in aquaculture
Guernsey (UK)	Fishing Ordinance, 1997
Indonesia	Decree of the Minister of Marine and Fisheries No. Kep.08/MEN/2004 on Procedures for importing fish of new kinds or varieties into the territory of the Republic of Indonesia
Italy	Regional Act No. 4 amending Regional Act No. 19 of 1998 setting out the rules for the protection of hydrobiological resources and of fish population and regulating internal and maritime waters fisheries in the Veneto Region
	Regional Act No. 7 making provision on inland fisheries
	Regional Act No. 19 setting up the Regional Park of River Trebbia
Kazakhstan	Order No. 290 of the Minister of Agriculture validating the Regulation introduction of animals, stock enhancement of waterbodies, and acclimatization of new species and fishery reclamation of waterbodies
Kenya	Water Resources Management Rules, 2007 (L.N. No. 171 of 2007)
	Environmental Management (Lake Naivasha Management Plan) Order, 2004 (Cap. 387)
	Agriculture, Fisheries and Food Authority Act, 2013 (No. 13 of 2013)
Kyrgyzstan	Ministerial Decree No. 161 validating the Regulation on protection of fish stocks and natural habitats thereof in the fishing waterbodies
Madagascar	Ordonnance n° 93–022 portant réglementation de la pêche et de l'aquaculture
	(Continued)

169

Table 10.1 Continued

Country/region	Legislation/policy including reference to non-native and/or invasive species
Malaysia	Fisheries (Prohibition of Import, etc., of Fish) (Amendment) Regulations, 2011
Namibia	Aquaculture (Licensing) Regulations: Aquaculture Act, 2002 (G.N. No. 245 of 2003)
Poland	Regulation on permissions for activities with the use of non-indigenous fish
Russian Federation	Regional Law No. 154-Z 'On protection and management of aquatic biological resources'
	Order No. 433 of the Federal Fisheries Agency validating the Regulation on the arrangements for acclimatization of aquatic biological resources
	Regional Decree No. 41 validating the Fisheries Regulation
	Regional Decree No. 199 validating Fisheries Regulation
	Order No. 211 of the Ministry of Natural Resources establishing fish protection requirements
	Methodical Instruction for determination of the level of natural resistance and valuation of immune status of fishes (No.13-4-2/1795 of 1999)
	Order No. 449 validating the Regulation on the modalities of distribution of quotas of catch (harvest) of aquatic biological resources for the purpose of stock enhancement and acclimatization thereof
Slovenia	Freshwater Fishery Act
Switzerland	Ordinance on fisheries
Tajikistan	Fishery Regulation No. 313 validated by the Ministry of Agriculture and Environmental Protection
Ukraine	Order No. 29 of the Ministry of Agrarian Policy validating the Regulation on the State Laboratory of Veterinary Medicine of Fish Disease and Biological Diversity Disease
United Kingdom	Aquaculture and Fisheries (Scotland) Act 2013 (Asp 7 of 2013)
Vietnam	Decree No. 128/2005/ND-CP providing for sanctioning of administrative violations in the fisheries domain

Source: Taken from supplementary material in Turbelin et al. (2017).

In some cases, quarantine and/or chemical treatment of transports or goods associated with introduction of freshwater non-natives takes place. As an example, the international trade in used tyres (among other containers) has played a significant role in introductions of the Asian tiger mosquito (*Aedes albopictus*), leading several countries to put in place inspection and chemical control programmes (e.g. spraying imported tyres or containers with insecticide) that have been successful in preventing mosquito establishment in some cases and at small scales but have sometimes proved impractical at larger scales (e.g. only c. 10% of used tyre imports to the US could be effectively screened). This has led to some countries (in South America, for example) placing embargoes on some imports to prevent introduction (or reintroduction) of the species (Leisnham, 2012).

Post-introduction, prevention of spread is often linked to organisational and personal behaviours. Various biosecurity and awareness-raising initiatives have been put in place in different countries to try to stop spread – for example, the Check-Clean-Dry campaign in the UK (NNSS, 2019) and the citizen-organised 'Toad Day Out' activities that attempt to hinder spread and reduce populations of cane toads (*Rhinella marina*) in Australia (Shine, 2012).

Detection techniques are becoming increasingly important for prevention and broader control efforts. In many cases these rely on basic monitoring and surveying efforts, but in freshwater ecosystems, eDNA can also be effective in establishing the presence of species of invasion concern and constructing both spatial and temporal patterns of distribution and spread (Davison et al., 2016; Stepien et al., 2019), although flow dynamics and connectivity (as discussed previously) can be complicating factors in establishing patterns (e.g. Jane et al., 2015).

Physical control

Physical control is the manual or automated removal of individuals from an invaded system, often by killing. For aquatic and riparian plants, common methods include hand pulling or cutting, mechanical harvesting such as the use of cutter boats (particularly for free-floating plants), mechanical excavation, maceration machines, water jets, suction dredging, covering of vegetation with sheeting to deprive it of sunlight and, in extreme cases, draining of waterbodies or flooding with salt or fresh water (Francis and Pyšek, 2012; Hussner et al., 2017). For animals, culling is the main form of physical control, either directly – e.g. by shooting or electrofishing – or indirectly by draining or fencing waterbodies or installing maceration devices in pipes.

Physical control methods are commonly applied in invaded freshwater ecosystems, but, as in other ecosystems, the main limitations are the amount of manual labour involved (Simberloff, 2008), especially if populations are abundant and/or widespread, and the cost of machinery that might be needed (Patil et al., 2016). Such efforts can be effective in the short term: Cockel et al. (2014) reported some success in reducing populations of Himalayan balsam (*Impatiens glandulifera*) by weeding while Ruiz-Avila and Klemm (1996) found physical removal of mats of floating pennywort (*Hydrocotyle ranunculoides*), followed by application of glyphosate, to be an effective short- to medium-term control strategy. Long-term control successes are rare (Havel et al., 2015, although see e.g. Perna et al., 2012), probably due to repeated invasions, meaning that any extirpations are temporary. Hussner et al. (2016) documented near eradication of water primrose (*Ludwigia grandiflora*) from an infested oxbow lake using hand weeding and predicted full eradication through subsequent applications. One notable eradication resulting from physical control was coypu/nutria (*Myocaster coypus*) in the UK. This was achieved through a combination of careful planning of trapping efforts and providing a fixed amount of funding that would expire after ten years – with the full amount payable to the trappers if they completed the eradication ahead of time (Bertolino et al., 2012.

Physical control is also often incorporated into river restoration efforts so that the removal of hard engineering or deregulation of flows, for example, often has the removal of non-native species as an objective, although successful interventions are relatively uncommon to date. Many non-natives are well suited to urban river conditions, sometimes more so than extirpated natives. For example, Arango et al. (2015) examined post-restoration responses in an urbanised stream in the Pacific Northwest (US) and found that less than a year after restoration, the fish community retained its dominance by non-native eastern brook trout (*Salvelinus fontinalis*). Suren (2009) planted a native macrophyte (*Myriophyllum triphyllum*) into streams dominated by non-natives, following weeding to reduce non-native populations, but these efforts were ultimately unsuccessful in removing the non-natives.

Chemical control

Chemical control is the application of herbicides and pesticides to kill or reduce competitive or reproductive fitness of organisms. The use of chemicals is a common form of control, and, when utilised correctly, such chemicals can be both cost-effective and have fewer collateral impacts than physical control methods (Hussner et al., 2017). They nevertheless do carry some risk of non-target impact, as in many cases they do not affect individual species alone. For plants, herbicides are usually applied where there are large monocultural stands of non-natives to maximise effectiveness. Common herbicides include 2,4-D, diquat, paraquat, glyphosate, fluridone and triclopyr (Coetzee and Hill, 2009; Hussner et al., 2017), but these have restricted availability in some regions (e.g. some European countries) due to their wider impacts (Hussner et al., 2017). For freshwater animals, common pesticides include various forms of piscicides (esp. rotenone, antimycin and saponins; Ling, 2009), insecticides (often organophosphates, carbamates and pyrethroids) and molluscicides (usually niclosamide/clonitralid).

Chemical control can be effective in some instances; for example, Britton and Brazier (2006) reported eradication of topmouth gudgeon (*Pseudorasbora parva*) from a UK fishery following repeated use of rotenone. Success using chemicals is usually achieved where waterbodies are small and relatively isolated, but even so chemical control has limitations. Many herbicides and pesticides are not target specific and therefore can have undesired environmental impacts, especially given the connected nature of freshwater ecosystems. Getting the right concentration of chemicals to ensure efficacy (yet hopefully avoiding substantial wider impacts) can be difficult; for plants especially, treated plants may develop resistance to herbicides (Getsinger et al., 2008).

There are other chemicals that can be used for control – for example, semiochemicals (such as pheromones) to entice animals into traps or disrupt animal behaviour such as mating (Dunn, 2012; Suckling, 2015) – but the use of semiochemicals is complex and has had limited application so far, despite its potential (Suckling, 2015).

Biological control

Biological control is the release of organisms (which may themselves be non-native) to kill, outcompete or otherwise compromise the fitness of a target organism. This tends to be something of a 'last resort' (Simberloff, 2008) as wider environmental impacts can result, especially if the released organism is not highly host specific, and because a large number of control agents have historically been released (in many cases) to attempt to control a single target species (Simberloff, 2008).

There are two main forms of biocontrol (Hussner et al., 2017): (1) classical (the use of a biocontrol agent that is itself non-native and often an organism that afflicts the target species in its native range) and (2) augmentive (where the biocontrol agent originates locally, i.e. is native to the invaded ecosystem). Within these two categories, there are both generalist and specific control agents, with generalists being biocontrol agents that may affect a variety of species (such as the use of grass carp, *Ctenopharyngodon idella*, to control submerged non-native plants; e.g. Shireman and Maceina, 1981) and specific biocontrol agents selected because they will usually affect only the target species (e.g. species-specific pathogens; Smither-Kopperl et al., 1999).

There have been some biocontrol successes in freshwater ecosystems, at least in terms of biomass reduction and compromising fitness, if not eradication: Hussner et al. (2017) document several cases where insects have been effectively deployed to reduce plant cover and biomass, particularly for emergent and floating aquatic plants (but not submerged plants). In such cases, insects can herbivorise stems and meristems to directly reduce biomass or compromise buoyancy and cause floating plants to sink. As with other forms of biological control, successful examples of freshwater non-native control are limited, and there is increasing recognition that success may rely on detailed knowledge of species life history traits and the ecology of the invaded ecosystem (Bajer et al., 2019).

There are also several cases in which the release of a biological control agent has itself become an IAS – for example, various mosquitofish (*Gambusia* spp.), which were originally introduced to various countries as control agents for mosquito larvae, in an attempt to reduce incidents of malaria (Walton et al., 2012). The risks associated with biocontrol have meant that these approaches are increasingly scrutinised and assessed prior to action. While biological control has been more extensively applied in (for example) North America and Australia, increasingly stringent regulations limit their use across Europe (Gerber and Schaffner, 2016).

Future directions for freshwater invasions research

Freshwater ecosystems are heavily influenced by species invasions, and several authors have noted that freshwater ecosystems have received less attention than their terrestrial counterparts (Havel et al., 2015). Most freshwater invasions research has tended to focus on temperate waterbodies in developed countries, leaving tropical, subtropical and boreal ecosystems relatively under-explored. Some research priorities for freshwater invasions echo those for other ecosystems: determining how communities respond to invasions, understanding what environmental and biological factors determine invasion success, establishing and quantifying the main impacts of species invasions and exploring the spatial and temporal scales over which such impacts unfold and how invasions might best be prevented or mitigated (e.g. Havel et al., 2015). However, particular opportunities lie in establishing the ways in which key characteristics of freshwater systems influence invasions. These include:

(1) Exploring how connectivity (and its loss) in all three spatial dimensions, including temporal changes, relates to patterns and rates of invasion, as well as the scale, extent, magnitude and duration of impacts, and how these may complicate control efforts.
(2) Establishing differences between lotic and lentic invasions.
(3) Investigating how invasions vary according to river types and sizes, in particular how hydrogeomorphic factors that drive many river processes may shape invasions or invasion response.
(4) Focusing more on freshwater invasions in less-developed countries and non-temperate freshwater ecosystems both to increase knowledge of these areas and potential awareness

of impacts and control and to allow contrasts to be drawn between these systems and the better-researched temperate waterbodies in developed countries.

(5) Reviewing the potential to expand physical, chemical and biological control programmes.

(6) Increasing understanding of ways to adapt to invasions if control and eradication efforts continue to have limited success.

(7) Establishing the extent and effectiveness of legislation and policy focused on freshwater invasions at national, regional and international scales.

(8) Critically evaluating how ecosystem restoration may more effectively relate to control of invasions, including the implications of increasing landscape connectivity.

Further investigation of these areas will be particularly important against the background of continued global freshwater ecosystem degradation, lack of restoration success for such degraded ecosystems and the pervasive and ongoing impacts of global climate change (Bellard et al., 2013).

References

Albertson, L.K. and Daniels, M.D. (2016) 'Effects of invasive crayfish on fine sediment accumulation, gravel movement, and macroinvertebrate communities', *Freshwater Science*, vol. 35, no. 2, pp. 644–653.

Allan, J.D. (1995) *Stream Ecology: Structure and Function of Running Waters*. Springer, Heidelberg.

Arango, C.P., James, P.W. and Hatch, K.B. (2015) 'Rapid ecosystem response to restoration in an urban stream', *Hydrobiologia*, vol. 749, no. 1, pp. 197–211.

Arnold, E. and Toran, L. (2018) 'Effects of bank vegetation and incision on erosion rates in an urban stream', *Water*, vol. 10, no. 4, p. 482.

Bailey, S.A. (2015) 'An overview of thirty years of research on ballast water as a vector for aquatic invasive species to freshwater and marine environments', *Aquatic Ecosystem Health & Management*, vol. 18, no. 3, pp. 261–268.

Bajer, P.G., Ghosal, R., Maselko, M., Smanski, M.J., Lechelt, J.D., Hansen, G.J.A. and Kornis, M.S. (2019) 'Biological control of invasive fish and aquatic invertebrates: A brief review with case studies', *Management of Biological Invasions*, vol. 10, in press.

Beletsky, D., Beletsky, R., Rutherford, E.S., Sieracki, J.L., Bossenbroek, J.M., Chadderton, W.L., Wittmann, M.E., Annis, G.M. and Lodge, D.M. (2017) 'Predicting spread of aquatic invasive species by lake currents', *Journal of Great Lakes Research*, vol. 43, no. 3, pp. 14–32.

Bellard, C., Thuiller, W., Leroy, B., Genovesi, P., Bakkenes, M. and Courchamp, F. (2013) 'Will climate change promote future invasions?', *Global Change Biology*, vol. 19, no. 12, pp. 3740–3748.

Bertolino, S., Guichón, M.L. and Carter, J. (2012) '*Myocastor coypus* Molina (coypu)', in R.A. Francis (ed.) *A Handbook of Global Freshwater Invasive Species*. Routledge, London.

Boers, A.M., Frieswyk, C.B., Verhoeven, J.T.A. and Zedler, J.B. (2006) 'Contrasting approaches to the restoration of diverse vegetation in herbaceous wetlands', in R. Bobbink, B. Beltman, J.T.A. Verhoeven and D.F. Whigham (eds.) *Wetlands: Functioning, Biodiversity Conservation, and Restoration*. Springer, Heidelberg.

Braga, R.R., Gómez-Aparicio, L., Heger, T., Vitule, J.R.S. and Jeschke, J.M. (2018) 'Structuring evidence for invasional meltdown: Broad support but with biases and gaps', *Biological Invasions*, vol. 20, no. 4, pp. 923–936.

Brauman, K.A. (2016) 'Freshwater', in M. Potschin, R. Haines-Young, R. Fish and R.K. Turner (eds.) *Routledge Handbook of Ecosystem Services*. Routledge, London.

Britton, J.R. and Brazier, M. (2006) 'Eradicating the invasive topmouth gudgeon, *Pseudorasbora parva*, from a recreational fishery in northern England', *Fisheries Management and Ecology*, vol. 13, no. 5, pp. 329–335.

Burlakova, L.E., Karatayev, A.Y. and Karatayev, V.A. (2012) 'Invasive mussels induce community changes by increasing habitat complexity', *Hydrobiologia*, vol. 685, no. 1, pp. 121–134.

Charles, H. and Dukes, J.S. (2008) 'Impacts of invasive species on ecosystem services', in W. Nentwig (ed.) *Biological Invasions: Ecological Studies (Analysis and Synthesis)*, vol. 193. Springer, Heidelberg.

Churchill, C.J. and Quigley, D.P. (2018) 'Downstream dispersal of zebra mussels (*Dreissena polymorpha*) under different flow conditions in a coupled lake-stream ecosystem', *Biological Invasions*, vol. 20, no. 5, pp. 1113–1127.

Cockel, C.P., Gurnell, A.M. and Gurnell, J. (2014) 'Consequences of the physical management of an invasive alien plant for riparian plant species richness and diversity', *River Research and Applications*, vol. 30, no. 2, pp. 217–229.

Coetzee, J.A. and Hill, M.P. (2009) 'Management of invasive aquatic plants', in M.A. Clout and P.A. Williams (eds.) *Invasive Species Management: A Handbook of Principles and Techniques*. Oxford University Press, Oxford.

Costanza, R., d'Arge, R., de Groot, R., Farber, S., Grasso, M., Hannon, B., Limburg, K., Naeem, S., O'Neill, R.V., Paruelo, J., Raskin, R.G., Sutton, P. and van den Belt, M. (1997) 'The value of the world's ecosystem services and natural capital', *Nature*, vol. 387, pp. 253–260.

Cowie, R.H. and Hayes, K.A. (2012) 'Apple snails', in R.A. Francis (ed.) *A Handbook of Global Freshwater Invasive Species*. Routledge, London.

Dallimer, M., Rouquette, J.R., Skinner, A.M.J., Armsworth, P.R., Maltby, L.M., Warren, P.H. and Gaston, K.J. (2012) 'Contrasting patterns in species richness of birds, butterflies and plants along riparian corridors in an urban landscape', *Diversity and Distributions*, vol. 18, pp. 742–753.

D'Amore, A., Kirby, E. and Hemingway, V. (2009) 'Reproductive interference by an invasive species: An evolutionary trap?', *Herpetological Conservation and Biology*, vol. 4, no. 3, pp. 325–330.

Davison, P.I., Créach, V., Liang, W.-J., Andreou, D., Britton, J.R. and Copp, G.H. (2016) 'Laboratory and field validation of a simple method for detecting four species of non-native freshwater fish using eDNA', *Fish Biology*, vol. 89, no. 3, pp. 1782–1793.

de Rouw, A., Ribolzi, O., Douillet, M., Tjantahosong, H. and Soulileuth, B. (2018) 'Weed seed dispersal via runoff water and eroded soil', *Agriculture, Ecosystems and Environment*, vol. 265, pp. 488–502.

De Ventura, L., Weissert, N., Tobias, R., Kopp, K. and Jokela, J. (2016) 'Overland transport of recreational boats as a spreading vector of zebra mussel *Dreissena polymorpha*', *Biological Invasions*, vol. 18, no. 5, pp. 1451–1466.

Drake, J.M. and Lodge, D.M. (2004) 'Global hot spots of biological invasions: Evaluating options for ballast – water management', *Proceedings of the Royal Society B*, vol. 271, no. 1539, pp. 575–580.

Dudgeon, D., Arthington, A.H., Gessner, M.O. Kawabata, Z.-I., Knowler, D.J., Lévêque, C., Naiman, R.J., Prieur-Richard, A.-H., Soto, D., Stiassny, M.L.J. and Sullivan, C.A. (2006) 'Freshwater biodiversity: Importance, threats, status and conservation challenges', *Biological Reviews*, vol. 81, no. 2, pp. 163–182.

Dufresnes, C., Pellet, J., Bettinelli-Riccardi, S., Thiébaud, J., Perrin, N. and Fumagalli, L. (2016) 'Massive genetic introgression in threatened northern crested newts (*Triturus cristatus*) by an invasive congener (*T. carnifex*) in Western Switzerland', *Conservation Genetics*, vol. 17, no. 4, pp. 839–846.

Dunn, J.C. (2012) '*Pacifastacus leniusculus* Dana (North American signal crayfish)', in R.A. Francis (ed.) *A Handbook of Global Freshwater Invasive Species*. Routledge, London.

Ellstrand, N.C. and Schierenbeck, K.A. (2000) 'Hybridization as a stimulus for the evolution of invasiveness in plants?', *PNAS*, vol. 97, no. 13, pp. 7043–7050.

Elton, C.S. (1958) *The Ecology of Invasions by Plants and Animals*. Methuen, London.

Facon, B., Jarne, P., Pointier, J.P. and David, P. (2005) 'Hybridization and invasiveness in the freshwater snail Melanoides tuberculata: Hybrid vigour is more important than increase in genetic variance', *Journal of Evolutionary Biology*, vol. 18, no. 3, pp. 524–535.

Faller, M., Harvey, G.L., Henshaw, A.J., Bertoldi, W., Bruno, M.C. and England, J. (2016) 'River bank burrowing by invasive crayfish: Spatial distribution, biophysical controls and biogeomorphic significance', *Science of the Total Environment*, vol. 569–570, pp. 1190–1200.

Fargen, C., Emery, S.M. and Carreiro, M.M. (2015) 'Influence of *Lonicera maackii* invasion on leaf litter decomposition and macroinvertebrate communities in an urban stream', *Natural Areas Journal*, vol. 35, no. 3, pp. 392–403.

Francis, R.A. and Chadwick, M.A. (2012) 'Invasive alien species in freshwater ecosystems: A brief overview', in R.A. Francis (ed.) *A Handbook of Global Freshwater Invasive Species*. Routledge, London.

Francis, R.A., Chadwick, M.A. and Turbelin, A.J. (2019) 'An overview of non-native species invasions in urban river corridors', *River Research and Applications*, vol. 35, no. 8, pp. 1269–1278.

Francis, R.A. and Pyšek, P. (2012) 'Management of freshwater invasive alien species', in R.A. Francis (ed.) *A Handbook of Global Freshwater Invasive Species*. Routledge, London.

Gallardo, B. and Aldridge, D.C. (2015) 'Is Great Britain heading for a Ponto–Caspian invasional meltdown?', *Journal of Applied Ecology*, vol. 52, no. 1, pp. 41–49.

Garvey, J.E. (2012) 'Bigheaded carps of the genus *Hypopthalmichthys*', in R.A. Francis (ed.) *A Handbook of Global Freshwater Invasive Species*. Routledge, London.

Gendron, A.D., Marcogliese, D.J. and Thomas, M. (2012) 'Invasive species are less parasitized than native competitors, but for how long? The case of the round goby in the Great Lakes-St. Lawrence Basin', *Biological Invasions*, vol. 14, no. 2, pp. 367–384.

Gerber, E. and Schaffner, U. (2016) *Review of Invertebrate Biological Control Agents Introduced into Europe.* CABI, Wallingford.

Getsinger, K.D., Netherland, M.D., Grue, C.E. and Koschnick, T.J. (2008) 'Improvements in the use of aquatic herbicides and establishment of future research directions', *Journal of Aquatic Plant Management*, vol. 46, pp. 32–41.

Goldschmidt, T., Witte, F. and Wanink, J. (1993) 'Cascading effects of the introduced Nile Perch on the detritivorous/phytoplanktivorous species in the sublittoral areas of Lake Victoria', *Conservation Biology*, vol. 7, no. 3, pp. 686–700.

Greenwood, P., Baumann, P., Pulley, S. and Kuhn, N.J. (2018) 'The invasive alien plant, *Impatiens glandulifera* (Himalayan Balsam), and increased soil erosion: Causation or association? Case studies from a river system in Switzerland and the UK', *Journal of Soils and Sediments*, vol. 18, no. 12, pp. 3463–3477.

Grosholz, E.D., Crafton, R.E., Fontana, R.E., Pasari, J.R., Williams, S.L. and Zabin, C.J. (2015) 'Aquaculture as a vector for marine invasions in California', *Biological Invasions*, vol. 17, no. 5, pp. 1471–1484.

Havel, J.E., Kovalenko, K.E., Thomaz, S.M., Amalfitano, S. and Kats, L.B. (2015) 'Aquatic invasive species: Challenges for the future', *Hydrobiologia*, vol. 750, no. 1, pp. 147–170.

Hicks, B., Ling, N. and Daniel, A.J. (2012) '*Cyprinus carpio* L. (common carp)', in R.A. Francis (ed.) *A Handbook of Global Freshwater Invasive Species*. Routledge, London.

Higgins, S.N. and Vander Zanden, M.J. (2010) 'What a difference a species makes: A meta–analysis of dreissenid mussel impacts on freshwater ecosystems', *Ecological Monographs*, vol. 80, no. 2, pp. 179–196.

Hong, S., Do, Kim, J.Y., Kim, D.-K. and Joo, G.-J. (2015) 'Distribution, spread and habitat preferences of nutria (*Myocastor coypus*) invading the lower Nakdong River, South Korea', *Biological Invasions*, vol. 17, no. 5, pp. 1485–1496.

Horsch, E.J. and Lewis, D.J. (2009) 'The effects of aquatic invasive species on property values: Evidence from a quasi-experiment', *Land Economics*, vol. 85, no. 3, pp. 391–409.

Hussner, A., Stiers, I., Verhofstad, M.J.J.M., Bakker, E.S., Grutters, B.M.C., Haury, J., van Valkenburg, J.L.C.H., Brundu, G., Newman, J., Clayton, J.S., Anderson, L.W.J. and Hofstrai, D. (2017) 'Management and control methods of invasive alien freshwater aquatic plants: A review', *Aquatic Botany*, vol. 136, pp. 112–137.

Hussner, A., Windhaus, M. and Starfinger, U. (2016) 'From weed biology to successful control: An example of successful management of *Ludwigia grandiflora* in Germany', *Weed Research*, vol. 56, no. 6, pp. 434–441.

International Maritime Organisation (IMO) (2019) 'Biofouling', www.imo.org/en/OurWork/Environment/Biofouling/Pages/default.aspx, accessed 18 November 2019.

Jackson, M.C. and Grey, J. (2013) 'Accelerating rates of freshwater invasions in the catchment of the River Thames', *Biological Invasions*, vol. 15, no. 5, pp. 945–951.

James, J., Nutbeam-Tuffs, S., Cable, J., Mrugała, A., Viñuela-Rodriguez, N., Petrusek, A. and Oidtmann, B. (2017) 'The prevalence of *Aphanomyces astaci* in invasive signal crayfish from the UK and implications for native crayfish conservation', *Parasitology*, vol. 144, no. 4, pp. 411–418.

Jane, S.F., Wilcox, T.M., McKelvey, K.S., Young, M.K., Schwartz, M.K., Lowe, W.H., Letcher, B.H. and Whiteley, A.R. (2015) 'Distance, flow and PCR inhibition: eDNA dynamics in two headwater streams', *Molecular Ecology Resources*, vol. 15, no. 1, pp. 216–227.

Johnson, L.E., Bossenbroek, J.M. and Kraft, C.E. (2006) 'Patterns and pathways in the post-establishment spread of non-indigenous aquatic species: The slowing invasion of North American inland lakes by the zebra mussel', *Biological Invasions*, vol. 8, no. 3, pp. 475–489.

Johnson, L.E., Ricciardi, A. and Carlton, J.T. (2001) 'Overland dispersal of aquatic invasive species: A risk assessment of transient recreational boating', *Ecological Applications*, vol. 11, no. 6, pp. 1789–1799.

Johnson, P.T.J., Olden, J.D. and Vander Zanden, M.J. (2008) 'Dam invaders: Impoundments facilitate biological invasions into freshwaters', *Frontiers in Ecology and the Environment*, vol. 6, no. 7, pp. 357–363.

Keller, R.P., Drake, J.M., Drew, M.B. and Lodge, D.M. (2011) 'Linking environmental conditions and ship movements to estimate invasive species transport across the global shipping network', *Diversity and Distributions*, vol. 17, no. 1, pp. 93–102.

Keller, R.P., Zu Ermgassen, P.S.E. and Aldridge, D.C. (2009) 'Vectors and timing of freshwater invasions in Great Britain', *Conservation Biology*, vol. 23, no. 6, pp. 1526–1534.

Killeen, G.F., Fillinger, U., Kiche, I., Gouagna, L.C. and Knols, B.G.J. (2002) 'Eradication of Anopheles gambiae from Brazil: Lessons for malaria control in Africa?', *The Lancet: Infectious Diseases*, vol. 2, no. 10, pp. 618–627.

Kornis, M.S. and Vander-Zanden, M.J. (2010) 'Forecasting the distribution of the invasive round goby (Neogobius melanostomus) in Wisconsin tributaries to Lake Michigan', *Canadian Journal of Fisheries and Aquatic Sciences*, vol. 67, no. 3, pp. 553–562.

Kouba, A., Tíkal, J., Císař, P., Veselý, L., Fořt, M., Příborský, J., Patoka, J. and Buřič, M. (2016) 'The significance of droughts for hyporheic dwellers: Evidence from freshwater crayfish', *Scientific Reports*, vol. 6, pp. 26569.

Kuglerová, L., García, L., Pardo, I., Mottiar, Y. and Richardson, J.S. (2017) 'Does leaf litter from invasive plants contribute the same support of a stream ecosystem function as native vegetation?', *Ecosphere*, vol. 8, no. 4, p. e01779.

Lee, D.J. and Gordon, R.M. (2006) 'Economics of aquaculture and invasive aquatic species: An overview', *Aquaculture Economics and Management*, vol. 10, no. 2, pp. 83–96.

Leisnham, P. (2012) '*Aedes albopictus* Skuse (Asian tiger mosquito)', in R.A. Francis (ed.) *A Handbook of Global Freshwater Invasive Species*. Routledge, London.

Lenzner, B., Leclère, D., Franklin, O., Seebens, H., Roura-Pascual, N., Obersteiner, M., Dullinger, S. and Essl, F. (2019) 'A framework for global twenty-first century scenarios and models of biological invasions', *BioScience*, vol. 69, no. 9, pp. 697–710.

Leppäkoski, E., Gollasch, S. and Olenin, S. (2002) 'Alien species in European waters', in E. Leppäkoski, S. Gollasch and S. Olenin (eds.) *Invasive Aquatic Species of Europe: Distribution, Impacts and Management*. Springer, Heidelberg.

Leuven, R.S.E.W., van der Velde, G., Baijens, I., Snijders, J., van der Zwart, C., Lenders, H.J.R. and bij de Vaate, A. (2009) 'The river Rhine: A global highway for dispersal of aquatic invasive species', *Biological Invasions*, vol. 11, p. 1989.

Levy, K. (2004) 'Neglected consequences: Role of introduced aquatic species in the spread of infectious diseases', *EcoHealth*, vol. 1. no. 3, pp. 296–305.

Ling, N. (2009) 'Management of invasive fish', in in M.A. Clout and P.A. Williams (eds.) *Invasive Species Management: A Handbook of Principles and Techniques*. Oxford University Press, Oxford.

Lisi, P.J., Childress, E.S., Gagne, R.B., Hain, E.F., Lamphere, B.A., Walter, R.P., Hogan, J.D., Gilliam, J.F., Blum, M.J. and McIntyre, P.B. (2018) 'Overcoming urban stream syndrome: Trophic flexibility confers resilience in a Hawaiian stream fish', *Freshwater Biology*, vol. 63, no. 5, pp. 492–502.

Loo, S.E., Keller, R.P. and Leung, B. (2007) 'Freshwater invasions: Using historical data to analyse spread', *Diversity and Distributions*, vol. 13, no. 1, pp. 23–32.

Lovell, S.J., Stone, S.F. and Fernandez, L. (2006) 'The economic impacts of aquatic invasive species: A review of the literature', *Agricultural and Resource Economics Review*, vol. 35, no. 1, pp. 195–208.

MacIsaac, H.J., Borbely, J.V.M., Muirhead, J.R. and Graniero, P.A. (2004) 'Backcasting and forecasting biological invasions of inland lakes', *Ecological Applications*, vol. 14, no. 3, pp. 773–783.

Macklin, M.G. and Lewin, J. (2015) 'The rivers of civilization', *Quaternary Science Reviews*, vol. 114, pp. 228–244.

Mandrak, N.E. and Cudmore, B. (2010) 'The fall of native fishes and the rise of non-native fishes in the great lakes basin', *Aquatic Ecosystem Health & Management*, vol. 13, no. 3, pp. 255–268.

Marbuah, G., Gren, I.-M. and McKie, B. (2014) 'Economics of harmful invasive species: A review', *Diversity*, vol. 6, no. 3, pp. 500–523.

Marmonier, P., Archambaud, G., Belaidi, N., Bougon, N., Breil, P., Chauvet, E., Claret, C., Cornut, J., Datry, T., Dole-Olivier, M.-J. Dumont, B., Flipo, N., Foulquier, A., Gérino, M., Guilpart, A., Julien, F., Maazouzi, C., Martin, D., Mermillod-Blondin, F., Montuelle, B., Namour, Ph., Navel, S., Ombredane, D., Pelte, T., Piscart, C., Pusch, M., Stroffek, S., Robertson, A., Sanchez-Pérez, J.-M., Sauvage, S., Taleb, A., Wantzen, M. and Vervier, Ph. (2012) 'The role of organisms in hyporheic processes: Gaps in current knowledge, needs for future research and applications', *Annales de Limnologie–International Journal of Limnology*, vol. 48, no. 3, pp. 253–266.

Marsden, J.E. and Ladago, B.J. (2017) 'The Champlain Canal as a non-indigenous species corridor', *Journal of Great Lakes Research*, vol. 43, no. 6, pp. 1173–1180.

Matsuzaki, S.S., Usio, N., Takamura, N. and Washitani, I. (2009) 'Contrasting impacts of invasive engineers on freshwater ecosystems: An experiment and meta-analysis', *Oecologia*, vol. 158, no. 4, pp. 673–686.

Mazza, G. and Tricarico, E. (eds.) (2018) *Invasive Species and Human Health*. CABI, Wallingford.

Mills, D.N., Chadwick, M.A. and Francis, R.A. (2019) 'Artificial substrate experiments to investigate potential impacts of invasive quagga mussel (*Dreissena rostriformis bugensis*, Bivalvia: Dreissenidae) on macroinvertebrate communities in a UK river', *Aquatic Invasions*, vol. 14, no. 2, in press.

Moorhouse, T.P. and Macdonald, D.W. (2015) 'Are invasives worse in freshwater than terrestrial ecosystems?', *WIREs Water*, vol. 2, no. 1, pp. 1–8.

Non-Native Species Secretariat (NNSS) (2019) 'Check-clean-dry', www.nonnativespecies.org/checkcle-andry/index.cfm, accessed 21 November 2019.

Nunes, A.L., Tricarico, E., Panov, V.E., Cardoso, A.C. and Katsanevakis, S. (2015) 'Pathways and gateways of freshwater invasions in Europe', *Aquatic Invasions*, vol. 10, no. 4, pp. 359–370.

Oreska, M.P.J. and Aldridge, D.C. (2011) 'Estimating the financial costs of freshwater invasive species in Great Britain: A standardized approach to invasive species costing', *Biological Invasions*, vol. 13, no. 2, pp. 305–319.

Padilla, D.K. and Williams, S.L. (2004) 'Beyond ballast water: Aquarium and ornamental trades as sources of invasive species in aquatic ecosystems', *Frontiers in Ecology and the Environment*, vol. 2, no. 3, pp. 131–138.

Pagnucco, K.S., Remmal, Y. and Ricciardi, A. (2016) 'An invasive benthic fish magnifies trophic cascades and alters pelagic communities in an experimental freshwater system', *Freshwater Science*, vol. 35, no. 2, pp. 654–665.

Patil, R., Itnare, R., Ahirrao, S., Jadhav, A. and Dhumal, A. (2016) 'Study of river harvesting and trash cleaning machine', *International Journal of Innovative Research in Science and Engineering*, vol. 2, no. 3, pp. 422–431.

Pearson, S.H., Avery, H.W. and Spotila, J.R. (2015) 'Juvenile invasive red-eared slider turtles negatively impact the growth of native turtles: Implications for global freshwater turtle populations', *Biological Conservation*, vol. 186, pp. 115–121.

Pejchar, L. and Mooney, H.A. (2009) 'Invasive species, ecosystem services and human well-being', *Trends in Ecology and Evolution*, vol. 24, no. 9, pp. 497–504.

Pergl, J., Perglová, I. and Pyšek, P. (2012) '*Heracleum mantegazzianum* Sommier & Levier (giant hogweed)', in R.A. Francis (ed.) *A Handbook of Global Freshwater Invasive Species*. Routledge, London.

Perna, C.N., Cappo, M., Pusey, B.J., Burrows, D.W. and Pearson, R.G. (2012) 'Removal of aquatic weeds greatly enhances fish community richness and diversity: An example from the Burdekin River flood-plain, tropical Australia', *River Research and Applications*, vol. 28, no. 8, pp. 1093–1104.

Pimentel, D., Zuniga, R. and Morrison, D. (2005) 'Update on the environmental and economic costs asso-ciated with alien-invasive species in the United States', *Ecological Economics*, vol. 52, no. 3, pp. 273–288.

Predick, K.I. and Turner, M.G. (2008) 'Landscape configuration and flood frequency influence invasive shrubs in floodplain forests of the Wisconsin River (USA)', *Journal of Ecology*, vol. 96, no. 1, pp. 91–102.

Rasmussen, J.L., Regier, H.A., Sparks, R.E. and Taylor, W.W. (2011) 'Dividing the waters: The case for hydrologic separation of the North American Great Lakes and Mississippi River Basins', *Journal of Great Lakes Research*, vol. 37, no. 3, pp. 588–592.

Ricciardi, A. and MacIsaac, H.J. (2000) 'Recent mass invasion of the North American Great Lakes by Ponto–Caspian species', *Trends in Ecology and Evolution*, vol. 15, no. 2, pp. 62–65.

Rood, S.B., Braatne, J.H. and Goater, L.A. (2010) 'Favorable fragmentation: River reservoirs can impede downstream expansion of riparian weeds', *Ecological Applications*, vol. 20, no. 6, pp. 1664–1677.

Ruiz-Avila, R.J. and Klemm, V.V. (1996) 'Management of *Hydrocotyle ranunculoides* L.f., an aquatic invasive weed of urban waterways in Western Australia', *Hydrobiologia*, vol. 340, nos. 1–3, pp. 187–190.

Savini, D., Occhipinti-Ambrogi, A., Marchini, A., Tricarico, E., Gherardi, F., Olenin, S. and Gollasch, S. (2010) 'The top 27 animal alien species introduced into Europe for aquaculture and related activities', *Journal of Applied Ichthyology*, vol. 26, no. s2, pp. 1–7.

Shaker, R.R., Yakubov, A.D., Nick, S.M., Vennie-Vollrath, E., Ehlinger, T.J. and Forsythe, K.W. (2017) 'Predicting aquatic invasion in Adirondack lakes: A spatial analysis of lake and landscape characteristics', *Ecosphere*, vol. 8, no. 3, p. e01723.

Shine, R. (2012) '*Rhinella marina* L. (cane toad)', in R.A. Francis (ed.) *A Handbook of Global Freshwater Invasive Species*. Routledge, London.

Shine, R. and Doody, J.S. (2011) 'Invasive species control: Understanding conflicts between researchers and the general community', *Frontiers in Ecology and the Environment*, vol. 9, no. 7, pp. 400–406.

Shireman, J.V. and Maceina, M.J. (1981) 'The utilization of grass carp, *Ctenopharyngodon idella* Val., for hydrilla control in Lake Baldwin, Florida', *Journal of Fish Biology*, vol. 19, no. 6, pp. 629–636.

Simberloff, D. (2008) 'We can eliminate invasions or live with them: Successful management projects', in D.W. Langor and J. Sweeney (eds.) *Ecological Impacts of Non-Native Invertebrates and Fungi on Terrestrial Ecosystems*. Springer, Heidelberg.

Simpson, A., Jarnevich, C., Madsen, J., Westbrooks, R., Fournier, C., Mehrhoff, L., Browne, M., Gra-ham, J. and Sellers, E. (2009) 'Invasive species information networks: Collaboration at multiple scales for prevention, early detection, and rapid response to invasive alien species', *Biodiversity*, vol. 10, nos. 2–3, pp. 5–13.

Smith, E.R.C. (2019) *Conduits of Invasive Species into the UK: The Angling Route?* Doctoral dissertation, University College London, UK.

Smither-Kopperl, M.L., Charudattan, R. and Berger, R.D. (1999) '*Plectosporium tabacinum*, a pathogen of the invasive aquatic weed *Hydrilla verticillata* in Florida', *Plant Disease*, vol. 83, no. 1, pp. 24–28.

Souty-Grosset, C., Anastácio, P., Reynolds, J. and Tricarico, E. (2018) 'Invasive freshwater invertebrates and fishes: Impacts on human health', in G. Mazza and E. Tricarico (eds.) *Invasive Species and Human Health*. CABI, Wallingford.

Stanley, E.H. and Doyle, M.W. (2003) 'Trading off: The ecological effects of dam removal', *Frontiers in Ecology and the Environment*, vol. 1, no. 1, pp. 15–22.

Stepien, C.A., Snyder, M.R. and Elz, A.E. (2019) 'Invasion genetics of the silver carp *Hypophthalmichthys molitrix* across North America: Differentiation of fronts, introgression, and eDNA metabarcode detection', *PLoS One*, vol. 14, no. 3, p. e0203012.

Suckling, D.M. (2015) 'Can we replace toxicants, achieve biosecurity, and generate market position with semiochemicals?', *Frontiers in Ecology and Evolution*, vol. 3, art.17.

Suren, A.M. (2009) 'Using macrophytes in urban stream rehabilitation: A cautionary tale', *Restoration Ecology*, vol. 17, no. 6, pp. 873–883.

Tolley-Jordan, L.R. and Chadwick, M.A. (2012) '*Centrocestus formosanus* Nishigori (Asian gill-trematode)', in R.A. Francis (ed.) *A Handbook of Global Freshwater Invasive Species*. Routledge, London.

Turbelin, A.J., Malamud, B.D.M. and Francis, R.A. (2017) 'Mapping the global state of invasive alien species: Patterns of invasion and policy responses', *Global Ecology and Biogeography*, vol. 26, no. 1, pp. 78–92.

UNCTAD (2018) 'Review of Maritime Transport 2018', https://unctad.org/en/PublicationsLibrary/rmt2018_en.pdf, accessed 18 November 2019.

Vilà, M. and Hulme, P.E. (2017) *Impact of Biological Invasions on Ecosystem Services*. Springer, Heidelberg.

Walsh, J.R., Carpenter, S.R. and Vander Zanden, M.J. (2016) 'Invasive species triggers a massive loss of ecosystem services through a trophic cascade', *PNAS*, vol. 113, no. 15, pp. 4081–4085.

Walton, W.E., Henke, J.A. and Why, A.M. (2012) '*Gambusia affinis* (Baird and Girard) and *Gambusia holbrooki* Girard (Mosquitofish)', in R.A. Francis (ed.) *A Handbook of Global Freshwater Invasive Species*. Routledge, London.

Ward, J.V., Malard, F. and Tockner, K. (2002) 'Landscape ecology: A framework for integrating pattern and process in river corridors', *Landscape Ecology*, vol. 17, no. s1, pp. 35–45.

Woodford, D.J., Impson, N.D., Day, J.A. and Bills, I.R. (2005) 'The predatory impact of invasive alien smallmouth bass, *Micropterus dolomieu* (Teleostei: Centrarchidae), on indigenous fishes in a Cape Floristic Region mountain stream', *African Journal of Aquatic Science*, vol. 30, no. 2, pp. 167–173.

World Ocean Review (2010) 'Living with the oceans: A report on the state of the world's oceans', https://worldoceanreview.com/en/wor-1/, accessed 18 November 2019.

Xing, Y., Zhang, C., Fan, E. and Zhao, Y. (2016) 'Freshwater fishes of China: Species richness, endemism, threatened species and conservation', *Diversity and Distributions*, vol. 22, no. 3, pp. 358–370.

Zabin, C., Davidson, I., Holzer, K., Smith, G., Ashton, G., Tamburri, M. and Ruiz, G. (2018) 'How will vessels be inspected to meet emerging biofouling regulations for the prevention of marine invasions?', *Management of Biological Invasions*, vol. 9, no. 3, pp. 195–208.

Zavaleta, E. (2000) 'The economic value of controlling an invasive shrub', *AMBIO*, vol. 29, no. 8, pp. 462–467.

Zedler, J.B. and Kercher, S. (2004) 'Causes and consequences of invasive plants in wetlands: Opportunities, opportunists, and outcomes', *Critical Reviews in Plant Sciences*, vol. 23, no. 5, pp. 431–452.

Zhan, A., Zhang, L., Xia, Z., Ni, P., Xiong, W., Chen, Y., Haffner, G.D. and MacIsaac, H.J. (2015) 'Water diversions facilitate spread of non-native species', *Biological Invasions*, vol. 17, no. 11, pp. 3073–3080.

11

'NEW' RECOMBINANT ECOLOGIES AND THEIR IMPLICATIONS – WITH INSIGHTS FROM BRITAIN

Ian D. Rotherham

Introduction

Discussions of the importance of native and non-native species take place against a backdrop of rapidly increasing extinctions and loss of biodiversity (e.g. Thomas et al., 2004; Rotherham, 2014a). Furthermore, in recent decades there has been a great increase in interest in the roles and impacts of invasive species and especially invasive non-natives on the functioning of ecological systems (Rotherham and Lambert, 2011; Davis et al., 2001; Grime and Pierce, 2012; Johnson, 2010; Monbiot, 2013; Jørgensen et al., 2013). Moreover, with rising levels of urbanisation, there is a marked increase in attention to urban ecologies (e.g. Bornkamm et al., 1982; Gilbert, 1989; Sukopp and Hejny, 1990; Barker et al., 1994; Sukopp et al., 1995; Niemelä, 2011; Gaston, 2011; Douglas et al., 2011). Urban ecology and urban nature have thus emerged as major themes for debate (e.g. Gilbert, 1989, 1992a, 1992b; Goode, 2014) alongside the impacts of non-native invaders. Recent works describing the latter include Stace and Crawley (2015) and, in terms of plant hybridisation, Stace et al. (2015).

The scale of human impacts on ecological systems is now widely accepted, although the potential solutions remain elusive (Rotherham, 2014a). However, it is increasingly the case that researchers and the popular media accept that we have now entered a new geological era called the Anthropocene (Steffen et al., 2007; Davies, 2016); in this environmental watershed moment, it is humanity that is driving the major changes. Against this backdrop of human-induced environmental stresses, the roles of invasive and non-native species become increasingly significant (Rosenzweig, 2003), and researchers move on from largely cataloguing the rise of the non-native as documented in earlier texts (e.g. Salisbury, 1961; Elton, 1958, 1966) towards more sophisticated analysis and debate (e.g. Bridgewater, 1990; Soulé, 1990; Simberloff, 2003, 2005, 2011). In the last two decades, this discussion has moved on to address issues of new ecologies or 'novel ecosystems' (Hobbs et al., 2009; Hobbs et al., 2006, 2013, 2014; Hulme, 2006; Hulme et al., 2008; Francis and Chadwick, 2012). Also in recent decades, there has been an emerging interest in 'future nature' and 'futurescapes' (e.g. Adams, 2003; Rotherham, 2014b, 2018).

It is important at the outset to address some key definitions and concepts. Some important concepts and definitions include what is 'native' and what is 'non-native', as questions of 'where' and 'when' become significant in terms of 'alien species' (see Mulcock and Trigger, 2008; see also Fall, this collection, and Lambert and Mark-Shadbolt, this collection). It is worth considering what the term 'alien' means. Dictionary definitions suggest:

1) 'Belonging to another person, place or family especially to a foreign nation or allegiance. Foreign in nature, character or origin' (1673)
2) 'A stranger or a foreigner. A resident foreign in origin and not naturalised' (1330)
3) 'One excluded from citizenship, privileges *etc*' (1549)
4) 'A plant originally introduced from other countries' (1847)

(from Rotherham, 2005a)

Clement and Foster (1994) used 'alien' in a broad sense to denote all plants whether or not they were believed to have arrived as a result of human activities. In this definition they included plants referred to by other authors as adventives, casuals, ephemerals, exotics, introductions and volunteers. At around the same time, Ellis (1993) wrote a very useful and pocket-sized introduction to invasive plants in Britain. He suggested that, for many, an 'alien plant' is essentially one that is not native. Furthermore, in this case, a native plant is one that arrived in Britain prior to the closure of the English Channel around 7,000 to 8,000 years ago. Therefore, an alien is a species arriving after this time.

Along with perceptions of native, ideas and concepts of 'naturalness' also affect the paradigms of invasive and alien species (see, for example, Rotherham, 2009a, 2014b) and influence conservation responses. The extent, then, to which humanity is a part of nature or separate from it becomes problematic. It is within this framework of definitions and perceptions that we need to consider matters of 'ecological novelty' and 'invasive species', both native and non-native.

Ellis (1993) regarded alien plants as those introduced by people, deliberately or accidentally, and observed that, in reality, and with longer time perspectives, most 'native' flora could be considered 'alien invaders'. This is, of course, due to the dynamic and fluctuating nature of vegetation in a landscape with long-term changes of key factors such as climate. Indeed, this latter point becomes increasingly important as climate appears to be one of a number of key environmental factors changing significantly and rapidly. From a nature conservation context, this complicates the underlying priorities for contemporary management. Furthermore, such recognition serves to emphasise that much important conservation work is based substantially on subjective human needs, opinions and priorities and not necessarily on ecological science. There is increasing awareness that some changes may be inevitable and, even more controversially, might provide answers and solutions to some critical environmental questions (e.g. Pearce, 2015; Thomas, 2013). There has also been a long-term interest in the history and natural history of introduced and exotic plants and animals (e.g. Lever, 1977; Robinson, 1870). The nature and characteristics of invasiveness have also been considered (Noble, 1989).

However, it is also clear that 'invasiveness' as an ecological characteristic is little different in non-native species and native ones (Davis et al., 2001). Grime develops this matter further in Grime (2005), Grime et al. (2007) and Grime and Pierce (2012).

Recombinant ecologies

From Soulé (1990), Barker (2000, 21–24) understood that 'recombinance' means the following:

The ecology of communities of plants and animals, the constituent members of which are drawn from a wide range of global biogeographic zones. These communities typically are mixtures of species long established in the wild in the area concerned and species relatively recently established there in the wild. In much of the literature, these are referred to respectively as 'natives' and 'aliens'. The cut-off between the two is defined differently in different places and from one piece of literature to another.

And furthermore,

The continental European convention is taken where newcomers are defined as species established since 1500AD (roughly the point at which European world-trading accelerated) although some, but not all, species established earlier as a consequence of human activity are also included in the literature and in everyday thought as 'aliens'.

Barker and those who contributed to the 2000 conference organised by George Barker ('Ecological Recombination in Urban Areas: Implications for Nature Conservation') and to the resultant proceedings also provided insights into the implications of recombination for nature conservation. They highlighted the issue that ecological science dispassionately describes and observes the processes and defines species interactions. Nature conservation, on the other hand, introduces value judgements and management practice. The 2000 meeting sought to identify what an understanding of these processes and the recombinant ecologies might mean for British nature conservation philosophy and practice.

One important suggestion from experts in the field at the time was 'that change in the species composition of ecotopes is continuous, unpredictable in direction and, effectively beyond human control'. Consequently,

This has implications for programmes which seek to maintain ecotopes developed in the past; to restore selected species locally extinct in the recent past; to prevent the local loss of any more species; and to prevent local gain of species not currently, or only recently, established there in the wild.

(Barker, 2000)

These observations remain pertinent today.

Changes are triggering the emergence of novel ecologies globally (Hobbs et al., 2013), which in the UK are generally 'communities' or 'habitat patches' rather than 'novel ecosystems' as such.

Additionally, in terms of recombinant ecology, there are important issues of how species are adapting to the challenges of environmental change and the opportunities and pressures that arise. Then, in relation to biosecurity, the changes that are happening increasingly widely may lead to impacts, risks, costs and consequences.

However, as noted earlier, finding precise and applicable definitions of 'natural', 'native' and 'non-native' is increasingly difficult and problematic. In terms of biosecurity, it is more relevant to consider 'problem' species (however defined or determined) rather than necessarily 'non-native' or 'alien' ones (see, for example, Rotherham and Lambert, 2011). This is particularly the case as climate change, other environmental fluxes and globalisation are changing the baseline condition of our ecological systems. Within this broad context, the emergence and spread of pests and diseases are probably the most serious adverse biosecurity impacts.

A question commonly asked concerns the differences between novel ecosystems and recombinant ones. However, from the work of Hobbs and colleagues (2013), we have a working definition of novel systems and a separation from historical and hybrid ones. Barker (2000) provided a broad working definition of recombinant ecology which, along with the work of Peter Shepherd and Oliver Gilbert, for example (in Barker, 2000), gives an approach that is workable in Britain and Europe.

However, there do seem to be differences in approach between the British and European researchers and those from Australia, New Zealand and the US. Nevertheless, Meurk (2011, in Douglas et al., 2011, 198) gives a thorough account of recombinant ecology but with a strongly urban perspective. His overview starts with a definition:

> Recombinant ecosystems comprise novel plant and animal associations that have been induced or created by people deliberately, inadvertently or indirectly. They are generally made up of indigenous and exotic species, but they may also involve associations of indigenous species alone, never before seen in nature, for example plant signature (Robinson, 1993), native landscape garden designs and pictorial meadows (Dunnett and Hitchmough, 2004), indigenous feature species introduced to areas beyond their natural range, or back-filled 'gaps' created by local extinctions.

In this context, Meurk explains the importance of a) human action, b) introduced species and c) environmental change. In urban environments, many of the adapted species cope well with high levels of disturbance and sometimes stress. However, the roles of relict and often-persistent species from the historic landscape are not discussed, and in the British setting, these are often a strong element of the urban matrix (see Rotherham, 1999).

Meurk notes the need to better understand the ecological envelopes of native or indigenous species entering into novel ecosystems, which is an important area for future research. Grime et al. (2007) help place species into a spatial frame of stress-disturbance-competition and so provide a useful tool to guide predictions. Alongside these aspects of recombinant communities, it is useful to consider origins, relative proportions of native and exotic species and issues of stability and sustainability. All these issues are in the context of successional changes in the ecological systems.

The factors as described may influence ideas for management interventions on the one hand and likely ecological outcomes of successions if these are transient or ephemeral stages. To help guide this, Meurk (2011) provides a simple table to aid the description of his recombinant ecosystems in a vision strongly influenced by New Zealand's experience. Here, European imports since the 1700s have caused mass extinctions of native species. Not all of Meurk's ideas transfer easily to the European or British situation, and in this sense the definition of community, ecology, ecosystem, etc. may also be problematic.

Is this discussion about novel elements and communities of local distribution and occurrence in a wider ecosystem context rather than 'ecosystems' per se? In many British and European situations, for many communities of plants, animals and fungi, it is difficult to recognise them as intact functioning 'ecosystems'. Following centuries of fragmentation and transformation, many biological systems are essentially habitat patches in a sea of agroindustrial farming, industrial extraction and processing or urban sprawl. The loss of ecosystem functions and associated services as widely reported are obvious consequences of this. In this situation, we may need to distinguish between 'novel ecologies' and 'novel ecosystems', and in Britain, the former predominate as recombinant communities (Rotherham, 2005a, 2005b).

The processes of eco-fusion

Many authors such as Ratcliffe (1984) and Rackham (1986) have described the way that British vegetation has been transformed through human agency over the historic period. New ecologies develop through various forms of ecological fusion (Rotherham, 2017a, 2017b; see Figure 11.1) or hybridisation to generate 'hybrid ecologies'.

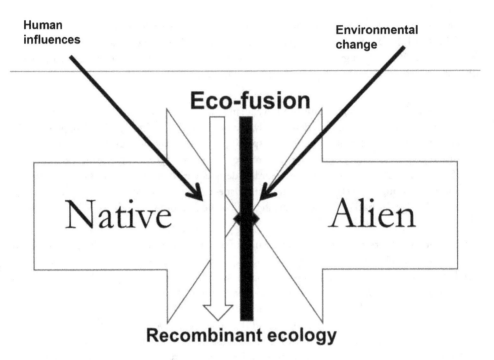

Figure 11.1   The process of eco-fusion to form recombinant communities.

Novelty results from:

1) **Invasion:** Non-native species arriving in the region in question through natural colonisation and/or human-induced movement. However, in the Anthropocene it becomes increasingly difficult to separate out the human- and non-human changes.

2) **Extinction:** Species removal by extinction from established 'natural' ecologies with the release of new ecological trajectories and the creation of vacant ecological niche spaces for potential new colonists. This process is especially significant when the species involved are 'keystone' components of the existing system.

3) **Hybridisation:** Hybridisation (and mutation) to form new 'species' which may have competitive advantages in the context of fluctuating environmental conditions.

Figure 11.2 shows the historic drivers of recombination in Great Britain.

We might also add to the three main processes the situation whereby changing baseline environmental conditions reduce a formerly widespread species and allow a rare one to become widespread and abundant. This is the case, for example, with the so-called 'pollution lichens' such as *Lecanora* which responded to 19th- and 20th-century air pollution so dramatically so as to become dominant on tree bark and other suitable surfaces (Hawksworth and Rose, 1970; Richardson, 1992; Rotherham, 2014a).

The results of the three aforementioned processes include the fusion or hybridisation of ecological communities to develop recombinant ecologies and, in some cases, novel ecosystems. Examples of species responding to these three drivers in the British context are given in Table 11.1.

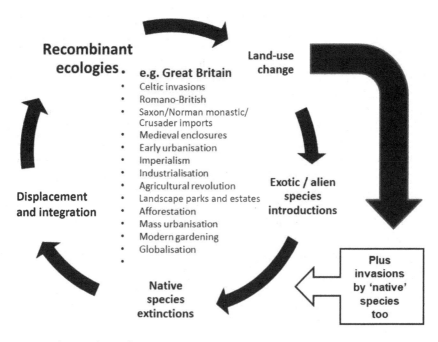

Figure 11.2   Historic drivers of recombination in Great Britain.

Table 11.1 Three key drivers of ecological fusion and examples of species responding to these drivers in the British context.

Driver	Species
Invasion	Himalayan balsam (*Impatiens glandulifera*), giant hogweed (*Heracleum mantegazzianum*), Japanese knotweed (*Fallopia japonica*), signal crayfish (*Pacifastacus leniusculus*), ring-necked parakeet (*Psittacula krameri*), muntjac deer (*Muntiacus muntjak*), fallow deer (*Dama dama*), sika deer (*Cervus nippon*), Chinese water-deer (*Hydropotes inermis*), grey squirrel (*Sciurus carolinensis*), rabbit (*Oryctolagus cuniculus*), brown hare (*Lepus europaeus*), mink (*Neovison vison*), ring-necked pheasant (*Phasianus colchicus*), red-legged partridge (*Alectoris rufa*), Canada goose (*Branta canadensis*), feral grey-lag goose (*Anser anser*), wild fuchsia (*Fuchsia magellanica*), tamarisk (*Tamarix ramosissima*).
Extinction	Wolf (*Canis lupus*), brown bear (*Ursus arctos*), beaver (*Castor fiber*), wild boar (*Sus scrofa*), and, at regional levels, wildcat (*Felis silvestris grampia*), pine marten (*Martes martes*), golden eagle (*Aquila chrysaetos*), osprey (*Pandion haliaetus*), common buzzard (*Buteo buteo*), red kite (*Milvus milvus*), otter (*Lutra lutra*), polecat (*Mustela putorius*).
Hybridisation	Hybrid bluebell (*Hyacinthoides non-scripta* x *hispanica* = *H.* x *massartiana*), montbretia (*Crocosmia* x *crocosmiiflora*), variegated yellow archangel (*Lamiastrum galeobdolon* subsp. *argentatum*), *Rhododendron ponticum* hybrid, feral pigeon (*Columba livia domestica*), wildcat hybrid (*Felis silvestris grampia* x *Felis silvestris catus*), polecat-ferret (*Mustela putorius* x *Mustela putorius furo*), sika deer x red deer (*Cervus nippon* x *Cervus elaphus*).

Historical context

Ecological systems and their species have always fluxed as they ebb and flow with changing environmental drivers and human-induced changes (Rotherham, 2014a). In this context, any attempt at precise definition of native and non-native becomes problematic. This issue becomes more difficult in a landmass as the core area, and thus its connectivity to neighbouring ecological systems increases. These issues also become very difficult when considering marine systems.

In an island (or archipelago) situation such as the British Isles, a starting point is often taken based on when the island is cut off from adjacent landmasses by rising sea levels (e.g. Ellis, 1993). In many ways this is problematic as a definitional point since sea level has risen and fallen over time. The key pragmatic cut-off point is the sea closure of the Channel following the last glaciation, but this really is a matter of convenience and a concept of separation by marine inundation. Bearing in mind that many taxa and species can cross a relatively narrow sea barrier anyway, this does present some issues that are generally overlooked. For mobile species that can fly, swim or else be dispersed on the wind, the presence or absence of a land bridge may be irrelevant.

However, in Britain, this date of approximately 7,500 years ago is generally taken as a starting point, and species that arrive after closure and through human agency are taken as non-native. Again, in the context of human-induced global environmental change, there may be problems in deciding when species spread is caused directly or indirectly by human actions (Rotherham, 2015, 2017a, 2017b). While this matter of definition may seem trivial, it does influence what we view as native or non-native and hence as novel or recombinant. Human agency may include the importation of species either deliberately or accidentally and the processes of both extinction on the one hand and hybridisation on the other (Rotherham, 2014a).

Globalisation of ecology and its consequences

Nature conservation is often concerned with halting or reversing the declines of habitats, communities and species. Within this spectrum of target actions, those species of ecosystems considered to be 'native' and/or 'natural' are generally taken as priorities. However, as we move into the Anthropocene, globalisation increasingly tips the balance towards widespread declines and, ultimately, mass extinctions. Associated with these broader trends are the increasing emergence of novel ecological systems and 'new' species. It has also been argued that the extinction events in the past have triggered rapid speciation and the evolution of new biodiverse ecologies (e.g. Thomas, 2011, 2013). However, it is worth noting that the emerging 'new' species generally do not provide like-for-like compensation for those lost, and while in a geological time frame new biodiverse systems might emerge, in any reasonably human-related historical time frame, this is not the case (Rotherham, 2015, 2017a, 2017b). In the British context, species lost include the wolf, brown bear, beaver, wild boar, wildcat, golden eagle, etc. Emerging species include variegated yellow archangel, montbretia, hybrid wildcat, red deer x sika deer hybrid, polecat-ferret, *Rhododendron ponticum* hybrid, etc.

Recombinant ecology and biosecurity issues

With social, economic and environmental globalisation there is dramatically increased mixing of species and ecological communities around the world. This process can be seen throughout human history; for example, in Europe, there is evidence of widespread species dispersal with Celtic migrations from east to west and settlements along the Atlantic coastline. However, despite the longevity of the basic processes, their scale and importance increase over time.

Some key watershed periods for human-induced species dispersal to Britain in history include (from Rotherham, 2014a, 2017a): 1) Celts; 2) Romano-British; 3) monastic settlements; 4) the Crusades; 5) medieval trade; 6) European imperialism, world discovery and exploitation; 7) the Victorian period; and 8) the 20th century. These key watershed periods go alongside climate change and socio-economic and political change and associated traumas in the countryside. The outcomes or manifestations of the latter include war, famine, disease, conquest, changed land ownership or management, etc.

In terms of biosecurity issues, there is a two-way relationship between recombinant ecologies and species considered to be biosecurity threats. On the one hand, species effectively integrating with 'native' ecologies might give cause for optimism in terms of functioning (albeit recombinant) communities. From a different perspective, some species may displace natives from their established niches or even divert a community into an altogether different ecological trajectory. Those invasive species which have the most significant impacts on native communities will be considered negatively in terms of biosecurity. Of these species, those acting as major pests and diseases that may displace, eradicate or significantly harm native species and communities, or even threaten human well-being, must be priorities for biosecurity control measures.

Some examples of historically significant invaders to Britain that have affected the countryside, human populations and the ecology are given in Table 11.2.

Effects on ecology may be direct, through eco-fusion impacts, or indirect, following changes to human land management. Furthermore, major human-related events, activities and actions have dramatically altered established ecological communities and species (see, for example, Rotherham, 2009b, 2013a, 2013b, 2014a). Through these interactions are created new opportunities for invaders and novel ecologies.

Examples of these critical processes (from Rotherham, 2014a) include:

1) Major fenland drainage and conversion to agriculture
2) Major removal of tree cover
3) Major urbanisation
4) Conversion of most open land in the lowlands to modern enclosed agriculture
5) Conversion and drainage of most upland to sheep-walk and grouse-moor
6) Widespread imposition of modern forestry mostly with exotic conifer tree species
7) Engineering of watercourses and creation of water-impoundment reservoirs
8) Engineering impacts on coastal flats, marshes, saltmarshes and estuarine habitats
9) Widespread pollution of air, land and water with especially intense impacts in many urban areas and heavy industrial zones
10) The widespread importation and cultivation of non-native trees and other plants
11) The widespread importation and establishment of non-native fauna

These human-induced impacts are combined with the removal of keystone native species and with processes numbered 10 and 11 above. The introduction of exotic plants, animals, fungi and other biota has increased through time.

Recombinance

In some cases, the new taxa become embedded in the existing ecology with little obvious impact or (subjectively induced) human concern. Examples where species have become widespread within native-dominated communities and ecosystems include mammals such as the brown hare (*Lepus europaeus*), which is a long-term exotic and now a Biodiversity Action Plan Species and

Table 11.2 Examples of invasive species that have had a significant impact on the countryside, human populations and ecology of Britain

Plants	Mammals	Birds	Invertebrates	Micro-organisms
Himalayan balsam (*Impatiens glandulifera*)	Black rat (*Rattus rattus*)	Ring-necked parakeet (*Psittacula krameria manillensis*)	Harlequin ladybird (*Harmonia axyridis*)	Bubonic plague (*Yersinia pestis*)
Wild rhododendron (*Rhododendron ponticum*)	Rabbit (*Oryctolagus cuniculus*)	Canada goose (*Branta canadensis*)	Asian hornet (*Vespa mandarinia*)	Malaria (*Plasmodium vivax*)
Portuguese laurel (*Prunus Lusitanica*)	Coypu (*Myocastor coypus*)	Mandarin duck (*Aix galericulata*)	Signal crayfish (*Pacifastacus leniusculus*)	Leprosy (*Mycobacterium leprae*)
Sitka spruce (*Picea sitchensis*)	Grey squirrel (*Sciurus carolinensis*)	Ruddy duck (*Oxyura jamaicensis*)	Australian flatworm (*Australoplana sanguinea*)	Myxomatosis (*Myxomatosis myxoma*)
Sycamore (*Acer pseudoplatanus*)	Brown rat (*Rattus norvegicus*)	Little owl (*Athene noctua*)	New Zealand flatworm (*Arthurdendyus triangulates*)	Dutch elm disease (*Ophiostoma ulmi*)
Norway maple (*Acer platanoides*)	Fallow deer (*Dama dama*)	Ring-necked pheasant (*Phasianus colchicus*)	Australian flatworm (*Australoplana sanguinea*)	
Sweet chestnut (*Castanea sativa*)	Sika deer (*Cervus nippon*)		New Zealand flatworm (*Arthurdendyus triangulates*)	
Giant hogweed (*Heracleum mantegazzianum*)	Muntjac deer (*Muntiacus reevesi*)			
Japanese knotweed (*Fallopia japonica*)	Mink (*Mustela lutreola*)			
New Zealand pygmyweed (*Crassula helmsii*)				
Canadian pondweed (*Elodea canadensis*)				
Buddleia (*Buddleia davidii*)				
Variegated yellow archangel (*Lamiastrum galeobdolon* subsp)				
Montbretia (*Crocosmia* × *crocosmiiflora*)				

valued highly for conservation reasons; the heath star moss (*Campylopus introflexus*), which is native to southern Asia but now widespread on moors, heaths and bogs in Britain; and many examples such as English oak (*Quercus robur*), which was widely planted as Dutch genetic stock from nurseries in the 1700s and 1800s and is quite distinctive from the genuinely British-grown stock. There are many non-native plants and animals that are simply absorbed into the native ecological systems with little obvious adverse impact. Furthermore, it is clear that while some species, such as Himalayan balsam (*Impatiens glandulifera*), took many decades to be recognised as a seriously problematic invader, others, such as Canadian pondweed (*Elodea canadensis*), for example, seem to move from notoriety to obscurity quite quickly. Whether this is a real-time ecological phenomenon or simply because they slip out of public view and are replaced by other new, high-profile invaders is unclear. Indeed, objectivity aside, many 'new' species, such as leopard's bane (*Doronicum orientale*) and sweet cicely (*Myrris odorata*), are welcomed by the public and botanists as 'honorary natives'. There may be localised displacement of established species through competition for niche-space, but overall in these cases the integration of old and new is accepted.

In other situations, the arrival and establishment of a newcomer is visually obvious and may involve clear and apparent negative impacts on the established ecological order. Some cases involve the arrival and spread of pests and diseases with clearly catastrophic impacts. The loss of most native elms (*Ulmus* spp.) to Dutch elm disease and current threats to oak and ash are high-profile examples. However, even here the problems and their impacts can be difficult to assess in terms of 'native' species in 'recombinant' ecologies or non-natives suffering disease or pest outbreaks. Examples of the latter include horse chestnut (*Aesculus hippocastanum*) and its blight, rabbit and myxomatosis and the various diseases and pests of introduced but commercially important conifer trees. The issues in these cases are real and significant but nothing at all to do with native species or natural communities. However, it can be argued that these instances are very much centred within the domain of the dynamics of recombinant ecologies and with both exotic hosts and their pests or diseases.

It seems that environmental stresses and traumas may create opportunities for new colonisers arriving directly through human agency or in some cases independently. Sometimes, such as in urban and industrial environments in the 20th century, entire native communities were eradicated. As conditions subsequently improved following either or both the decline of the polluting industries and environmental controls enforced by legislation, re-colonisation was possible. This was generally by a mix of native and non-native species (both animals and plants) to form hybrid recombinant ecologies, and over time these communities have developed complexity and species richness. While involving significant degrees of ecological novelty, these communities are still able to deliver substantial ecosystem services.

However, from the heartlands of urban rivers and 'urban commons' (Gilbert, 1992a, 1992b), for example, it is clear that some species are capable of spreading more widely to form hybrid communities in the wider countryside. This can then become a major conservation concern, although realistically (Rotherham, 2014a, 2015) the process is, to some extent at least, inevitable. Furthermore, it has been happening for many centuries and in many countries. We can discuss the scale of colonisation and the acceptability or not of its likely impacts, but by and large they will happen. Additionally, in Britain the ecological systems of the wider countryside have already been traumatised by widespread landscape transformations such as agriculture, forestry, transport infrastructure and major extractive industries. There is also widespread pollution of land, air and water from all of these activities. While some atmospheric impacts include the almost ubiquitous deposition of nitrogen, in the past there was widespread fallout of sulphur and associated smoke and soot. In areas around and downwind of urban and industrial centres, this was especially extreme.

Along with these environmental stresses, British ecology has been affected by the introductions of numerous exotic species into the wild, many of which are potential invaders. Some of these are non-native species, some exotic genotypes of natives, some hybrids grown in cultivation or captivity and others hybrids in the wild between natives and non-natives. Sources include deliberate introductions to the wild, garden escapees, forestry introductions and others arriving with limited direct human intervention. However, in the wider countryside, this rather mixed bag of possible invaders meets with an already transformed landscape and a strongly hybrid ecology. In this melting pot, there is inevitably a shuffling of the ecological pack and a sorting and sifting of species within these increasingly recombinant communities.

Implications for biosecurity

It is suggested that many of the changes described are to some degree inevitable and, indeed, have already been happening for many centuries. Only today, the processes have sped up to match the scale of human interventions in the environment. However, with regard to the issues of biosecurity, it seems that the following are key considerations and concerns:

1) **Species displacement, reduction, extinction:**

   a) Impacts on conservation-significant species and communities – especially on keystone native species – through competition and through human persecution and pollution, etc.
   b) Loss of or reduction in key species through diseases and pests and the creation of ecological niche opportunities for invaders
   c) Loss or reduction in species through genetic invasion, i.e. hybridisation

2) **Ecosystem functions:** Impacts on ecosystem services and functioning
3) **Economic consequences:** Impacts on economic activities, food production, forestry and horticulture
4) **Consequences for people:** Impacts on human benefits through disruption to ecosystem services (no. 2 above) and on human welfare, etc.

However, most these issues can be attributed to invasive species, especially non-native invaders. The questions then focus on the implications of the processes of eco-fusion and recombinance in this context. First, in Britain and in many other countries, we already have an ecology which is strongly hybrid and recombinant, and this is concentrated in areas of maximum human interference with 'natural' processes. Second, with increasing impacts of stresses such as climate change, habitat fragmentation, eutrophication and urbanisation, we can expect the impacts to increase rather than decrease. Third, globalisation and the scale and speed of worldwide transportation are undoubtedly accelerating the processes. So, in this context, what are the expected implications of recombinant ecology for biosecurity issues?

One issue that needs to be recognised is that most ecological systems in the British case study, especially in developed areas but also in the wider countryside, are in part at least recombinant and over time are becoming more so. If recombination reduces resilience in ecological systems, there may be increased susceptibility to invasive pests and diseases, although these are more likely to be driven by globalisation and environmental stresses like climate change. Many invasive species, including pests and diseases, are spread directly by human agency and others indirectly by human-induced changes. Land-use changes and landscape transformations stress existing ecological systems, and the processes of invasion, extinction and hybridisation move them towards increasing recombinance. The vulnerability of such altered ecological systems to the loss of ecosystem services and to further declines of native species is as yet unclear. In some cases, at least, the non-native invaders become 'absorbed' into the existing systems, which continue to function effectively. Importantly, as Grime and associates have observed, the ecological strategies and behaviour of native and non-native invaders follow the same pathways and so being 'exotic' does not affect the ecological outcome (e.g. Grime, 2002, 2003; Grime et al., 2007).

However, it is also clear that with increasing human influences and many environmental changes having reached tipping points beyond which a return to a previous stasis is not possible, the future ecological systems will be recombinant; biosecurity issues need to be considered within this context. It is therefore important to gain a better insight into what these communities might be and how they might develop and function.

Wider implications

The case study is based largely on Britain but in a context of global environmental change. At a worldwide level, the same basic observations apply as ecologies are transformed by globalisation, human activities and environmental stresses. Displacement and extinction of island ecologies is well known throughout the world, and many areas (such as Jamaica, for example) already have systems dominated by imported species. European-style agriculture and forestry have been spread around the globe and have transformed landscapes, and increasingly the global human population lives in urban areas dominated by exotic and recombinant ecological communities. In the wider environment, removal of native vegetation, erosion of soils, pollution by fertilisers, etc. combine with species displacement and extinctions to transform the remaining ecological systems. Here, too, the future ecology seems to be increasingly a recombinant one.

References

Adams, W. (2003) *Future Nature: A Vision for Conservation*. Earthscan, London.

Barker, G. (ed.) (2000) *Ecological Recombination in Urban Areas: Implications for Nature Conservation*. English Nature, Peterborough, pp. 21–24.

Barker, G., Luniak, M., Trojan, P. and Zimny, H. (eds.) (1994) 'Proceedings of the Second European Meeting of the International Network for Urban Ecology', *Memorabilia Zoologica*, vol. 49, Warsaw.

Bornkamm, R., Lee, J.A. and Seaward, M.R.D. (eds.) (1982) *Urban Ecology*. Blackwell Scientific Publications, Chichester.

Bridgewater, P.B. (1990) 'The role of synthetic vegetation in present and future landscapes of Australia', *Proceedings of the Ecological Society of Australia*, vol. 16, pp. 129–134.

Clement, E.J. and Foster M.C. (1994) *Alien Plants of the British Isles*. Botanical Society of the British Isles, London.

Davies, J. (2016) *The Birth of the Anthropocene*. University of California Press, Oakland, California.

Davis, M.A., Thompson, K. and Grime, J.P. (2001) 'Charles S. Elton and the dissociation of invasion ecology from the rest of ecology', *Diversity and Distributions*, vol. 7, pp. 97–102.

Douglas, I., Goode, D., Houck, M.C. and Wang, R. (eds.) (2011) *The Routledge Handbook of Urban Ecology*. Routledge, London.

Dunnett, N. and Hitchmough, J. (eds.) (2004) *The Dynamic Landscape: Design, Ecology and Management of Urban Planting*. Spon Press, London.

Ellis, G.R. (1993) *Aliens in the British Flora*. National Museum of Wales, Cardiff.

Elton, C.S. (1958) *The Ecology of Invasions by Animals and Plants*. Methuen, London.

Elton, C.S. (1966) *The Pattern of Animal Communities*. Methuen, London.

Francis, R.A. and Chadwick, M.A. (2012) 'What makes a species synurbic?', *Applied Geography*, vol. 32, no. 2, pp. 514–521.

Gaston, K.J. (ed.) (2011) *Urban Ecology*. Cambridge University Press, Cambridge.

Gilbert, O.L. (1989) *The Ecology of Urban Habitats*. Chapman and Hall, London.

Gilbert, O.L. (1992a) *The Flowering of the Cities . . . . the Natural Flora of 'Urban Commons'*. English Nature, Peterborough.

Gilbert, O.L. (1992b) 'The ecology of an urban river', *British Wildlife*, vol. 3, pp. 129–136.

Goode, D. (2014) *Nature in Towns and Cities*. Collins New Naturalist, London.

Grime, J.P. (2002) 'Declining plant diversity: Empty niches or functional shifts?', *Journal of Vegetation Science*, vol. 13, pp. 457–460.

Grime, J.P. (2003) 'Plants hold the key: Ecosystems in a changing world', *Biologist*, vol. 50, pp. 87–91.

Grime, J.P. (2005) 'Alien plant invaders; threat or side issue?', *ECOS*, vol. 40, pp. 33–40.

Grime, J.P., Hodgson, J.G. and Hunt, R. (2007) *Comparative Plant Ecology: A Functional Approach to Common British Species*. 2nd edition. Castlepoint Press, Dalbeattie, Scotland.

Grime, J.P. and Pierce, S. (2012) *The Evolutionary Strategies that Shape Ecosystems*. Wiley-Blackwell, Chichester.

Hawksworth, D.L. and Rose, F. (1970) 'Qualitative scale for estimating sulphur dioxide air pollution in England and Wales using epiphytic lichens', *Nature*, vol. 227, pp. 145–148.

Hobbs, R.J., Arico, S., Aronson, J., Baron, J.S., Bridgewater, P., Cramer, V.A., Epstein, P.R., Ewel, J.J., Klink, C.A., Lugo, A.E., Norton, D., Ojima, D., Richardson, D.M., Sanderson, E.W., Valladares, F., Vilà, M., Zamora, R. and Zobel, M. (2006) 'Novel ecosystems: Theoretical and management aspects of the new ecological world order', *Global Ecology & Biogeography*, vol. 18, pp. 1–7.

Hobbs, R.J., Higgs, E.S. and Hall, C.M. (eds.) (2013) *Novel Ecosystems: Intervening in the New Ecological World Order*. Wiley-Blackwell, Chichester.

Hobbs, R.J., Higgs, E.S., Hall, C.M., Bridgewater, P., Chapin III, F.S., Ellis, E.C., Ewel, J.J., Hallett, L.M., Harris, J., Hulvey, K.B., Jackson, S.T., Kennedy, P.L., Kueffer, C., Lach, L., Lantz, T.C., Lugo, A.E., Mascaro, J., Murphy, S.D., Nelson, C.R., Perring, M.P., Richardson, D.M., Seastedt, T.R., Standish, R.J., Staromski, B.M., Suding, K.N., Tognetti, P.M., Yakob, L. and Yung, L. (2014) 'Managing the whole landscape: Historical, hybrid, and novel ecosystems', *Frontiers in Ecology and the Environment*, vol. 12, no. 10, pp. 557–564.

Hobbs, R.J., Higgs, E.S. and Harris, J.A. (2009) 'Novel ecosystems: Implications for conservation and restoration', *Trends in Ecology & Evolution*, vol. 24, pp. 599–605.

Hulme, P.E. (2006) 'Beyond control: Wider implications for the management of biological invasions', *Journal of Applied Ecology*, vol. 43, no. 5, pp. 835–847.

Hulme, P.E., Bacher, S., Kenis, M., Klotz, S., Kühn, I., Minchin, D. and Nentwig, W. (2008) 'Grasping at the routes of biological invasions: A framework for integrating pathways into policy', *Journal of Applied Ecology*, vol. 45, no. 2, pp. 403–414.

Johnson, S. (ed.) (2010) *Bioinvaders*. White Horse Press, Cambridge.

Jørgensen, D., Jørgensen, F.A. and Pritchard, S.B. (eds.) (2013) *New Natures: Joining Environmental History with Science and Technology Studies*. University of Pittsburgh Press, Pittsburgh.

Lever, C. (1977) *The Naturalized Animals of Britain and Ireland*. Hutchinson & Co (Publishers) Ltd, London.

Little, W., Fowler, H.W., Coulson, J., Onions, C.T. and Friedrichsen, G.W.S. (1993) *The New Shorter Oxford English Dictionary*. Oxford University Press, Oxford.

Meurk, C. (2011) 'Recombinant ecology of urban areas: Characterisation, context, and creativity', in I. Douglas, D. Goode, M.C. Houck and R. Wang (eds.) *The Routledge Handbook of Urban Ecology*. Routledge, London, pp. 198–220.

Monbiot, G. (2013) *Feral: Searching for Enchantment on the Frontiers of Rewilding*. Allen Lane, London.

Mulcock, J. and Trigger, D. (2008) 'Ecology and identity: A comparative perspective on the negotiation of "nativeness"', in D. Wylie (ed.) *Toxic Belonging? Identity and Ecology in Southern Africa*. Cambridge Scholars Publishing, Newcastle upon Tyne.

Niemelä, J. (ed.) (2011) *Urban Ecology*. Oxford University Press, Oxford.

Noble, I.R. (1989) 'Attributes of invaders and the invading process; terrestrial and vascular plants', in J.A. Drake, H.A. Mooney, F di Castri, F., R.H., Groves, F.J. Kruger, M. Rejmanek and M. Williamson (eds.) *Biological Invasions: A Global Perspective*. Wiley, Chichester, pp. 301–313.

Pearce, F. (2015) *The New Wild: Why Invasive Species Will Be Nature's Salivation*. Icon Books, London.

Rackham, O. (1986) *The History of the Countryside*. Dent, London.

Ratcliffe, D.A. (1984) 'Post-medieval and recent changes in British vegetation: The culmination of human influence', *New Phytologist*, vol. 98, pp. 73–100.

Richardson, D.H.S. (1992) *Pollution Monitoring with Lichens*. Naturalists' Handbooks No. 19, Richmond Publishing Co. Ltd, Slough, England.

Robinson, N. (1993) 'Place and planting design – plant signatures', *The Landscape*, vol. 53, pp. 26–28.

Robinson, W. (1870) The Wild Garden. J. Murray, London.

Rosenzweig, M.L. (2003) 'Reconciliation ecology and the future of species diversity', *Oryx*, vol. 37, no. 2, pp. 194–295.

Rotherham, I.D. (1999) 'Urban environmental history: The importance of relict communities in urban biodiversity conservation', *Practical Ecology and Conservation*, vol. 3, no. 1, pp. 3–22.

Rotherham, I.D. (2005a) 'Invasive plants: Ecology, history and perception', *Journal of Practical Ecology and Conservation Special Series*, vol. 4, pp. 52–62.

Rotherham, I.D. (2005b) 'Alien plants and the human touch', *Journal of Practical Ecology and Conservation Special Series*, vol. 4, pp. 63–76.

Rotherham, I.D. (2009a) 'Exotic and alien species in a changing world', *ECOS*, vol. 30, no. 2, pp. 42–49.

Rotherham, I.D. (2009b) 'The importance of cultural severance in landscape ecology research', in A. Dupont and H. Jacobs (eds.) *Landscape Ecology Research Trends*. Nova Science Publishers Inc, Hauppauge, New York.

Rotherham, I.D. (2013a) *The Lost Fens: England's Greatest Ecological Disaster*. The History Press, Stroud.

Rotherham, I.D. (ed.) (2013b) *Cultural Severance and the Environment: The Ending of Traditional and Customary Practice on Commons and Landscapes Managed in Common*. Springer, Dordrecht, The Netherlands.

Rotherham, I.D. (2014a) *Eco-History: An Introduction to Biodiversity and Conservation*. The White Horse Press, Cambridge.

Rotherham, I.D. (2014b) 'The Call of the Wild: Perceptions, history people & ecology in the emerging paradigms of wilding', *ECOS*, vol. 35, no. 1, pp. 35–43.

Rotherham, I.D. (2015) '"Times they are a changin" – Recombinant Ecology as an emerging paradigm', *International Urban Ecology Review*, vol. 5, pp. 1–19.

Rotherham, I.D. (2017a) *Recombinant Ecology: A Hybrid Future?* Springer Briefs, Springer, Heidelberg.

Rotherham, I.D. (2017b) 'Eco-fusion of alien and native as a new conceptual framework for historical ecology', in E. Vaz, C.J. de Melo and L. Pinto (eds.) *Environmental History in the Making, Vol. 1: Explaining*. Springer, Dordrecht, The Netherlands, pp. 73–90.

Rotherham, I.D. (2018) 'The implications of ecological fusion and cultural severance for re-wilding', *Aspects of Applied Ecology*, vol. 139, pp. 91–101.

Rotherham, I.D. and Lambert, R.A. (eds.) (2011) *Invasive and Introduced Plants and Animals: Human Perceptions, Attitudes and Approaches to Management*. Earthscan, London.

Salisbury, E. (1961) *Weeds and Aliens*. Collins, London.

Simberloff, D. (2003) 'Confronting introduced species: A form of xenophobia?', *Biological Invasions*, vol. 5, pp. 179–192.

Simberloff, D. (2005) 'Non-native species do threaten the natural environment', *Journal of Agricultural and Environmental Ethics*, vol. 18, pp. 595–607.

Simberloff, D. (2011) 'The rise of modern invasion biology and American attitudes towards introduced species', in I.D. Rotherham and R. Lambert (eds.) *Invasive and Introduced Plants and Animals: Human Perceptions, Attitudes and Approaches to Management*. Earthscan, London.

Soulé, M.E. (1990) 'The onslaught of alien species, and other challenges in the coming decades', *Conservation Biology*, vol. 4, no. 3, pp. 233–239.

Stace, C.A. and Crawley, M.J. (2015) *Alien Plants*. HarperCollins, London.

Stace, C.A., Preston, C.D. and Pearman, D.A. (2015) *Hybrid Flora of the British Isles*. Botanical Society of Britain & Ireland, Bristol.

Steffen, W., Crutzen, P.J. and McNeill, J.R. (2007) 'The Anthropocene: Are humans now overwhelming the great forces of nature', *AMBIO*, vol. 36, no. 8, pp. 614–621.

Sukopp, H. and Hejny, S. (eds.) (1990) *Urban Ecology: Plants and Plant Communities in Urban Environments*. SPB Academic Publishing, The Hague, The Netherlands.

Sukopp, H., Numata, M. and Huber, A. (eds.) (1995) *Urban Ecology as the Basis of Urban Planning*. SPB Academic Publishing, The Hague, The Netherlands.

Thomas, C. (2011) 'Britain should welcome climate refugee species', *New Scientist*, 2nd November, pp. 29–30.

Thomas, C. (2013) 'The Anthropocene could raise biological diversity', *Nature*, vol. 502, p. 7.

Thomas, C.D., Cameron, A., Green, R.E., Bakkenes, M., Beaumont, L.J., Collingham, Y.C., Erasmus, B.F.N., Ferreira de Siqueira, M., Grainger, A., Hannah, L., Hughes, L., Huntley, B., van Jaarsveld, A.S., Midgley, G.F., Miles, L., Ortega-Huerta, M.A., Peterson, A.T., Phillips, O.L. and Williams, S.E. (2004) 'Extinction risk from climate change', *Nature*, vol. 427, pp. 145–148.

12

INDUSTRIAL AGRICULTURAL ENVIRONMENTS

Robert G. Wallace, Alex Liebman, David Weisberger,
Tammi Jonas, Luke Bergmann, Richard Kock
and Rodrick Wallace

Introduction

Modern agriculture is proving to be a leader in the next mass extinction (Broswimmer, 2002; Dawson, 2016). Along with agriculture's advance, primary natural habitat and non-human populations are contracting across locales at record rates (Foley et al., 2011; Valladares et al., 2012; Sinclair and Dobson, 2015; Ferreira et al., 2015; Wolf and Ripple, 2017). At the same time, agriculture is founding new ecologies in the place of the lost landscapes. Agricultural production and trade promote invasive species and alternate xenospecific relationships, permitting emergent pathogens, pests and other previously marginalised populations to disrupt long-term ecosystemic function (Crowl et al., 2008; Paini et al., 2016; Chapman et al., 2017; Wyckhuys et al., 2019). The crops and livestock in and of themselves represent the greatest of such introgression (Hoffman, 2010).

Succession, one community of species replacing another, is indubitably a pervasive ecological process (Bast, 2016). All of the earth's biomes originated in part by just such serendipity, with taxa mixing and matching across the globe since life's origins. But there is something foundationally different in the scale and speed at which capital-led production has transformed environment and community alike. In prioritising profits, an abstraction with foundational real-world impact, industrial agriculture risks undercutting the very regenerative capacity of the biological sum upon which even the most human-centric activities, food production included, depend (Lefebvre, 1974 [2000]; Foster, 1999; Malm, 2016; Okamoto et al., 2019; Wallace et al., 2020).

In this chapter we explore how such an agriculture has selected for new epizoologies in livestock and plant epidemiologies in crops, producing potential threats to human health and economy alike. Along the way, we address how capital monetises the process of controlling those very same diseases that are of its own making. We propose that, however ingenious, the disease control strategies enacted to protect food animals and plants provide nominal defence, acting more as a self-exculpating scientism wielded against alternate food systems. That is, *biosecurity* is an imposition in *biogovernance*, how capital and its allies in the public sector rule societies by intervening into human populations – and the food sources upon which they depend – from individual bodies to broader demographics. We argue that biosecurity is deployed first and foremost to protect the most lucrative markets in invasive agriculture.

Livestock and poultry: industrialising pathogen evolution

Megafarm conditions select for a wide array of deadly diseases that belie the livestock and poultry industries' self-absolving invocations of 'biosecurity' (Graham et al., 2008; Hinchliffe, 2013; Allen and Lavau, 2015; Wallace, 2016a; Leibler et al., 2017). Among recent emergent and re-emergent farm and foodborne pathogens of increasing deadliness and spatial extent are African swine fever, *Campylobacter*, Covid-19, *Cryptosporidium*, *Cyclospora*, Ebola Reston, *E. coli* O157:H7, foot and mouth disease, hepatitis E, *Listeria*, Nipah virus, Q fever, *Salmonella*, *Vibrio*, *Yersinia* and a variety of novel influenza A variants, including H1N1 (2009), H1N2v, H3N2v, H5N1, H5N2, H5Nx, H6N1, H7N1, H7N3, H7N7, H7N9 and H9N2 (Tauxe, 1997; Guinat et al., 2016; Wallace et al., 2016; Marder et al., 2018; Wallace et al., 2020; Wallace, 2020).

Raising vast monocultures removes immunogenetic firebreaks that in more diverse populations cut off transmission booms (Garrett and Cox, 2008; Vandermeer, 2010; Thrall et al., 2011; Denison, 2012; Gilbert et al., 2017). Pathogens routinely just evolve around the now near-ubiquitous host immune genotypes. Industry densities can also depress immune response (Houshmand et al., 2012; Gomes et al., 2014; Yarahmadi et al., 2016; Li et al., 2019). Larger herd sizes and greater densities select for increases in rates of transmission and recurrent infections across pathogen strains (Pitzer et al., 2016; Gast et al., 2017; Diaz et al., 2017; EFSA Panel on Biological Hazards, 2019). High throughput offers a continually renewed supply of susceptibles at barn, farm and regional levels, removing the demographic cap on the evolution of pathogen deadliness (Atkins et al., 2011; Allen and Lavau, 2015; Pitzer et al., 2016; Rogalski et al., 2017). Housing such concentrations rewards those strains that can burn through them fastest. As the *persistence* of acute pathogens is also selected for in industrial livestock at population sizes orders of magnitude *less* than in humans, such production can amplify disease spillover into human populations (Rogalski et al., 2017).

Mathematical modelling and a proliferation of data suggest additional epizoological perversities (Rozins and Day, 2017). Decreasing the age at sacrifice – to 6 weeks in chickens and 22 in hog – may select for greater pathogen virulence and viremia able to survive more robust immune systems (Wallace, 2009; Atkins et al., 2013; Wallace, 2016b; Mennerat et al., 2017). 'All-in/all-out' production, an attempt to control outbreaks by growing out cohorts in batches, may introduce transmission optima at the level of the barn or farm. The practice may select for a population infection threshold per barn that lines up with the finishing times farmers set for their herds or flocks (Atkins et al., 2011; Kennedy et al., 2017). That is, successful strains evolve life histories that kill farm animals grown out and near slaughter, when stock are most valuable to farmers. With no reproduction on-site and breeding conducted offshore largely for morphometric traits alone – big breasts in chickens, for instance – food animal populations are also unable to pass on resistance to circulating infections (Wallace, 2016c; Gilbert et al., 2017). Natural selection is removed as an ecosystem service, no longer able to conduct near-free work for farmers. The modelling is finding support in the historical record. Increases in avian influenza virulence have been documented nearly exclusively in larger commercial operations (Wallace, 2009; Atkins et al., 2011; Dhingra et al., 2018).

Global trade compounds these production effects. Beyond the farm gate, lengthening the geographic extent of burgeoning live animal trade has expanded the diversity of genomic segments their pathogens trade, increasing the rate and combinatorials over which disease agents explore their evolutionary possibilities (Nelson et al., 2011; Fuller et al., 2013; Wallace and Wallace, 2015; Mena et al., 2016; O'Dea et al., 2016; Dee et al., 2018; Gorsich et al., 2019; Nelson et al., 2019). The greater the variation in their genetics, the faster pathogen populations evolve.

Even successful gambits in biosecurity carry their costs. At this scale of operations, by both everyday sanitation and emergency culling, population disease resistance can be filtered out, less virulent serovars that exclude their more virulent competitors can be removed and exposures that natural immune development requires can be excluded, including pathogen epitopes that elicit cross-reactive immune protection (Rabsch et al., 2000; Shim and Galvani, 2009; Nfon et al., 2012; Yang et al., 2018). Vaccines – which certainly have their place in food animal health – can also mask or select for the evolution of virulence under industrial conditions (Smith et al., 2006; Pasquato and Seidah, 2008; Lauer et al., 2015). Farms that require stringent biosecurity are typically spatially clustered for *economies of scale*, producing super-epidemiologies that extend beyond any one farm's capacity to control an outbreak, with decided impacts upon human spillover and pathogen fitness (Wallace, 2016d; Lantos et al., 2016; Ma et al., 2019). Some epizoological dynamics are so irregular that the possibility of eradication is eliminated, whatever controls are developed and adhered to (Kennedy et al., 2018).

Other results imply breakages in scientific thinking. Modelling of Marek's disease virus, an alphaherpesvirus that produces a variety of cancers in poultry, indicates the effects of stocking densities on pathogen mortality may depend on whether layers moult and enter a natural abeyance in laying eggs (Rozins and Day, 2017; Rozins et al., 2019). Whereas Rozins et al. (2019) argue such a result indicates improving hen welfare need not be at odds with industry economics, the high stocking densities that led to *reduced* egg loss in the model required the seasonal interruptions moulting imposes (and which the layer sector attempts to *circumvent* with counterseasonal lighting for year-round laying). The researchers' modelling rationale here – searching for room for industrial practices – is particularly loaded given that blaming smallholders wholesale for outbreaks and demanding biosecurity protocols that pasture producers are unable to afford for diseases rarely of their origin are now part of the industry's standard outbreak crisis management package (Bryant and Garnham, 2014; Wallace, 2016e; Wallace, 2017; Wallace, 2018a). Forster and Charnoz (2013) and Wallace (2018a) argue such imposition extends beyond a shock doctrine by which outbreaks are used to capital's passing financial advantage. As Ingram (2013) and Dixon (2015) describe it, biosecurity offers a mode of governance by which global capital accumulates through nature at smallholders' expense.

The dichotomy between intensive and extensive production upon which industry arguments are based is itself problematic (Wallace, 2018a). The essentialist distinction that has been made between industrial farms exercising biosecurity and small farmers whose herds and flocks are exposed to the epidemiological elements is belied by complexities in ownership and contractual obligation (Wallace, 2009; Atkins et al., 2011; Leonard, 2014). In many industrial countries, agribusiness ships day-old chicks to be raised piecework by contract farmers. Once grown (and exposed to migratory birds and environmental sources of disease), the flocks are shipped back to the factory for processing. The violation of biosecurity is built directly into the industrial model.

The whole of these constraints – the loss of immune diversity and responsive evolution, the selection for virulence and persistence – are almost entirely self-imposed. Beyond passing off blame, the sector has responded to these disease traps of its own making by modifying food animals. The logistical lengths undertaken in the lab and on the farm, however, are based on expedient readings of the nature of livestock biologies (Lulka, 2004; Lorimer and Driessen, 2013) that are also proving gothic in their gore.

'Sterile' hog sheds, for instance, are populated with piglets obtained by 'snatch farrowing' – collected directly at birth and reared in isolation (Harris, 2000). Under the HYPAR or HCDC variations, the piglets are 'hysterectomy-procured' or '-derived' and 'artificially reared' or deprived of colostrum (Henry, 1965; Harris, 2000; Stibbe, 2012). That is, sows that were on the cusp of farrowing are euthanised before or after delivery by terminal hysterectomy (Muñoz et al., 1999; Zimmerman et al., 2019). Their uteri are removed and placed in a humidicrib or doused in antiseptic before the piglets are removed from their uterus casing. In some cases, the piglets are latched onto first-generation HYPAR sows so they can obtain colostrum. Some are medically induced into early weaning. In others, the piglets are fully formula reared. The aim in breaking the mother-piglet bond is to produce what the industry calls a 'minimal-disease' (MD), 'high health status', or 'specific-pathogen free' (SPF) herd that breaks the vertical trans-mission of pathogens (and a beneficial microbiome) from sow to progeny (Cameron, 2000; Huang et al., 2013). MD-brand pigs are expected to be free of industrially specific brucellosis, enzootic pneumonia, pleuropneumonia, swine dysentery, external parasites such as sarcoptic mange and internal parasites, including large round worm, nodule worm, whip worm and stomach worm *Hyostrongylus*.

Post-partum development is characterised by another series of surgical and environmental interventions (Zimmerman et al., 2019). Piglet tails are routinely docked by cautery iron or cutters, eye teeth removed to stop bored piglets from biting each other and castration chemi-cally induced in juveniles (Sutherland et al., 2008; Van Beirendonck et al., 2012). In an effort to conserve farm space and labour, finances again favoured first, sows are placed in small gestation and farrowing crates in which they are unable to turn; they chew bars out of bore-dom and develop sores attempting to stand up and lie down (Schrey et al., 2017; Pedersen, 2017; Baxter et al., 2017). Welfare is almost entirely presented in terms of lesser mortalities for sow and piglet alike than animal experience (e.g. Chantziaras et al., 2018). Industrial livestock are raised more as always-already meat than living animals (Jonas, 2015a, 2015b). The presumption carries over across veterinary care. Subtherapeutic antibiotics in hog, other livestock and poultry – 80% of total US usage in human and non-human alike, summing to 34 million pounds a year stateside – were until 2017 nearly universally deployed first as growth promoters and prophylaxes for diseases that for the most part would be otherwise preventable by a change in the production model (Spellberg et al., 2016; Wallace, 2016a; FDA, 2019). The agricultural applications still help contribute to antibiotic resistance across the bacterial infections killing 23,000–100,000 Americans a year and, with increasing application globally today, 700,000 people worldwide (Spellberg et al., 2016; Robinson et al., 2014; Mughini-Gras et al., 2019).

Such herd densities, as noted, are characterised by little background immunity, presenting the very risk of disease the industrial model of production presumes to codify out by imposing sterile initial conditions in barns full of bodily fluid and manure. The Danish entry/exit is now prescribed: a contaminated area where work boots and clothing are left, leading through an intermediate area to a clean zone where outbarn wear is provided (Lambert and D'Allaire, 2009; Pitkin et al., 2009; Janni et al., 2016; Dewulf and Van Immerseel, 2018). Downtime is imposed before visitors are allowed. Showers are provided on-site upon entry and exit. Foot-baths are on offer. Production is organised around 'all-in/all-out', wherein a herd or flock is brought into a barn only upon the removal of the previous cohort. Barns are outfitted with double-door, 'airlock' systems and filters for air circulating in and out. Perimeter fencing and wild bird- and rodent-proofing aim to keep dirty nature at bay. Machinery and tools – trucks, forklifts, ear taggers and the like – are dedicated by barn. Dead livestock are incinerated, buried or composted on-site. A separate lunchroom is made available. Processed animal products are banned. Staff are prevented from contact with similar animals – domestic, commercial and wild – even as sector economics also select for squads of poorly paid itinerant farmhands work-ing farm to farm (Blanchette, 2015; Wallace, 2016g; Wickramage et al., 2018; Wickramage and Annunziata, 2018; Moyce and Schenker, 2018). Interventions extend off-site. Companies have inspected farmworkers' homes as if they are the source of the problem (Blanchette, 2015; Wallace, 2016g).

However assiduously followed, even the most stringent industrial biosecurity programmes are proving ineffectual against an increasing array of deadly pathogens evolving in part as a matter of their own agency. Despite months of warning in advance, wide media coverage and state and extension programme publicity campaigns, highly pathogenic avian influenza A H5N2 burned through turkey and layer operations in the US Midwest, killing 50 million birds by direct infection or culling (Wallace, 2016c). The strain appeared spread by fomites on the wind for which the Danish model offered little protection and the sector's spatial concentration favoured. Disease, alongside pollution, farmworker abuse and all the other rifts touched on in this chapter, exemplifies a little-alluded-to and intrinsic *diseconomy* of scale. The greater the size of operations at both farm and region levels, the *worse* the problem.

The disease problematique extends beyond the logistical capacity the sector can bear on the outbreaks (Lee et al., 2019). In 2016, France's duck and goose sectors were hit across 18 southwest departments by simultaneous outbreaks of highly pathogenic avian influenzas H5N1, H5N2, H5N3 and H5N9 (Wallace, 2017; Briand et al., 2018). In response, poultry farmers moved to end industrial production on the basis that the biosecurity practices long pursued were insufficient for biocontrol (Belaich, 2016). In contrast to US production, sector-wide regulation is introducing four-month breaks between flocks during which farmers are expected to clean out and disinfect their barns, a length of time that would effectively end France's place in the global race-to-the-bottom in production ecology.

The quandary is actualised by more than the small likelihood the usual stable of interventions can succeed in staunching these newly emergent infections. Agribusiness pathogens are also evolving through the core of the model of production as living refutations. The new pathogens, H5Nx and African swine fever among them, circumvent the culturally bounded notions of what 'biosecurity' *must* mean to the sector as both a matter of economic necessity and as a 'master signifier' on which to ground the story of food for the greater public (Hill, 2015; Wallace, 2017; Maclean et al., 2019). Industrial cascades of 'sterile' livestock are proving no such thing. The resulting damage, extending beyond herd and flock loss, is infusing agribusiness leadership with an existential anxiety (Collier and Lakoff, 2008; Hinchliffe, 2013; Allen and Lavau, 2015; Gowdy and Baveye, 2019). Dirty diseases that escape putatively clean food 'contaminate' industry's narrative during an already ongoing crisis in thick legitimacy and public trust (Akram-Lodhi, 2015; Montenegro de Wit and Iles, 2016; Murray, 2018; Wallace et al., 2020). Factions of agricultural capital, already clashing over whether to supply highly competitive markets or protect value by planned scarcity, are beginning to lose the kind of class discipline needed to resolve the sector's multiplying predicaments in agribusiness's favour.

Some industrial factions appear adventitious in their response, beyond even the grab-and-go of terminal hysterectomies. The impression left is that it is hoped that some solution will stick this financial quarter. Face recognition software for monitoring thousands of livestock for disease or lasers for driving migrating wild waterfowl off farms offer few prospects and involve little in the way of transforming the production schedules at the heart of the business model selecting for virulence and persistence (Powell, 2017; Benjamin and Yik, 2019; Ma, 2019). 'Preprogramming' disease resistance into food animals as if a transgenic subject settles the matter before the production line starts addresses nothing of pathogens that across millions of hosts daily evolve solutions to prophylaxes many times over before the pharmaceuticals are even introduced (Wallace, 2016h; Proudfoot et al., 2019).

Agribusiness's capacity to adapt shouldn't be underestimated, however. The sector has long parlayed its intrinsic failure modes in disease control (Leonard, 2014; Wallace, 2016e;

Adams et al., 2018). The ingenuity is astonishing if we miss that the materialism upon which industrial agriculture works extends beyond flesh and machine and into the social and the semiotic (Fracchia, 2017; Wallace, 2018b). Rather than rethinking the model of production, decades ago intensive husbandry in the US and other industrial countries spun off growing out animals – what many imagine as farming itself – to contract farmers to bear the worst of disease's fiscal losses. Contractors are hired on a gig basis while responsible for millions in US dollars in loans to buy the land, barns, equipment and other inputs to raise food animals to company specifications. The companies 'grow' the contractors, and the capital debts to which these farmers are indentured, to sop up the worst of the damage outbreaks cause.

Until now the fail-safe has worked as designed. For the 2015 outbreak of avian influenza H5N2 in the US Midwest, the direct costs of birds killed by the virus, for which a vaccine produced during the outbreak proved ineffective, fell on contract farmers, for whom no outbreak insurance is on offer (Wallace, 2016d). The costs of culling flocks as yet uninfected by H5N2 but in danger were paid for by federal taxpayers. In short, to the benefit of the deadliest pathogens, allowed to continue to circulate across a vast network of barns, farms and national borders, capitalism here, against its own characterisation, failed to punish the sector's market failure. The system moves the damage off balance sheets and upon food animals, wildlife, famers, consumers and communities local and abroad instead. Foundational failures in biosecurity are tucked into production, to be offset by farmers and governments first, before a single hog or chicken batch makes it off the truck. One finds similar end-runs in crops.

Crops: Palmer amaranth and biopolitical governance

Industrialised crop agriculture offers its own dysfunctional analogues to spreading zoonosis. In this section, we examine problems within US commodity crop production related to the expansion of an invasive plant, Palmer amaranth (*Amaranthus palmeri* S. Watson). The weed and efforts to control it embody how invasion and its management are strategically framed in contemporary agronomic discourse.

Palmer amaranth is an annual forb native to northern Mexico and the southwestern US (Moerman, 1986). It has moved outside of its native range, becoming a weed of concern to major grain and fibre producing regions of the Southeast US, and has begun a northward expansion (Ward et al., 2013; Chahal et al., 2015). The reasons for this expansion are myriad and centre around Palmer amaranth's plasticity and adaptive ability in relation to eco-evolutionary selection pressures, namely changing crop management practices and weather patterns. Palmer amaranth is a dioecious obligate outcrosser – the movement of pollen from distinct male to female plants is needed for fertilisation and production of offspring. Its reproductive strategy helps maintain a broad genetic background and is aided by key biological advantages, such as high fecundity – with high levels of pollen and seed production – even relative to other ruderal, primary-successional weed species (Ward et al., 2013; Chahal et al., 2015). These traits allow the weed to maintain high levels of genetic diversity within populations, permitting the transfer of adaptive traits across geographic spaces that are agnostic with respect to legal delineations.

Industrial agricultural crop production systems have directly facilitated the evolution and spread of Palmer amaranth. Nowhere has this been truer than the Southeast's Cotton Belt. In the mid-1990s, both the adoption of conservation tillage practices – specifically no-tillage and strip tillage systems that limit soil disturbance throughout the cropping season – and the elimination of more diversified weed management programmes – including a reduction in the diversity of herbicide sites of action – selected for more robust and competitive biotypes of Palmer amaranth (Webster and Sosnoskie, 2010). The reduction of such diversity in application arose concurrently with the adoption of cotton varieties engineered to tolerate glyphosate and was most pronounced in the ten years following the roll-out of this singular, broad-spectrum herbicide (Kniss, 2018). Indeed, these two management practices should be considered co-facilitating factors, each allowing the other to become more ubiquitous over the landscape. While goals of reduced soil erosion, carbon oxidation and energy use were and are valid, conservation tillage systems and the reduction of weed management practices, facilitated through the mass planting of glyphosate-tolerant cotton, imposed strong selection pressures on Palmer amaranth. Undisturbed soil conditions created more advantageous conditions for germination, growth and development. Seed from this species was now primarily left on the soil surface instead of being buried deeper in the soil profile, generating larger initial populations of the weed and resulting in a much greater probability of the presence of herbicide-resistant traits and individuals (Price et al., 2011; Menalled et al., 2016).

These traits were strongly selected for as repeated applications of glyphosate, upon which farmers in the region were soon heavily reliant, became commonplace (Webster and Sosnoskie, 2010; Ward et al., 2013; Chahal et al., 2015). To compound the problem, Palmer amaranth weed seeds, many of which carried traits conferring resistance to one or more herbicide sites of action, are mobile across environments of varied distance. Weed seeds, and consequently future generations of resistant individuals, are harvested along with a given commodity crop. This occurs within field boundaries via tractors and harvesters, generating local infestations of various levels of intensity and distribution. Grain or fibre is then transported to both local and distant locations through transnational commodity circulation, turning semi-trailers and rail cars into vectors of dispersal and helping resistant populations colonise fields across both proximal and distant geographies.

Much as in the aforementioned epizootic examples, weed invasions are not some ancillary feature of such production but rather the direct result of the industrial agriculture logics imposed on the landscape. Palmer amaranth has become a highly problematic weed in the Southeast US and has been working its way north, with populations confirmed in areas as far as the Canadian border. In almost every state within the invaded region, both single- and multiple-herbicide-resistant biotypes of Palmer amaranth have been confirmed, showing that both individual herbicides and mixtures are ceasing to work (Heap, 2018). That this list is growing suggests a bright future for Palmer amaranth. The fallout associated with this level of human-induced mismanagement and disregard for ecological processes has been socially catastrophic. Invasions of Palmer amaranth are responsible for devastating effects on agronomic production and economic returns related to major commodity crops such as corn, cotton, peanut and soybean (Ward et al., 2013; Chahal et al., 2015).

Considerable effort has been spent describing Palmer amaranth's development, growth habits and impact upon cropping systems (see reviews Ward et al., 2013; Chahal et al., 2015), as well as in introducing chemical and non-chemical forms of management (e.g. Culpepper et al., 2010; Culpepper et al., 2011; Price et al., 2011; Gaines et al., 2012). Such efforts presently dominate weed management science, extension literature and popular farm press, the latter two representing major arbiters of information reaching farmers. Current status quo recommendations for the management of Palmer amaranth involve the continued use of herbicides, with increasingly expensive products and mixtures. While such programmes may provide temporary respite to farmers, this management paradigm continues to fuel what has become an evolutionary arms race, leading to more herbicide-resistant Palmer amaranth. Those that stand to gain the most in the long-term from this trajectory are the agribusiness giants involved in chemical and biotechnology development and sales. As popular herbicides such as glyphosate have become less efficacious in controlling Palmer amaranth over the past ten-plus years, companies have returned to the promotion and sale of older, more volatile herbicide chemistries (group 4 herbicides, specifically 2,4-D and dicamba), concomitantly breeding transgenic crops with tolerance to these chemistries in addition to glyphosate. Unsurprisingly, farmer adoption has quickly followed in the Cotton Belt. This 'solution' merely reproduces the same logic that facilitated Palmer amaranth's exceptional adaptation in the first place. Palmer amaranth plants with resistance to these new mixtures of glyphosate, 2,4-D and dicamba have already been discovered in Kansas, suggesting a limited timeline for this technology at the very outset of its implementation (Kumar et al., 2019).

The chemical volatility of these types of herbicide active ingredients has also sparked production issues and social tensions. Those choosing not to adopt these new transgenic crops face the potential for economically destructive crop damage due to herbicide drift across farms. The use of these 'newer' technologies has also sparked physical violence between neighbouring farmers due to reported crop damage, capsizing long-standing social norms (McCune, 2017). Widespread litigation among industry and individual farmers, as well as among members of the same rural communities, is expected (Alexander, 2017; Hakim, 2017; Bunge, 2020). Researchers have noted how risk is borne unevenly in the adoption of basic production mandates or adaptation to regulatory changes among smallholder strawberry or contract poultry production compared to large-scale commodity shippers and traders (Moodie, 2017; Guthman, 2017). Similarly, Palmer amaranth and its management may simultaneously devastate individual farmers and social relations while generating enormous profits for industry through the adoption of new iterations of herbicide-resistant crops and their associated management programmes.

Palmer amaranth is commonly understood and managed as an alien phenomenon, a foreign threat to the productive material base of the American heartland, requiring all the contemporary tools of ecocide and industry. To better encapsulate how the plant has been managed and discussed, we turn to Michel Foucault's *biopolitics* framework. Foucault, the French political theorist best known for his studies of institutions – of mental hospitals, prisons and schools – also studied the ways in which human and non-human life is governed and managed. Tracing the evolution of sovereign forms of government in the 16th and 17th centuries into modern states, Foucault noted that power, exercised spectacularly and publicly under the rule of kings, had become increasingly focused on managing the molecular, biological conditions of life itself, a capacity he dubbed 'biopower'. Leaders no longer exercised power over a territory based only on divine and violent rule legitimated through political sovereignty but dealt with 'a nature . . . the perpetual conjunction, the perpetual intrication of a geographical, climatic, and physical milieu' (Foucault, 2007, 23). As territory could no longer be controlled through tough borders and military power alone, governance strategies also folded in the management of domestic populations. Statistical studies of risk, health and agriculture, alongside intensive, population-wide schooling, burgeoned throughout Europe during the 18th and 19th centuries, eventually shaping norms of governance in the US context today. Industrialising agriculture and its socio-spatial flow served as a core of these shifts in political strategies.

Biogovernance is explicitly defined as 'a matter of organising circulation, eliminating its dangerous elements, making a division between good and bad circulation, and maximising the good circulation by diminishing the bad' (Foucault, 2007, 18). As applied here, such a hermeneutic is informed by recent challenges to the credo of invasive species ecology and conservation biology (Larson, 2007; Bierman and Mansfield, 2014; Larson, 2016; Srinivasan, 2017) and recent reformulations in the Foucauldian framework to study the post-9/11 proliferation of biosecurity (Barker, 2015). In the case of Palmer amaranth, the herbicidal regime mimics the population-level management of plague and the 'abhorrence of contagion' (Stanescu, 2013). Palmer amaranth's state is a specific consequence of mechanisms in biopolitical governance of contemporary industrial grain and fibre production, production mandates and 'invasive species' control, ideologically and materially supported by agronomic science and industry.

The control of invasive species in highly capitalised, industrial agriculture environments reflects the formation of partitions and securitisation. The management of Palmer amaranth – eradication vis-a-vis herbicidal arms race, selective regulation of gene flows, disciplined mono-culture, export zones wrought out of unruly nature – constitutes a regime of governance that enforces utilitarian circulations, among them global crop production, sales of agrichemicals and mitigation of capital-suppressing pathogens and weeds, all the while studiously neglecting the ways in which industrial agriculture cultivates the very conditions for these incursions (Barker, 2015).

The proximate problem is well-studied. There is empirical clarity as to the underlying causes and consequences of 'invasive plants' in the present industrial agricultural paradigm. Agronomic scientists have stressed the preponderance of herbicide-resistant weeds as a direct consequence of the herbicide industry's impact upon both farm production and the intellectual output of weed scientists (Davis and Frisvold, 2017; Harker et al., 2017). A next step, reframing the co-evolutionary interactions between Palmer amaranth and the logics of industrial capitalist agri-culture as a particular form of biopolitical management, finds where different forms of life and evolutionary processes are newly *valorised*, frequently with disastrous social and ecological results. Hindmarsh (2005) theorised the dialectic between containment and uncontrolled gene flow across genetically modified and non-GMO crops as a Foucauldian 'escape' from the attempted mechanistic, ordered management of natural resources. Palmer amaranth's escape embodies a resistance to the industrial orderings that drove its emergence.

Understanding invasive species control and agronomic sciences as deeply biopolitical forms of ordering helps clarify the contradictory directions of regulation and management. On the one hand, crop production helps drive agri-capital's 'spatial fixes' (Harvey, 2001). Through shifts in technology, land price and locational competition, among other factors, a locale may suddenly become transiently conducive to cheap crop production and advantageous exchange (Harvey, 1982 [2006]; Walker and Stroper, 1991; Wallace et al., 2015). On the other hand, the borders of farm fields are increasingly securitised and surveilled. Furthermore, such a biopolitics depends upon the epistemological disconnect required to create and promote certain invaders while ignoring others, proceeding 'not through connections and contagions, but rather [producing] subjectivity through separation and disavowal' (Stanescu, 2013).

A singular focus on the expansion and optimisation of crop populations and export-oriented landscapes has co-produced the actuality and the spectre of Palmer amaranth. These 'invasions' can be spotted in their Janus face, each side, farmer and agribusiness, reflecting a version of the co-determined 'wild' and 'cultured' processes of crop production's evolution and ecology while in their clash each is also at odds with the other.

Conclusion: locating the outside of a global invasion

How might a critical reading of industrial agriculture and biosecurity help us make sense of emerging 'invasion' phenomena across material, economic, discursive and relational domains? Upon what do the seemingly disparate influenza A H5Nx in France and Palmer amaranth in Iowa converge?

In this chapter, we have argued that 'invasion' is embodied by capitalist agricultural logics and the monocultures, extraction ideologies and rippling forms of variegated local agricultures tied to global industrial trade. On its face, there seems no 'outside' (Morton, 2013). The threat of invasion is simultaneously internal and external to our agricultural systems. It 'comes from afar' from futures markets, foreign direct investment and offshore terminal cross-breeding programmes while it is also produced 'at home' through intensive monocultures, tillage regimes and increases in antimicrobiotics, chemical fertiliser and herbicide application (Chen et al., 2013; Robinson et al., 2013). The constituent components together geologically transform vast swaths of the earth's surface into solar factories, carbon mines and manure lagoons, an alien landscape hostile to most life forms outside the interest of capital save a subset of suddenly opportunistic pathogen and pest stowaways (Despommier et al., 2006; Wallace, 2016a).

These changes are filtered by locale. Megafarms are sites of mutual construction across livestock, crops, farmer, labour, architecture and political economy (Lewontin, 1998; Coppin, 2003; Baird, 2011). Their operations effectively divest from the historically ecological and social integration of local landscapes in favour of new programmes in biodiversity, water use, outwaste, labour discipline and economic extraction. Farms are engineered into veritable spaceships, capable of articulating with any biogeography on capital's terms first (and launching free upon market failure). The earth's ecosystems and local cultures are treated as alien territory, with any resistance to accumulating value to be filtered out. Capital has acted with no compunction in destroying the very agroecological and social resilience needed to control regional disease systems before public health or medical intervention (Rotz and Fraser, 2015; Wallace et al., 2019). Indeed, nature and all other non-market subjectivities are competitors to be *defeated*, a contrast exceeding even Franklin's (1979) chilling juxtaposition that under such a system imagining the end of the world is easier than imagining the end of capitalism (Stuart and Gunderson, 2018; Wallace et al., 2018).

The present programme is a losing proposition in the short term, too. The pathogens that arise – H5Nx and Palmer amaranth, among hundreds of others that invade the invasion – are integral byproducts of its operations, dependent upon the diverse and proximal entwining with local biogeographies and health ecologies (Hinchliffe, 2013; Karatani, 2014; Wallace et al., 2015; Foster and Burkett, 2016; Harvey, 2018; Kallis and Swyngedouw, 2018; Wallace et al., 2019). The feral and the factory interplay atop global capital's peripatetic crest – where a suddenly tightened schedule in poultry production, for instance, selects for the just-in-time ontogeny of avian influenza or where Palmer amaranth evolves virulence or resistance in response to a spike in herbicide (Keck, 2019; Tsing, 2017; Tsing et al., 2020). Even without newly moving in, endemic pathogens invade agriculture by evolving out from underneath even the most conscientious biocontrol (Wallace and Wallace, 2015; Gowdy and Baveye, 2019). 'Invasives' also spill back out into the destitute spaces that by spatial fixes capital has abandoned, transforming these 'ghost' landscapes into new forms (Meiners et al., 2002; Jordan et al., 2012; Stein, 2020).

Invasions also assault the epistemic and the normative, how we think and what we believe (Simberloff, 2012; Code, 2013; Doan, 2016). Scientific approaches have selectively overlooked the role of capitalist production – its mechanisation, simplification, geographical re-ordering and incessant spatio-temporal movement – as a causal factor in the production of invasive species. Indeed, whole classes of modelling, down into their mathematical formalisms, recapitulate such a politics of omission as a matter of first principle (Levins, 1998; Levins, 2006; Winther, 2006; Schizas, 2012; Nikisianis and Stamou, 2016). Cost-effectiveness analysis in environmental studies and public health, for instance, aiming to minimise costs in benefits' favour, implicitly accepts the premises of a social system of extreme inequity (Wallace et al., 2018). Given widespread socio-economic destitution, the modelling asks, what should we do? The models are organised around an ethics of economism that prudently minimises expenditures for *some* (less powerful) institutions rather than addressing the structural expropriation that produces the very artificial scarcities the analyses ostensibly target (Farmer, 2008; Sparke, 2009; Chiriboga et al., 2015; Sparke, 2017).

The scientific, governmental and industrial responses to pathogenic invasive species together constitute new forms of biopolitical governance and new spaces for the expansion of capital – as far down as the molecular level and as ethereally abstract as mathematical proofs. Research is filtered through the interests of available extramural funding with agribusiness increasingly replacing the state as the primary source of funding for studies in agriculture in capital centres and peripheries alike (Chapura, 2007; Food and Water Watch, 2012; Pardey et al., 2018; Kalaitzandonakes et al., 2018). While mainstream scientific literature acknowledges the role globalisation and rising global inequality play in the increasing proliferation of invasive species, its conclusions usher in biosecurity logics based in ever-more surveillance and control of an order that enforces those dynamics – ag-gag laws, inspecting farmworker homes and microscopic colonialism domestic and abroad (Hulme, 2009; Blanchette, 2015; Bagnato, 2017).

The new order extends beyond scientific practice at the level of the investigator. The epidemiological commons we need to protect ourselves from pathogens and pests are being sold off as just another series in proprietary inputs. The rawest of epidemiological data – outbreak locales, premise identification numbers, shipping records, pathogen genetics and transgenic histologies – are being wrapped in a confidentiality that is taking precedence over public health as a basic domain of intervention (Lezaun and Porter, 2015; Wallace, 2016f; Borkenhagen et al., 2019; Gorsich et al., 2019; Okamoto et al., 2019). Scientists are increasingly unable to assess even where outbreaks are exactly happening.

Are there alternate programmes that can address this triple invasion – industrial agriculture, pathogens and commoditised investigation? Is there an outside to such a global invasion? Is another world possible? Yes, there is, and it is already underway. Capital, as described here, aims no more than to articulate with the landscapes it arrives upon as if an otherworldly spaceship. Multinationals and their local subsidiaries aim to plant and grow out without interference from or co-option by local peoples and non-human populations alike, save as a base of resources and labour. The reality is a different matter, with agroecosystemic players co-evolving or interweaving together even under the best of capital's circumstances (Norgaard, 1984; Noailly, 2008; Tsing, 2009; Moreno-Peñaranda and Kallis, 2010; Coq-Huelva et al., 2017; Giraldo, 2019). Other alternatives are more consciously oppositional, pursuing returns of the agricultural commons (IPES-Food, 2016, 2018; Chappell, 2018; Manu, 2019; Arias et al., 2019; Vivero-Pol et al., 2019; Anderson et al., 2019).

Agroecological approaches, such as the 'push-pull' of semiochemical intercropping and trap crops in Africa, diversified cropping systems and 'many little hammers' weed management approaches in the US Midwest and the farmer-negotiated probiotic ecologies of livestock landscape health, can foundationally reconfigure agricultural systems in biocontrol's favour (Liebman et al., 1997; Enticott, 2008; Wallace, 2016i; Midega et al., 2018; Wallace et al., 2019). The practices are contingent upon local contexts that refuse to meet global capitalism on anything like its terms as both a matter of principle and practical survival, thwarting the logics of ever-increasing homogenisation, scale and expropriation. Crop diversification, integration of animals into production and farmer-controlled systems of commercialisation and trade also repopulate rural landscapes with human community, undercutting industrial extractivism and biopolitical governance from afar.

While these agroecologies as they are practised in the Global North have been increasingly cut from their foundationally political bases (Holt-Giménez and Altieri, 2016), there is now a new push at presenting the approaches as an epistemological, ontological, cultural and political break from capital-led efforts to industrialise natural economies (Giraldo, 2019). Agroecology is more than about healthy soil and diverse livestock. Critical evaluations of biopolitics and capitalist geographies need to be folded in as a part of helping reconstitute the community-controlled agricultures that can stem industrial food's imperial ecology.

References

Adams, T., Gerber, J.-D., Amacker, M. and Haller, T. (2018) 'Who gains from contract farming? Dependencies, power relations, and institutional change', *Journal of Peasant Studies*, vol. 46, no. 7, pp. 1435–1457.

Akram-Lodhi, A.H. (2015) 'Land grabs, the agrarian question and the corporate food regime', *Canadian Food Studies*, vol. 2, no. 2, pp. 233–241.

Alexander, A. (2017) 'Court finds spraying of dicamba by third-party farmers an intervening cause [Bader Farms, Inc. v. Monsanto Co.]', National Agricultural Law Center, http://nationalaglawcenter.org/court-finds-spraying-dicamba-third-party-farmers-intervening-cause-bader-farms-inc-v-monsanto-co/

Allen, J. and Lavau, S. (2015) '"Just-in-time" disease: Biosecurity, poultry and power', *Journal of Cultural Economy*, vol. 8, no. 3, pp. 342–360.

Anderson, C.R., Brull, J., Chappell, M.J., Kiss, C. and Pimbert, M.P. (2019) 'From transition to domains of transformation: Getting to sustainable and just food systems through agroecology', *Sustainability*, vol. 11, no. 19, p. 5272.

Arias, P.F., Jonas, T. and Munksgaard, K. (eds.) (2019) *Farming Democracy: Radically Transforming the Food System from the Ground Up*. Australian Food Sovereignty Alliance, Melbourne.

Atkins, K.E., Read, A.F., Savill, N.J., Renz, K.G., Islam, A.F., Walkden-Brown, S.W. and Woolhouse, M.E. (2013) 'Vaccination and reduced cohort duration can drive virulence evolution: Marek's disease virus and industrialized agriculture', *Evolution*, vol. 67, no. 3, pp. 851–860.

Atkins, K.E., Wallace, R.G., Hogerwerf, L., Gilbert, M., Slingenbergh, J., Otte, J. and Galvani, A. (2011) *Livestock Landscapes and the Evolution of Influenza Virulence: Virulence Team Working Paper No. 1*. Animal Health and Production Division, Food and Agriculture Organization of the United Nations, Rome.

Bagnato, A. (2017) 'Microscopic colonialism', *E-Flux*, www.e-flux.com/architecture/positions/153900/microscopic-colonialism/

Baird, I.G. (2011) 'Turning land into capital, turning people into labour: Primitive accumulation and the arrival of large-scale economic land concessions in the Lao People's Democratic Republic', *New Proposals: Journal of Marxism and Interdisciplinary Inquiry*, vol. 5, pp. 10–26.

Barker, K. (2015) 'Biosecurity: Securing circulations from the microbe to the macrocosm', *The Geographical Journal*, vol. 181, no. 4, pp. 357–365.

Bast, F. (2016) 'Primary succession recapitulates phylogeny', *Phylogenetics & Evolutionary Biology*, vol. 4, pp. 1.

Baxter, E.M., Andersen, I.L. and Edwards, S.A. (2017) 'Sow welfare in the farrowing crate and alternatives', in M. Špinka (ed.) *Advances in Pig Welfare*. Elsevier, Duxford, UK, pp. 27–72.

Belaich, P.C. (2016) 'Face à la grippe aviaire, les éleveurs du Sud-Ouest se remettent "en ordre de marche"', *Le Monde*, April 28, www.lemonde.fr/economie/article/2016/04/29/face-a-la-grippe-aviaire-les-eleveurs-du-sud-ouest-se-remettent-en-ordre-de-marche_4910900_3234.html

Benjamin, M. and Yik, S. (2019) 'Precision livestock farming in swine welfare: A review for swine practitioners', *Animals*, vol. 9, p. 133.

Biermann, C. and Mansfield, B. (2014) 'Biodiversity, purity, and death: Conservation biology as biopolitics', *Environment and Planning D: Society and Space*, vol. 32, pp. 257–273.

Blanchette, A. (2015) 'Herding species: Biosecurity, posthuman labor, and the American industrial pig', *Cultural Anthropology*, vol. 30, no. 4, pp. 640–669.

Borkenhagen, L.K., Salman, M.D., Mai-Juan, M. and Gray, G.C. (2019) 'Animal influenza virus infections in humans: A commentary', *International Journal of Infectious Diseases*, vol. 88, pp. 113–119.

Briand, F.X., Niqueux, E., Schmitz, A., Hirchaud, E., Quenault, H. et al. (2018) 'Emergence and multiple reassortments of French 2015–2016 highly pathogenic H5 avian influenza viruses', *Infection, Genetics and Evolution*, vol. 61, pp. 208–214.

Broswimmer, F.J. (2002) *Ecocide: A Short History of the Mass Extinction of Species*. Pluto Press, London.

Bryant, L. and Garnham, B. (2014) 'Economies, ethics and emotions: Farmer distress within the moral economy of agribusiness', *Journal of Rural Studies*, vol. 34, pp. 304–312.

Bunge, J. (2020) 'Bayer, BASF ordered to pay $265 million in weedkiller crop-damage suit', *Wall Street Journal*, February 26, www.wsj.com/articles/bayer-basf-ordered-to-pay-265-million-in-weedkiller-crop-damage-suit-11581795711

Cameron, R.D.A. (2000) 'A review of the industrialisation of pig production worldwide with particular reference to the Asian region', www.fao.org/ag/againfo/themes/documents/pigs/A%20review%20of%20the%20industrialisation%20of%20pig%20production%20worldwide%20with%20particular%20reference%20to%20the%20Asian%20region.pdf

Chahal, P.S., Aulakh, J.S., Jugulam, M. and Jhala, A.J. (2015) 'Herbicide-resistant Palmer amaranth (*Amaranthus palmeri* S. Wats.) in the United States–mechanisms of resistance, impact, and management', in A. Price,

J. Kelton and L. Sarunite (eds.) *Herbicides: Agronomic Crops and Weed Biology*. InTech, Rijeka, Croatia, pp. 1–29.

Chantziaras, I., Dewulf, J., Van Limbergen, T., Klinkenberg, M. and Palzer, A. (2018) 'Factors associated with specific health, welfare and reproductive performance indicators in pig herds from five EU countries', *Preventive Veterinary Medicine*, vol. 159, pp. 106–114.

Chapman, D., Purse, B.V., Roy, H.E. and Bullock, J.M. (2017) 'Global trade networks determine the distribution of invasive non-native species', *Global Ecology and Biogeography*, vol. 26, no. 8, pp. 907–917.

Chappell, M.J. (2018) *Beginning to End Hunger: Food and the Environment in Belo Horizonte, Brazil, and Beyond*. University of California Press, Berkeley.

Chapura, M. (2007) *Actor Networks, Economic Imperatives and the Heterogeneous Geography of the Contemporary Poultry Industry*. MA thesis, Department of Geography, University of Georgia, https://getd.libs.uga.edu/pdfs/chapura_mitchell_e_200708_ma.pdf

Chen, G.Q., He, Y.H. and Qiang, S. (2013) 'Increasing seriousness of plant invasions in croplands of Eastern China in relation to changing farming practices: A case study', *PLoS One*, vol. 8, no. 9, p. e74136.

Chiriboga, D., Buss, P., Birn, A.E., Garay, J., Muntaner, C. and Nervi, L. (2015) 'Investing in health', *The Lancet*, vol. 383, no. 9921, p. 949.

Code, L. (2013) 'Doubt and denial: Epistemic responsibility meets climate change skepticism', *Onati Socio-Legal Series*, vol. 3, no. 5, pp. 838–853.

Collier, S.J. and Lakoff, A. (2008) 'The problem of securing health', in A. Lakoff and S.J. Collier (eds.) *Biosecurity Interventions: Global Health and Security in Question*. Columbia University Press, New York.

Coppin, D. (2003) 'Foucauldian hog futures: The birth of mega-hog farms', *The Sociological Quarterly*, vol. 44, no. 4, pp. 597–616.

Coq-Huelva, D., Higuchi, A., Alfalla-Luque, R., Burgos-Morán, R. and Arias-Gutiérrez, R. (2017) 'Co-evolution and bio-social construction: The Kichwa agroforestry systems (*Chakras*) in the Ecuadorian Amazonia', *Sustainability*, vol. 9, no. 10, p. 1920.

Crowl, T.A., Crist, T.O., Parmenter, R.R., Belovsky, G. and Lugo, A.E. (2008) 'The spread of invasive species and infectious disease as drivers of ecosystem change', *Frontiers in Ecology and the Environment*, vol. 6, no. 5, pp. 238–246.

Culpepper, A.S., Richburg, J.S., York, A.C., Steckel, L.E. and Braxton, L.B. (2011) 'Managing glyphosate- resistant Palmer amaranth using 2,4-D systems in DHT cotton in GA, NC, and TN', *Proceedings of the 2011 Beltwide Cotton Conference*. National Cotton Council of America, Cordova, TN, p. 1543.

Culpepper, A.S., Webster, T.M., Sosnoskie, L.M. and York, A.C. (2010) 'Glyphosate-resistant Palmer amaranth in the US', in V.K. Nandula (ed.) *Glyphosate Resistance: Evolution, Mechanisms, and Management*. J. Wiley, Hoboken, pp. 195–212.

Davis, A.S. and Frisvold, G.B. (2017) 'Are herbicides a once in a century method of weed control?', *Pest Management Science*, vol. 73, pp. 2209–2220.

Dawson, A. (2016) *Extinction: A Radical History*. OR Books, New York.

Dee, S.A., Bauermann, F.V., Niederwerder, M.C., Singrey, A., Clement, T. et al. (2018) 'Survival of viral pathogens in animal feed ingredients under transboundary shipping models', *PLoS One*, vol. 14, no. 3, p. e0214529.

Denison, R.F. (2012) *Darwinian Agriculture: How Understanding Evolution Can Improve Agriculture*. Princeton University Press, Princeton.

Despommier, D., Ellis, B.R. and Wilcox, B.A. (2006) 'The role of ecotones in emerging infectious diseases', *Ecohealth*, vol. 3, no. 4, pp. 281–289.

Dewulf, J. and Van Immerseel, F. (2018) *Biosecurity in Animal Production and Veterinary Medicine: From Principles to Practice*. Acco, Leuven.

Dhingra, M.S., Artois, J., Dellicour, S., Lemey, P., Dauphin, G. et al. (2018) 'Geographical and historical patterns in the emergences of novel Highly Pathogenic Avian Influenza (HPAI) H5 and H7 viruses in poultry', *Front. Vet. Sci.*, June 5, https://doi.org/10.3389/fvets.2018.00084

Diaz, A., Marthaler, D., Corzo, C., Muñoz-Zanzi, C., Sreevatsan, S., Culhane, M. and Torremorell, M. (2017) 'Multiple genome constellations of similar and distinct influenza A viruses co-circulate in pigs during epidemic events', *Scientific Reports*, vol. 7, p. 11886.

Dixon, M.W. (2015) 'Biosecurity and the multiplication of crises in the Egyptian agri-food industry,' *Geoforum*, vol. 61, pp. 90–100.

Doan, M.D. (2016) 'Responsibility for collective inaction and the knowledge condition', *Social Epistemology*, vol. 30, pp. 532–554.

EFSA Panel on Biological Hazards (EFSA BIOHAZ Panel), Koutsoumanis, K., Allende, A. Alvarez-Ordóñez, A., Bolton, D. et al. (2019) '*Salmonella* control in poultry flocks and its public health impact', *EFSA Journal*, vol. 17, no. 2, p. e05596.

Enticott, G. (2008) 'The spaces of biosecurity: Prescribing and negotiating solutions to bovine tuberculosis', *Environment and Planning A*, vol. 40, pp. 1568–1582.

Farmer, P. (2008) 'Challenging orthodoxies: The road ahead for health and human rights', *Health and Human Rights*, vol. 10, no. 1, p. 519.

FDA (2019) *FACT SHEET: Veterinary Feed Directive Final Rule and Next Steps*, www.fda.gov/animal-veterinary/development-approval-process/fact-sheet-veterinary-feed-directive-final-rule-and-next-steps

Ferreira, S., Martínez-Freiría, F., Boudot, J.-P., El Haissoufi, M. and Bennas, N. (2015) 'Local extinctions and range contraction of the endangered *Coenagrion mercuriale* in North Africa', *International Journal of Odonatology*, vol. 18, no. 2, pp. 137–152.

Foley, J., Ramankutty, N., Brauman, K.A., Cassidy, E.S., Gerber, J.S. et al. (2011) 'Solutions for a cultivated planet', *Nature*, vol. 478, pp. 337–342.

Food and Water Watch (2012) 'Public research, private gain: Corporate influence over university agricultural research', www.foodandwaterwatch.org/sites/default/ files/Public%20Research%20Private%20Gain%20Report%20April%202012.pdf

Forster, P. and Charnoz, O. (2013) 'Producing knowledge in times of health crises: Insights from the international response to avian influenza in Indonesia', *Revue d'anthropologie des connaissances*, vol. 7, no. 1, pp. w–az.

Foster, J.B. (1999) 'Marx's theory of metabolic rift: Classical foundations for environmental sociology', *American Journal of Sociology*, vol. 105, no. 2, pp. 366–405.

Foster, J.B. and Burkett, P. (2016) *Marx and the Earth: An Anti-Critique*. Brill Academic Publishers, Leiden.

Foucault, M. (2007) *Security, Territory, Population: Lectures at the College de France, 1977–78*. Palgrave Macmillan, Basingstoke.

Fracchia, J. (2017) 'Organisms and objectifications: A historical-materialist inquiry into the "Human and the Animal"', *Monthly Review*, vol. 68, no. 10, pp. 1–16.

Franklin, H.B. (1979) 'What are we to make of J.G. Ballard's apocalypse?', in T.D. Calreson (ed.) *Voices for the Future: Essays on Major Science Fiction Writers, Volume Two*. Bowling Green State University Popular Press, Bowling Green.

Fuller, T.L., Gilbert, M., Martin, V., Cappelle, J., Hosseini, P., Njabo, K.Y., Abdel Aziz, S., Xiao, X., Daszak, P. and Smith, T.B. (2013) 'Predicting hotspots for influenza virus reassortment', *Emerging Infectious Diseases*, vol. 19, no. 4, pp. 581–588.

Gaines, T.A., Ward, S.M., Bekun, B., Preston, C., Leach, J.E. and Westra, P. (2012) 'Interspecific hybridization transfers a previously unknown glyphosate resistance mechanism in Amaranthus species', *Evolutionary Applications*, vol. 5, pp. 29–38.

Garrett, K.A. and Cox, C.M. (2008) 'Applied biodiversity science: Managing emerging diseases in agriculture and linked natural systems using ecological principles', in R.S. Ostfeld, F. Keesing and V.T. Eviner (eds.) *Infectious Disease Ecology: Effects of Ecosystems on Disease and of Disease on Ecosystems*. Princeton University Press, Princeton, pp. 368–386.

Gast, R.K., Guraya, R., Jones, D.R., Anderson, K.E. and Karcher, D.M. (2017) 'Frequency and duration of fecal shedding of *Salmonella* Enteritidis by experimentally infected laying hens housed at different stocking densities', *Frontiers in Veterinary Science*, doi: 10.3389/fvets.2017.00047

Gilbert, M., Xiao, X. and Robinson, T.P. (2017) 'Intensifying poultry production systems and the emergence of avian influenza in China: A "one health/ecohealth" epitome', *Archives of Public Health*, vol. 75, p. 48.

Giraldo, O.F. (2019) *Political Ecology of Agriculture: Agroecology and Post-Development*. Springer, Cham.

Gomes, A.V.S., Quinteiro-Filho, W.M., Ribeiro, A., Ferraz-de-Paula, V., Pinheiro, M.L. et al. (2014) 'Overcrowding stress decreases macrophage activity and increases Salmonella Enteritidis invasion in broiler chickens', *Avian Pathology*, vol. 43, no. 1, pp. 82–90.

Gorsich, E.E., Miller, R.S., Mask, H.M., Hallman, C., Portacci, K. and Webb, C.T. (2019) 'Spatio-temporal patterns and characteristics of swine shipments in the U.S. based on Interstate Certificates of Veterinary Inspection', *Scientific Reports*, vol. 9, p. 3915.

Gowdy, J. and Baveye, P. (2019) 'An evolutionary perspective on industrial and sustainable agriculture', in G. Lemaire, P.C.D.F. Carvalho, S. Kronberg, and S. Recous (eds.) *Agroecosystem Diversity: Reconciling Contemporary Agriculture and Environmental Quality*. Academic Press, London, pp. 425–433.

Graham, J.P., Leibler, J.H., Price, L.B., Otte, J.M., Pfeiffer, D.U., Tiensin, T. and Silbergeld, E.K. (2008) 'The animal: Human interface and infectious disease in industrial food animal production: Rethinking biosecurity and biocontainment', *Public Health Reports*, vol. 123, pp. 282–299.

Guinat, C., Gogin, A., Blome, S., Keil, G., Pollin, R. et al. (2016) 'Transmission routes of African swine fever virus to domestic pigs: Current knowledge and future research directions', *Veterinary Record*, vol. 178, no. 11, pp. 262–267.

Guthman, J. (2017) 'Life itself under contract: Rent-seeking and biopolitical devolution through partnerships in California's strawberry industry', *The Journal of Peasant Studies*, vol. 44, no. 1, pp. 100–117.

Hakim, D. (2017) 'Monsanto's weed killer, dicamba, divides farmers', *New York Times*, September 21, www.nytimes.com/2017/09/21/business/monsanto-`dicamba-weed-killer.html

Harker, K.N., Mallory-Smith, C., Maxwell, B.D., Mortensen, D.A. and Smith, R.G. (2017) 'Another view', *Weed Science*, vol. 65, pp. 203–205.

Harris, D.L. (2000) *Multi-Site Pig Production*. Iowa State University, Ames.

Harvey, D. (1982 [2006]) *The Limits to Capital*. Verso, New York.

Harvey, D. (2001) 'Globalization and the "spatial fix"', *Geographische Revue*, vol. 2, pp. 23–30.

Harvey, D. (2018) 'Marx's refusal of the labour theory of value: Reading Marx's Capital with David Harvey', 14 March, http://davidharvey.org/2018/03/marxsrefusal-of-the-labour-theory-of-value-by-david-harvey/

Heap, I. (2018) 'The international survey of herbicide-resistant weeds', www.weedscience.com

Henry, D.P. (1965) 'Experiences during the first eight weeks of life of HYPAR piglets', *Australian Veterinary Journal*, vol. 41, no. 5.

Hill, A. (2015) 'Moving from "matters of fact" to "matters of concern" in order to grow economic food futures in the Anthropocene', *Agriculture and Human Values*, vol. 32, no. 3, pp. 551–563.

Hinchliffe, S. (2013) 'The insecurity of biosecurity: Remaking emerging infectious diseases', in A. Dobson, K. Baker and S.L. Taylor (eds.) *Biosecurity: The Socio-Politics of Invasive Species and Infectious Diseases*. Routledge, New York, pp. 199–214.

Hindmarsh, R. (2005) *Green Biopolitics and the Molecular Reordering of Nature*. Paper presented at the 'Mapping Biopolitics: Medical-Scientific Transformations and the Rise of New Forms of Governance' Workshop, European Consortium for Political Research Conference, Granada, Spain, 14–19 April.

Hoffmann, I. (2010) 'Livestock biodiversity', *Review Scientifique et Technique*, vol. 29, no. 1, pp. 73–86.

Holt-Giménez, E. and Altieri, M. (2016) *Agroecology "Lite": Cooptation and Resistance in the Global North*. Food First – Institute for Food and Development Policy, https://foodfirst.org/agroecology-lite-cooptation-and-resistance-in-the-global-north/

Houshmand, M., Azhar, K., Zulkifli, I., Bejo, M.H. and Kamyab, A. (2012) 'Effects of prebiotic, protein level, and stocking density on performance, immunity, and stress indicators of broilers', *Poultry Science*, vol. 91, pp. 393–401.

Huang, Y., Haines, D.M. and Harding, J.C.S. (2013) 'Snatch-farrowed, porcine-colostrum-deprived (SF-pCD) pigs as a model for swine infectious disease research', *Canadian Journal of Veterinary Research*, vol. 77, no. 2, pp. 81–88.

Hulme, P.E. (2009) 'Trade, transport and trouble: Managing invasive species pathways in an era of globalization', *Journal of Applied Ecology*, vol. 46, pp. 10–18.

Ingram, A. (2013) 'Viral geopolitics: Biosecurity and global health governance', in A. Dobson, K. Baker and S.L. Taylor (eds.) *Biosecurity: The Socio-Politics of Invasive Species and Infectious Diseases*. Routledge, New York, pp. 137–150.

IPES-Food (2016) *From Uniformity to Diversity: A Paradigm Shift from Industrial Agriculture to Diversified Agroecological Systems*. Louvain-la-Neuve, Belgium, www.ipes-food.org/_img/upload/files/UniformityToDiversity_FULL.pdf

IPES-Food (2018) *Breaking Away from Industrial Food and Farming Systems: Seven Case Studies of Agroecological Transition*. Louvain-la-Neuve, Belgium, www.ipes-food.org/_img/upload/files/CS2_web.pdf

Janni, K.A., Jacobson, L.D., Noll, S.L., Cardona, C.J., Martin, H.W. and Neu, A.E. (2016) 'Engineering challenges and responses to the pathogenic avian influenza outbreak in Minnesota in 2015', 2016 American Society of Agricultural and Biological Engineers Annual International Meeting.

Jonas, T. (2015a) 'The vegetarian turned pig-farming butcher', in N. Rose (ed.) *Fair Food: Stories from a Movement Changing the World*. University of Queensland Press, St Lucia.

Jonas, T. (2015b) 'How to respond to vegan abolitionists', *Tammi Jonas: Food Ethics*, 24 March, www.tammijonas.com/2015/03/24/how-to-respond-to-vegan-abolitionists/

Jordan, N.R., Aldrich-Wolfe, L., Huerd, S.C., Larson, D. and Muehlbauer, G. (2012) 'Soil–occupancy effects of invasive and native grassland plant species on composition and diversity of mycorrhizal associations', *Invasive Plant Science and Management*, vol. 5, no. 4, pp. 494–505.

Kalaitzandonakes, N., Carayannis, E.G., Grigoroudis, E. and Rozakis, S. (2018) *From Agriscience to Agribusiness: Theories, Policies and Practices in Technology Transfer and Commercialization*. Springer, Cham.

Kallis, G. and Swyngedouw, E. (2018) 'Do bees produce value? A conversation between and ecological economist and a Marxist geographer', *Capitalism Nature Socialism*, vol. 29, no. 3, pp. 36–50.

Karatani, K. (2014) *The Structure of World History: From Modes of Production to Modes of Exchange*. Duke University Press, Durham, NC.

Keck, F. (2019) 'Livestock revolution and ghostly apparitions: South China as a sentinel territory for influenza pandemics', *Current Anthropology*, vol. 60, no. S20, pp. S251–S259.

Kennedy, D.A., Cairns, C., Jones, M.J., Bell, A.S., Salathe, R.M. et al. (2017) 'Industry-wide surveillance of Marek's disease virus on commercial poultry farms', *Avian Diseases*, vol. 61, pp. 153–164.

Kennedy, D.A., Dunn, P.A. and Read, A.F. (2018) 'Modeling Marek's disease virus transmission: A framework for evaluating the impact of farming practices and evolution', *Epidemics*, vol. 23, pp. 85–95.

Kniss, A.R. (2018) 'Genetically engineered herbicide-resistant crops and herbicide-resistant weed evolution in the United States', *Weed Science*, vol. 66, no. 2, pp. 260–273.

Kumar, V., Liu, R., Boyer, G. and Stahlman, P.W. (2019) 'Confirmation of 2, 4-D resistance and identification of multiple resistance in a Kansas Palmer amaranth (*Amaranthus palmeri*) population', *Pest Management Science*, https://doi.org/10.1002/ps.5400

Lambert, M.E. and D'Allaire, S. (2009) 'Biosecurity in swine production: Widespread concerns?', *Advances in Pork Production*, vol. 20, pp. 139–148.

Lantos, P.M., Hoffman, K., Höhle, M., Anderson, B. and Gray, G.C. (2016) 'Are people living near modern swine production facilities at increased risk of influenza virus infection?', *Clinical Infectious Diseases*, vol. 63, no. 12, pp. 1558–1563.

Larson, B.M.H. (2007) 'Who's invading what? Systems thinking about invasive species', *Canadian Journal of Plant Science*, vol. 87, pp. 993–999.

Larson, B.M.H. (2016) 'New wine and old wineskins? Novel ecosystems and conceptual change', *Nature and Culture*, vol. 11, no. 2, pp. 148–164.

Lauer, D., Mason, S., Akey, B., Badcoe, L., Baldwin, D. et al. (2015) *Report of the Committee on Transmissible Diseases of Poultry and Other Avian Species*. United States Animal Health Association, www.usaha.org/upload/Committee/TransDisPoultry/report-pad-2015.pdf

Lee, J., Schulz, L. and Tonsor, G. (2019) 'Swine producers' willingness to pay for Tier 1 disease risk mitigation under ambiguity', Selected Paper prepared for presentation at the 2019 Agricultural & Applied Economics Association Annual Meeting, Atlanta, GA, 21 July– 23 July, https://ageconsearch.umn.edu/record/290908/files/Abstracts_19_05_13_19_46_48_31__129_186_252_93_0.pdf

Lefebvre, H. (1974 [2000]) *The Production of Space*. Blackwell Publishers, Oxford.

Leibler, J.H., Dalton, K., Pekosz, A., Gray, G.C. and Silbergeld, E.K. (2017) 'Epizootics in industrial livestock production: Preventable gaps in biosecurity and biocontainment', *Zoonoses Public Health*, vol. 64, no. 2, pp. 137–145.

Leonard, C. (2014) *The Meat Racket: The Secret Takeover of America's Food Business*. Simon & Schuster, New York.

Levins, R. (1998) 'The internal and external in explanatory theories', *Science as Culture*, vol. 7, no. 4, pp. 557–582.

Levins, R. (2006) 'Strategies of abstraction', *Biology & Philosophy*, vol. 21, pp. 741–755.

Lewontin, R. (1998) 'The maturing of capitalist agriculture: Farmer as proletarian', *Monthly Review*, vol. 50, no. 3, p. 72.

Lezaun, J. and Porter, N. (2015) 'Containment and competition: Transgenic animals in the One Health agenda', *Social Science & Medicine*, vol. 129, pp. 96–105.

Li, W., Wei, F., Xu, B., Sun, Q., Deng, W. et al. (2019) 'Effect of stocking density and alpha-lipoic acid on the growth performance, physiological and oxidative stress and immune response of broilers', *Asian-Australasian Journal of Animal Studies*, https://doi.org/10.5713/ajas.18.0939.

Liebman, M., Gallandt, E.R. and Jackson, L.E. (1997) 'Many little hammers: Ecological management of crop–weed interactions', *Ecology in Agriculture*, vol. 1, pp. 291–343.

Lorimer, J. and Driessen, C. (2013) 'Bovine biopolitics and the promise of monsters in the rewilding of Heck cattle', *Geoforum*, vol. 48, pp. 249–259.

Lulka, D. (2004) 'Stabilizing the herd: Fixing the identity of nonhumans', *Environment and Planning D*, vol. 22, no. 3, pp. 439–463.

Ma, A. (2019) 'China uses AI, facial recognition, and blockchain to monitor its farms–but it still can't stop the gruesome swine fever that will leave 200 million pigs dead', *Business Insider*, May 22, www.businessinsider.com/china-pig-facial-recognition-african-swine-fever-still-hits-2019-5

Ma, J., Shen, H., McDowell, C., Liu, Q., Duff, M. et al. (2019) 'Virus survival and fitness when multiple genotypes and subtypes of influenza A viruses exist and circulate in swine', *Virology*, vol. 532, pp. 30–38.

Maclean, K., Farbotko, C. and Robinson, C.J. (2019) 'Who do growers trust? Engaging biosecurity knowledges to negotiate risk management in the north Queensland banana industry, Australia', *Journal of Rural Studies*, vol. 67, pp. 101–110.

Malm, A. (2016) *Fossil Capital: The Rise of Steam Power and the Roots of Global Warming.* Verso, New York.

Manu (2019) 'Agroecology: Real innovation from and for the people', *Nyéléni Newsletter*, p. 36, https://nyeleni.org/DOWNLOADS/newsletters/Nyeleni_Newsletter_Num_36_EN.pdf

Marder, E.P., Griffin, P.M., Cieslak, P.R., Dunn, J., Hurd, S. et al. (2018) 'Preliminary incidence and trends of infections with pathogens transmitted commonly through food – Foodborne Diseases Active Surveillance Network, 10 U.S. sites, 2006–2017', *MMWR*, vol. 67, no. 11, pp. 324–328.

McCune, M. (2017) 'A pesticide, a pigweed, and a farmer's murder', *Planet Money*, National Public Radio, www.npr.org/2017/06/14/532879755/a-pesticide-a-pigweed-and-a-farmers-murder

Meiners, S.J., Pickett, S.T.A. and Cadenasso, M.L. (2002) 'Exotic plant invasions over 40 years of old field successions: Community patterns and associations', *Ecography*, vol. 25, pp. 215–223.

Mena, I., Nelson, M.I., Quezada-Monroy, F., Dutta, J., Cortes-Fernández, R. et al. (2016) 'Origins of the 2009 H1N1 influenza pandemic in swine in Mexico', *Elife*, vol. 5, pp. ii, e16777.

Menalled, F., Peterson, R., Smith, R., Curran, W., Páez, D. and Maxwell, B. (2016) 'The eco-evolutionary imperative: Revisiting weed management in the midst of an herbicide resistance crisis', *Sustainability*, vol. 8, no. 12, p. 1297.

Mennerat, A., Ugelvik, M.S., Jensen, C.H. and Skorping, A. (2017) 'Invest more and die faster: The life history of a parasite on intensive farms', *Evolutionary Applications*, vol. 10, no. 9, pp. 890–896.

Midega, C.A., Pittchar, J.O., Pickett, J.A., Hailu, G.W. and Khan, Z.R. (2018) 'A climate-adapted push-pull system effectively controls fall armyworm, *Spodoptera frugiperda* (J E Smith), in maize in East Africa', *Crop Protection*, vol. 105, pp. 10–15.

Moerman, D.E. (1986) *Medicinal Plants of Native America.* Museum of Anthropology, Ann Arbor.

Montenegro de Wit, M. and Iles, A. (2016) 'Toward thick legitimacy: Creating a web of legitimacy for agroecology', *Elementa: Science of the Anthropocene*, vol. 4, p. 115.

Moodie, A. (2017) 'Fowl play: The chicken farmers being bullied by big poultry', *The Guardian*, April 22, www.theguardian.com/sustainable-business/2017/apr/22/chicken-farmers-big-poultry-rules

Moreno-Peñaranda, R. and Kallis, G. (2010) 'A coevolutionary understanding of agroenvironmental change: A case-study of a rural community in Brazil', *Ecological Economics*, vol. 69, no. 4, pp. 770–778.

Morton, T. (2013) *Hyperobjects: Philosophy and Ecology after the End of the World.* University of Minnesota Press, Minneapolis.

Moyce, S.C. and Schenker, M. (2018) 'Migrant workers and their occupational health and safety', *Annual Review of Public Health*, vol. 39, pp. 351–365.

Mughini-Gras, L., Dorado-García, A., van Duijkeren, E., van den Bunt, G., Dierikx, C.M. et al. (2019) 'Attributable sources of community-acquired carriage of *Escherichia coli* containing β-lactam antibiotic resistance genes: A population-based modelling study', *The Lancet Planetary Health*, vol. 3, pp. e357–e369.

Muñoz, A., Ramis, G., Pallarés, F.J., Martínez, J.S., Oliva, J. et al. (1999) 'Surgical procedure for specific pathogen free piglet by modified terminal hysterectomy', *Transplantation Proceedings*, vol. 31, pp. 2627–2629.

Murray, A. (2018) 'Meat cultures: Lab-grown meat and the politics of contamination', *BioSocieties*, vol. 13, no. 2, pp. 513–534.

Nelson, M.I., Lemey, P., Tan, Y., Vincent, A., Lam, T.T. et al. (2011) 'Spatial dynamics of human-origin H1 influenza A virus in North American swine', *PLoS Pathogens*, vol. 7, no. 6, p. e1002077.

Nelson, M.I., Souza, C.K., Trovão, N.S., Diaz, A., Mena, I. et al. (2019) 'Human-origin influenza A (H3N2) reassortant viruses in swine, Southeast Mexico', *Emerging Infectious Diseases*, vol. 25, no. 4, pp. 691–700.

Nfon, C., Berhane, Y., Pasick, J., Embury-Hyatt, C., Kobinger, G. et al. (2012) 'Prior infection of chickens with H1N1 or H1N2 Avian Influenza elicits partial heterologous protection against Highly Pathogenic H5N1', *PLoS One*, vol. 7, no. 12, p. e51933.

Nikisianis, N. and Stamou, G.P. (2016) 'Harmony as ideology: Questioning the diversity-stability hypothesis', *Acta Biotheoretica*, vol. 64, no. 1, pp. 33–64.

Noailly, J. (2008) 'Coevolution of economic and ecological systems: An application to agricultural pesticide resistance', *Journal of Evolutionary Economics*, vol. 18, pp. 1–29.

Norgaard, R.B. (1984) 'Coevolutionary agricultural development', *Economic Development and Cultural Change*, vol. 32, no. 3, pp. 525–546.

O'Dea, E.B., Snelson, H. and Bansal, S. (2016) 'Using heterogeneity in the population structure of U.S. swine farms to compare transmission models for porcine epidemic diarrhea', *Scientific Reports*, vol. 6, p. 22248.

Okamoto, K.W., Liebman, A. and Wallace, R.G. (2019) 'At what geographic scales does agricultural alienation amplify foodborne disease outbreaks? A statistical test for 25 U.S. States, 1970–2000', *MedRxiv*, www.medrxiv.org/content/10.1101/2019.12.13.19014910v2

Paini, D.R., Sheppard, A.W., Cook, D.C., De Barro, P.J., Worner, S.P. and Thomas, M.B. (2016) 'Global threat to agriculture from invasive species', *PNAS*, vol. 113, no. 27, pp. 7575–7579.

Pardey, P.G., Alston, J.M., Chang-Kang, C., Hurley, T.M., Andrade, R.S. et al. (2018) 'The shifting structure of agricultural R&D: Worldwide investment patterns and payoffs', in N. Kalaitzandonakes, E.G. Carayannis, E. Grigoroudis and S. Rozakis (eds.) *From Agriscience to Agribusiness: Theories, Policies and Practices in Technology Transfer and Commercialization*. Springer, Cham.

Pasquato, A. and Seidah, N.G. (2008) 'The H5N1 influenza variant Fujian-like hemagglutinin selected following vaccination exhibits a compromised furin cleavage: Neurological consequences of highly pathogenic Fujian H5N1 strains', *Journal of Molecular Neuroscience*, vol. 35, no. 3, pp. 339–343.

Pedersen, L.J. (2017) 'Overview of commercial production systems and their main welfare challenges', in M. Špinka (ed.) *Advances in Pig Welfare*. Elsevier, Duxford, UK, pp. 3–25.

Pitkin, A., Otake, S. and Dee, S. (2009) *Biosecurity Protocols for the Prevention of Spread of Porcine Reproductive and Respiratory Syndrome Virus*. Swine Disease Eradication Center, University of Minnesota College of Veterinary Medicine, https://datcp.wi.gov/Documents/PRRSVBiosecurityManual.pdf

Pitzer, V.E., Aguas, R., Riley, S., Loeffen, W.L.A., Wood, J.L.N. and Grenfell, B.T. (2016) 'High turnover drives prolonged persistence of influenza in managed pig herds', *Journal of the Royal Society Interface*, vol. 13, p. 20160138.

Powell, J. (2017) 'Poultry farm sets up lasers to guard its organic hens from bird flu', *The Poultry Site*, March 6, https://thepoultrysite.com/news/2017/03/poultry-farm-sets-up-lasers-to-guard-its-organic-hens-from-bird-flu

Price, A.J., Balkcom, K.S., Culpepper, S.A., Kelton, J.A., Nichols, R.L. and Schomberg, H. (2011) 'Glyphosate-resistant Palmer amaranth: A threat to conservation tillage', *Journal of Soil and Water Conservation*, vol. 66, pp. 265–275.

Proudfoot, C., Lillico, S. and Tait-Bukard, C. (2019) 'Genome editing for disease resistance in pigs and chickens', *Animal Frontiers*, vol. 9, no. 3, pp. 6–12.

Rabsch, W. Hargis, B.M., Tsolis, R.M., Kingsley, R.A., Hinz, K.H., Tschäpe, H. and Bäumler, A.J. (2000) 'Competitive exclusion of *Salmonella enteritidis* by *Salmonella gallinarum* in poultry', *Emerging Infectious Diseases*, vol. 6, no. 5, pp. 443–448.

Robinson T.P., Wint, G.R.W., Conchedda, G., Van Boeckel, T.P., Ercoli, V., Palamara, E., Cinardi, G., D'Aietti, L., Hay, S.I. and Gilbert, M. (2014) 'Mapping the global distribution of livestock.' *PLoS One*, vol. 9, p. e96084.

Rogalski, M.A., Gowler, C.D. Shaw, C.L., Hufbauer, R.A. and Duffy, M.A. (2017) 'Human drivers of ecological and evolutionary dynamics in emerging and disappearing infectious disease systems', *Philosophical Transactions of the Royal Society B: Biological Sciences*, vol. 372, no. 1712, p. 20160043.

Rotz, S. and Fraser, E.D.G. (2015) 'Resilience and the industrial food system: Analyzing the impacts of agricultural industrialization on food system vulnerability', *Journal of Environmental Studies and Sciences*, vol. 5, pp. 459–473.

Rozins, C. and Day, T. (2017) 'The industrialization of farming may be driving virulence evolution', *Evolutionary Applications*, vol. 10, no. 2, pp. 189–198.

Rozins, C., Day, T. and Greenhalgh, S. (2019) 'Managing Marek's disease in the egg industry', *Epidemics*, vol. 27, pp. 52–58.

Schizas, D. (2012) 'Systems ecology reloaded: A critical assessment focusing on the relations between science and ideology', in G.P. Stamou (ed.) *Populations, Biocommunities, Ecosystems: A Review of Controversies in Ecological Thinking*. Bentham Science Publishers, Sharjah.

Schrey, L., Kemper, N. and Fels, M. (2017) 'Behaviour and skin injuries of sows kept in a novel group housing system during lactation', *Journal of Applied Animal Research*, vol. 46, no. 1, pp. 749–757.

Shim, E. and Galvani, A.P. (2009) 'Evolutionary repercussions of avian culling on host resistance and influenza virulence', *PLoS One*, vol. 4, no. 5, p. e5503.

Simberloff, D. (2012) 'Nature, natives, nativism, and management: Worldviews underlying controversies in invasion biology', *Environmental Ethics*, vol. 34, no. 1, pp. 5–25.

Sinclair, A.R.E. and Dobson, A. (2015) 'Conservation in a human-dominated world', in A.R.E. Sinclair, K.L. Metzger, S.A.R. Mduma and J.M. Fryxell (eds.) *Serengeti IV: Sustaining Biodiversity in a Coupled Human-Natural System*. University of Chicago Press, Chicago, pp. 1–10.

Smith, G.J., Fan, X.H., Wang, J., Li, K.S., Qin, K. et al. (2006) 'Emergence and predominance of an H5N1 influenza variant in China', *Proceedings of the National Academy of Sciences of the United States of America*, vol. 103, no. 45, pp. 16936–16941.

Sparke, M. (2009) 'Unpacking economism and remapping the terrain of global health', in A. Kay and O.D. Williams (eds.) *Global Health Governance: Crisis, Institutions and Political Economy*. Springer, Cham.

Sparke, M. (2017) 'Austerity and the embodiment of neoliberalism as ill-health: Towards a theory of biological sub-citizenship', *Social Science & Medicine*, vol. 187, pp. 287–295.

Spellberg, B., Hansen, G.R., Kar, A., Cordova, C.D., Price, L.B. and Johnson, J.R. (2016) *Antibiotic Resistance in Humans and Animals*. National Academy of Medicine. Discussion Paper, https://nam.edu/antibiotic-resistance-in-humans-and-animals/

Srinivasan, K. (2017) 'Conservation biopolitics and the sustainability episteme', *Environment and Planning A*, vol. 49, no. 7, pp. 1458–1476.

Stanescu, J. (2013) 'Beyond biopolitics: Animal studies, factory farms, and the advent of deading life', *PhaenEx.*, vol. 8, no. 2, pp. 135–160.

Stein, S. (2020) 'Witchweed and the ghost: A parasitic plant devastates peasant crops on capital-abandoned plantations in Mozambique', in A. Tsing, J. Deger, A.K. Saxena and E. Gan (eds.) *Feral Atlas: The More-than-Human Anthropocene*. Stanford University Press, Palo Alto.

Stibbe, A. (2012) *Animals Erased: Discourse, Ecology, and Reconnection with the Natural World*. Wesleyan University Press, Middletown.

Stuart, D. and Gunderson, R. (2018) 'Nonhuman animals as fictitious commodities: Exploitation and consequences in industrial agriculture', *Society & Animals*. doi: 10.1163/15685306-12341507

Sutherland, M.A., Bryer, P.J., Krebs, N. and McGlone, J.J. (2008) 'Tail docking in pigs: Acute physiological and behavioural responses', *Animal*, vol. 2, no. 2, pp. 292–297.

Tauxe, R.V. (1997) 'Emerging foodborne diseases: An evolving public health challenge', *Emerging Infectious Diseases*, vol. 3, no. 4, pp. 425–434.

Thrall, P.H., Oakeshott, J.G., Fitt, G., Southerton, S., Burdon, J.J. et al. (2011) 'Evolution in agriculture: The application of evolutionary approaches to the management of biotic interactions in agro-ecosystems', *Evolutionary Applications*, vol. 4, pp. 200–215.

Tsing, A. (2009) 'Supply chains and the human condition', *Rethinking Marxism*, vol. 21, no. 2, pp. 148–176.

Tsing, A. (2017) 'The buck, the bull, and the dream of the stag: Some unexpected weeds of the Anthropocene', *Soumen Anthropologi*, vol. 42, no. 1, pp. 3–21.

Tsing, A., Deger, J., Saxena, A.K. and Gan, E. (eds.) (2020) *Feral Atlas: The More-than-Human Anthropocene*. Stanford University Press, Palo Alto.

Valladares, G., Cagnolo, L. and Salvo, A. (2012) 'Forest fragmentation leads to food web contraction', *Oikos*, vol. 121, no. 2, pp. 299–305.

Van Beirendonck, S., Driessen, B., Verbeke, G., Permentier, L., Van de Perre, V. and Geers, R.(2012) 'Improving survival, growth rate, and animal welfare in piglets by avoiding teeth shortening and tail docking', *Journal of Veterinary Behavior*, vol. 7, no. 2, pp. 88–93.

Vandermeer, J. (2010) *The Ecology of Agroecosystems*. Jones and Bartlett Publishers, Sudbury.

Vivero-Pol, J.L., Ferrando, T., De Schutter, O. and Mattei, U. (eds.) (2019) *Routledge Handbook of Food as a Commons*. Routledge, New York.

Walker, R. and Stroper, M. (1991) *The Capitalist Imperative: Territory, Technology and Industrial Growth*. Wiley.

Wallace, R., Bergmann, L., Hogerwerf, L., Kock, R. and Wallace, R.G. (2016) 'Ebola in the hog sector: Modeling pandemic emergence in commodity livestock', in R.G. Wallace and R. Wallace (eds.) *Neoliberal Ebola: Modeling Disease Emergence from Finance to Forest and Farm*. Springer, Cham.

Wallace, R., Chaves, L.F., Bergmann, L.R., Ayres, C., Hogerwerf, L., Kock, R. and Wallace, R.G. (2018) *Clear-Cutting Disease Control: Capital-Led Deforestation, Public Health Austerity, and Vector-Borne Infection*. Springer, Cham.

Wallace, R. and Wallace, R.G. (2015) 'Blowback: New formal perspectives on agriculturally-driven pathogen evolution and spread', *Epidemiology and Infection*, vol. 143, no. 10, pp. 2068–2080.

Wallace, R.G. (2009) 'Breeding influenza: The political virology of offshore farming', *Antipode*, vol. 41, pp. 916–951.

Wallace, R.G. (2016a) *Big Farms Make Big Flu: Dispatches on Infectious Disease, Agribusiness, and the Nature of Science*. Monthly Review Press, New York.

Wallace, R.G. (2016b) 'Flu the farmer', in *Big Farms Make Big Flu: Dispatches on Infectious Disease, Agribusiness, and the Nature of Science*. Monthly Review Press, New York, pp. 316–318.

Wallace, R.G. (2016c) 'A pale, mushy wing', in *Big Farms Make Big Flu: Dispatches on Infectious Disease, Agribusiness, and the Nature of Science*. Monthly Review Press, New York, pp. 222–223.

Wallace, R.G. (2016d) 'Made in Minnesota', in *Big Farms Make Big Flu: Dispatches on Infectious Disease, Agribusiness, and the Nature of Science*. Monthly Review Press, New York, pp. 347–358.

Wallace, R.G. (2016e) 'Collateralized farmers', in *Big Farms Make Big Flu: Dispatches on Infectious Disease, Agribusiness, and the Nature of Science*. Monthly Review Press, New York, pp. 336–340.

Wallace, R.G. (2016f) 'Protecting H3N2v's privacy', in *Big Farms Make Big Flu: Dispatches on Infectious Disease, Agribusiness, and the Nature of Science*. Monthly Review Press, New York, pp. 319–321.

Wallace, R.G. (2016g) 'Banksgiving', *Farming Pathogens*, November 30, https://farmingpathogens.wordpress.com/2016/11/30/banksgiving/

Wallace, R.G. (2016h) 'Cave/Man', in *Big Farms Make Big Flu: Dispatches on Infectious Disease, Agribusiness, and the Nature of Science*. Monthly Review Press, New York, pp. 277–278.

Wallace, R.G. (2016i) 'A probiotic ecology', in *Big Farms Make Big Flu: Dispatches on Infectious Disease, Agribusiness, and the Nature of Science*. Monthly Review Press, New York, pp. 250–256.

Wallace, R.G. (2017) 'Industrial production of poultry gives rise to deadly strains of bird flu H5Nx', Institute for Agriculture and Trade Policy blog, January 24, www.iatp.org/blog/201703/industrial-productionpoultry-gives-rise-deadly-strains-bird-flu-h5nx

Wallace, R.G. (2018a) *Duck and Cover: Epidemiological and Economic Implications of Ill-Founded Assertions That Pasture Poultry Are an Inherent Disease Risk*. Australian Food Sovereignty Alliance, https://afsa.org.au /wpcontent/uploads/2 018/10/WallaceDuck-and-CoverReport-September2018.pdf

Wallace, R.G. (2018b) 'Book review: "Paul Richards""Ebola: How a People's Science Helped End an Epidemic" (Zed Books, 2016)', *Antipode* [Internet], January 25, https://antipodefoundation.org/2018/01/25/wallace-book-review-ebola/

Wallace, R.G. (2020) *Dead Epidemiologists: On the Origins of COVID-19*. Monthly Review Press, New York.

Wallace, R.G., Alders, R., Kock, R., Jonas, T., Wallace, R. and Hogerwerf, L. (2019) 'Health before medicine: Community resilience in food landscapes', in M. Walton (ed.) *One Planet, One Health: Looking After Humans, Animals and the Environment*. Sydney University Press, Sydney.

Wallace, R.G., Bergmann, L., Kock, R., Gilbert, M., Hogerwerf, L., Wallace, R. and Holmberg, M. (2015) 'The dawn of Structural One Health: A new science tracking disease emergence along circuits of capital', *Social Science & Medicine*, vol. 129, pp. 68–77.

Wallace, R.G., Okomoto, K. and Liebman, A. (2020) 'Earth, the alien planet', in D.B. Monk and M. Sorkin (eds.) *Between Catastrophe and Redemption: Essays in Honor of Mike Davis*. UR Books, New York.

Ward, S.M., Webster, T.M. and Steckel, L.E. (2013) 'Palmer Amaranth (*Amaranthus palmeri*): A review', *Weed Technology*, vol. 27, no. 1, pp. 12–27.

Webster, T.M. and Sosnoskie, L.M. (2010) 'Loss of glyphosate efficacy: A changing weed spectrum in Georgia cotton', *Weed Science*, vol. 58, no. 1, pp. 73–79.

Wickramage, K. and Annunziata, G. (2018) 'Advancing health in migration governance, and migration in health governance', *The Lancet*, vol. 392, no. 10164, pp. 2528–2530.

Wickramage, K., Gostin, L.O., Friedman, E., Prakongsai, P., Suphanchaimat, R. et al. (2018) 'Missing: Where are the migrants in pandemic influenza preparedness plans?', *Health and Human Rights*, vol. 20, no. 1, pp. 251–258.

Winther, R.G. (2006) 'On the dangers of making scientific models ontologically independent: Taking Richard Levins' warnings seriously', *Biology & Philosophy*, vol. 21, pp. 703–724.

Wolf, C. and Ripple, W.J. (2017) 'Range contractions of the world's large carnivores', *Royal Society Open Science*, vol. 4, p. 170052.

Wychkhuys, K.A.G., Hughes, A.C., Buamas, C., Johnson, A.C., Vasseur, L., Reymondin, L., Deguine, J.-P. and Sheil, S. (2019) 'Biological control of an agricultural pest protects tropical forests', *Communications Biology*, vol. 2, p. 10.

Yang, Y., Tellez, G., Latorre, J.D., Ray, P.M., Hernandez, X. et al. (2018) 'Salmonella excludes salmonella in poultry: Confirming an old paradigm using conventional and barcode-tagging approaches', *Frontiers in Veterinary Science*, vol. 5, p. 101.

Yarahmadi, P., Miandare, H.K., Fayaz, S., Marlowe, C. and Caipang, A. (2016) 'Increased stocking density causes changes in expression of selected stress- and immune-related genes, humoral innate immune parameters and stress responses of rainbow trout (*Oncorhynchus mykiss*)', *Fish & Shellfish Immunology*, vol. 48, pp. 43–53.

Zimmerman, J.J., Karriker, L.A., Ramirez, L., Schwartz, K.J., Stevenson, G.W. and Zhang, J. (eds.) (2019) *Diseases of Swine*. John Wiley & Sons, Hoboken.

13

URBANISATION AND GLOBALLY NETWORKED CITIES

Meike Wolf

In May 2019, a 37-year-old woman returned from her vacation in a holiday resort in Alicante – a port city located on the Spanish Costa Blanca – to her home in Reykjavik. When she suddenly developed a high fever, she sought out medical help only to be diagnosed with chikungunya, a formerly tropical disease transmitted by tiger mosquitoes (*Aedes albopictus*). Shortly after, her sister and niece were also diagnosed with the infection. The three Icelandic residents received unwanted fame: they were the first confirmed cases of autochthonous chikungunya infections contracted in Spain.

The presence of *Aedes albopictus* mosquitoes in Central Europe is a story of entanglements across a multitude of fields. It is also a story of invasion. The Asian tiger mosquito is one of the most recent infamous additions to wildlife in Europe: from its origins in Southeast Asia, the insect managed to spread to Africa, the Americas, Oceania and Europe. As a container breeder, the tiger mosquito's eggs can endure prolonged periods of drought, allowing it to passively spread all over the world. In Europe, the tiger mosquito made its first appearance in Albania in 1979. In 1990, it was reported in the city of Genoa in Italy. From there it slowly migrated north along the Mediterranean coast either through trade in used tyres or plants (especially lucky bamboo) or via road traffic, when a female mosquito in search of a blood meal entered cars or trucks.

From the 1990s on, tiger mosquitoes were detected in more and more European countries. Medical entomologists managed to transform tiger mosquitoes into objects of governance: by developing the *Guidelines for the Surveillance of Invasive Mosquitoes in Europe* at the European Centre for Disease Prevention and Control (ECDC, 2014), they transformed the mosquito's appearance in Europe from 'insects out of place' into the register of biosecurity (see Shaw et al., 2010 on the difficulty of capturing the ontological nature of the mosquito and the diversity of management practices). Ideally the mosquitoes would be detected right after entering the country through the inspection of goods originating in infested countries. The guidelines construct tiger mosquitoes as a problem of national border control by referring to the concepts of exotic and invasive species. Exotic species are 'not native to an ecosystem and, if present, [have] been introduced' to said ecosystem. An invasive species – in addition to being an exotic species that has established itself and is successfully proliferating in the ecosystem – is also believed to cause 'economic or environmental impact or harm to human health' (ECDC, 2014, iv). Within these guidelines, the national territory is conceptualised as a container for native biodiversity, making borders matter as the detection of the invasive mosquitoes is closely linked to national territories (see also Olwig, 2003).

However, the concept of invasion, rather than simply mirroring a 'natural' order of the biological world, is political in nature and rich in political as well as scientific assumptions. Geographer Juliet Fall (2013; see Chapter 2, this collection) stresses that current Western ways of thinking about nature in terms of biodiversity are fairly recent. Biodiversity is never a matter of mere numbers. Rather, the concept mobilises a valuation of the degree of variation of life forms within a given ecosystem. Paradoxically, not all species are considered to be a valuable addition to a particular ecosystem (and clearly, tiger mosquitoes are not). But the notions of biodiversity and invasion also entangle our scientific concepts of place, nature and environment and the question of 'who is responsible for defining and solving specific problems' (ibid., 169; see also Robbins, 2004; Waterton, 2002 on the performative nature of classifications).

In this context, the tiger mosquito is attributed political importance as it is considered an invasive species with relevance to public health due to its ability to transmit a series of vector-borne diseases such as dengue fever, chikungunya fever and Zika, along with a number of other viruses. In 2007, it caused an outbreak of chikungunya fever in Italy with approximately 330 cases, followed by significantly smaller outbreaks in France in 2010, 2014 and 2017. The last outbreak of chikungunya in Italy in 2017 was of 245 cases (ECDC, 2017, 8). Due to their potential to transmit a series of tropical diseases, tiger mosquitoes are classified as invasive rather than exotic in relation to European national territories.

Whether chikungunya in 2017, Zika in 2016, SARS in 2003/2004, H1N1 (more commonly known as swine flu) in 2009 or Ebola in 2014, many of these recently emerged pathogens successfully spread in and through urban environments. Cities are often thought of as nodal points in the distribution and accumulation of infectious disease risks (see Ali and Keil, 2008). Moreover, underlying this is the assumption that urban environments have an impact on the health and lives of their inhabitants that includes both positive and negative health outcomes. This assumption is mirrored in initiatives such as the WHO Healthy Cities programme, in Urban Health as a scholarly discipline and in journals focusing on topics relevant to urban health. In what follows, I offer a brief reflection on the negative outcomes of city life in terms of infectious disease entanglements and biosecurity reasoning. While more recently, cities have commonly been associated with an increased disease burden caused by chronic conditions like asthma, diabetes, cardiovascular disease or obesity, the focus here will be on infectious diseases and infectious disease risks.

Nowadays, infectious diseases still constitute some of the leading causes of death worldwide (with respiratory diseases, diarrhea and tuberculosis being the most prominent examples; see WHO, 2018),[1] and they are associated with a high morbidity burden. They are, however, different from conditions described as chronic in character in several aspects. Infectious diseases by definition are dependent on infectious agents such as bacteria, viruses, prions or vectors to be transmitted. I stretch the definition of microbe beyond the usual reference to bacteria to include other kinds of infectious agents such as viruses or prions. Whereas many chronic conditions can be prevented, predicted, calculated or inherited (within certain limits, of course), this is not true for many diseases caused by microbes or vectors. Infectious diseases tend to show a specific temporal dynamic. Their most prominent manifestation is the outbreak, a sudden and unpredictable appearance of disease within a confined population. Outbreaks possess a beginning and an ending (they may, however, last several years).

Chronic conditions, in contrast, develop slowly, possess a rising tide character and in most cases end only if the diseased individual dies – they usually are lifelong conditions. Both types of diseases also show differences in terms of their relevance (to whom do they matter?) and their visibility (how are they embedded in the medical discourse?). Chronic conditions have sometimes been discussed as characteristic of modern societies or even of the modernisation process as such. While concepts such as the epidemiological transition model (Omram, 2005) have been rightfully criticised for being overly simplistic in their attempt to determine disease risks and mortality on a global scale, infectious diseases are indeed bound to those conditions that the involved infectious agents need to survive and thrive.

Multispecies scholars (see Kirksey and Helmreich, 2010) rightfully claim that in order to understand human life – in urban areas or elsewhere – it is crucial to develop an understanding of those biological actors whose lives are closely entangled with ours: animals, plants, fungi and microbes.[2] Infectious diseases by definition are transgressive in nature – they not only transgress the borders of nation states, urban and rural areas, but also bodily boundaries. They involve messy and malleable configurations of humans, environments and other biological organisms. For public health interventions and also for social science approaches, it might be useful to differentiate between the various causes of infectious diseases that are increasingly problematised as urban health risks.

Vector-borne diseases (for example, malaria, dengue, Zika or plague) are caused by bacteria, viruses or parasites, and they depend on vector animals such as fleas, ticks or mosquitoes to be transmitted between humans or between humans and vertebrate animals. It is not uncommon for them to be restricted to certain geographical areas and climatic conditions, often in tropical and subtropical areas. Malaria, for example, currently is not much of a threat to people living in Novosibirsk. As they depend on the vector organism to adapt to their environment, these diseases tend to spread slowly. However, they affect a growing number of the human population on a global scale, especially as some vector species successfully accommodate themselves to more moderate climate zones and successfully conquer new habitats – the three Icelandic chikungunya cases described at the beginning of this chapter are a striking example of a remarkably successful vector species (see also Flacio et al., 2016). Disease management practices enacted to regulate the spread of vectorborne diseases might target the human population, the vector itself, the environment or the pathogen.

Zoonotic diseases, in contrast, can be transmitted between animals and humans. These include, for example, influenza, rabies, bovine tuberculosis and many viral hemorrhagic fevers such as Ebola or hanta. When compared to vector-borne diseases, they spread quickly and possess a higher mobility as they are not confined to certain geographical areas or climatic conditions but depend on their host organisms' survival. H5N1 could very much post a threat to people living in Novosibirsk. The management of zoonotic diseases needs to engage with human as well as with animal health and with the environments the species in question inhabit and co-produce.

In the context of biosecurity studies, these infectious agents are interesting exactly because of their liveliness and agency and their ignorance of human-made borders, policies and categories. As biological matter, they are unruly in nature; as unruly agents, they escape attempts to be disciplined and controlled. They make visible, as Sydney Dekker argues (2011, see also Wynne, 1988), a gap between practices of regulation, certification and design on the one hand and the messy liveliness of practices – be they of human, microbial or animal origin – on the other hand.

To further complicate matters, a total elimination of infectious disease threats is an impossible task to perform given the vast number of potentially contagious contact zones, infectious agents and their mutagenic potential and impressive adaptivity. Consequently, humans are thrown back to disease management practices, and to this day these are open-ended and potentially eternal: infectious diseases are here to stay. Other than humans (or any vertebrate animal), microbes are likely to survive global warming and even nuclear war. Rightfully, multispecies scholars have argued for cohabitation between humans and non-humans instead of designating distinct spaces to each of them.

In this chapter, I focus on the complex interrelation between urban environments and infectious disease risks as posed by the unwanted mobility of biological agents and organisms (see Barker et al., 2013). Interestingly, not all infectious agents are problematised as biosecurity risks, and even those that are classified as such constitute a very heterogeneous group. From a social science perspective, no infectious agent is a biosecurity risk in itself until it is classified as such (see also Lupton, 1997). It is in this context that some pathogens have come under close scrutiny (prominent examples include H5N1, anthrax and smallpox). Their routes of transmission and their genetic make-up have been analysed. They are embedded in political rationales and transnational surveillance systems. Following Mary Douglas's famous account, an emerging virus within urban environments represents 'matter out of place' (Douglas, 2011 [1966]) and therefore elicits a wide range of socio-technical, political and biomedical countermeasures. Pathogens such as H5N1 are moved into the domain of the biosecurity apparatus. Other pathogens, however, have the potential to pass unnoticed as they are neither classified as emerging nor diagnosed or made visible to surveillance systems (for example, Chagas disease or Leishmaniasis).

Social science scholarship on urban infections has explored both how urban environments are configured as areas of risk and how the concept of emergence itself plays a key role within infectious disease management practices and is enacted within specific analytical frameworks (consisting, for example, of surveillance systems, classification schemes, reports, funding bodies and so on; see Washer, 2014). These frameworks determine what can be seen, known or said within the biomedical and political context of emerging infectious diseases (see Wolf, 2016). These frameworks bear strong links to the underlying assumption that emerging infections are 'natural' consequences of ongoing processes of urbanisation and globalisation and that pathogens move from infected spaces into those that are disease-free (see also Lavau, 2014).

If we take seriously critical accounts on biosecurity and disease emergence (Farmer, 1996; Grisotti and Dias de ÁvilaPires, 2010; see also Füller, 2014), pathogens do not suddenly 'emerge' somewhere in a pre-modern rural setting or in the backyard of a southeastern poultry farmer (see also Hinchliffe and Bingham, 2008). Rather, for a pathogen to be classified as 'emerging', it needs to be connected meaningfully to a specific disease (for example, bird flu), to a vulnerable population (for example, children and newborns), to surveillance systems and to a territory (for example, Europe). It is exactly through this framework that pathogens – living, resisting, unruly organisms – are translated into microbial risks that can be known, calculated and managed accordingly. For our purposes here, the emergence, development and spread of urban infections is not the result of given global orders. Rather, urban infections co-produce these orders (see Wolf, 2016).[3]

Resuscitating borders

Seeing cities through the perspective of multispecies approaches, the liveliness and organic nature of urban environments, is brought into the foreground. Rather than understanding them as static, bound and local entities, cities might be described as more-than-human spaces (Whatmore, 2006; see also Hinchliffe and Whatmore, 2006), inhabited by a multitude of different organisms and organised within metabolic networks of interaction and disease transmission. Animal and microbial organisms' remarkable capacity to shape human bodies and human health has been scrutinised in recent years by multispecies scholars, attending to the manifold ways animals and microbes make their appearance in urban life: as sentinels for infectious disease outbreaks (Keck, 2011), as part of rooftop agriculture (Hinchliffe and Bingham, 2008), as companion species (Haraway, 2003), as flying public health tools (Beisel and Boëte, 2013), as food (Porter, 2012) or as household pests (Biehler, 2013), to mention only some of the central fields of study. These studies show a particular interest in the becoming and unboundedness of cities: how spaces are made and remade; how new modes of governance are brought up; how organisms circulate and come to live, connect and mingle.

Seen from this perspective, we share space, metabolic flows and susceptibilities with manifold other biological beings as our bodies are porous and open to other forms of biological life through practices of eating, breathing, exuding, sneezing and touching (see Greenhough, 2008). Against this backdrop, it might be useful to understand precisely how these co-evolutionary and metabolic linkages between different biological agents such as humans, plants, animals and microbes benefit from environmental, political or social dynamics – and what this means for our concepts of urban health.

Among the many urban dwellers such as rodents, birds and microbes that mostly remain invisible to the human eye, insects have recently come under close scrutiny. Among them, mosquitoes play an important role for a growing number of outbreaks, as species such as *Aedes aegypti* and *Aedes albopictus* have successfully accommodated to human habitation patterns and thus become endemic in many urban areas (RigauPérez et al., 1998). The insects emerge, for example, from neglected swimming pools in the US (Reisen et al., 2008) or find breeding sites in urban household wastes such as pitchers, coconut shells, vases or plastic containers (Banerjee et al., 2013), questioning and transgressing any assumed dialectical relationships between 'natural' sites and urban environments.

Multispecies scholar Alex Nading (2012, 2014) suggests the application of entanglement perspectives to people, pathogens and spaces that are interconnected in a process of 'mutual becoming' (2012, 69). In his ethnographic study of dengue management programmes in urban Nicaragua, he explores how female community workers are integrated through participatory approaches into mosquito management strategies (similar attempts are currently being tested in Europe where Asian tiger mosquitoes have become a local nuisance). Within these campaigns, visions are invoked of disciplined households managed by their female members who struggle to keep human and insect worlds strictly apart. However, Nading encountered profound discrepancies between public health rationales that aim at separating humans from insects and the community workers' belief in accepting their households as places of multiple interspecies encounters. As a consequence, his research shows how attempts to implement better education programmes and 'more science' are doomed to failure and, accordingly, how urban environments might never be ordered rationally. Underlying this is a concept of cities as open landscape, as vicissitudinous in character, as unstable in form, but Nading also provides insights into how the community is shaped by the downsizing of public health investments. Entanglement perspectives, however, do not enable us to confine or accurately locate infectious disease risks within urban environments (see Wolf, 2016).

Anthropologist Ann Kelly and Science and Technology Studies scholar Javier Lezaun (2014, 2013) analyse a different kind of vectorborne disease in urban contexts, namely public health intervention strategies in Dar es Salaam aimed at disentangling humans from malariatransmitting mosquitoes. The *Anopheles* mosquito finds numerous breeding opportunities in the mobile and shifting nature of Dar es Salaam's fabric: holes, pits, drains, construction sites, sewage ponds and containers, often restricted by fences and private premises. Similar to the situation in Nicaragua, the mosquitoes have succeeded in adapting to the rhythms and surfaces of the urban environment. The city here is characterised by contingency, by shifting surfaces and underlying political, social and moral dimensions of city life that have an impact on both insects and humans. Dichotomous constructions of diseasefree and diseaseridden areas, underlying many biosecurity approaches, might be called into question as the city and its rural hinterland are both organised through a malaria geography (see also Brown and Kelly, 2014, on the material proximities of viral hemorrhagic fevers). Here, boundaries are porous, cities are expansive and urban dwellers suffer from an increased vulnerability to vectorborne diseases (compared to their rural counterparts).

Approaches such as these, concerned with the vitality of urban-rural networks, share a concern with bios, with the organic nature of urban environments that are understood as biosocial spaces, multispecies communities or simply life spaces. Their unruly geographies are organised through the activities of many species and metabolic-organic flows. As it becomes clear from these perspectives, health is always unpredictable and vicissitudinous, but it cannot be achieved or regulated by disentangling humans from other urban inhabitants, but only through living with urban natures in practices of cohabitation (Wolf, 2016). Urban infectious diseases enact ontologies of connectivity (Latimer, 2013) and relations between humans and other organisms. As Latimer reminds us, in theorising the relationship between human and other biological beings, it might be useful 'to keep open the nature and form of such orderings' (ibid., 22), not reducing them to hybrids in a Latourian sense, but rather 'reconstituting humans as being . . . part of nature' (ibid., 23). When looking at the concept of emerging infectious diseases, their event-like and unpredictable character can be called into question as they have always been an intrinsic and intra-active part of urban (and rural) landscapes, brought into being through a process of mutual becoming (see Ingold, 2004, 2000).

Managing borders

Biosecurity as a mode, a socio-technological apparatus and a mindset of regulation cannot be discussed independently of its geographical context. Within biosecurity reasoning, physical space is both transgressed and invoked, weakened and strengthened, made visible and opened up to regulation and governance. Thus, outbreak events are translated into local areas of intervention. From a public health perspective, controlling pathogen exchange – for example, within global travel or agricultural practices – has been understood as a key way of acting upon infectious agents, carefully designating disease-free and infected zones (on the underlying spatial logics of regions, see Lavau, 2014).

Studies on biosecurity practices foreground the topological dimensions of urban infections and their management. As Barker et al. (2013) remind us, biosecurity practices can never be reduced to simply constitute response to outbreak events but are embedded in larger frameworks of politics and problematisation. In analysing biosecurity as a practice, we might understand how pathogens (or other biological beings) are translated into risks to be acted upon and how this process impacts geopolitical boundaries.

Priscilla Wald reminds us that global networks simultaneously serve as threat and solution (2008, 8; she also points out the powerful nature of outbreak narratives underlying many current attempts of infectious disease management). While globalised flows of people, microbes, goods, traffic and transport pose a threat to urban health by connecting manifold organisms and places to the bodies of urban residents (see Braun, 2008 on 'thinking the city through SARS'), they also create transnational collaborations, enable surveillance networks and connect expertise and experts (see Enticott, Chapter 4, this volume).

Understanding infectious diseases as globalised and globalising in nature is based upon the assumptions that microbes live in a borderless world. The focus here is on those pathogens that are highly visible, often airborne viruses, embedded in complex political rationales and believed to be of global significance. 'It is in this context that SARS has marked a historical change – the pandemic highlighted how cities once again have become places of increased vulnerability and risk', as Ali and Keil convincingly argue (2008, 5). In their concept of 'networked disease', they paint this sketch:

> Another reality becomes visible: the global cities hierarchy is really a complex network of topological relations both externally (cities among one another) and internally (the capillary system of the globalized metropolis). What is central to such conceptualizations of both the relationship amongst global cities, and within global cities, are the notions of mobility, flow, and dynamism, and the consideration of such factors is perhaps best understood through more networked and topological understandings of the city.
>
> *(ibid., 5–6)*

Through the global cities network, microbes are enabled to transgress the urban-global dialectic. While the authors foreground the topological nature of emerging infectious diseases and their intimate relation with urban environments and global travel and trade networks, the biological and social reality of emerging viruses (and other pathogens) gives them their own dynamic, rather than being organised in a predictable pattern. In spite of thorough and elaborate attempts to predict microbial mobility patterns, we still do not know which city will be affected most by the next pandemic. Similarly, sociologist Evelyn Lu Yen Roloff (2007), in her empirical work on the SARS crisis in Hong Kong, sketches the urban governance of the epidemic as part of a larger restructuring process of established political hierarchies which, in turn, enabled citizens to articulate (bio)political needs of security and bodily integrity, if only temporarily.

British geographers Steve Hinchliffe and Nick Bingham (2008), in their research on avian flu governance in Cairo, portray a city shaped by multispecies interfaces. Contrary to official depictions of Cairo as a clean and modern global city, multispecies interfaces (such as rooftop poultry) are here believed to be an integral part of a globalised modernity and its accompanying patterns of consumption, trade and traffic – rather than belonging to a 'pre-modern elsewhere'. In 2005–2006 the Egyptian government identified rooftop and backyard poultry as culprits in the ongoing spread of avian flu virus and thus an estimated 34 million birds were killed and disposed of as a means of biosecurity. At the same time, an agricultural crisis struck Egypt and caused the breakdown of small local economies, causing food prices to increase and rooftop poultry to proliferate. As a consequence, food shortages emerged, and people tried to hide seemingly healthy chickens or hurried to bring infected birds to the market. As Hinchliffe and Bingham show, while the applied biosecurity measures reduced the number of potential hosts, they also brought about new risks – and new niches for the virus to hide (in dead or hidden birds). Biosurveillance approaches such as this, the argument goes, underestimate both humans and non-humans in building upon 'a logic of control and instrumentalism'. Consequently, they don't succeed in adapting to the indeterminate nature of networks (ibid., 1547; see Wolf, 2016): impure, complex, topologically distributed.

French anthropologist Frédéric Keck (2011), on the basis of fieldwork in Hong Kong, analyses avian flu prevention as the outcome of specific human-animal relations. Hong Kong is believed to be of central importance to the emergence of new flu viruses as these are often located as originating in South China's ecology, which is characterised by a close proximity of numerous species, many of which are part of larger food chains. Keck describes how Hong Kong is embedded in a web of relationships and human-animal interactions, bringing about the city as a sentinel for avian flu and a biosecurity hotspot. He depicts it as 'a place where live poultry is produced, exchanged, and consumed, or as a city where nature is at the same time controlled and proliferating' (2011, 57). To Keck, animals – mainly birds, in this case – and their immune systems are transformed into sentinels for the emergence of pandemic flu viruses. Paradoxically, while at the same time being part of assumingly 'pre-modern' patterns of consumption, the birds also contribute to attempts to modernise Hong Kong into a bulwark of disease control.

As approaches such as these show, cities are both embedded in globalised flows and circulatory movements and affected differently by modernisation processes – both of which impact the implementation and enactment of biosecurity measures. As Ash Amin and Nigel Thrift remind us, cities are

> A complex imbroglio of actors with different goals, methods and ways of practice. Then, precisely because of this complexity, the city can never be wholly fathomed. There always remain parts that can never be reached . . . gaps, blind spots, mistakes, unreliable paradoxes, ambiguities, anomalies, invisibilities.
>
> *(Amin and Thrift, 2002, 92)*

By translating health into the domain of biosecurity, contradictory patterns emerge: on the one hand, health seems to result from successful governance practices; on the other hand, it is embedded in numerous flows, traverses organic boundaries and escapes scalar logics.

Conclusions

When taken together, the accounts discussed here foreground new paths of connection, liveliness and association, as these are integral to urban infectious diseases and their management. In this context, microbial and human actors enact different kinds of border-making practices. For microbial and insect actors, borders are a means to proliferate successfully, exposing human populations to infectious disease risks. Human borders, however, are part of larger (bio)security practices, enabling or restricting mobility, belonging, policies and political rights. In Gitte du Plessis's (2018, 392) analysis of microbes' attempt to territorialise spaces, she argues:

> Microbial cartography is based on ecological factors for which separations between nature and culture do not make sense. Acknowledging non-human beings such as pathogenic microbes as having real political leverage helps to establish a new, more realistic, playing field for security politics. It is important to note that microbes are distinct political actors precisely because they are in relation with everything else.

Taking the manifold infectious disease entanglements in urban and other environments seriously might mean acknowledging that we as humans are embedded in material networks that exceed human control.

Finally, are infectious disease risks really an inherent quality of globalised urban environments as anchored in many biosecurity approaches? Just as Austin Zeiderman argues in his approach to the management of environmental risk in Bogotá (2012, 1572), we should not investigate how the urban poor came to settle within landscapes of risk. Rather, as he proposes, it might be useful to ask how zones of risk came to inhabit the territories of the urban poor. When transferred to the topic of emerging infectious diseases, it might be wise to ask how the surveillance, anticipation, visualisation, politicisation and management – the biosecuritisation – of infectious disease risks came to choose urban environments as one of their major targets (Wolf, 2016). Within critical biosecurity studies, infectious disease risks are understood as enactments of complex socio-technical assemblages, consisting of biomedical expertise, biological organisms, policies and infrastructures. With the practices of emergency planning performed as tools of urban governance, it can be observed how new areas of intervention emerge alongside the pathogen.

Indeed, the classification, comparison and visualisation of infectious disease risks bear political dimensions, as the manufacturing of emerging infectious disease risks might result in sociospatial transformation processes and specific urban futures (ibid.).

Acknowledgements

The author would like to acknowledge the German Research Association (Deutsche Forschungsgemeinschaft, DFG) for the funding that made this chapter possible. Some of the thoughts discussed in this chapter have previously been published in Meike Wolf, 2016, 'Rethinking Urban Epidemiology: Natures, Networks and Materialities', *International Journal of Urban and Regional Research* 40: 958–982.

Notes

1 In the African region, however, the top ten causes of death include the following communicable diseases: respiratory infections, HIV/AIDS, diarrhoeal diseases, malaria and tuberculosis.
2 Different from (other) animals, however, vector species and microbes rarely seem to claim any ethical standing – or, in other words, 'it may be difficult to love the mosquito' (Spielman and D'Antonio, 2001, xviii).
3 Cities and urban environments cannot be reduced to a universal form or structure, they are not self-evident or predetermined formations and they are affected by infectious diseases in various ways (for a detailed discussion see Brenner and Schmidt, 2014; Wolf, 2016).

References

Ali, S.H. and Keil, R. (eds.) (2008) *Networked Disease: Emerging Infections in the Global City*. Wiley-Blackwell, Oxford.
Amin, A. and Thrift, N. (2002) *Cities: Reimagining the Urban*. Polity, Cambridge.
Banerjee, S., Aditya, G. and Saha, G.K. (2013) 'Household disposables as breeding habitats of dengue vectors: Linking wastes and public health', *Waste Management*, vol. 33, no. 1, pp. 233–239.
Barker, K., Taylor, S.L. and Dobson, A. (2013) 'Introduction: Interrogating bio-insecurities', in A. Dobson, K. Barker and S.L. Taylor (eds.) *Biosecurity: The Socio-Politics of Invasive Species and Infectious Diseases*. Routledge, Oxon, New York, pp. 3–27.
Beisel, U. and Boëte, C. (2013) 'The flying public health tool: Genetically-modified mosquitoes and malaria control', *Science as Culture*, vol. 22, no. 1, pp. 38–60.
Biehler, D.D. (2013) *Pests in the City: Flies, Bedbugs, Cockroaches & Rats*. University of Washington Press, Seattle.
Braun, B. (2008) 'Thinking the city through SARS: Bodies, topologies, politics', in S.H. Ali and R. Keil (eds.) *Networked Disease: Emerging Infections in the Global City*. Wiley-Blackwell, Oxford.
Brenner, N. and Schmidt, C. (2014) 'The "urban age" in question', *International Journal of Urban and Regional Research*, vol. 38, no. 3, pp. 731–755.
Brown, H. and Kelly, A. (2014) 'Material proximities and hotspots: Towards an anthropology of viral hemorrhagic fevers', *Medical Anthropology Quarterly*, vol. 28, no. 2, pp. 280–303.
Dekker, S. (2011) *Drift into Failure: From Hunting Broken Components to Understanding Complex Systems*. Ashgate, Farnham.
Douglas, M. (2011 [1966]) *Purity and Danger: An Analysis of Concepts of Pollution and Taboo*. Routledge, London and New York.
European Centre for Disease Prevention and Control (2014) *Guidelines for the Surveillance of Invasive Mosquitoes in Europe*. ECDC, Stockholm.
European Centre for Disease Prevention and Control (2017) 'Rapid Risk Assessment', Clusters of autochthonous chikungunya cases in Italy. First update, 9 October. ECDC, Stockholm.
Farmer, P. (1996) 'Social inequalities and emerging infectious diseases', *Emerging Infectious Diseases*, vol. 2, no. 4, pp. 259–269.
Flacio, E., Engeler, L., Tonolla, M. and Müller, P. (2016) 'Spread and Establishment of Aedes Albopictus in Southern Switzerland between 2003 and 2014: An Analysis of Oviposition Data and Weather Conditions', *Parasites & Vectors*, vol. 9, p. 304.

Füller, H. (2014) 'Global health security? Questioning an "emerging diseases worldview"', in T. Cannon and L. Schipper (eds.) *World Disasters Report 2014: Focus on Culture and Risk*. International Federation of Red Cross and Red Crescent Societies, Geneva.

Greenhough, B. (2008) 'Where species meet and mingle: The common cold unit 1946–1990', Paper presented at the RG/IBG, London, September.

Grisotti, M. and Dias de Ávila-Pires, F. (2010) 'The concept of emerging infectious disease revisited', in A. Mukherjea (ed.) *Understanding Emerging Epidemics: Social and Political Approaches*. Advances in medical sociology, Vol. 11, Emerald Group Publishing Limited, Bingley.

Haraway, D. (2003) *The Companion Species Manifesto: Dogs, People and Significant Otherness*. Prickly Paradigm, Chicago.

Hinchliffe, S. and Bingham, N. (2008) 'Securing life: The emerging practices of biosecurity', *Environment and Planning A*, vol. 40, no. 7, pp. 1534–1551.

Hinchliffe, S. and Whatmore, S. (2006) 'Living cities: Towards a politics of conviviality', *Science as Culture*, vol. 15, no. 3, pp. 123–138.

Ingold, T. (2000) *The Perception of the Environment: Essays on Livelihood, Dwelling and Skill*. Routledge, London.

Ingold, T. (2004) 'Two reflections on ecological knowledge', in G. Sanga and G. Ortalli (eds.) *Nature Knowledge: Ethnoscience, Cognition, and Utility*. Berghahn Books, Oxford.

Keck, F. (2011) 'How Hong Kong became a sentinel for avian flu', Paper presented at the Influenza Workshop, Rennes, 24–26 August.

Kelly, A. and Lezaun, J. (2013) 'Walking or waiting? Topologies of the breeding ground in Malaria control', *Science as Culture*, vol. 22, pp. 86–107.

Kelly, A. and Lezaun, J. (2014) 'Urban mosquitoes, situational politics, and the pursuit of interspecies separation in Dar es Salaam', *American Ethnologist*, vol. 41, no. 2, pp. 363–383.

Kirksey, S.E. and Helmreich, S. (2010) 'The emergence of multispecies ethnography', *Cultural Anthropology*, vol. 25, no. 4, pp. 545–576.

Latimer, J. (2013) 'Being alongside: Rethinking relations among different kinds', *Theory, Culture & Society*, vol. 1, pp. 1–18.

Lavau, S. (2014) 'Viruses: From travel companion to companion species', in P. Adey, D. Bissell, K. Hannamm, P. Merriman and M. Sheller (eds.) *Routledge Handbook of Mobilities*. Routledge, New York, NY, pp. 298–305.

Lupton, D. (1997) *The Imperative of Health: Public Health and the Regulated Body*. SAGE, London.

Nading, A. (2012) 'Dengue mosquitoes are single mothers: Biopolitics meets ecological aesthetics in Nicaraguan community health work', *Cultural Anthropology*, vol. 27, no. 4, pp. 572–596.

Nading, A. (2014) *Mosquito Trails: Ecology, Health, and the Politics of Entanglement*. University of California Press, Berkeley, CA.

Olwig, K.R. (2003) 'Natives and aliens in the national landscape', *Landscape Research*, vol. 28, pp. 61–74.

Omran, A.R. (2005 [1971]) 'The epidemiological transition: A theory of the epidemiology of population change', *The Milbank Quarterly*, vol. 83, no. 4, pp. 731–757.

Plessis, G. du (2018) 'When pathogens determine the territory: Toward a concept of non-human border', *European Journal of International Relations*, vol. 24, no. 2, pp. 391–413.

Porter, N. (2012) 'Risky zoographies: The limits of place in Avian Flu Management', *Environmental Humanities*, vol. 1, pp. 103–121.

Reisen, W.K., Takahashi, R.M., Carroll, B.D. and Quiring, R. (2008) 'Delinquent mortgages, neglected swimming pools, and West Nile virus, California', *Emerging Infectious Diseases*, vol. 14, no. 11, pp. 1747–1749.

Rigau-Pérez, J.G., Clark, G.G., Gubler, D.J., Reiter, P., Sanders, E.J. and Vorndam, A.V. (1998) 'Dengue and dengue haemorrhagic fever', *The Lancet*, vol. 352, pp. 971–977.

Robbins, P. (2004) 'Comparing invasive networks: Cultural and political biographies of invasive species', *The Geographical Review*, vol. 94, pp. 139–156.

Roloff, E.L.Y. (2007) Die SARS-Krise in Hongkong: zur Regierung von Sicherheit in der Global City, Transcript, Bielefeld.

Shaw, I.G.R., Robbins, P.F. and Jones III, J.P. (2010) 'A bug's life and the spatial ontologies of mosquito management', *Annals of the Association of American Geographers*, vol. 100, no. 2, pp. 373–392.

Spielman, A. and D'Antonio, M. (2001) *Mosquito: A Natural History of Our Most Persistent and Deadly Foe*. Hyperion, New York.

Wald, P. (2008) *Cultures, Carriers, and the Outbreak Narrative*. Duke University Press, Durham.

Washer, P. (2014) *Emerging Infectious Diseases and Society*. Palgrave, London.

Waterton, C. (2002) 'From field to fantasy: Classifying nature, constructing Europe', *Social Studies of Science*, vol. 32, no. 2, pp. 177–204.

Whatmore, S. (2006) 'Materialist returns: Practising cultural geography in and for a more-than-human world', *Cultural Geographies*, vol. 13, no. 4, pp. 600–609.

WHO (2018) 'Top 10 causes of death', www.who.int/gho/mortality_burden_disease/causes_death/top_10/en/, accessed 25 July 2019.

Wolf, M. (2016) 'Rethinking urban epidemiology: Natures, networks and materialities', *International Journal of Urban and Regional Research*, vol. 40, pp. 958–982.

Wynne, B. (1988) 'Unruly technology: Practical rules, impractical discourses and public understanding', *Social Studies of Science*, vol. 18, pp. 147–167.

Zeiderman, A. (2012) 'On shaky ground: The making of risk in Bogotá', *Environment and Planning A*, vol. 44, no. 7, pp. 1570–1588.

14

GARDENERS' PERSPECTIVES AND PRACTICES IN RELATION TO PLANTS IN MOTION

Katarina Saltzman, Carina Sjöholm and Tina Westerlund

Introduction

In their gardens, people inevitably affect, and are affected by, the growth, life and death of other, non-human actors. In this chapter we focus on gardeners' perspectives and practical involvement in such interactions. We want to highlight how everyday actions, when it comes to the spreading of plants and other organisms in the garden context, relate to the making and crossing of borders and boundaries such as those between nature and culture. More specifically, we discuss how private gardeners share their garden plants in the form of cuttings, seedlings and divided plants, based on examples from Sweden. Sharing garden plants is an everyday practice, driven by gardeners' curiosity, generosity and desire to try something new, as well by their wish to get rid of excess plants. Some gardeners are aware that sharing may have undesirable consequences in their own garden and elsewhere if a specific plant begins to spread beyond control. Other gardeners are not much bothered by such risks. Most gardeners see it as a benefit when plants they find attractive and useful are willingly spreading in the garden.

Gardeners have always used plants of different origins, moving them around to new environments and working hard to make the plants survive and thrive under new conditions. During the late Renaissance, around the second half of the 16th century, an interest in 'new' plants developed within scientific work, which led to an increased diversity of plants in gardens (Hobhouse, 2002). This history is reflected in the development of botanical gardens, which have a long tradition of collecting plants from different parts of the world for research and education. Subsequently, they have also played an important role in the spreading of 'new' plants to a wider audience and in the preservation of species and varieties, something that in today's debate on invasiveness is recognised as an activity that has contributed to an undesirable spreading of plants (e.g. Hulme, 2015).

Certain plant varieties have proved to be particularly vital, have a long history as cultivars and are today understood as heritage plants. Many have spread over centuries and are regarded as 'belonging' and as more or less 'natural'. Others again have spread too much and have increasingly become acknowledged as a risk, potentially threatening local biodiversity, and hence have been labelled invasive. With increasing focus on threats to biodiversity, 'war' has been declared towards the most persistent invaders. In this context gardeners and their circulation of garden plants have often been blamed, by nature protection authorities as well as in the media, for not taking adequate responsibility for the effects of unintentional spreading. In this context, as we were told in an interview with the head of the Botanical Garden in Gothenburg, the role of botanical gardens has been extended to also take responsibility for the spreading of knowledge about how plants that are potentially invasive could be controlled or removed.

The chapter is based on ethnographic material that includes in-depth interviews with gardeners and professional horticulturalists in Sweden, as well as written stories about people's individual relationship to gardens collected through the Swedish folklore archives. Our material reflects how gardeners in Sweden talk and think about what grows and lives in their gardens, whether it is there due to their own actions or just by chance. How the plants and other organisms people share their gardens with are described in these stories reflects the ways in which gardeners understand their non-human neighbours and companions.

There are different views among gardeners on what belongs within the garden and what should stay on the other side of the garden fence (cf. Head and Muir, 2006). Understandings of boundaries between nature and culture, and between garden and the surrounding environment, have changed over time with shifting gardening ideals and are continuously changing. In most gardens, there are different degrees of openness towards the wild and spontaneous in different parts of the garden, and to some extent these attitudes also change with the time of year. What is seen as a weed in one context can be regarded as beautiful and useful in another. Birds, mammals and insects that are defined as pests by many gardeners may be welcomed on some occasions; for example, the connection between the large white butterfly (*Pieris brassicae*) and its gluttonous caterpillars is not obvious to everyone. In the context of biosecurity, we need to consider how contemporary gardeners think about which plants and animals are welcome in their gardens. Which species and varieties are welcomed when and where, and which are feared, detested and even killed by gardeners?

Plants on the move – roots en route

In Sweden, as well as in other parts of the world, many plants that are now known to be invasive have spread through cultivation in private gardens. The establishment and spreading of such introduced species in new environments are today acknowledged as potential threats to biodiversity in large parts of the world. The discussion about invasive and alien species in the garden has perhaps been most intense in a number of countries with a colonial history where a large number of non-native species followed the establishment of new, Western ways of using the landscape. In Australia and New Zealand, for example, the spread of invasive species through the introduction of non-native horticultural plants is a widely recognised problem. Among gardeners in these countries, there is a relatively large awareness of this issue, where non-native and native gardening are mentioned as clearly separated gardening ideals (Head and Muir, 2004, 2006; Barker, 2008a). Despite the focus on Sweden in this chapter, it is influenced also by research on invasive species in antipodal settings (e.g. Barker, 2008b; Head et al., 2015; Head, 2017).

In Scandinavia, the issue of invasive alien species has begun to attract more general attention in the last decades. In Iceland, for example, the rapid spread of blue sand lupins (*Lupinus nootkatensis*) on barren fields and mountainsides and in the open Icelandic landscape has come to be noted as a threat but also as a tourist attraction (Benediktsson, 2015; *New York Times*, 2018). In Norway, just like in Sweden, environmental authorities have faced challenges when trying to spread knowledge about invasive alien species, not least to domestic gardeners. As Qvenild and colleagues have shown, Norwegian home gardeners are generally not talking about their gardens and plants in terms of native or alien. From the Norwegian perspective, this way of categorising vegetation is rooted within science and environmental policymaking rather than in the situated practice of gardening (Qvenild, 2013; Qvenild et al., 2014).

Generally in Europe, movement of species has been going on for a long time, and it is rather complicated to set clear boundaries for what is to be regarded as foreign. The European network on invasive alien species, NOBANIS, was initiated in the early 2000s by the Nordic Council of Ministers and now gathers countries in northern Europe from Iceland to Austria and from the Netherlands to Russia. Within the EU, the regulation 1143/2014 *On Invasive Alien Species* has increased the focus on these issues. The EU's list of invasive alien species of concern lists species that have been determined as severe threats, to which specific restrictions on keeping, import-ing, selling, breeding and growing apply. The list first entered into force in 2014 and has since been updated, in 2019 composed of 36 plant and 30 animal species within the EU. According to Naturvårdsverket, the Swedish Environmental Protection Agency (EPA), 12 of these species are established in Sweden, and another 8 have been observed occasionally. By November 2019 NOBANIS have in total listed 392 invasive alien species in Sweden.

In Sweden, a more widespread discussion about invasive alien species as a threat has only recently emerged. Until just a few years ago, invasive alien plants were not a well-known issue in the context of private gardening. However, this has now changed, and these issues have received quite a lot of attention in media, with headlines such as 'Watch Out for the Plants That Threaten Our Nature' (SvD, 2019) and 'Five Species That Are Threatening Sweden' (SVT, 2019). The background is that the Swedish EPA and the Swedish Species Information Centre have published lists of invasive species in Sweden, along with instructions on how gardeners should manage these (e.g. Naturvårdsverket and Havs- och Vattenmyndigheten, 2017a, 2018; Strand et al., 2018). These publications seem to have contributed to raising public awareness of invasive alien species in Sweden.

Some of the plant species that have been listed as invasive and alien are ones that are very common in Swedish gardens, including shrubs such as lilac (*Syringa vulgaris*), perennials such as the garden lupine (*Lupinus polyphyllus*) and bulbs such as daffodils (*Narcissus pseudonarcissus*). These species are generally regarded as 'belonging' within the heritage of Swedish garden cul-ture, and seeing them listed as potential threats has led to upset debates among gardeners as well as in media. Not least the garden lupine has turned into a 'hot' issue. This North American species has been cultivated in European gardens since the 1920s and has gradually been spread-ing into the environment, along roads and railways, and more widely in areas where agriculture is declining (Tschan, 2018). In many places in Sweden the pink, blue and white flowers of the lupines colour vast areas in early summer. While increasingly being acknowledged as a threat to local biodiversity, the spreading of the lupine has also turned it into a much-loved flower that is at the same time regarded as part of national and regional heritage. This has led to affected con-flicts when eradication campaigns have been initiated and carried out (Frihammar et al., 2018).

As Qvenild et al. (2014) have shown, Norwegian gardeners have claimed that discussions on invasive species do not sufficiently consider the varying geographical conditions for cultivation in their country. Similarly in Sweden, experienced gardeners have questioned to what extent the lists and rules have been sufficiently adapted to the specific conditions in different parts of the country. They have also claimed the importance of enriching the Swedish flora through the introduction of new species over centuries (e.g. Söderlund, 2018). It has also been noted that the instructions from the authorities in some cases have led to confusion. Despite such instructions, it can be difficult for individual gardeners to assess the impact of specific plants and to know how to avoid unwanted spreading.

As people are in many different ways involved in the spreading of plants, both desired and undesired, the expansion and movement of vegetation have to be understood as cultural and social issues. Such movements can be interpreted in several ways, not least in the context of the Anthropocene (Head, 2016). Some experts advocate a more positive interpretation of the 'threat' from invasive plants and even claim that they can be our rescue in a rapidly changing world (e.g. Pearce, 2015). When vulnerable indigenous species are threatened, we should perhaps be happy that other species are more vigorous, regardless of their background. In our study we have encountered similar approaches among professional gardeners and horticultural researchers, who are constantly looking at species from other parts of the world when searching for species that can better adapt to climate change and to human-made growing conditions – for example, in paved urban environments (Sjöman et al., 2012).

Simply too much?

A key feature of all gardens is that they accommodate change and movement over time. Some of these movements are rapid, some lead to a sudden transformation, while others are slow and unnoticed, such as when trees grow and species spread. Sometimes plants migrate by themselves and pop up in new settings as welcome additions or unwanted weeds. The processes of change in the garden are complex and include many different, mutually interwoven movements over time: the cycles of the day and year, the flow of materials and biochemical processes involved in growth and degradation – all linked to decisions and choices made by human actors as well as strategies of non-human actors. All life, anthropologist Tim Ingold reminds us, is social as well as biological. According to Ingold, all organisms, human and non-human, are best understood as 'biosocial becomings' of a relational, material and continuously changing nature (Ingold, 2013). With such perspectives gardens can be seen not as compounds of individuals and objects but rather as a world characterised by material flows and movements, where organisms such as humans and plants are best understood as 'biosocial becomings' rather than as static and separate 'beings'.

The garden is a place where human and non-human movements and transformations are intermingled in many different ways. 'On gardening I have a lot to say; to me, it means life', writes a Swedish woman who looks back on her many years of gardening to see that it is almost impossible to separate her own life from the life of her garden, with its plants and animals. Her reflections are a typical example of the relation to non-human co-species expressed by devoted home gardeners: 'I love all my flowers and plants and I am delighted with everything that comes up. It is nature's wonder that growth continues, and it is just as new and fantastic every year' (DAG, 2011). For many gardeners the repetitive tasks in the garden become a way of linking to nature's cycles and processes. Everyday routines such as weeding, and rituals such as daily walks around the garden, provides possibilities to follow day by day and week by week what grows, blooms and eventually withers (Saltzman and Sjöholm, 2016).

The dynamics and vitality of plants are often regarded as assets but can sometimes be turned into a problem when plants simply grow too much. Even experienced gardeners are repeatedly surprised by plants that expand more and quicker than expected and therefore need to be limited in order not to outcompete other plants. Trees and shrubs tend to grow larger, and some perennials to spread more widely, than you could ever imagine. For example, when we visited Ingela, who is a keen home gardener, she showed us a Japanese maple (*Acer palmatum*) that she had bought in the belief that it was miniature variety and planted in a small flowerbed. The maple, however, turned out to have a completely different capacity, and it kept on growing: 'So I need to cut, continuously cut it down and remove branches in order to keep it small'. Indeed, many of our informants reflect on what happens when specific plants grow more vigorously than expected, at the expense of other plants. Sometimes, one of them notes, it gets necessary to remove such a plant, and then suddenly others may become visible and get new life. The limits for how much growth one can accept are often exceeded yearly and keep hoes, weeders, saws and pruners busy in the hands of gardeners.

Dividing and sharing

Let's now look into some of the strategies gardeners use when their plants grow too much. Evy, a homeowner in her 70s with a well-established interest in gardening, reflects on this when we talk about the plants in her garden:

> Plant lovers are quite generous and enjoy sharing. It is not hard when you have plenty of it. I give a lot to the neighbours, yes those who want. Well, you have to have a limit because it must not get too much. And it does give you a bit of heartache to throw away a nice good plant.

Evy is one of several gardeners that we interviewed who testify that a particular kind of generosity is expressed when growers share plants and parts of plants with each other. In our research on life in contemporary private gardens, many gardeners have told us that they are happy to share what they grow – plants and seeds, as well as fruit and other kinds of harvest from the garden; they give their excess to friends, neighbours and relatives and in some cases spread it via more or less informal markets.

There are many different reasons, and many different ways, to divide and share garden plants. During our field studies among Swedish gardeners, we were often reminded that this is a common practice, especially among more experienced gardeners. One of these was Barbro, who devotes a lot of time and commitment to the small garden of her semi-detached house. When we asked if she shares her perennials, she immediately replied: 'Would you like a piece?' and then explained: 'I like sharing, because I need the space here'. Sharing takes place in many different ways, and the reason Barbro mentions seem to be recurring among gardeners. Many share their plants because they feel they need more space – often to create room for other plants. Sharing what grows in one's garden can be considered a simple, everyday gesture of friendship that creates and consolidates social bonds. It is also a way to manage excess and get rid of things you do not want or need. Through the practice of sharing, many different kinds of values and relationships are created. Plant sharing may today be regarded as a specific form of the 'sharing economy'. Thus, it can be seen as an expression of a sustainability trend, but the ideals of frugality also have roots far back in time (Saltzman and Sjöholm, 2019).

Plants, and parts of plants, are also spread on informal markets such as plant flea markets and plant exchange forums online where gardeners meet, share their own excess, bargain and reuse. The significant development of these markets is related to the fact that gardening today is a large economic sector in terms of materials (plants, soil, tools, etc.) and services as well as dreams, trends and ideals. Although contemporary garden trends emphasise aspects such as 'local produce', 'ecological' and 'heritage', garden plants are extensively traded at garden centres linked to large-scale business chains, where the local connection is increasingly limited (Gunnarsson et al., 2017).

Previous research on the division and sharing of plant material shows that material and social links are created when plant material is shared through various more or less informal networks. Much of this research is based on examples relating to the handling of seeds, not least among dedicated 'seed savers'. These are often driven by the commitment to preserve and spread rare varieties as a way to promote and preserve biodiversity as well as traditional knowledge and skills in cultivation. Many have an explicit ambition to maintain an alternative to established market channels and the growing influence of commercial interests (Pottinger, 2018a, 2018b; Phillips, 2013).

Similarly, our informants dig up, divide and move plants; collect and dry seeds; and share them in many different ways. In our studies, sharing appears to be a low-key and everyday practice which in various ways relates to people's ambitions to take care of their physical and social environment.

Share to get more – or less

Plants have the ability to multiply in various ways, both on their own and with the help of humans. As plants spread both sexually, through seeds, and vegetatively – for example, through stolons and runners – gardeners can use different methods to propagate their plants. When it comes to cultivated varieties of horticultural plants, vegetative propagation is often considered preferable as it is the only way to create new plants that retain the genetic characteristics of the parent plant. Seeds result in new plants of the same species, but properties such as colour, shape, taste and growth patterns may vary, whereas vegetative propagation offers an opportunity to create identical new plants. Traditional propagation methods used by professional gardeners include a variety of techniques for vegetative propagation, and many of these techniques are also applied by hobby gardeners. A common method is to simply dig up and divide a larger root clump into several smaller parts, each forming a new plant. Other ways are to separate smaller pieces, called cuttings, from specific parts of the plant (for example, root, leaf or stem) and let them develop into new plants. In this way, a single plant can give rise to a large number of new plants (Westerlund, 2017, 2021 in press).

Gardening can be a costly hobby, and dividing plants can sometimes be an economic necessity in order to achieve your garden ideals. If, for example, you would like 100 plants of lavender to cover a new flower bed, it may simply be too expensive to go and buy them from a garden centre and much more affordable if you can manage to make them from cuttings. And perhaps these can be taken from a few plants that were, in turn, grown from cuttings shared by another gardener. The driving forces behind sharing can be of many kinds, and for some an important motive is linked to ideologies of sustainability. Others, such as this gardener, may rather be attracted by the challenges involved and the economic benefits: 'For me it is a bit of a sport to develop the garden without expensive investments. Buying new on the market will soon become very expensive'. This woman describes herself as 'self-sufficient' when it comes to plants and obviously sees an intrinsic value in being able to get new plants without buying them (LUF, 2010). Many of our informants brought up the economic aspect and effect of 'making their own plants', and Lisbet was one of those who talked about plants that are especially suitable for propagation. She proudly told us about her self-sown white columbines:

> There are a lot of columbines here; white columbines. It was my daughter who, when they got a garden and they bought a house we drove to Löddeköpinge, some store there, and she bought white columbine. And they were so damn expensive, so [laughter], I said, I'll take seeds from you. And I have taken seeds, I have lots of white columbines, I have given away lots of white columbine. She still has her two plants, they do not seed there.

The quote shows how something can go from being an economic reality ('they were so damn expensive', as she puts it) to becoming an end in itself through the joy of being able to make your own and a pride in this skill. Lisbet, like many gardeners, has learned through experience what works to propagate and how to do it. She has devoted a lot of time and dedication to completely redesigning her home garden. In her yard, she has replaced the lawn with a rock garden with a multitude of small trees and bushes, perennials and artistic decorations. In her work on remaking the garden, she has clearly benefited from multiplying, sharing and exchanging plants.

In our study, gardeners have mentioned the cutleaf coneflower (*Rudbeckia laciniata*) and snow-drops (*Galanthus*) as plants they have tried to propagate by dividing them. Snowdrops are con-sidered to be a bit difficult to grow from the bulbs that are purchased and planted in autumn. A recurring alternative advice to gardeners is to move and multiply snowdrops 'in the green' – i.e. by digging them up and dividing the clumps of bulbs while the leaves are green – if you want to plant snowdrops in new places. Dividing and moving plants can thus be a method to build the garden you want. Kristina and Örjan have created an ambitious and relatively large home garden, and they even open their garden to the public a few times a year. They have a focus on colour composition and thus want a large number of the same kind of plants for the sake of effect. They are members of both a local and a national garden association and see the economic benefits of this, especially in the plant flea markets arranged by the garden associations: 'Then you can go and buy some cheap plants, yes it is very practical', says Kristina. Buying shared plants from other members simply made it easier for them to shape their garden, and they have also been dividing and moving around existing plants within their own garden.

Most private gardens are relatively small, and when gardeners propagate their plants at the same time the plants themselves are growing and spreading, the desire to want more can quickly be replaced by problems of managing too much. In the garden, as in many other contexts, abundance and overflow can be quite problematic assets (Czarniawska and Löfgren, 2012, 2013). When our informants talk about sharing, exchanging and selling, it becomes clear that this can go through different phases: from having too few plants to having too many. Kristina and Örjan have come to the stage in their gardening that they no longer exchange plants with their neighbours: 'No, we rather just give away', says Örjan, explaining that they no longer need more plants themselves.

To take care and make use

While Kristina and Örjan are quite fussy about what they want in their garden and make active choices based on a plan on how they want it to develop, other gardeners prefer to let the availabil-ity of plants decide how their gardens get composed. There are contexts where people may feel encouraged to take care of plants and move them to their own garden. One of our informants, for example, told us about an informal compost, close to their own garden, where neighbours had thrown away garden waste, including living plants. She described this compost heap as 'a gold mine' where she could get hold of perennials of various kinds free of charge (LUF, 2010). Such sites and other communal spaces can sometimes contain plants that have spread on their own from adjoining gardens but also excess that gardeners have thrown or sometimes even planted there. Certain plants may have been placed there with the hope that someone else wants to take care of them, and this can also be regarded as a means of sharing. Picking and digging among discarded plants presents a different kind of social situation, as donors and recipients do not meet or even have any contact. Such spaces and sharing practices show that there are no clear-cut boundaries between desired and despised (Saltzman, 2006; Saltzman and Sjöholm, 2016).

Taking care and making use of things that would otherwise be wasted often connects to an ambition to save resources, closely linked to ideals of sustainability. In Laura Pottinger's studies of seed sharing, the 'generous exchange' of seed activists is often a deliberate political act (Pottinger, 2018a, 2018b). For most of the home gardeners in our study, it seems more appropriate to interpret these practices as based on ideals of thrifty husbandry. Such ideals may be, but are not always, grounded in limited financial resources. Many say, like Barbro, that 'it is hard to throw something that is alive', or like Lisbet, 'when there is life in it I can't throw it away. No'. In their view there is a double value in sharing: the joy of making something grow is paired with the joy of giving and thus keeping the plants alive. In their view, discarding a living plant is definitely the last resort, even in the case of plants you do not like at all.

Taking care of and cultivating plants that someone else has thrown away can also be seen as a challenge, and it can be the start of a new garden plot or even a new business. One example of this was when the professional gardener Hermann Krupke found some peony plants thrown on a customer's compost pile. The peonies had been lying there unprotected over the winter, but this did not stop Hermann from trying to save them. Looking back, he could see that it was these wasted plants that became the start of the peony cultivation run at Guldsmedsgården's nursery by Hermann and his son since the mid-1970s. Over the years it had developed into one of the most well known in Sweden (Westerlund, 2014).

Peonies (*Paeonia*) on a compost pile in Sweden pose no great risk to turn invasive and start spreading into the surroundings, as peonies do not spread with runners or long distance with seeds, and their seeds take many years to develop into full-grown plants. In other cases, however, piles of garden waste can cause undesirable spreading of plants. The Swedish EPA expresses warnings concerning 'run-away' plants and particularly points out composts and piles of garden waste as sources for dispersal in the immediate environment, outside gardens (Naturvårdsverket and Havs- och Vattenmyndigheten, 2018). When a communal compost pile is not well-managed, there is a risk that both seeds and plant parts will survive in the pile, and particularly plant species with the ability to multiply from pieces of roots and runners are prone to spread from such sites. The need to investigate such practices in relation to invasive species is clear, and research on the role of garden waste is currently being carried out in Norway (Bajada and Setten, 2019).

Plants as gifts and memories

As we have seen, one important driving force when it comes to dividing plants seems to be the joy of succeeding in getting plants to thrive and to succeed in propagating them, especially if they are known to be difficult, but also with more common and 'easy' ones. Lisbet, for example, says that she never buys geraniums any more but preserves and propagates the ones she has: 'And it's so much fun to make them survive, and make them come again. Just that. And also, to give away to friends who live in apartments where they cannot keep such plants over winter'.

Plants can be perceived as valuable through the histories they are associated with. Many of our informants told us about plants that carry a particular history, quite often linked to persons or places where the plant came from. In some cases the stories that individual plants carry can be just as important to preserve as the plant itself. Evy presented an example:

> I just looked outside the door before and saw a high snowdrop that I got from a patient, many years ago, who in turn had received it from some old lady who had told that it had grown at some monastery here in Skåne many, many years ago.

By preserving plants that carry a history, stories can be kept alive, and the stories also seem to help keep the plants alive. Another reason to preserve plants with history may be that old varieties are appreciated as hardy and good varieties:

> Many of the plants I have in the garden are legacy from previous generations. These are durable plants, which are not the least tricky for a little lack of care every now and then. In addition, it is a nice task to preserve these old varieties.
>
> *(LUF, 2010)*

Passing on heritage plants seems to be one important driving force behind plant sharing. Growing and passing on plants with history becomes a way of building relationships that extend over time and space and thus also a way of actively taking responsibility for the survival of the plants and the memories and stories associated with them (Saltzman et al.,2020). The willingness to take care of things and pass them on also has ethical dimensions, especially when it comes to living materials such as plants. This ethical dimension has been given new relevance since the discussion about invasive species gained momentum: 'Pay attention to what kind of plants you buy or accept into your garden', the Swedish EPA declares in their information material (Naturvårdsverket and Havs- och Vattenmyndigheten, 2018). Their obvious ambition is to teach the public how invasive alien species can be identified and managed, and they particularly stress that plants also change their behaviour in connection to climate change:

> Increased travel, growing world trade and climate change are factors that accelerate the negative development. So far, we have been spared in Sweden, but as climate zones shift, it will be easier for plants and animals from warmer areas to establish here and some of them will become invasive.
>
> *(translation from Naturvårdsverket, 2020)*

As we have seen, many give plants away to create space for something new. Others do it just because they enjoy cultivating, being able to share with others and keeping old varieties alive. This is indeed very similar to what plant collectors throughout history seem to have thought and done. Sometimes it is also a mere coincidence that someone happens to be sharing and giving away plants. Receiving from others, giving to friends, sharing and exchanging are all about maintaining relationships and creating new biosocial networks.

Coexistences and confrontations

We have seen how humans and other organisms interact and benefit each other in many different ways in gardens. However, gardening can be interpreted as the art of refining, on the one hand, and, on the other hand, fighting nature. The word 'nature' may refer to a garden aesthetic or environmental ideal, perhaps particularly prominent in some parts of the garden, but it may also refer to the 'wild' weeds and vermin that threaten the crops and also to a larger physical and metaphysical context that includes the garden (cf. Pollan 1991). Many gardeners struggle with issues concerning the extent of control in their garden. Vigorous plants can simultaneously be an asset and a problem. In between plant vitality as a positive value and as a problem, such plants highlight the difficulties in cultivating and at the same time protecting 'nature'. Sometimes it is clear which plants to care for and which to clear away, but this can also be a matter where taste differs, as one gardener shows: 'I let the plants establish wherever they like and don't care whether they are in neat rows or not. Some move in by themselves – I have never sown or planted tansy and foxgloves, but there they are' (LUF, 2010).

Many associate gardening with dreams of living a good and simple life, close to nature. But, in practice, managing nature in your garden can be a real struggle (Saltzman and Sjöholm, 2016). These seemingly contradictory relations to 'nature' become especially clear in cases where neighbours have different approaches to invasive weeds and pests. The conditions for what is welcomed and what is seen as a threat in the garden are determined by cultural and social processes. It is quite common for gardeners to use war metaphors in their description of the fight against unwanted plants and creatures. In Sweden, such language is, for example, often used when talking about ground elder (*Aegopodium podagraria*), which can spread quickly and does not care about property limits. Evy, who has been diligently looking after her garden to keep it weed-free for almost 40 years, told us about the one time ground elder appeared in a corner:

> I was on it with both spit and shovel and God knows what. Big alarm and I told my neighbours that they had some in the hedge and said, 'Can I take it?' 'Yes, of course', they said. . . . So, it's a scourge, beautiful leaves and beautiful flowers, but awful.

Even plant lovers who are otherwise happy to welcome self-sown additions to the garden use metaphors of war and extermination when talking about invasive species, and the word 'invasive' does, of course, have military connotations (cf. Qvenild et al., 2014; Ernwein and Fall, 2015). However, as we have seen, what is considered a weed in one garden may be very welcome in another. Ground elder is edible, and so are nettles (*Urtica dioica*), another common weed in Swedish gardens. Anneli, for example, told us that the nettles in her garden are regarded as a crop and a delicacy and that she also turns dandelion buds into pickles, similar to capers. But it is clear that plants that grow in abundance in one setting may be less prone to spread in another. This gardener has, for example, gradually learned that sweet cicely (*Myrrhis odorata*) and oregano (*Origanum vulgare*) thrive almost too well in her garden: 'You need to be tough on the cicely, otherwise it occupies the whole garden, and so does the oregano, but oh how butterflies love the latter' (LUF, 2010).

Some gardeners are actively working to get specific plants to spread. One such example is the snake head (*Fritillaria meleagris*) that now grows in abundance in Kristina and Örjan's garden. Kristina says that they originate from a few small pots that she planted: 'They simply seed if you just leave them alone'. When it comes to spring flowering bulbs, gardeners often seem to tolerate and enjoy unrestrained and boundless spreading. The early spring flowers are often allowed to grow beyond borders and flowerbeds, and many of these spread easily with seeds, bulbs and stolons. One gardener, for example, seems very happy to reveal how 'a corner of the lawn is like a pink carpet, full of corydalis in spring' (DAG, 2011). One explanation of this tolerance might be that flowers are particularly welcome early in the season and another that these plants tend to wither and 'disappear' before summer comes. However, several of the species that form coloured carpets in Swedish gardens for a short time in spring are included in the Swedish Species Information Centre's list of potentially invasive alien species, such as squill (*Scilla*) and striped squill (*Puschkinia scilloides*; see Strand et al., 2018).

Even gardeners who are otherwise concerned about everything 'natural' have their limits and express antipathy to specific plant species. Nina, for example, talks about hops 'that are all around, making you crazy'. For the safety of the children, Nina and her husband also remove poisonous plants such as yew and foxgloves. On the other hand, they make use of wild plants such as willowherb (*Chamaenerion angustifolium*). During our visits, a number of gardeners showed us plants in their garden that just popped up unannounced or spread to new places, perhaps with the help of wind or birds. Nina commented on a variety of maiden pink (*Dianthus deltoides*) that she used to cultivate as a border plant; it later spread while the original plants disappeared. Nowadays, they are growing in the gravel on the driveway: 'These have spread here, and they are wonderful. I love them'.

Another example of a plant that can be considered attractive but is also feared as invasive is the creeping bellflower (*Campanula rapunculoides*), a bluebell species that spreads with seeds as well as underground stolons. One gardener says:

> I have a hard time deciding whether it is a beautiful flower that may be left to thrive there, or an extremely troublesome weed that can hardly be eradicated. It probably ends with me at least trying to delineate the extent of it.
>
> *(LUF, 2010)*

In our encounters with Swedish gardeners, a recurrent theme has been questions about whether a plant species should be considered a weed or not – or, as Malin put it when showing us a small plant with yellow flowers, 'Is this something one could want to have?' As researchers, we are in such situations reminded that our interaction with the informants inevitably affects what they tell us and possibly also affects their relationship with the plants in their gardens. In one garden we made a brief comment on the leaves of creeping bellflower that were growing next to a rose. It turned out that the gardener, who had just told us about her intense campaign against ground elder, did not know this was creeping bellflower. As an unintended result of our attention to the bellflower's presence in her garden, she got another enemy to fight.

Animal friends and foes

For insects, birds and other creatures, a garden can be their entire habitat or an essential part of it. The presence of animals can sometimes, but not always, be appreciated by gardeners. As with plants, what is welcomed can vary from garden to garden and from one occasion to another. Kristina, for example, talks fondly about the birds in her garden and feeds them in winter: 'They give back what they get in the winter by catching insect pests in summer'. This view – that birds and humans help each other in the garden – reflects an attitude we have often seen among gardeners. Everyday interaction with different kinds of living organisms seems to make many gardeners aware that cooperation can extend across species boundaries. This reminds us of the concept of 'co-species' that Donna Haraway established for similar kinds of multispecies relations (Haraway, 2008).

One of the most controversial species in contemporary Swedish gardens is the Spanish slug (*Arion vulgaris*), in Swedish known as *mördarsnigel*, which translates as 'the murderer slug'. This clearly invasive species has been spreading in Europe during the second half of the 20th century and in Sweden since the 1970s (cf. Ginn, 2014). According to the Swedish EPA, Spanish slugs cause significant damage in private gardens as well as in agriculture and commercial gardens and also pose a threat to the domestic slug species (Naturvårdsverket and Havs- och Vattenmyndigheten, 2017b). In private gardens, attempts to stop the progress of the slugs has led to new routines and rituals:

> I used to sow flowers, vegetables and more, but when everything gets eaten by the beasts you give up. Every night during the summer, if it rains, I go out with a planting shovel and a milk carton. I chop them [the slugs] in half and put them in the carton. Sometimes I count them, one evening I got 150.
>
> *(LUF, 2010)*

There are many different methods for handling the slugs, and gardeners are inventive. Sven and Margareta told us that one year their garden was 'flooded' with slugs and admitted that they kill them with salt even if they know this method is not accepted by everyone. 'It is animal cruelty', says Margareta, but they saw no other solution when their garden was invaded and their plants were threatened. Others freeze, poison or cut the slugs, but not Barbro, who is tender to animals of all varieties and does not kill the slugs she finds in her garden:

> I carry them away. I can't kill them. It has not been many this year, but I have picked a lot of slug eggs. They will often come under pots and such, I take them and then I throw them away [in public areas]. I know you shouldn't, but it is, I cannot. And if I can't, I can't. But there haven't been that many this year.

Garden co-becomings

The garden is often described as a symbol of people's encounter with and cultivation of nature, but attempts to establish and consolidate boundaries between nature and culture are constantly exceeded. In the everyday interaction with fellow biosocial becomings – animals as well as plants and other living organisms – many gardeners seem to experience connections that extend across boundaries. As we have seen in this chapter, many boundaries are mental as well as physical, and they are also dynamic and changing over time. Common understandings of which species are welcome and which ones are not can be challenged by differentiated and constantly trans-forming individual approaches, as well as changing conditions – for example, if a loved garden plant turns out to be too willing to spread. As we have seen, 'wild' plants are often regarded as invaders in the garden just as horticultural plants popular within the garden's boundaries can turn invasive when they escape over the fence into the surrounding landscape. Even in a nature lover's garden there are values concerning desired and undesirable plants and animals. People's attitudes towards living co-species in the garden make the boundaries and categorisations in everyday gardening visible.

What we call gardening is largely about managing and trying to control the forces and dynamics of nature. Plants move and are moved between different places in the garden and from one garden to another. Such movements illustrate how the constructions of boundaries – in and between gardens as well as between countries and continents – are to some extent con-stantly under negotiation. In this context, cutting and weeding can be seen as boundary-making activities designed to keep vegetation that has been introduced under control and limit and even eradicate plants considered undesirable.

Gardens are important places for people's everyday engagement with the complex dynam-ics of life, commonly understood as 'nature'. Gardeners' perspectives on plants and animals in their own backyard provide insights into how people relate to other biosocial becomings in a shared environment and indeed to contemporary processes of environmental change. As other researchers have noted, it is important to pay attention to the garden's non-human organisms not only as objects but also as acting subjects in order to understand these complex processes (e.g. Head et al., 2015)

From neighbour to neighbour, and from garden to garden, plants are exchanged and shared as well as inherited between generations. What began as a way to dispose of one's own abundance, perhaps in a plant flea market, can lead to meetings with like-minded people with whom one can also exchange knowledge, services and experiences. For gardeners, there are many reasons to divide, and share, their plants. Sharing can be seen as a meaning-making practice, where different kinds of knowledge, ideals and motivation are required depending on the context of sharing. Gardeners share plants, among other things, to maintain relationships. They also do it in order to take care of a plant, to get rid of it, to get more of the same, to keep things alive, or simply because they enjoy sharing and/or are in the habit of doing so. Sometimes the motives are economic as sharing can provide opportunities to get hold of plants in a cheaper way than the plant shop. It can also be a way to get hold of unusual, old, local or in other ways special varieties, which are often followed by the story about their origin. In this sense, exchanging garden plants with your friends can be a cultural heritage management act. Furthermore, as we have seen, sharing can also be part of a more or less active form of resistance to commodification and mainstreaming of garden culture; it can be done for ideological reasons, as an endeavour for sustainability or to promote biodiversity.

The forms of generous exchange of plant material that are so obviously recurring in the gar-den context show that material and (bio)social relationships are closely interconnected. In these contexts, plant vitality is often regarded as a value. Practical aspects such as how to deal with lack and abundance are in this context intertwined with ideals of sustainability and heritage management, as well as with cultural and ethical perspectives on gifts and responsibilities. In practice, the generosity of gardeners, as we have seen, is mixed with other feelings and priorities, such as wanting to get rid of excess, resistance to throwing living things away or simply an inability to stop collecting, growing and sharing. The curiosity to try new plants as well as the joy of managing to propagate old ones and share them with others are important factors to take into consideration when looking for sustainable ways to manage changing conditions and increasing mobility of species.

In the discussion of invasive alien species – as well as in the parallel and related discussions about endangered species and loss of biodiversity – cultural, historical and emotional connections between people and plants tend to be overlooked. In order to manage the threat posed by invasive species, we need to improve our understanding of what happens in everyday biosocial encounters among people, plants and other species. By looking closely into people's stories about how they share and spread plants, we have, in this chapter, tried to highlight the complexity of the question of how to limit the spread of certain species. The everyday practices by which private as well as professional gardeners divide, multiply and share plants are an important reminder that the spreading of species is closely connected to cultural contexts and social networks: practices that set plants in motion.

References

Bajada, F. and Setten, G. (2019) 'Garden waste and the life and death of invasive alien plants', Presentation at the Nordic Geographers Meeting, Trondheim.

Barker, K. (2008a) *Cultivating Biosecurity: Governance, Citizenship and Gardening in Aotearoa New Zealand.* University of London, London.

Barker, K. (2008b) 'Flexible boundaries in biosecurity: Accommodating gorse in Aotearoa New Zealand', *Environment and Planning A*, vol. 40, pp. 1598–1614.

Benediktsson, K. (2015) 'Floral hazards: Nootka lupin in Iceland and the complex politics of invasive life', *Geografiska Annaler: Series B, Human Geography*, vol. 97, no. 2, pp. 139–154.

Czarniawska, B. and Löfgren, O. (eds.) (2012) *Managing Overflow in Affluent Societies.* Routledge, London.

Czarniawska, B. and Löfgren, O. (eds.) (2013) *Coping with Excess: How Organizations, Communities and Individuals Manage Overflows.* Edward Elgar Publishing, Cheltenham.

DAG (2011) Department of Dialectology, Onomastics and Folklore Research in Gothenburg, Sweden: Responses to the question list DAG 22 The Home Garden (Trädgården), distributed in 2011. Archival series F.

Ernwein, M. and Fall, J.J. (2015) 'Communicating invasion: Understanding social anxieties around mobile species', *Geografiska Annaler: Series B, Human Geography*, vol. 97, no. 2, pp. 155–167.

Frihammar, M., Kaijser, L. and Lagerqvist, M. (2018) 'Hotet i vägrenen. Blomsterlupinens plats i ett svenskt kulturarvslandskap', *Kulturella perspektiv*, vol. 28, nos. 3–4.

Ginn, F. (2014) 'Sticky lives: Slugs, detachment and more-than-human ethics in the garden', *Transactions of the Institute of British Geographers*, vol. 39, no. 4, pp. 532–544.

Gunnarsson, A., Saltzman, K. and Sjöholm, C. (2017) *Ett eget utomhus. Perspektiv på livet i villaträdgården.* Makadam, Göteborg/Stockholm.

Haraway, D.J. (2008) *When Species Meet.* University of Minnesota Press, Minneapolis.

Head, L. (2016) *Hope and Grief in the Anthropocene.* Routledge, London.

Head, L. (2017) 'The social dimensions of invasive plants', *Nature Plants*, vol. 3, p. 17075.

Head, L., Atchison, J. and Phillips, C. (2015) 'The distinctive capacities of plants: Re-thinking difference via invasive species', *Transactions of the Institute of British Geographers*, vol. 40, pp. 399–413.

Head, L. and Muir, P. (2004) 'Nativeness, invasiveness, and nation in Australian plants', *The Geographical Review*, vol. 94, no. 2, pp. 199–217.

Head, L. and Muir, P. (2006) 'Suburban life and the boundaries of nature: Resilience and rupture in Australian backyard gardens', *Transactions of the Institute of British Geographers*, vol. 31, no. 4, pp. 505–524.

Hobhouse, P. (2002) *The Story of Gardening*. DK Publishing, New York.

Hulme, P.E. (2015) 'Resolving whether botanic gardens are on the road to conservation or a pathway for plant invasions', *Conservation Biology*, vol. 29, pp. 816–824.

Ingold, T. (2013) 'Prospect', in T. Ingold and G. Pálsson (eds.) *Biosocial Becomings: Integrating Social and Biological Anthropology*. Cambridge University Press, Cambridge.

LUF (2010) The Folklore Archives in Lund, Sweden: Responses to the question list LUF 233 The Home Garden (Trädgården), distributed in 2010. Archival series M.

Naturvårdsverket (2020) 'Sprid inte invasiva främmande arter', www.naturvardsverket.se/Var-natur/Djur-och-vaxter/Invasiva-frammande-arter/

Naturvårdsverket and Havs- och Vattenmyndigheten (2017a) Fact sheet: *Har du något förbjudet som simmar eller växer i din trädgårdsdamm?*

Naturvårdsverket and Havs- och Vattenmyndigheten (2017b) Fact sheet: *Spansk skogssnigel. Arion vulgaris.*

Naturvårdsverket and Havs- och Vattenmyndigheten (2018) Fact sheet: *Råd om hur du hindrar spridning av invasiva främmande arter i och från din trädgård.*

New York Times (2018) 'Beauty or Beast? Iceland Quarrels Over an Invasive Plant', November 9.

Phillips, C. (2013) *Saving More than Seeds*. Ashgate, Farnham.

Pearce, F. (2015) *The New Wild: Why Invasive Species Will Be Nature's Salvation*. Icon Books Ltd., London.

Pollan, M. (1991) *Second Nature: A Gardeners Education*. Grove Press, New York.

Pottinger, L. (2018a) 'Growing, guarding and generous exchange in an analogue sharing economy', *Geoforum*, vol. 96, sid. 108–118.

Pottinger, L. (2018b) '"It feels connected in so many ways" circulating seeds and sharing garden produce', in A. Ince and S.M. Hall (eds.) *Sharing Economies in Times of Crisis: Practices, Politics and Possibilities.* Routledge, London/New York.

Qvenild, M. (2013) *Wanted and Unwanted Nature: Invasive Plants and the Non-Native Dichotomy*. Norwegian University of Science and Technology, Trondheim.

Qvenild, M., Setten, G. and Skår, M. (2014) 'Politicising plants: Dwelling and invasive alien species in domestic gardens in Norway', *Norsk Geografisk Tidskrift*, vol. 68, no. 1, pp. 22–33.

Saltzman, K. (2006) 'Composting', *Ethnologia Europaea*, vol. 35, nos. 1–2, pp. 63–68.

Saltzman, K. and Sjöholm, C. (2016) 'Managing nature in the home garden', in L. Head, K. Saltzman, G. Setten and M. Stenseke (eds.) *Nature, Time and Environmental Management: Scandinavian and Australian Perspectives on Peoples and Landscapes*. Routledge, Abingdon, Oxon.

Saltzman, K. and Sjöholm, C. (2019) 'Att dela och dela med sig. Trädgårdsväxter, kulturarv och ekonomi', in M.P. McIntyre, B. Johansson and N. Sörum (eds.) *Konsumtionskultur. Innebörder och praktiker*. Makadam, Göteborg.

Saltzman, K., Sjöholm, C. and Westerlund, T. (2020) 'Sellable histories: The use of stories in marketing of heritage plants and gardens', *Ethnologia Scandinavica*, vol. 50, pp. 121–144.

Sjöman, H., Östberg, J. and Bühler, O. (2012) 'Diversity and distribution of the urban tree population in ten major Nordic cities', *Urban Forestry & Urban Greening*, vol. 11, no. 2012, pp. 31–39.

Strand, M., Aronsson, M. and Svensson, M. (2018) *Klassificering av främmande arters effekter på biologisk mångfald i Sverige – ArtDatabankens risklista*. ArtDatabanken Rapporterar 21. ArtDatabanken SLU, Uppsala.

SvD *Svenska Dagbladet* (2019) 'Se upp för växterna som hotar vår natur', 4 May.

SVT *Sveriges Television* (2019) 'Fem arter som hotar Sverige', 17 June.

Söderlund, T. (2018) 'EU förbjuder invasiva växter', *Trädgårdsamatören*, vol. 1, pp. 30–31.

Tschan, G.F. (2018) *Invasiva arter och transportinfrastruktur: en internationell kunskapsöversikt med fokus på vägar och växter*. Statens väg- och transportforskningsinstitut, Linköping.

Westerlund, T. (2014) 'Pionodlaren på Guldsmedsgården', in G. Almevik, S. Höglund and A. Winbladh (eds.) *Hantverkare emellan*. Hantverkslaboratoriet, Göteborgs universitet, Mariestad.

Westerlund, T. (2017) *Trädgårdsmästarens förökningsmetoder. Dokumentation av hantverkskunskap*. Diss. Acta Universitatis Gothoburgensis, Göteborgs universitet, Göteborg.

Westerlund, T. (2021, in press) 'Classification of plant propagation practice', in T. Westerlund, C. Groth and G. Almevik (eds.) *Craft Sciences*. Hantverkslaboratoriet, University of Gothenburg, Mariestad.

PART 3

Practices

15

NATIONAL BIOSECURITY REGIMES

Plant and animal biopolitics in the UK and China

Damian Maye and Kin Wing (Ray) Chan

Introduction

'Biosecurity' is a contested concept, but in broad terms it refers to mitigation strategies that prevent disease incursion and the spread of infectious pathogens among humans, animals, plants and other natural environments (Koblentz, 2010). Since the onset of African swine fever (ASF) and highly pathogenic avian influenza (HPAI), reducing the risk to health from the transnational spread of infectious diseases across nations and regions is a global priority. Strategies to manage disease operate at multiple scales, including regulatory frameworks set by national governments and intergovernmental organisations (Wilkinson et al., 2011; Maye et al., 2012). The economic interests of countries can 'deflect international economic concerns requiring biosecurity solutions to be reworked at local scales' (Enticott, 2014a, 42). Politics at different scales shape the production, distribution and consumption of livestock and plant products, involving global organisations (e.g. the Food and Agriculture Organization, FAO) and national and local government departments (Chan, 2015). This chapter examines national biosecurity regimes to contain and manage the impact of infectious animal and plant diseases, including control over the implementation and monitoring of biosecurity measures and relations with international and local actors.

Climate change and globalisation are key processes implicated in the spread of animal and plant diseases (Waage and Mumford, 2008; Wilkinson et al., 2011; Hinchliffe et al., 2016; House of Lords, 2018; Tsouvalis, 2019). Globalisation, including the increasing movement of goods and people, is of particular relevance here given the impact these processes have on border controls, supply chains and trade, as well as the development of national and international systems of regulation. Globalisation and trade limit 'the autonomy of traditional, nation-state-based systems of authority' (Wilkinson et al., 2011, 1935), requiring additional measures and collective and individual responsibilities of farms and businesses, as well as transnational systems of regulation.

The chapter compares two national biosecurity regimes, the UK and China. Biosecurity is ostensibly apolitical, about managing risk, but in practice it is highly political and entangled with broader tensions between biosecurity and trade priorities (Braun, 2007; Brasier, 2008; Maye et al., 2012; Enticott, 2014b). National biosecurity regimes are instruments of governance, and understanding the 'biopolitics' that underpins them is important. A common distinction is between *state-centred* and *neoliberal regimes*, which are contrasted in terms of relations between state and private capital and science for disease planning, with the former more top-down and state centric in contrast to neoliberal models that devolve responsibilities and costs to private actors. This chapter shows the influence of political and socio-economic context in shaping national biosecurity regimes. In practice, they represent forms of institutional blending. The next section reviews natural and social science biosecurity studies to conceptualise what we mean by 'national biosecurity regimes', including critiques of risk-based approaches to biosecurity governance.

National biosecurity regimes: disease classifications, risk, politics and borderlands

To examine state regulations and institutional regimes that govern against infectious animal and plant disease incursions, it is helpful to first consider classifications of disease threat and how that translates into international standards and agreements. A basic but fundamental distinction is between plant and animal diseases. This distinction is important from a governance perspective because plant and animal diseases are managed very differently by countries. Epidemiological characteristics (defined relative to national context) further distinguish animal and plant disease threats. For example, 'endemic' and 'exotic' animal and plant diseases and pests; 'notifiable' and 'non-notifiable' animal and plant diseases; 'zoonotic' and 'non-zoonotic' animal diseases. Various international regulations govern plants and animals as part of a 'global biosecurity regime'. For example, the Office International des Epizooties (OIE), the world organisation for animal health, has a list of diseases that should be reported internationally (see OIE, 2019). Most diseases on the list are epizootic, highly transmissible or zoonotic (Lerner, 2017). OIE also sets standards for international trade in animal commodities – e.g. beef, pork, poultry meat and milk. Plant pathogens are regulated by the International Plant Protection Convention (IPPC), an international treaty of the FAO (MacLeod et al., 2010) that has 184 contracting parties (FAO, 2021). International standards set by the IPPC and OIE are complemented by the World Trade Organization's (WTO) 'SPS Agreement', which is a provision for plant and animal protection in free trade agreements. A key objective of a national biosecurity regime is to keep disease threats out, but WTO members must also allow animal and plant material importations (i.e. market access/penetration of national products into other countries' domestic markets) unless meaningful risk is demonstrated (Maye et al., 2012, 155; Tsouvalis, 2019).

International standards and lists of disease are applied by nation states to *enable* international trade. Standards and lists define plant and animal health, with 'health' defined as 'the absence [of markers] of disease' (Lerner, 2017, 291). Plant health, for example, refers to 'the legislative and administrative procedures used by governments to prevent plant pests from entering and spreading within their territories' (MacLeod et al., 2010, 51). Disease status informs governance response. Animal diseases such as foot and mouth are classified in the UK, for example, as exotic, notifiable and non-zoonotic, while avian influenza is notifiable, exotic and potentially zoonotic and cryptosporidiosis is endemic and zoonotic. Each requires specific governance arrangements (Fish et al., 2011, 2024).

While each plant or animal disease exemplifies particular risks and uncertainties and requires specific governance responses, we can identify general parameters for containment in terms of how nation states organise their biosecurity regimes in terms of regulatory frameworks and risk management procedures. The significant difference is the way plant and animal health regimes are governed nationally. Plants are generally less well regulated. National regimes are not usually well integrated in terms of animal and plant health governance (exceptions being New Zealand and Australia; Maye et al., 2012). This is linked to biology – e.g. there are many more plant crops than farmed animals, and crop threats vary geographically. There are also 'historical political factors affecting the ways plant and animal diseases are dealt with' (Wilkinson et al., 2011, 1935), with greater protection afforded to animal diseases because of higher economic value, lobbying, public health concerns and historical links in protecting a country's reputation (see Woods, 2011).

Nation states employ common instruments and protocols in the name of 'biosecurity', including: non-tariff trade barriers to control disease spread; regulating the movement of people, animals and plants across borders; national policy strategies for disease containment; and plant and animal health biosecurity standards. These national biosecurity standards diffuse to local scales. Farm animal health planning on a poultry production unit in the UK, for instance, responds to national guidelines regarding on-farm biosecurity by adopting biosecurity practices such as washing and cleansing; tracking farm inputs, outputs and movements; and following procedures for reporting disease outbreaks. In China, the central state has drafted a National Action Plan and delivered administrative measures to local county government officials to eradicate African swine fever, including measures to control interprovincial live pig transportation, mandatory disease reporting and culling (The State Council, 2019).

However, 'making life safe' is not as straightforward as this. National biosecurity regimes at an operational level are messy assemblages, with disease part of entangled logics and practices that produce food such as poultry (Hinchliffe et al., 2013, 2016). In the biological invasions literature, frameworks have been developed that characterise the phases which need to be passed in order for a disease or pest to successfully 'invade' (see Williamson, 2006). This includes 'import into a country, release or escape in to the wild, establishing a population, spreading from an established population and becoming a problem that some people want to deal with' (ibid., 1561). This classifies biological invasions along ordered chronological lines. Although useful heuristically, this linear retelling of 'invasion' fails to capture messiness, uncertainty, relational qualities of disease and the role of human behaviour.

Fish et al. (2011) provide a general framework for the governance of animal disease. Analysis of uncertainty in containment practice encompasses, they argue, three 'arenas of action':

- *Prevention*: reducing the occurrence of animal disease and taking pre-emptive forms of action to reduce the chances of a disease outbreak – e.g. changing livestock management practices;
- *Anticipation*: acknowledging a potential animal disease threat (e.g. bluetongue outbreak in the UK) and predicting/preparing for disease outbreaks – e.g. modelling of disease scenarios; and
- *Alleviation*: the process of responding to disease occurrence. The focus here is on procedures adopted to control and eradicate disease in real-world circumstances.

The three arenas of action have different expressions according to the level of policy practice, differentiating among the *strategic level* (structures and processes that shape principles of containment, such as legislation on disease risks); the *tactical level* (strategic level goals are translated into practical rules, procedures and tools for decision-making – i.e. rationales are given procedural expression, such as criteria for intervening in a particular disease outbreak); and the *operational level* (practical contexts of disease containment and systems of technological and human practice, such as processes of vaccinating birds and livestock). They emphasise the need to look for interaction between arenas and levels to reveal uncertainties associated with strategies of containment. This is helpful in terms of identifying ways that scientific uncertainty can be included in decision-making about biosecurity control measures. However, the emphasis on containment still gives an impression of 'biosecurity threats as external rather than *lying with* its very system' (Enticott, 2014b, 125; emphasis added) or what Hinchliffe et al. (2013, 534) describe as 'a will to closure where success is measured by their ability to limit the flow of undesirables across territories and bodies'. This requires an approach to biosecurity that focuses less on 'borders' and 'breach points' and more on the internal vulnerability and complexity of relationships among disease, environment and host.

Conceptualising the governance of biosecurity regimes in line with the relational qualities of disease is significant because nation states are increasingly aware of the costs of state-led eradication programmes (foot and mouth disease [FMD] in the UK, for example) and are shifting costs and responsibility to agricultural sectors. This includes nurturing new forms of biosecurity subjectivity and citizenship to benefit, for instance, the 'national herd' (Enticott, 2014b) or to protect 'national identity' (Barker, 2010). Risk-based approaches to biosecurity, particularly for plant and tree diseases, are also integrated with neoliberalism (Tsouvalis, 2019), even though global trade accelerates the pace, emergence and spread of pests and pathogens. We elaborate on these tensions between national trade priorities and biosecurity in the next section, where we examine particular national biosecurity regimes, including how biosecurity is defined and operationalised, relations with international frameworks, approaches to govern plant and animal health risks and the influence of politics and socio-cultural factors.

Contrasting national biosecurity regimes: the UK and China

This section examines biosecurity in the UK and China respectively. For each regime, we introduce where 'biosecurity' fits within public policy discourse, review the overall regulatory framework for animals and plants, examine relations among the national regime and international standards and local practices and use key disease case studies to exemplify the evolution and tensions within each regime. This analysis extends earlier work on national biosecurity (Maye et al., 2012), which contrasted the Australian experience with the UK and revealed a complex geopolitics. Both cases were avowedly neoliberal and supportive of free trade, but while Australia used biosecurity practices to protect agricultural production, the UK was precautionary in its approach but conflicted by Single Market requirements. We review attempts by the UK to shift disease planning policy from state centric towards cost sharing and private responsibility. However, we argue that this policy objective is not well reflected in practice, with state intervention still very evident. Implications for biosecurity post-Brexit are also reviewed. Turning to China's biosecurity regime, at a national level it is highly state centric and centralised. In practice, however, we highlight a lack of independent monitoring and regulation, which means that biosecurity risk is passed to local stakeholders, resulting in practices that are flexible and renegotiated by socio-cultural factors.

The UK's biosecurity regime

The bovine spongiform encephalopathy (BSE) crisis in 1996 was a significant event in terms of influencing a more precautionary stance to animal disease in the UK. However, it was not until the FMD outbreak in 2001 that biosecurity emerged as a term in public policy, when was employed to explain measures that could be implemented to reduce disease spread on farms (Donaldson, 2008; Fish et al., 2011). Biosecurity policy in the UK is currently aligned with EU governance. In practice this means that 'most of the UK's legislative frame on biosecurity, and the systems and processes in place to manage biosecurity risks, comes from the EU' (House of Lords, 2018, 6). For example, EU legislation covers measures designed to prevent spread (surveillance and movement controls) and agreed lists of species that are potentially harmful to the EU (Maye et al., 2012). This means that 'plants and animals (and plant and animal products) can move freely within the EU Single Market; stricter controls are then in place for imports from non-EU countries' (House of Lords, 2018, 6). Through its current membership to the EU, the UK also shares disease notification systems and monitoring systems for plant and animal movements. Risk assessments and risk management decisions for plant and animal health, food and food safety are also undertaken at an EU level. EU-level legislation is based on international agreements established by the WTO. In other words, European international frameworks are nested within international frameworks, which determine what organisms and materials can be denied access without going against WTO rules (Wilkinson et al., 2011).

Within international frameworks set by the EU and WTO, the UK has had fairly limited flexibility to establish its own biosecurity measures (MacLeod et al., 2010), which are usually emergency measures to deal with biosecurity threats – e.g. the UK and Ireland were the only EU members to adopt national measures for *Epitrix*, a flea beetle that threatens potato crops (House of Lords, 2018). Future changes to EU biosecurity governance will potentially allow more flexibility at the member state level. In the UK the regulatory framework for biosecurity is currently structured around three key animal, plant and invasive regulations, all of which were set by EU institutions (ibid., 8–14):

- *Regulation (EU) 2016/429 on Transmissible Animal Diseases* [animal health law]: this regulation came into force in April 2016 (fully implemented in April 2021). It merges various existing acts into a single law regarding what is needed for disease prevention, disease awareness, biosecurity, animal traceability, intra-EU movement, animal and animal products entering the EU, etc.
- *Regulation (EU) 2016/2031 on Protective Measures against Pests of Plants* [plant health law]: introduced on 13 December 2016 (and fully implemented at the end of 2019), this prevents destructive plant pests from entering the EU and also prevents spread if they are found to be present, including 'priority pests', which are subject to enhanced measures. Plants and plant products can be potentially prohibited (if the risk is high) or plant materials must be accompanied by a phytosanitary certificate. All plants moving within the EU must have a 'plant passport' that certifies their health and traceability (see also Defra, 2014).
- *EU Regulation 1143/2014 on Invasive Alien Species* [invasive non-native species law]: introduced on 1 January 2015, this sets out prevention, early detection/rapid eradication and management measures to be taken in relation to a list of invasive non-native species.

UK biosecurity legislation thus depends on cooperation with the EU. From a national biosecurity perspective, one major implication is that the EU has open internal borders, which means any European nation state cannot on its own keep out animals and plants (Brasier, 2008). In other words, new biosecurity risks must be assessed by the EU and 28 member states who agree on an appropriate action.

If we look more specifically at the UK's biosecurity regime for animals and plants, there are two key points to note. First, attention is still mostly on border controls and the desire to keep out 'exotics' and 'notifiables' in conjunction with EU legislation and agreed lists of restricted species (i.e. a borders rather than borderlands approach; Defra, 2014, 2018). These containment strategies employ the strategic, tactical and operational levels described in Fish et al.'s (2011) framework. Figure 15.1, for example, is taken from Defra's (2018, 18) *Contingency Plan for Exotic Notifiable Diseases of Animals in England*. The contingency plan is classic emergency planning, using militarised language like 'command, control and communication' and 'tactical and

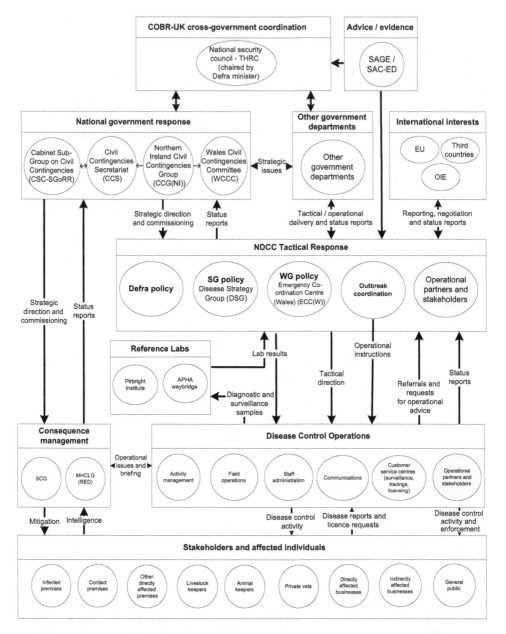

Figure 15.1 Contingency planning for exotic notifiable diseases of animals in England – command, control and communications structure.

Source: Defra, 2018, 18.

operational command structures', and describes in detail how the government will respond to and manage and prepare to respond to exotic disease outbreaks, with an emphasis on borders and exclusion. The second point is to recognise different governance regimes for plant and animal disease in the UK. Table 15.1 shows this point and reiterates the prominence of animal disease

Table 15.1 Comparing plant and animal disease biosecurity regimes in the UK.

	Plant	Livestock
Government intervention	Government does not compensate affected producers but covers costs of testing, surveillance, etc.	Government currently covers costs of disease control for exotic diseases plus compensation for some endemic diseases.
Industry cohesion	Cohesion is strong in agricultural sector, with industry-led trade agreements and market structures to discourage bad practices among producers; less cohesion in horticultural sector.	Individualistic approach to endemic disease control and poor communication and 'free-loading' by producers.
Disease surveillance	Routine testing and surveillance of regulated plants and plant products (e.g. potatoes); no surveillance of unregulated endemic pests and pathogens.	Routine testing for government-controlled (exotic) diseases, poor surveillance for endemic diseases.
Welfare	Apart from some aspects of biodiversity, plant welfare is not a public concern.	Zoonotic risks, animal welfare and biodiversity are important factors in disease control policy.
Professional expertise	Plant pathologists and plant health inspectors have a low profile.	Veterinarians are a well-organised profession and have a relatively high political profile; the chief veterinary officer holds considerable legal responsibilities.

Source: Wilkinson et al., 2011, 1936.

in UK government policy, largely because animal diseases have more significant detrimental impacts on the economy and society (Wilkinson et al., 2011). The recent 'ash dieback' outbreak has reduced this bias to some extent, but the emphasis remains towards animals and plans to prevent border breaches from 'outside threats'.

We turn now to look at specific disease examples. In relation to trees and plants, the UK has experienced a rapid growth in silviculture and horticulture in recent years (Potter et al., 2011; Tsouvalis, 2019). In particular, trade between East Asia and other regions has resulted in a number of new disease and pest introductions in forestry, including *Phytophthora ramourm* (also referred to as 'sudden oak death' pathogen), a fungus that is a major threat to indigenous trees and shrubs (Wilkinson et al., 2011, 1935). The fungus entered the UK around 1990, and emergency measures were enacted in the EU in 2000, but not all members complied and the disease spread throughout Europe via the nursery trade. This shows the weakness of UK plant biosecurity as part of a harmonised EU-wide plant health regime (of standards and passports), which replaced national biosecurity when the Single Market was formed (Maye et al., 2012). Analysis of the ash dieback (*Chalara fraxinea*) outbreak in 2012 is also critical of the EU plant health regime (Tsouvalis, 2019). In particular, this study shows how the risk-based approach to biosecurity is too narrow. The emphasis was expert-led governance, via the Tree Health and Plant Biosecurity Export Taskforce, set up in the wake of the outbreak. Defra (2014), in their appraisal of how ash dieback and Asian longhorn beetle (*Anoplophora glabripennis*) were managed, were keen to emphasise the success of collaborative approaches involving the government, industry, NGOs, landowners and the public. However, Tsouvalis (2019) shows, via in-depth interviews with taskforce members, that one of the key vectors, global trade, was not adequately discussed, even though they were very aware of this factor. It was considered pointless to articulate concerns about trade officially because 'the political will to do anything about them was lacking' (ibid., 3). Plant biosecurity and neoliberalism are thus interlinked to such an extent that trade as a vector of disease is not addressed politically, with a narrow focus on risk-based approaches that do not allow for sufficient discussion about alternative regimes, such as the 'pathways approach' (see Brasier, 2008).

Scientific and technical risk assessments also dominate policymaking for livestock diseases in the UK, as noted in Duckett et al.'s (2015) analysis of cryptosporidiosis, FMD and highly pathogenic avian influenza. All three cases, which have very distinct pathways, scales of impact and animal/human relationships, are managed through positivist epistemologies (i.e. approaches that favour quantitative modelling). A major historical difference between plant and animal disease biosecurity regimes is in terms of government intervention, with no compensation for affected plant producers but costs covered for exotic livestock disease control and compensation for some endemic diseases. However, studies of animal health governance in the UK indicate a redistribution of 'disease responsibility away from state actors and onto private companies and, in particular, onto farms' (Hinchliffe et al., 2013, 533). One core objective is for disease prevention costs to be borne more by the farm and less by the state.

The 'cost and responsibility' approach is taken from New Zealand (Enticott, 2014b), and we see the influence of this approach in different animal disease examples. In 2007–2008, for example, there was a significant outbreak of bluetongue in the UK (BTV-8). The disease was introduced into East Anglia via windborne transmission of infected midges from continental Europe. Defra adopted a partnership approach to manage the disease that involved a core group of industry stakeholders and veterinary professionals (O'Rourke, 2009). This involved working with the core group, devolved administrations and veterinary professionals to develop an emergency vaccination plan, with mass vaccination achieved through a voluntary approach and supported by an industry-led communications campaign that promoted the benefits of vaccination. In 2012 the Schmallenberg virus was confirmed as present in the UK, and there was no government compensation or direct assistance provided by the government and no vaccine was provided. We can also find evidence of cost-sharing approaches to deal with one of the most troublesome endemic livestock diseases in the UK: bovine TB. This includes schemes developed for badger vaccination and attempts by government to establish local farmer-owned syndicates to manage badger culls (Maye et al., 2014). At a policy level, the UK is moving towards a neoliberal approach to animal biosecurity, emphasising individual responsibilities, industry-led partnerships and cooperation to share costs. However, in reality government intervention remains crucial because it is expected and/or farmers do not respond well to neoliberal styles of governing which conflict with long-held belief systems about risk and nature, as seen through responses to badger vaccination contracts (ibid.) and elements of the badger cull, which remains a huge financial burden on the state (Godfray et al., 2018).

In summary, the case study shows that despite the rhetoric of the UK government to adopt New Zealand's and Australia's style of biosecurity policy regimes in relation to animal health, the varied uptake and success of cost-sharing models suggests this approach is currently limited. On the plant health side, biosecurity is significantly undermined by trade politics. Now that the UK has left the EU, biosecurity laws will be adopted through the EU Withdrawal Act so that the UK replicates functions previously undertaken by EU institutions. Post-Brexit, the UK may establish its own lists of restricted species and could potentially implement stricter biosecurity controls. It could also move away from the EU's biosecurity system (based on lists of restricted species) towards a more unified policy across sectors – e.g. adopting a more restrictive biosecurity regime that combines legislation with public awareness campaigns (Barker, 2010), although any changes will likely remain in tension with trade priorities. As we will see later, tensions among trade, politics and biosecurity regimes are also becoming more evident in China.

China's biosecurity regime

There are a number of reasons why it is important to understand China's biosecurity regime. First, China has been the source of numerous infectious diseases, including severe acute respiratory syndrome (SARS) and H7N9 influenza. Second, China's rapid urbanisation and rural-urban migration increases the risk of disease outbreaks (Alirol et al., 2011, 132). Third, the emergence of zoonotic disease is linked to rapid intensification of animal production, land-use changes, the live-poultry market and the emergence of trade networks between China and other countries (Gilbert et al., 2014). Finally, China engages in global health diplomacy (e.g. supporting health programmes in Africa) and employs 'soft power' to mobilise its influence and shape the global health agenda (Barr, 2009; Feldbaum and Michaud, 2010).

Biosecurity in China also sits within and against a wider set of competing and interrelated national tensions, including population pressure and food security. This context is important. China has a population of 1.3 billion, and this exerts significant pressure on internal food production. This has driven rapid restructuring in meat and crop production sectors in recent years through intensification and vertical integration, with the aim to improve productivity and standards (Day and Schneider, 2017; Wu, 2019). Rapid agricultural restructuring has fostered corporate farming (Scott et al., 2014), and the scale and concentration of agricultural production have increased the risk of disease epidemics (Meadows et al., 2018). China also increasingly outsources foreign food crops and animals to maintain food security. This is in response to the rapid loss of arable land due to urbanisation and environmental pollution (Morton, 2017) but is also a political decision. This increases the risk of importing infected animals and plant products. For example, Panama disease[1] (i.e. Fusarium wilt [FW] of banana) was imported to China via the importation of bananas from Taiwan in 2001 (Lin and Shen, 2017), and a devastating outbreak of African swine flu in September 2018 has been linked to a strain likely to have entered via Russian pork imports (Normile, 2018; SCMP, 2018).

Developing an effective regulatory framework for animal and plant disease management in China is therefore a national and international priority. The governance of China's biosecurity system is top-down and technocratic (Wei et al., 2015; Salter and Waldby, 2011), embracing the use of scientific information to evaluate biosecurity risks with limited involvement of lay knowledge (Barr, 2009). The term 'biosecurity' is employed in China in relation to laboratory procedures and policies, education, industrial animal farming, migratory birds, public health systems and border studies (Barr, 2009; Barr and Zhang, 2010; He et al., 2018; Kaufman, 2008; Smart and Smart, 2012). In public policy discourse, however, the term is more limited: 'biosecurity' refers to 'the protection and control of pathogens and toxins to prevent their deliberate theft, misuse, or diversion for the purposes of biological warfare or terrorism' (Barr and Zhang, 2010, 155). Biosecurity problems are attributed to human negligence, including poor training and a lack of management skills and expert knowledge (ibid., 119). China works with the FAO and OIE to fulfil international standards for biosecurity so that China's animal health standards align with global requirements. In this sense China maintains a 'state-centric' biosecurity system under neoliberalism (Schneider, 2017; Chan and Flynn, 2018).

The regulatory framework to monitor and control the importation of foods and plants is dynamic and complex. Similar to the UK, China has two biosecurity regimes to manage plant and animal diseases respectively that can be compared in terms of (1) state-funded agricultural support, (2) agricultural insurance coverage, (3) disease surveillance systems and (4) professional expertise (Table 15.2). The Ministry of Finance (MoF) provides more financial subsides for animal breeders than cash crop producers in terms of compensation for endemic diseases. China has standardised its disease management, risk analysis and quarantine practices to control invasive pests and diseases for animals and plants. However, managing zoonotic diseases is the

Table 15.2 Comparing plant and animal health biosecurity regimes in China

	Plant	Livestock
State-funded agricultural support	The MoF mostly provides subsidies for major cash crops (e.g. paddy rice, corn, wheat, cotton, rapeseed, bean and peanut) to compensate for endemic diseases.	The MoF provides financial subsides for sow, dairy cattle, pig, sheep, poultry, shrimp and fish breeding to compensate for epidemic diseases.
Agricultural insurance coverage	Mainly covers horticulture, which comprise 60% of agricultural insurance.	Limited.
Disease surveillance system	Most attention given to the surveillance of regulated exotic plants and products.	Good surveillance for pathogenic animal diseases, but surveillance is poor for endemic diseases.
Professional expertise	More professional training is needed.	More professional training is needed.

Sources: Bermouna and Li (2014), Krychevska et al. (2017, 5), Wilkinson et al. (2011, 1936).

main priority because zoonosis outbreaks cause higher socio-economic impacts, violate national reputations and threaten global trade and citizens' health (Qiu et al., 2018).

We now look at national biosecurity controls for plants and animals (pigs and poultry) in China. International trading of plants and global tourism are the main ways in which invasive species enter China (Qiang, 2007). There are currently 283 non-native invasive plant species in China, which have caused economic losses of around 14,985 billion USD per year, led to a decline in biological diversity and increased biological and genetic pollution (Ou et al., 2008; Mu et al., 2010). The number of invasive plant species in China reached 614 in 2015, of which 283 are non-native and 50 are among the world's 'worst' invasive species (Jiang et al., 2017, 240; Wang et al., 2017). To prevent and restrict the importation of invasive species, China has introduced quarantine services, set up an invasive species elimination programme, established cetology of risk supervision and enforced risk assessment of imported species before their introduction (Mu et al., 2010; He et al., 2018). The risk assessment approach in China is based on risk identification, estimation and evaluation (Chen et al., 2015).

However, while the preventative paradigm is acknowledged in principle, in practice it is relatively weak, in part because a technical approach to risk is favoured to control the problem. The approach adopted for border control is also technical and procedural. The General Administration of Customs of China (GACC) controls plant quarantine at ports in China. The Department of Animal and Plant Quarantine (DAPQ) is responsible for formulating regulations and procedures for the inspection and quarantine of animal and plant species (Figure 15.2). China has formulated more than 300 national and industry plant biosecurity standards to control and prevent non-native pests and diseases (Li et al., 2014, 57). There is a strong commitment to control disease spread, even though, as noted earlier, China now has a significant number of invasive alien species in agriculture, forestry and natural ecosystem environments. In terms of domestic pests and diseases, the Ministry of Agriculture and Rural Affairs (MARA) and the National Forestry and Grassland Administration (NFGA) control domestic agriculture and forest plant quarantine (Li et al., 2014, 47–48; NFGA, 2019). This includes establishing a database to aid in the control of and prevent the spread of non-native species. To improve the management and administrative capacity of invasive alien species (IAS), MARA established the Office for Alien Species Management in 2003 and the Rural

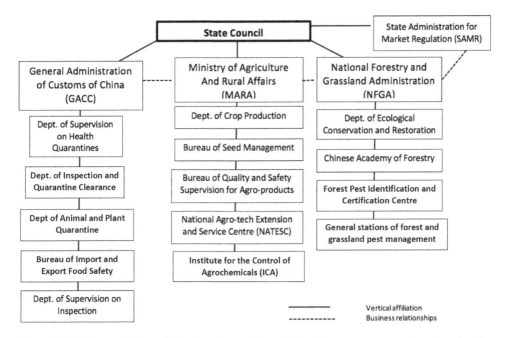

Figure 15.2   The organisation of alien invasive species and plant disease management in China after the 2018 structural reform.

Sources: GACC, 2019; MARA, 2019; MMTA, 2018; MWE China Law Offices, 2018; NFGA, 2019; Swiss Business Hub China, 2018a, 2018b.

Energy and Environment Agency in 2012 to monitor and implement nationwide IAS eradication programmes (Wang et al., 2017, 152). MARA also manages and monitors plant products prior to their circulation in Chinese markets and regulates agricultural inputs, including seed quality and agrochemical usage (Figure 15.2). State-funded scientific research and risk assessments are conducted by research institutions to reduce and contain non-native species, including developing new technology (e.g. early warning and risk assessment technology) to increase plant biosecurity (Li et al., 2014).

The national system of biosecurity governance for animals is also complex (Figure 15.3). In 2004 the state implemented a series of policies in rural China, including 'Building a New Socialist Countryside',[2] with a view to resolving farmers' livelihood problems, revitalising agricultural activity and regenerating rural areas (and thus solving the *Sanlong Wenti*).[3] China is the largest producer of live pigs (54.5 million tonnes were produced in 2016) and the largest consumer of pork (31.6 kg per capita annually; DBS Asian Insight, 2017, 6). China is the second largest consumer of poultry after the US. Animal biosecurity is therefore of high priority, with a vast animal health administration system created to manage risks that involves 300,000 government officials and 73,000 veterinarians (Wu, 2019, 7). Veterinary technical support institutions include three public institutes directly affiliated with MARA (MARA, 2019). These are: (1) the China Animal Disease Control Centre, responsible for surveillance and control; (2) the China Institute of Veterinary Drug Control, charged with assessing drug quality and regulating use; and (3) the China Animal Health and Epidemiology Centre, which investigates specific outbreaks and more generally identifies risk factors. At the county level, local government has established county veterinary authorities to inspect livestock and

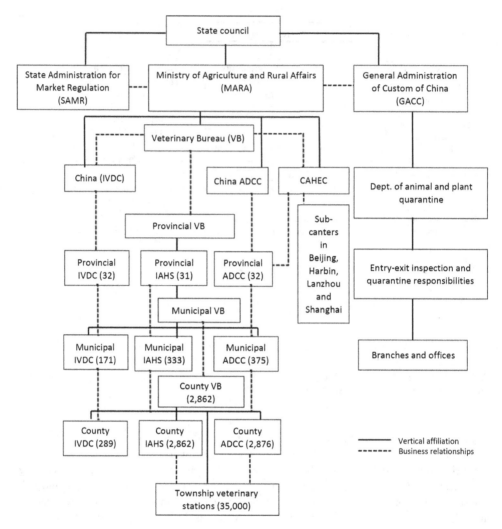

Figure 15.3 The organisation of China's (animal health) veterinary service after the 2018 structural reform.

Abbreviations: ADCC = Animal Disease Control Centre, CAHEC = China Animal Health and Epidemiology Centre, IAHS = Institute of Animal Health Supervision, IVDC = Institute of Veterinary Drug Control.

Sources: The number of institutions is cited from Animal Health in China 2012 (Veterinary Bureau, 2014, 4) and a study of 'Biosecurity and Disease Management in China's Animal Agriculture Sector' (Wei et al., 2015, 57).

livestock products for disease prevention (Wei et al., 2015, 57). After the 2018 reform[4] to China's Food and Drug Administration, GACC is responsible for the inspection and quarantine of animals and animal products that enter and exit China (GACC, 2019). In 2018, the state council launched the institutional reform plan to restructure government departments. For instance, the General Administration of Quality Supervision, Inspection and Quarantine (AQSIQ) was reorganised into the separate entities of the State Administration for Market Regulation (SAMR)[5] and the GACC in order to improve administrative efficiency and speed up the process of animal health inspection. The integration of the inspection and quarantine bureaux and GACC was undertaken to facilitate disease testing and allow for the development of platforms for joint examinations, as well as to improve administrative processes for animal importation (GACC, 2019).

Brief analysis of the pig and poultry sectors is useful at this point to illustrate how the national regime for animal health translates in practice. The Chinese state has been attempting to modernise pig production via subsidies for sow insurance, disease-prevention services, improving pig breeds and standardising the scale of production (The State Council, 2007). Measures implemented by MARA include a nationwide immunisation plan against animal epidemic diseases through the enforcement of compulsory vaccinations against three major pig diseases: porcine reproductive respiratory syndrome (PRRS), foot and mouth disease and swine fever (MARA, 2019). The scheme provides vaccines to farmers as well as training courses to improve pig farmers' animal health procedures. The appropriate use of vaccines is also promoted to address the challenges of antimicrobial resistance, as well as reduce disease incidence. However, despite these vaccine interventions, there have been continued disease outbreaks, such as the PRRS epidemic in 2006–2007. Observers believe that the PRRS outbreak started in 'backyards, smallholdings and medium-scale pig farms and then spread to intensive [commercial] pig farms' (Zhou and Yang, 2010, 32).

Turning to the biosecurity management of poultry, it is again important to recognise structural changes that have happened in the industry in recent decades (Pi et al., 2014, 11). The Chinese state has, since the 1990s, sought to industrialise and increase the scale of production in the poultry industry. This has been done by providing subsidies to farmers to convert backyard poultry farms into large concentrated poultry feeding operations, alongside strict drugs usage and waste management controls (Hu et al., 2017). The Chinese state introduced measures to increase the application of rapid diagnostics, enhance surveillance systems and improve the training of poultry workers (Swayne et al., 2000) and also implemented culling and distributed free H5N1 avian influenza vaccines.

In summary, China has enhanced its biosecurity measures and improved its disease management. However, several major challenges remain regarding the control of invasive alien species and infectious diseases. First, the state-centric biosecurity approach may not effectively improve biosecurity outcomes because the central state does not exercise a high degree of control over implementation and monitoring processes (Kostka, 2016; Kostka and Nahm, 2017). This produces wide gaps between policy goals and the implementation outcomes at local levels. Second, and related, the implementation of biosecurity measures can be difficult because of fragmented governance in terms of 'non-specific and unclear languages' (Ran, 2017, 639) and contradictions between policy objectives and outcomes. The weak and limited authority of biosecurity departments at the local level means they have difficulty implementing regulations (Ran, 2017). Third, the local government at county and township levels adjusts central state policies by evaluating trade-offs between environmental goals and local livelihoods (Zinda et al., 2017, 120) and renegotiating rules and regulations with local actors based on interpersonal relationships and cultural norms (Chan and Enticott, 2019; Mertha, 2009). Local government may pay more attention to short-term economic gains by increasing the intensity of animal farming at the expense of increasing the risks of disease outbreaks. Fourth, weak legal enforcement and a number of inconsistencies in implementation of biosecurity measures pose major challenges to China's biosecurity system, including insufficient veterinary training, overlapping roles at a governmental level, poor coordination among governing agencies in charge of the regulation of alien plants (e.g. MORA, NFGA, GACC) and poor monitoring and early warning mechanisms (Jiang et al., 2017; Wang et al., 2017; Wei et al., 2015). Fifth, there is a lack of institutional capacity to implement educational outreach and raise public awareness of disease prevention (Kaufman, 2008).

Conclusions

This chapter examined biosecurity systems in the UK and China to (1) compare implementation and monitoring of biosecurity measures and (2) consider relations with international and local actors. The UK's biosecurity regime involves experts, citizens and farmers to govern disease. At a policy level, there is a desire to shift costs and responsibilities away from the state. In contrast, China's biosecurity regime is characterised as a more centralised 'command and control' enforcement approach. Biosecurity policies are delegated centrally, and local states are enforced by the central state to execute biosecurity policy mandates. This state-centric approach may not effectively improve biosecurity outcomes because the central state does not exercise a high degree of control over implementation and monitoring (Kostka, 2016). This creates a wide gap between policy goals and the implementation outcomes at a local level. In the UK, there are opportunities for citizen participation and independent third parties (i.e. the involvement of public and independent parties) to monitor biosecurity practices, and collaborative models of governance are supported, even though, in practice, they are sometimes challenged because of an expectation of continued state involvement.

In China socio-cultural dynamics highlight the innate complexity of China's biosecurity system, with a gap between the 'rigid', 'top-down' national-scales and 'flexible' and 'negotiable' local levels of biosecurity practices. Interpersonal relationships influence the actual practices of biosecurity policies, and there is a lack of independent third-party presence to monitor the effectiveness of biosecurity practices at a local level. The local state can manipulate biosecurity data without genuinely complying with the central state policy goals and mandates. Pig farmers, for example, renegotiate with state veterinary officials. Smallholders may not be able to access the state's subsidies or other supporting policies because they are unable to develop close relationships with government officials (Chan and Enticott, 2019). This has widened the social and economic gaps between smallholders and corporate animal breeders and state funding support.

Political imperatives have significant consequences for national governance structures that are established to combat plant and animal disease threats, including how trade is regulated. Both the UK and China adopt 'precautionary approaches' to biosecurity, and both are influenced by neoliberalism. However, China actively develops its own biosecurity measures to fulfil international standards, ensuring its animal health standards align with global requirements to maintain transnational trading of meat products while maintaining an essentially state-centric biosecurity system (Chan and Flynn, 2018). For the UK – or any other European Union member country – national biosecurity is nested within EU legislation rather than defined by the state. The decision-making process for Chinese national biosecurity controls on fauna and flora is a state-driven 'precautionary' approach. The central and local state collaborates with scientists to collect scientific evidence and uses technical knowledge to identify 'risk factors' in order to manipulate and reduce disease outbreaks. The tendency is to favour risk-based approaches and border control management. However, this ignores the relational qualities of disease. Infectious diseases exist everywhere. Comparative analysis of national biosecurity regimes helps problematise normalised ontologies of disease that draw a boundary to wall off against disease. Generalised understandings of 'state-centred' and 'neoliberal' styles of governance related to animal and plant disease management are also flawed. Political, socio-economic and cultural factors shape national animal and plant biosecurity regimes and represent forms of institutional blending, with policy rhetoric playing out differently in specific contexts, spaces and places.

Notes

1 A lethal fungal disease caused by the soil-borne fungus *Fusarium oxysporum* f. sp. *cubense* (Foc).
2 Since 2004, the central state has implemented the 'building a new socialist countryside policy' to abolish agricultural tax and school fees, establish cooperative medical systems and extend electricity networks in rural areas as well as organise farmers into farmers' cooperatives.
3 In 2006, the National People's Congress named the *Sanlong Wenti* (Three Rural Problems), which refers to the problems of (1) low agricultural productivity, (2) rural-urban disparity in terms of income and culture and (3) farmers' burdens and low quality of life.
4 The 13th National People's Congress released the plan to launch the structural reform to China's food and drug administration. Under this plan, the State Administration for Market Regulation will take all responsibilities previously held by the State Administration for Industry and Commerce (SAIC); the General Administration of Quality Supervision, Inspection and Quarantine (AQSIQ); and the State Administration of Food and Drug Administration (CFDA).
5 SAMR is in charge of food and industrial product safety, quality inspection, certification and accreditation in China (Swiss, 2018a, 2018b).

References

Alirol, E., Getaz, L., Stoll, B., Chappuis, F. and Loutan, L. (2011) 'Urbanisation and infectious diseases in a globalised world', *The Lancet Infectious Diseases*, vol. 11, no. 2, pp. 131–141.

Barker, K. (2010) 'Biosecure citizenship: Politicising symbiotic associations and the construction of biological threat', *Transactions of the Institute of British Geographers*, vol. 35, pp. 350–363.

Barr, M. (2009) 'The importance of China as a biosecurity actor', in B. Rappert and C. Gould (eds.) *Biosecurity: Origins, Transformations and Practice*. Palgrave Macmillan, Basingstoke, pp. 121–132.

Barr, M. and Zhang, J.Y. (2010) 'Bioethics and biosecurity education in China: Rise of a scientific superpower', in B. Rappert (ed.) *Education and Ethics in the Life Sciences: Strengthening the Prohibition of Biological Weapons*. Australian National University Press, Canberra, pp. 115–129.

Bermouna, S. and Li, J. (2014) 'China's agricultural project finance and support policies: The framework of China's major agricultural subsidies', *European Food and Feed Law Review*, vol. 9, p. 171.

Brasier, C. (2008) 'The biosecurity threat to the UK and global environment from international trade in plants', *Plant Pathology*, vol. 57, pp. 792–808.

Braun, B. (2007) 'Biopolitics and the molecularization of life', *Cultural Geographies*, vol. 14, no. 1, pp. 6–28.

Chan, K.W. (2015) *Contesting Urban Agriculture: The Politics of Meat Production in the License-Buy-Back Scheme (2006–2007) in Hong Kong*. Routledge, London, pp. 294–314.

Chan, K.W. and Enticott, G. (2019) 'The Suzhi farmer: Constructing and contesting farming Subjectivities in post-Socialist China', *Journal of Rural Studies*, vol. 67, pp. 69–78.

Chan, K.W. and Flynn, A. (2018) 'Food production standards and the Chinese local state: Exploring new patterns of environmental governance in the Bamboo Shoot Industry in Lin'an', *The China Quarterly*, pp. 1–27.

Chen, C., Huang, D. and Wang, K. (2015) 'Risk assessment and invasion characteristics of alien plants in and around the agro-pastoral ecotone of northern China', *Human and Ecological Risk Assessment: An International Journal*, vol. 21, no. 7, pp. 1766–1781.

Day, A.F. and Schneider, M. (2017) 'The end of alternatives? Capitalist transformation, rural activism and the politics of possibility in China', *The Journal of Peasant Studies*, pp. 1–25.

DBS Asian Insights (2017) 'China pork: A meaty task to meet demand', *DBS Asian Insights: Sector Briefing*, vol. 45, pp. 1–67.

Defra (2014) *Protecting Plant Health: A Plant Biosecurity Strategy for Great Britain*. Defra, London.

Defra (2018) *Contingency Plan for Exotic Notifiable Diseases of Animals in England*. Defra, London.

Donaldson, A. (2008) 'Biosecurity after the event: Risk politics and animal disease', *Environment and Planning A*, vol. 40, pp. 1552–1567.

Duckett, D., Wynne, B., Christley, R.M., Heathwaite, A.L., Mort, M., Austin, Z., Wastling, J.M., Latham, S.M., Alcock, R. and Haygarth, P. (2015) 'Can policy be risk based? The cultural theory of risk and the case of livestock disease containment', *Sociologia Ruralis*, vol. 55, pp. 379–398.

Enticott, G. (2014a) 'Relational distance, neoliberalism and the regulation of animal health', *Geoforum*, vol. 52, pp. 42–50.

Enticott, G. (2014b) 'Biosecurity and the bioeconomy: The case of disease regulation in the UK and New Zealand', in T.K. Marsden and A. Morley (eds.) *Sustainable Food Systems: Building a New Paradigm.* Earthscan, London, pp. 122–142.

FAO – Food and Agriculture Organization of the United Nations (2021) 'International Plant Protection Convention (IPPC)'s list of countries', https://www.ippc.int/en/countries/all/list-countries/, accessed 5 January 2021.

Feldbaum, H. and Michaud, J. (2010) 'Health diplomacy and the enduring relevance of foreign policy interests', *PLoS Medicine*, vol. 7, no. 4, p. e1000226.

Fish, R., Austin, Z., Christley, R., Haygarth, P., Heathwaite, L., Latham, S., Medd, W., Mort, M., Oliver, D., Pickup, R., Wastling, J. and Wynne, B. (2011) 'Uncertainties in the governance of animal disease: An interdisciplinary framework for analysis', *Philosophical Transactions of the Royal Society B: Biological Sciences*, vol. 366, pp. 2023–2034.

GACC – General Administration of Customs People's Republic of China (2019) 'Overview of GACC', http://english.customs.gov.cn/about/mission, accessed 19 January 2019.

Gilbert, M., Golding, N., Zhou, H., Wint, G.W., Robinson, T.P., Tatem, A.J., . . . and Huang, Z. (2014) 'Predicting the risk of avian influenza A H7N9 infection in live-poultry markets across Asia', *Nature Communications*, vol. 5, p. 4116.

Godfray, C., Donnelly, C.A., Hewinson, G., Winter, M. and Wood, J. (2018) *Bovine TB Strategy Review.* Defra, London.

He, S., Yin, L., Wen, J. and Liang, Y. (2018) 'A test of the Australian Weed Risk Assessment system in China', *Biological Invasions*, vol. 20, pp. 2061–2076.

Hinchliffe, S., Allen, J., Lavau, S., Bingham, N. and Carter, S. (2013) 'Biosecurity and the topologies of infected life: From borderlines to borderlands', *Transactions of the Institute of British Geographers*, vol. 38, no. 4, pp. 531–543.

Hinchliffe, S., Bingham, N., Allen, J. and Carter, S. (2016) *Pathological Lives: Disease, Space and Biopolitics.* Wiley-Blackwell, Oxford.

House of Lords (2018) *Brexit: Plant and Animal Biosecurity.* House of Lords European Union Committee 21st Report of Session 2017–19. HL Paper 191. Authority of the House of Lords, London.

Hu, Y., Cheng, H. and Tao, S. (2017) 'Environmental and human health challenges of industrial livestock and poultry farming in China and their mitigation', *Environment International*, vol. 107, pp. 111–130.

Jiang, M., Zhan, A., Guo, H. and Wan, F. (2017) 'Research and management of biological invasions in China: Future perspectives', in *Biological Invasions and Its Management in China.* Springer, Singapore, pp. 239–247.

Kaufman, J.A. (2008) 'China's heath care system and avian influenza preparedness', *The Journal of Infectious Diseases*, vol. 197, no. Supplement_1, pp. S7–S13.

Koblentz, G.D. (2010) 'Biosecurity reconsidered: Calibrating biological threats and responses', *International Security*, vol. 34, no. 4, pp. 96–132.

Kostka, G. (2016) 'Command without control: The case of China's environmental target system', *Regulation & Governance*, vol. 10, no. 1, pp. 58–74.

Kostka, G. and Nahm, J. (2017) 'Central–local relations: Recentralization and environmental governance in China', *The China Quarterly*, vol. 231, pp. 567–582.

Krychevska, L., Shynkarenko, I. and Shynkarenko, R. (2017) 'Agricultural insurance in China: History, development and success factors', http://agroinsurance.com/wp-content/uploads/2017/04/Agricultural-insurance-in-China-Agroinsurance-International.pdf, accessed 20 October 2019.

Lerner, H. (2017) 'Conceptions of health and disease in plants and animals', in T. Schramme and S. Edwards (eds.) *Handbook of the Philosophy of Medicine.* Springer, Dordrecht, pp. 287–301.

Li, Z.H., Zhu, S.F. and Wan, F.H. (2014) 'Domestic regulatory framework and invasive alien species in China', in T. Schramme and S. Edwards (eds.) *Handbook of the Philosophy of Medicine.* Springer, Dordrecht, pp. 45–72.

Lin, B. and Shen, H. (2017) 'Burrowing Nematode Radopholus similis (Cobb)', in F. Wang, M. Jiang and A. Zhan (eds.) *Biological Invasions and Its Management in China.* Springer, Singapore, pp. 23–31.

MacLeod, A., Pautasso, M. and Jeger, M. (2010) 'Evolution of the international regulation of plant pests and challenges for future plant health', *Food Security*, vol. 2, pp. 49–70.

MARA – Ministry of the Agriculture of the People's Republic of China (2019) 'The organization structure and departments of the MARA', http://english.agri.gov.cn/aboutmoa/departments/, accessed 20 October 2019.

Maye, D., Dibden, J., Higgins, V. and Potter, C. (2012) 'Governing biosecurity in a neoliberal world: Comparative perspectives from Australia and the United Kingdom', *Environment and Planning A*, vol. 44, no. 1, pp. 150–168.

Maye, D., Enticott, G., Fisher, R., Ilbery, B. and Kirwan, J. (2014) 'Animal disease and narratives of nature: Farmers' reactions to the neoliberal governance of bovine Tuberculosis', *Journal of Rural Studies*, vol. 36, pp. 401–410.

Meadows, A.J., Mundt, C.C., Keeling, M.J. and Tildesley, M.J. (2018) 'Disentangling the influence of livestock vs. farm density on livestock disease epidemics', *Ecosphere*, vol. 9, no. 7, p. e02294.

Mertha, A. (2009) '"Fragmented authoritarianism 2.0": Political pluralization in the Chinese policy process', *The China Quarterly*, vol. 200, pp. 995–1012.

MMTA – Minor Metal Trade Association (2018) 'Latest updates on China's trade policy', https://mmta.co.uk/2018/08/03/11717/, accessed 19 January 2019.

Morton, K. (2017) 'Learning by doing: China's role in the global governance of food security', in S. Kennedy (ed.) *Global Governance and China – The Dragon's Learning Curve*. Routledge, Abingdon, Oxon, pp. 207–227.

Mu, X., Hu, Y., Song, H., Wang, P. and Luo, J. (2010) 'Damage and management of Alien Species in China', *Journal of Agricultural Science*, vol. 2, no. 1, p. 188.

MWE China Law Office (2018) 'Tracking China's structural reform of food and drug', www.mwechinalaw.com/en/publicationeventlanding/publications/2018/china-structural-reform-fda, accessed 19 January 2019.

NFGA – National Forestry Grassland Administration – Department Structure, www.forestry.gov.cn/, accessed 20 October 2019.

Normile, D. (2018) 'Can China, the world's biggest pork producer, contain a fatal pig virus? Scientists fear the worst', *Science*, www.sciencemag.org/news/2018/08/can-china-world-s-biggest-pork-producer-contain-fatal-pig-virus-scientists-fear-worst, accessed 20 November 2018.

OIE (2019) 'OIE-listed diseases, infections and infestations in force in 2019', www.oie.int/animal-health-in-the-world/oie-listed-diseases-2019/, accessed 18 October 2019.

O'Rourke, J. (2009) 'Bluetongue-Defra policy in action', *Government Veterinary Journal*, vol. 20, no. 2, pp. 11–15.

Ou, J., Lu, C. and O'Toole, D.K. (2008) 'A risk assessment system for alien plant bio-invasion in Xiamen, China', *Journal of Environmental Sciences*, vol. 20, no. 8, pp. 989–997.

Pi, C., Rou, Z. and Horowitz, S. (2014) 'Fair or fowl? Industrialization of poultry production in China', *Institute for Agriculture and Trade Policy*, vol. 9, pp. 15–16.

Potter, C., Harwood, T., Knight, J. and Tomlinson, I. (2011) 'Learning from history, predicting the future: The UK Dutch Elm disease outbreak in relation to contemporary tree disease threats', *Philosophical Transactions of the Royal Society B*, vol. 366, pp. 2045–2053.

Qiang, W. (2007) 'Efforts to strengthen biosafety and biosecurity in China', Department of Arms Control and Disarmament of China's Ministry of Foreign Affairs, www.nonproliferation.org/wp-content/uploads/2014/02/070917_wang.pdf, accessed 19 October 2019.

Qiu, W., Chu, C., Mao, A. and Wu, J. (2018) 'Studying communication problems for emergency management of SARS and H7N9 in China', *Journal of Global Infectious Diseases*, vol. 10, no. 4, p. 177.

Ran, R. (2017) 'Understanding blame politics in China's decentralized system of environmental governance: Actors, strategies and context', *The China Quarterly*, vol. 231, pp. 634–661.

Salter, B. and Waldby, C. (2011) 'Biopolitics in China: An introduction', *East Asian Science, Technology and Society: An International Journal*, vol. 5, no. 3, pp. 287–290.

Schneider, M. (2017) 'Wasting the rural: Meat, manure, and the politics of agro-industrialization in contemporary China', *Geoforum*, vol. 78, pp. 89–97.

SCMP (South China Morning Post) (2018) 'China's US$128 billion pork industry is under threat by a deadly, mysterious virus', www.scmp.com/business/commodities/article/2164631/chinas-us128-billion-pork-industry-under-threat-deadly, accessed 21 November 2018.

Scott, S., Si, Z., Schumilas, T. and Chen, A. (2014) 'Contradictions in state-and civil society-driven developments in China's ecological agriculture sector', *Food Policy*, vol. 45, pp. 158–166.

Smart, A. and Smart, J. (2012) 'Biosecurity, quarantine and life across the border', *A Companion to Border Studies*, vol. 26, p. 354.

The State Council (2007) Opinions of the State Council on Promoting the Production and Development of Live Pigs and Stabilizing the Market Supply No. 22 C.F.R.

The State Council (2019) Opinions of the State Council on Strengthening Animal Diseases Prevention System and Providing Clear Demands, www.gov.cn/xinwen/2019-07/04/content_5406111.htm, accessed 11 October 2019.

Swayne, D.E., Garcia, M., Beck, J.R., Kinney, N. and Suarez, D.L. (2000) 'Protection against diverse highly pathogenic H5 avian influenza viruses in chickens immunized with a recombinant fowlpox vaccine containing an H5 avian influenza hemagglutinin gene insert', *Vaccine*, vol. 18, no. 11–12, pp. 1088–1095.

Swiss Business Hub China (2018a) 'The State Administration for Market Regulation (SAMR): A clear structure unveiled', www.s-ge.com/sites/default/files/publication/free-form/samr-a_clearer_structure_unveiled-s-ge.pdf, accessed 19 January 2019.

Swiss Business Hub China (2018b) 'China's New State Market Regulatory Administration: State Administration for Market Regulation (SAMR)', www.s-ge.com/sites/default/files/publication/free-form/overview-of-effects-on-swiss-companies-s-ge.pdf, 19 January 2019.

Tsouvalis, S. (2019) 'The post-politics of plant biosecurity: The British Government's response to ash dieback in 2012', *Transactions of the Institute of British Geographers*, vol. 44, no. 1, pp. 195–208.

Veterinary Bureau (2014) *Ministry of Agriculture: Animal Health in China 2012*. China Agricultural Publishing House, Beijing.

Waage, J.K. and Mumford, J.D. (2008) 'Agricultural biosecurity', *Philosophical Transactions of the Royal Society B: Biological Sciences*, vol. 363, no. 1492, pp. 863–876.

Wang, R., Wan, F. and Li, B. (2017) 'Roles of Chinese government on prevention and management of invasive Alien Species', in F. Wang, M. Jiang and A. Zhan (eds.) *Biological Invasions and Its Management in China*. Springer, Dordrecht, pp. 149–156.

Wei, X., Lin, W. and Hennessy, D.A. (2015) 'Biosecurity and disease management in China's animal agriculture sector', *Food Policy*, vol. 54, pp. 52–64.

Wilkinson, K., Grant, W., Green, L., Hunter, S., Jeger, M., Lowe, P., Medley, G., Mills, P., Phillipson, J., Poppy, G. and Waage, J. (2011) 'Infectious diseases of animals and plants: An interdisciplinary approach', *Philosophical Transactions of the Royal Society B: Biological Sciences*, vol. 366, pp. 1933–1942.

Williamson, M. (2006) 'Explaining and predicting the success of invading species at different stages of invasion', *Biological Invasions*, vol. 8, pp. 1561–1568.

Woods, A. (2011) 'A historical synopsis of farm animal disease and public policy in twentieth century Britain', *Philosophical Transactions of the Royal Society B: Biological Sciences*, vol. 366, no. 1573, pp. 1943–1954.

Wu, Z. (2019) *Antibiotic Use and Antibiotic Resistance in Food-Producing Animals in China* (No. 134). OECD Publishing, Paris. https://doi.org/10.1787/18156797

Zhou, L. and Yang, H. (2010) 'Porcine reproductive and respiratory syndrome in China', *Virus Research*, vol. 154, nos. 1–2, pp. 31–37.

Zinda, J.A., Trac, C.J., Zhai, D. and Harrell, S. (2017) 'Dual-function forests in the returning farmland to forest program and the flexibility of environmental policy in China', *Geoforum*, vol. 78, pp. 119–132.

16

THE FUTURE OF BIOSECURITY SURVEILLANCE

Evangeline Corcoran and Grant Hamilton

Biosecurity surveillance

Biosecurity surveillance broadly refers to the collection and analysis of biosecurity data. While the International Standards for Phytosanitary Measures defines surveillance as 'an official process which collects and records data on pest occurrence or absence by survey, monitoring or other procedure', this omits explicit mention of important aspects of the analysis that inform the *capture* of data and the *analysis* of that data. Both these components critically inform biosecurity data collection. An effective surveillance system can help prevent environmental and economic impacts of new pests and diseases and mitigate the impacts of those that are already present.

Quarantine, containment and eradication are all vital in the biosecurity system; however, none of these could function properly without effective biosecurity surveillance (Colunga-Garcia et al., 2013; Jarrad et al., 2015; Poland and Rassati, 2019). While sometimes missing from definitions, a key qualification is that such surveillance should be guided by specific biosecurity objectives to ensure that data collection and analysis are well targeted and therefore efficient. In a narrow sense, the intention of biosecurity surveillance is the early warning, detection and identification of threats to agriculture, horticulture and silviculture, although more recently (and importantly), the environment has been added to this list in a number of countries. The greatest economic and environmental benefits will accrue with early detection. Biosecurity surveillance data can be collected to demonstrate pest absence, occurrence, abundance and distribution (Baxter and Hamilton, 2016). The practices used to collect and analyse this data are classified based on a number of factors including invasion stage, target species or species group and mode of action (Hulme, 2014).

Biosecurity surveillance can be divided into three categories based on the invasion stage they target: pre-border, border and post-border surveillance (Hulme, 2014; Poland and Rassati, 2019). Pre-border surveillance practices aim to prevent invasion of non-native species entirely and relate to the development of policies to enable safe importing of commodities (Colunga-Garcia et al., 2013). Border surveillance aims to identify invasive species and pathogens at the arrival stage, and post-border surveillance targets the early establishment stage in order to aid the implementation of management strategies while population levels of non-native species remain low (Westphal et al., 2008). The latter can be carried out in specific infested areas, sometimes referred to as containment surveillance, or over larger spatial scales (Poland and Rassati, 2019).

Biosecurity surveillance can be further divided by target species or species group. Specific surveillance practices target a single species or very narrow species group while generic surveillance targets a broader range (Froud et al., 2008). Surveillance practices can be considered active if they are carried out regularly, often as part of a monitoring plan by pest specialists with direct consequences for management, or passive if they rely on non-expert and community reporting (Poland and Rassati, 2019).

The current state of biosecurity surveillance practices

Surveillance practices to date have primarily been used to determine the 'health status' of countries or regions based on whether a plant or animal pest is present or absent; there has therefore been a heavy focus on directing surveillance efforts towards early detection of incursions and confirmation of eradication (Jarrad et al., 2015). However, surveillance has also been used to a lesser extent to inform control and containment programmes for ongoing incursions at intermediate establishment phases (Noordhuizen et al., 2013). These two objectives have typically been addressed using a combination of detection, monitoring and delimiting surveys (Stephenson et al., 2003). *Detection surveys* are used at the earliest stages to determine if a pest is present in an area, *monitoring surveys* are typically conducted on ongoing incursions to gain insight into characteristics of pest populations such as abundance or rate of spread and *delimiting surveys* aim to establish boundaries of what areas are considered to be free or infested by a pest (Committee of Experts on Phytosanitary Measures, 1996). In practice, detection surveys will often be conducted after notification of an anomaly from a member of the community. The biosecurity information gathered from these three types of surveys is then used to run predictive models to better understand the movement of emerging pests, judge official pest status and guide decisions on risk in international trade.

Surveys for biosecurity surveillance are commonly conducted in the field, often with trained personnel on the ground directly identifying invasive species or diseased plants and animal pathogens by sight (Kalaris et al., 2014). Invasive animals are also often surveyed indirectly through signs such as scat, scratches and auditory signals, or traps may be set up in field sites to catch insects, collect samples of hair or catch animals on camera (Jarrad et al., 2010). Surveillance has also heavily relied on passively collected data supplied from growers, agricultural industry members, researchers and the public reporting pest detections or pathogen outbreaks through avenues such as toll-free telephone lines (Froud and Bullians, 2015).

Despite how commonly these survey techniques have been adopted, a number of issues concerning their effectiveness have been documented. For example, in cases where there are no viable traps or lures available for insect species of interest, officials are forced to rely on visual surveys, which have been found to be inaccurate in many cases (Kalaris et al., 2014). There are also significant costs for conducting pathogen surveillance, with reviewers suggesting current detection technologies are inadequate, resulting in suboptimal outcomes for plant and animal disease eradication (Kalaris et al., 2014). To date, many pests have been introduced to regions and remained undetected until it was too late for them to be easily eradicated, which suggests current surveillance practices are in need of improvement (Noordhuizen et al., 2013). A number of methods and technologies have been developed to support survey efforts involving statistics and geographical information sciences (GIS) methods and data management technologies. However, high costs and limited resources to conduct surveys that result in capture of high quality data remain barriers to exploiting the full potential of these analysis tools (Kalaris et al., 2014; Poland and Rassati, 2019).

With the growing conceptual and operational importance of biosecurity making it a priority for many government agencies, most surveillance practices are carried out by national authorities

(Goldson et al., 2010). However, surveillance programmes at the national level have previously been criticised for a lack of thorough and conclusive information on pest status or adequate data management (Convention on Biological Diversity Secretariat [CBD] 2001). National trade decisions have also been found to be biased by political will, rather than based on empirical data from surveillance (CBD, 2001). Conversely, regional and international entities have been critical in ensuring surveillance practices are successful by providing advanced alert systems and evidence-based guidelines that can be applied to biosecurity management at the national level. For example, the European and Mediterranean Plant Protection Organisation and North American Plant Protection Organisation have provided early warning systems that have successfully facilitated identification of pest risks in many countries (Maye et al., 2012). In terms of animal health, regional entities such as the European Food Safety Authority have guided effective management of international food safety crises such as incursions of bovine spongiform encephalopathy (commonly known as mad cow disease) and dioxins in food products (Deluyka and Silano, 2012).

In recent years many organisations have attempted to enhance surveillance for plant and animal health. These efforts have included exploring the factors most likely to result in increased effectiveness of surveillance practices in detecting pest species, guiding control and confirming eradication of incursions (International Plant Protection Convention (IPPC) Secretariat, 2012). The top priority factors have been determined as increased financial and human resources, increased training, awareness and inter-organisational cooperation, with improved survey operating procedures, equipment and information technology also deemed to be of high importance (IPPC Secretariat, 2012). The role of risk analysis and management decision support tools has been emphasised, with some progress having been made in this area through special funding of research and development programmes (Jarrad et al., 2015). One example of an advance that resulted from this is the development of a phytosanitary risk rating for individual countries (Jarrad et al., 2015).

Future needs for surveillance practices

As domestic and international trade increase among growing global populations, there are more opportunities for biological incursions and therefore an increasing need to predict, prevent and mitigate the impacts of these events through good surveillance practice (Jarrad et al., 2015). With the potential increase in incursions, there is also a need for greater efficiency in surveillance practices in order to respond to these events, so trialling new surveillance technologies to make faster, better informed decisions is clearly important (Baxter and Hamilton, 2016). Adoption of new technologies needs to be contingent on their accuracy, reliability and cost benefits.

Advanced technologies cannot improve detection and management in isolation as there will always be a need to manipulate the data they capture in order to extract the relevant information. More sophisticated and specialised modelling methods will therefore be needed to aid this process and collect the most useful insights from new surveillance devices. Novel decision-making tools should then be developed from these insights to guide management actions and ensure the most beneficial biosecurity outcomes. With this in mind, throughout the rest of the chapter, we evaluate and discuss future directions for surveillance practices over the entire data life cycle from data capture to analysis to informing decisions and the possible constraints on these emerging technologies and methods.

Emerging data capture methods

Collection of data is critical to many surveillance objectives, from initial detection to tracking the spread of pests and finally confirming their eradication. Increased opportunities for more frequent and widespread incursions in the future mean that data capture methods will need to keep up, facilitating detection as early as possible, responding more quickly to change and covering more expansive areas. Research investment has led to the development of equipment and technologies that may enable surveillance practices to match this demand for data, although it is important to remember that increased reliance on these will necessitate regular and significant investment in their development, maintenance and handling.

Diagnostics

The development of mobile and rapid diagnostic techniques had enabled early in situ detection of many pests, providing capability for surveillance to reach more remote areas and faster response times to contain incursions (Baxter and Hamilton, 2016). For instance, advancements in molecular testings such as real-time polymerase chain reaction (PCR) assays have allowed identification and quarantine of imported fruit infested with invasive fly long before species identification by morphology is possible (Schaad et al., 2003; Li et al., 2018). The recent development of standardised multispecies DNA barcoding methods also provides greater flexibility in molecular diagnostics to respond to changing pest priorities and adaptation to various countries or regions without the need to develop a specific test for every area or species (Armstrong and Ball, 2005; Piper et al., 2019). Multispecies DNA barcoding has proven of particular use for monitoring invasive insects, with the technique providing the opportunity for surveillance plans to be dramatically scaled up as a very large quantity of insect traps can be processed in a small amount of time (Piper et al., 2019).

Outside of molecular testing, there has been success in trialling the use of other organisms to detect pest species in what are termed 'biosurveillance' practices (Poland and Rassati, 2019). Thus far, two approaches have been tested using the perception of dogs (Hoyer-Tomiczek et al., 2016) and predatory wasps (Careless et al., 2014). Dogs are capable of detecting minute traces of target scents, including those of many insects, and have been used to actively survey points of entry for specific species such as the Asian longhorned beetle (Hoyer-Tomiczek et al., 2016) and the red palm weevil (*Rhynchophorus ferrugineus*; Suma et al., 2014). Predatory wasps have been shown to be more useful for broadscale surveillance as they are less limited in terms of work hours and are capable of detecting insect scents in a wider range of settings (Careless et al., 2014; Rutledge et al., 2011). By analysing the community of prey species found in the nest of wasp species such as the buprestid-hunting wasp (*Cerceris fumipennis*), insights can be gained into the presence and abundance of cryptic pest species (Careless et al., 2014; Nalepa et al., 2015).

Electronic devices have also been developed to detect the presence of volatile organic compounds released by damaged or stressed plants and pheromones emitted by insects (Cellini et al., 2017; Rock et al., 2008). These 'electronic noses' may be suitable for border surveillance, with portable models being used to distinguish damaged plants or insect-infested goods at points of entry, as well as potentially being useful for post-border containment with regularly placed devices monitoring spread (Wilson, 2013).

Citizen science

The increased distribution of mobile devices provides an opportunity to capture large quantities of passive surveillance data over wide areas (Dehnen-Schmutz et al., 2016). This can be used to inform biosecurity management when supplied directly, such as when landowners and members of the public were encouraged to report the occurrence of acute oak decline disease on their property (Baker et al., 2018). Alternatively, it can be gathered indirectly from social media trends through methods such as text mining, demonstrated in a study that showed how migratory moth incursions could be successfully identified by searching for their common names in Twitter data (Caley et al., 2019; Welvaert et al., 2017). The number of citizen science initiatives that monitor ecological data have risen steadily over the past decade and have been found to be capable of detecting invasive species at early stages, allowing for a timely management response (Baker et al., 2018)

With values shifting towards protection of the natural environment being seen as a responsibility that should be shared by all, the role of non-experts in contributing data for biosecurity is likely to grow. Industry organisations will have a particularly important role to play in engaging members of agribusiness in contributing to surveillance. This can be achieved through developing interfaces that are straightforward and easy to use and promoting the direct and financial benefits of improved biosecurity to growers and agronomists (Baxter and Hamilton, 2016). Wider participation of the public could also be fostered through user-friendly interfaces but will also rely on garnering public support for and buy-ins to biosecurity measures (Baxter and Hamilton, 2016). Indigenous communities can also provide valuable information, as in many remote areas they may be the first responders to pest incursions (Black et al., 2019; see Lambert and Mark-Shadbolt, this collection). First nations communities also offer a promising avenue of societal support for biosecurity, although relationships with pest species may be complex (Black et al., 2019).

Remote sensing

In recent years there have been rapid advances in technologies that enable collection of high-resolution data from a distance, particularly in relation to aerial imaging (Kellenberger et al., 2018). For broadscale surveillance, enhanced satellite coverage and resolution of up to 30 centimetres ground distance offers the potential for improved predictions of species distribution (Kim and Beresford, 2009). For example, integrated satellite images and geographic information systems data have been used to map the risk of dwarf bunt disease across the entirety of New Zealand (Kim and Beresford, 2009). Satellites are also capable of picking up signals outside the visible spectrum. This can improve quantification of soil quality, water availability, meteorological events and vegetation health. For example, a recent study demonstrated it is feasible to detect and differentiate between plant diseases including cassava brown streak disease, cassava mosaic disease and sooty mould in satellite images (Sims et al., 2018).

Unmanned aerial vehicles, also known as UAVs or drones, may improve data capture for more specific surveillance. These platforms are portable and easy to manoeuvre and can be fitted with a wide variety of sensors that can detect signals across visible and non-visible spectra for a relatively low cost compared to satellite surveys (Hollings et al., 2018). Unlike satellites they can sample at very close range but without the direct contact involved in ground surveys. This means that the potential for spread of contaminants from infested sites is reduced, as well as making surveys safer for researchers who do not need to navigate unfavourable terrain. Image capture from UAVs could also be supplemented by ground-based cameras or drones, to add complimentary near-scale or under-leaf data (Shi et al., 2016). UAV technology, while yet to reach its full potential, has been the subject of intense interest around the world, which has led to a rapid expansion in experimentation and use. Further research is needed to investigate the most accurate and efficient ways to conduct biosecurity surveillance surveys with these platforms (Baxter and Hamilton, 2018). While individual trials are currently being carried out for detection of a variety of pest species, there is yet to be a coherent understanding of the utility of this technology in many areas.

Automation and artificial intelligence

While biosecurity surveillance data are conventionally collected manually, these methods can be very time-consuming, labour intensive and expensive to carry out (Jurdak et al., 2015). These practices are therefore often restricted to small scales and occur infrequently, meaning that biosecurity surveillance may not be able to keep up with an increased likelihood of incursions in the future (Jurdak et al., 2015). Autonomous sensors, robotic systems and algorithms could reduce cost, time and bias in biosecurity data collection and widen the scale and enhance the temporal resolution at which data can be gathered (Velusamy et al., 2012).

Automated sensors can be set up in field sites or at points of entry to capture, store and transmit data in response to specific stimuli with minimal human involvement and have been shown to provide finer-grain spatio-temporal event information for less battery power than sensors that collect data at regular timed intervals (Jurdak et al., 2013). This kind of technology was recently used to detect and track flying foxes, which can be vectors of zoonotic disease, over time and space (Jurdak et al., 2013). Automated cameras that activate in response to movement have also been tested inside insect traps, allowing in situ species identification and detection of fruit fly pests (Jurdak et al., 2015). In terms of robotics, autonomous drones that can be pre-programmed to conduct surveys can aid quick detection of invasive species and pathogens over large and inaccessible areas (CSIRO, 2014). For example, these aerial platforms were used to detect early stages of *Miconia* invasions in tropical forests (CSIRO, 2014).

Examination of remotely sensed data is typically done manually to see if target organisms or pathogens can be detected. However, automated methods have been shown to be capable of detecting target species in audio and images faster and with a reduced likelihood of false negative results (Hodgson et al., 2018). Progress in machine learning (otherwise known as artificial intelligence) such as deep-learning neural networks means that automated detection is less resource intensive than ever to implement as pre-existing general purpose algorithms can be retrained to detect specific species or signs of disease with a relatively small amount of relevant data (Rey et al., 2017). Such automated detection methods have been shown to detect wildlife in ecological surveys more efficiently and with less false negative results (Corcoran et al., 2019; Seymour et al., 2017; Witczuk et al., 2018) and could be applied to the detection of pest species in colour or thermal imagery collected from satellites or UAVs. These methods are not limited to image data; algorithms can also be trained to recognise audio signals (Demertzis et al., 2018). For example, the machine hearing framework has begun development to enable identification of invasive marine species by the sounds they produce under water (Demertzis et al., 2018)

New ways to understand data

Statistical modelling and analytics

As datasets are likely to get bigger and messier due to the increased implementation of passive and general surveillance methods, advanced statistical modelling methods may be needed to sort meaningful signals from noise (Baxter and Hamilton, 2016). Recent technological advances providing more computational power have enabled development of more efficient algorithms capable of extracting useful insights from large and complex datasets from multiple sources. This is important as models for biosecurity need to be adaptable to changing environmental conditions and unforeseen events in order to guide response to highly dynamic incursion events. Being able to input and exploit data on ongoing incursions rapidly means that managers can then predict when attempting eradication of a pest species is most feasible and cost-effective (Stanaway et al., 2011).

Bayesian statistical models are well designed for this purpose as they allow incorporation of existing data in the form of 'priors' with new information (Cook et al., 2007). They have also been shown to be very effective at accounting for uncertainty in observational data, such as that derived from active and passive biosecurity surveillance (Cressie et al., 2009). For example, Bayesian modelling frameworks were recently used to evaluate the effectiveness of early detection programmes in Queensland, Australia, and to map the probability of infestation across space to determine where surveillance resources could most effectively be deployed (Lopes et al., 2019).

Bayesian belief networks (BNNs), a specific kind of Bayesian modelling, could also prove very useful in the context of biosecurity surveillance (Froese et al., 2019; Miraballes et al., 2019). They enable modelling of very complex relationships or 'networks' in which one factor may influence and be influenced by many others to differing extents (Fenton and Neil, 2013). Biosecurity risks are driven by a multitude of underlying factors, so models that can account for all these factors and their relationship with each other are useful, especially for visualisation of how pests spread (Cope et al., 2019; Wiltshire et al., 2019). BNNs are also relatively easy for non-statisticians to interpret, with the arrows representing the direction of causal relationships between different variables and simple probability values representing the likelihood of different outcomes (i.e. risk of spread to certain areas) given different inputs (i.e. seasonal and environmental conditions, what biosecurity are in place; Fenton and Neil, 2013).

There is often a separation between the on-the-ground need for rapid action to contain an incursion once identified and the data requirements of complex statistical models used for prediction. Models tend to be bespoke, created using the information gathered for a particular species. The act of gathering this information, and then constructing the model, takes considerable time. Balancing these two competing demands is challenging, but creating tools that are generic and freely available online is a step towards realising this aim (Froese et al., 2019).

Network models can be applied to a number of spatial resolution scales across all stages of past invasion, as demonstrated in several recent studies (Cope et al., 2019; Miraballes et al., 2019; Wiltshire et al., 2019). For instance, BNNs have been used for specific post-border surveillance to assess the likelihood of farms in Uruguay becoming infested with cattle ticks through the introduction of infested cattle. This guides farmers' decisions on when to attempt eradication rather than waste resources when conditions are likely to result in the reintroduction of the pest (Miraballes et al., 2019). They have also been used assess the vulnerability of production hogs to the spread of diseases in the US (Wiltshire et al., 2019). BNNs can also be applied to very broadscale general surveillance. Global invasion risk into Australia was recently modelled based on the ecological similarities and level of transport connectivity between different regions (Cope et al., 2019).

Supporting better decision-making

As data capture technologies and diagnostic research advances, there remains a need to integrate new findings to provide up-to-date support for decision-makers in surveillance programmes. Quantitative tools, such as the statistical modelling methods previously discussed, can be incorporated directly into these decision processes to evaluate the possible results of decisions before they are made or to guide strategies by enabling data to be visualised in a clear and relevant context.

Simulation

Statistical simulations or optimisation models allow evaluation of many possible scenarios before they occur. They can therefore be of particular use to biosecurity surveillance by providing guidance for how, where and when to look for pests or pathogens (Barnes et al., 2019). This includes optimising the allocation of resources for early detection programmes, as was investigated in the Torres Strait Islands (Barnes et al., 2019). Another advantage is that simulation allows the impact of technological constraints and varying environmental or ecological factors to be taken into account so the most appropriate data capture method can be identified for different biosecurity objectives. For instance, simulation has provided insight into the optimal speed and height UAVs should be flown at in order to detect species with different distributions and modes of dispersal (Baxter and Hamilton, 2018). For current and ongoing incursions, simulation models can be carried out even with minimal data and used to explore and identify the best strategies for ongoing monitoring and predict the further spread of the pest species and outcomes of proposed interventions (Merrill et al., 2019). An example of the latter is when optimisation models were used to investigate how best to communicate messages of infection risks to personnel in animal livestock industries in order to ensure compliance with biosecurity practices (Merrill et al., 2019).

Value of Information approaches

Although the act of collecting surveillance data itself reduces uncertainty around biosecurity issues, the adoption of Value of Information (VOI) approaches could be beneficial as they can be used to determine exactly what data is likely to make the most helpful addition to current understandings (Maxwell et al., 2015). VOI analysis was developed according to the theory of information economics and assesses the benefits of collecting data while considering the context of the decision that needs to be made. It is particularly useful for high-risk decisions with a great deal of uncertainty surrounding them (Yokota and Thompson, 2004). In biosecurity this could apply to when a manager needs to decide between many possible interventions for an incursion and the choice is affected by an uncertain variable such as risk of spread or likelihood of compliance with biosecurity procedures (Runge et al., 2017). VOI could calculate the expected improvement in the outcome of that decision – in this case how likely a pest is to be eradicated from an area with the chosen intervention – *if* the uncertainty is resolved by collecting data from different areas or sources (Runge et al., 2017). The capacity for VOI to take into account the goal of decision-making means these approaches are well suited for biosecurity, as there are often significant risks to collecting surveillance data, such as potential contamination during surveying. It is therefore important to understand whether the benefits will outweigh the possible negative impacts of surveillance practices before they are carried out so the most valuable data can be collected to make the most beneficial decision.

Spatial modelling and risk maps

It is intuitive for risk in biosecurity to be represented spatially, whether it be the risk of the spread of a pest across a landscape, the risk of introduction between different properties or pathways showing the connections between global points of entry. Thus, spatial models and their associated risk maps, which depict the geographic patterns of susceptibility to pest or pathogen invasion, are becoming a crucial component of decision-making tools for surveillance across all stages of invasion. After myrtle rust (*Austropuccinia psidii*) was detected for the first time on mainland New Zealand, maps were generated weekly showing the potential risks of infection, latent periods and sporulation in response to changing weather conditions across the country (Beresford et al., 2018). Follow-up analysis of surveillance data revealed the maps displayed accurate predictions of geographic risk, and they are now being used for ongoing planning of surveillance and management efforts (Beresford et al., 2018). In another recent example, maps generated from geographic information system data to predict the risk of highly pathogenic avian influenza outbreaks in Korea enabled targeted surveillance zones to be established to prevent transmission of the disease (Lee and Pak, 2017). In order to be of most use to future biosecurity surveillance, spatial risks will need to be modelled and mapped even more rapidly so that dynamic changes can be displayed during response to ongoing incursions, suggesting the need for generic interactive visualisation that can be rapidly tailored to individual incursions. This is important not only to understand the dynamics of incursions and allocate resources but also to communicate to stakeholders, including the public. The use of freely available products such as Google Earth will be an essential component of such models, as will the capacity for these models to be used by biosecurity staff who may not have high-level modelling skills. While challenging, this is a powerful approach, and software such as Biospark (Baxter et al., 2017) have paved the way in this area.

Constraints on future possibilities

When planning future biosecurity surveillance, it is important to consider the possible roadblocks to implementing emerging technologies and methods. No single approach will be appropriate for all situations, so careful consideration of the context of proposed surveillance will help identify which practices will provide the most beneficial outcomes. This context includes what kind of data needs to be collected; what is the goal of surveillance; what species, group or pathogen is being monitored; and the social and moral implications of certain practices.

Data-specific considerations

As data capture technologies improve, the quantity of data available for biosecurity surveillance is growing exponentially. Although this provides useful additional information to complement pre-existing data, it also presents challenges in terms of how that sheer volume of data can be handled, stored, assessed for quality control and analysed. These obstacles to putting 'big data' to its best use have undergone considerable scrutiny and are documented extensively in current literature (Duncan et al., 2019; Kozminski, 2015; Verma et al., 2020). The quantity of data generated from many surveillance sources, particularly passive sources, in many cases exceeds the maximum amount that can be stored, computed and retrieved, and there are limited experts available to wrangle and analyse big data when it is possible (Duncan et al., 2019). This problem is likely to be reduced over time, but only if substantial investment is made in the development of software and the training of data scientists (Duncan et al., 2019). Data-sharing agreements between bodies that collect data can also be complex and can potentially prevent managers from being able to see the full picture when making biosecurity decisions; therefore, global cooperation will be necessary to most effectively coordinate surveillance and biosecurity measures (Scott et al., 2019). In terms of quality control, passive and citizen science datasets tend to contain a significant amount of noise and may be biased by self-selection and positive reporting (Baker et al., 2018). Therefore, it is likely that the usefulness of these kinds of data will decrease from early detection stages onwards, and other forms of surveillance may need to be considered for later stages of invasion (Baker et al., 2018; Baxter and Hamilton, 2016).

Objective-specific considerations

When deciding how best to conduct biosecurity surveillance, it is critical to keep the end goal in mind so the best return on investment can be obtained. This is particularly relevant when it comes to applying new technologies and techniques to later invasion stages, as decisions will need to be made on where to focus search effort. Depending on whether the goal of later stage surveillance is to detect new incursions, map the extent of the current spread of a pest, confirm eradication success or provide evidence for area freedom, the best spatio-temporal deployment will differ vastly and will impact what new and existing surveillance approaches are applicable. For instance, citizen science data that only indicates presence would not be very appropriate for determining if a pest is truly eradicated and absent from an area (Baker et al., 2018). However, shortcomings in the current fitness for purpose of emerging tools and methods can also provide insight into future opportunities for research and development to broaden their applicability and adoption.

Organism-specific considerations

Important contexts to consider in the development of future surveillance practices are the biology and ecology of the target species, species group or pathogen. These considerations include factors that vary over time, such as latent or asymptomatic periods, seasonal patterns and phenology, and spatial factors such as mode of spread, mobility, distribution of host plants and range of spatial extent (Baxter and Hamilton, 2016). Traits such as rate of reproduction, survivorship and abundance are also important to consider. For example, there has been difficulty in scaling up molecular diagnostic tests that work by testing a single specimen per reaction. Very large numbers of insects are typically found in traps in biosecurity surveillance situations, so identifying each one by one would be impractical (Piper et al., 2019). The wider environmental context is also worth considering as factors such as host–plant diversity, abundance and distribution of vectors and microclimatic conditions are likely to effect likelihood of invasion of infection (Beresford et al., 2018; Cope et al., 2019; Wiltshire et al., 2019). This wide range of considerations precludes any one technological innovation being the best solution to conducting surveillance on all species, but, as previously discussed, many are adaptable to diverse scenarios, particularly statistical modelling results that can be strategically generalised (Lopes et al., 2019).

Ethical considerations

Ethical challenges are also important to consider when planning how to implement new tools and techniques for biosecurity surveillance (Devitt et al., 2019). Improved biosecurity surveillance has many potential humanitarian benefits such as ensuring adequate access to food and providing support for sustainable agricultural practices and international trade, but this improvement may come at the cost of invasion of privacy (Johnson, 2015; Simpson and Srinivasan, 2014). This is a particular concern when surveillance data is acquired from private citizens, such as indirectly and directly gathered citizen science data or remote sensing data collected from private property (Zuboff, 2019). If this data collection is too pervasive and indiscriminately captured from growers, the risk of that data being misused by third parties becomes significant (Zuboff, 2019). Secure data storage is therefore critical, as breaches of confidentiality could have major negative repercussions for future participation, confidence, trust and buy-in for future surveillance and could lead to non-compliance with recommended biosecurity measures (Vinatzer et al., 2019). This has become a greater challenge in recent years as reliance on large databases has increased, making them more likely to be a target for cybersecurity attacks (Peccoud et al., 2018).

Linking of data from multiple integrated sources is undeniably useful for biosecurity managers wanting to get a clearer understanding of incursions but also raises concerns as it increases the potential for identifying individuals or groups that could be targeted for malicious activities or political reasons (Caswell et al., 2019; Daly et al., 2019; Vinatzer et al., 2019). Identifiable data can also be exploited by biosecurity providers for financial gain – by selling the data to third parties in order to target advertising and affect behaviours to improve business opportunities – rather than being used to protect and support farming communities (Zuboff, 2019).

Public perceptions including whether or not certain new surveillance technologies are considered acceptable or likely to result in invasion of privacy must also be managed to maximise participation and access to data (Ogilvie et al., 2019). For example, many farmers have expressed concerns about drone operators for biosecurity surveillance infringing on their privacy by flying over private property (Devitt et al., 2019). Ensuring consistent standard operating procedures that pay respect to the privacy of farmers and the general public, and raising awareness in the community and encouraging co-ownership of biosecurity operations, could help increase trust and compliance with surveillance practices (Devitt et al., 2019; Ogilvie et al., 2019).

Overall, it is important to note that while new technologies may greatly enhance biosecurity data capture capabilities, these advancements may pose a threat to maintaining a free and just society if they remain unchecked. There is a potential for conflict between the need to conduct thorough biosecurity surveillance to protect native ecosystems and agricultural services and the rights of growers and members of the public not to be put under undue surveillance or entered into unethical data relationships. Careful consideration must be given to how innovative surveillance techniques and technologies are developed and implemented to reduce negative impacts on human freedoms while ensuring high standards of ecological and food security are maintained.

Conclusions

There is great potential for biosecurity surveillance practices to improve thanks to new technologies and advanced quantitative analyses. In the future a more efficient and effective surveillance scheme for global biosecurity is likely to rely on statistical models that are capable of handling large and complex datasets while remaining accessible to a wide range of users. These models can exploit the results of improved data capture so that updated, clear information can be provided to growers, industry and other stakeholders in real time and can be used to make increasingly timely and relevant decisions. Context will also be a critical component of an improved surveillance strategy. To get maximum benefit out of these new technologies and analyses, the goal of surveying, the targeted organism and the tools and human resources available need to be considered. Decisions on surveillance practice should therefore be nested in a holistic or hierarchical approach so the combination of available techniques can be deployed at the spatio-temporal resolution with the best capacity to detect the organism of interest and so the most beneficial trade-off between the benefits of data collected and the costs of carrying out surveillance can be gained. Multi-criteria decision support tools such as simulation, VOI approaches and risk mapping should be widely adopted to help compare and evaluate different strategies and create a more transparent framework for future directions and investments in biosecurity.

Calls have been made in many countries for a cooperative information highway or cyberinfrastructure for sharing and compiling biosecurity data (Ministry for Primary Industries, 2016). To maintain this, contributions and support from a wide range of stakeholder groups including growers, researchers, industries, governments and private citizens will be required. While successful biosecurity efforts are likely to lead to sustained or increased agricultural productivity through preventing incursions, more efficient eradication of pest species and confirmation of pest absence, unsuccessful biosecurity can lead to severe negative economic, social and environmental impacts in places where incursions occur. Therefore, there is an ever-present need for partnership among stakeholder groups to invest in biosecurity surveillance research financially or through contributing knowledge or time. It would be best to consider advancing surveillance practices as a shared responsibility so that with our information, technology and expertise we can practise biosecurity surveillance in the best, most efficient and mutually rewarding way.

References

Armstrong, K.F. and Ball, S.L. (2005) 'DNA barcodes for biosecurity: Invasive species identification', *Philosophical Transactions of the Royal Society B*, vol. 360, pp. 1813–1823.

Baker, E., Jeger, M.J., Mumford, J.D. and Brown, N. (2018) 'Enhancing plant biosecurity with citizen science monitoring: Comparing methodologies using reports of acute oak decline', *Journal of Geographical Systems*, vol. 21, pp. 111–131.

Barnes, B., Giannini, F., Arthur, A. and Walker, J. (2019) 'Optimal allocation of limited resources to biosecurity surveillance using a portfolio theory methodology', *Ecological Economics*, vol. 161, pp. 153–162.

Baxter, P.W.J., Woodley, A. and Hamilton, G. (2017) 'Modelling the spatial spread risk of plant pests and pathogens for strategic management decisions', in G. Syme, D. Hatton MacDonald, B. Fulton and J. Piantadosi (eds.) *MODSIM2017, 22nd International Congress on Modelling and Simulation*. Modelling and Simulation Society of Australia and New Zealand, Hobart, Australia, pp. 209–215.

Baxter, P.W.J. and Hamilton, G. (2016) *An Analysis of Future Spatiotemporal Surveillance for Biosecurity*. Plant Biosecurity CRC, Brisbane, Australia.

Baxter, P.W.J. and Hamilton, G. (2018) 'Learning to fly: Integrating spatial ecology with unmanned aerial vehicle surveys', *Ecosphere*, vol. 9, p. e02194.

Beresford, R.M. et al. (2018) 'Predicting the climatic risk of myrtle rust during its first year in New Zealand', *New Zealand Plant Protection*, vol. 71, pp. 332–347.

Black, A. et al. (2019) 'How an indigenous community responded to the incursion and spread of myrtle rust (*Austropuccinia psidii*) that threatens culturally significant plant species: A case study from New Zealand', *Pacific Conservation Biology*, vol. 25, pp. 348–354.

Caley, P., Welvaert, M. and Barry, S.C. (2019) 'Crowd surveillance: Estimating citizen science reporting probabilities for insects of biosecurity concern', *Journal of Pest Science*, vol. 93, pp. 543–550.

Careless, P., Marshall, S.A. and Gill, B.D. (2014) 'The use of Cerceris fumipennis (Hymenoptera: Crabronidae) for surveying and monitoring emerald ash borer (Coleoptera: Buprestidae) infestations in eastern North America', *The Canadian Entomologist*, vol. 146, pp. 90–105.

Caswell, J. et al. (2019) 'Defending our public biological databases as a global critical infrastructure', *Frontiers in Bioengineering and Biotechnology*, vol. 7.

Cellini, A., Blasioli, S., Biondi, E., Bertaccini, A., Braschi, I. and Spinelli, F. (JOUR et al.) (2017) 'Potential applications and limitations of electronic nose devices for plant disease diagnosis', *Sensors*, vol. 17, p. 2596.

Colunga-Garcia, M., Haack, R.A., Magarey, R.D. and Borchert, D.M. (2013) 'Understanding trade pathways to target biosecurity surveillance', *NeoBiota*, vol. 18, pp. 103–118.

Committee of Experts on Phytosanitary Measures (1996) *Report of the Third Meeting of the Committee of Experts on Phytosanitary Measures*. Rome, Italy.

Convention on Biological Diversity Secretariat (2001) *Open-Ended Inter-Sessional Meeting on the Strategic Plan, National Reports, and Implementation of the Convention on Biological Diversity*. Montreal, Canada.

Cook, A., Marion, G., Butler, A. and Gibson, G. (2007) 'Bayesian inference for the spatio-temporal invasion of alien species', *Bulletin of Mathematical Biology*, vol. 69, pp. 2005–2025.

Cope, R.C., Ross, J.V., Wittmann, T.A., Watts, M.J. and Cassey, P. (2019) 'Predicting the risk of biological invasions using environmental similarity and transport network connectedness', *Risk Analysis*, vol. 39, pp. 35–53.

Corcoran, E., Denman, S., Hanger, J., Wilson, B. and Hamilton, G. (2019) 'Automated detection of koalas using low-level aerial surveillance and machine learning', *Scientific Reports*, vol. 9, p. 3208.

Cressie, N., Calder, C.A., Clark, J.S., Hoef, J.M.V. and Wikle, C.K. (2009) 'Accounting for uncertainty in ecological analysis: The strengths and limitations of hierarchical statistical modeling', *Ecological Applications*, vol. 19, pp. 553–570.

CSIRO (2014) *Robots to ResQu Our Rainforests*. Canberra, Australia.

Daly, A., Devitt, K. and Mann, M. (eds.) (2019) *Good Data*. Institute for Network Cultures, Amsterdam.

Dehnen-Schmutz, K., Foster, G.L., Owen, L. and Persello, S. (2016) 'Exploring the role of smartphone technology for citizen science in agriculture', *Agronomy for Sustainable Development*, vol. 36, p. 25.

Deluyka, H. and Silano, V. (2012) 'Editorial: The first ten years of activity of EFSA, a success story', *EFSA Journal*, vol. 10, special issue, pp. 1–6.

Demertzis, K., Iliadis, L.S. and Anezakis, V.-D. (2018) 'Extreme deep learning in biosecurity: The case of machine hearing for marine species identification', *Journal of Information and Telecommunication*, vol. 2, pp. 492–510.

Devitt, S.K., Paxter, P.W.J. and Hamilton, G. (2019) 'The ethics of biosurveillance', *Journal of Agricultural and Environmental Ethics*, vol. 32, pp. 709–740.

Duncan, S.E. et al. (2019) 'Cyberbiosecurity: A new perspective on protecting U.S. food and agricultural system', *Frontiers in Bioengineering and Biotechnology*, vol. 7.

Fenton, N. and Neil, M. (2013) *Risk Assessment and Decision Analysis with Bayesian Networks*. CRC Press, Taylor & Francis Group, Boca Raton, FL.

Froese, J., Pearse, A. and Hamilton, G. (2019) 'Rapid spatial risk modelling for management of early weed invasions: Bridging the gap between ecological complexity and operational needs', *Methods in Ecology and Evolution*, vol. 2019, no. 00, pp. 1–13.

Froud, K.J. and Bullians, M.S. (2015) 'Investigation of biosecurity risk organisms for the plant and environment domains in New Zealand for 2008 and 2009', *New Zealand Plant Protections*, vol. 63, pp. 262–269.

Froud, K.J., Oliver, T.M., Bingham, P.C., Flynn, A.R. and Rowswell, N.J. (2008) *Surveillance for Biosecurity: Pre-Border to Pest Management*. The New Zealand Plant Protection Society, Christchurch.

Goldson, S., Frampton, E.R. and Ridley, G. (2010) 'The effects of legislation and policy in New Zealand and Australia on biosecurity and arthropod biological control research and development', *Biological Control*, vol. 52, pp. 241–244.

Hodgson, J.C. et al. (2018) 'Drones count wildlife more accurately and precisely than humans', *Methods in Ecology and Evolution*, vol. 9, pp. 1160–1167.

Hollings, T. et al. (2018) 'How do you find the green sheep? A critical review of the use of remotely sensed imagery to detect and count animals', *Methods in Ecology and Evolution*, vol. 9, pp. 881–892.

Hoyer-Tomiczek, U., Sauseng, G. and Hoch, G. (2016) 'Scent detection dogs for the Asian longhorn beetle, Anoplophora glabripennis', *EPPO Bulletin*, vol. 46, pp. 148–155.

Hulme, P.E. (2014) *The Handbook of Plant Biosecurity: Principles and Practices for the Identification, Containment and Control of Organisms That Threaten Agriculture and the Environment Globally* (eds. G. Gordh and S. McKirdy). Springer, Netherlands, pp. 1–25.

International Plant Protection Convention Secretariat (2012) *IPPC Strategic Framework 2012–2019*. Food and Agriculture Organization of the United Nations, Rome, Italy.

Jarrad, F., Barrett, S., Murray, J., Parkes, J., Stoklosa, R., Mengersen, K. and Whittle, P. (2010) 'Improved design method for biosecurity surveillance and early detection of non-indigenous rats', *New Zealand Journal of Ecology*, vol. 35, pp. 1–13.

Jarrad, F., Low-Choy, S. and Mengerson, K. (2015) *Biosecurity Surveillance: Quantitative Approaches*. 1st edition. CAB International, Boston, MA.

Jurdak, R., Elfes, A., Kusy, B., Tews, A., Hu, W., Hernandez, E., Kottege, N. and Sikka, P. (2015) 'Autonomous surveillance for biosecurity', *Trends in Biotechnology*, vol. 33, pp. 201–207.

Jurdak, R., Sommer, P., Kusy, B., Kottege, N., Christopher, C., McKeown, A. and Westcott, D. (2013) 'Camazotz: Multimodal activity-based GPS sampling', *Proceedings of the 12th International Conference on Information Processing in Sensor Networks (ISPN)*, IEEE, pp. 67–78.

Johnson, H. (2015) 'Eating for health and the environment: Australian regulatory responses for dietary change', *QUT Law Review*, vol. 15, p. 139.

Kalaris, T. et al. (2014) *The Handbook of Plant Biosecurity: Principles and Practices for the Identification, Containment and Control of Organisms That Threaten Agriculture and the Environment Globally* (eds. G. Gordh and S. McKirdy). Springer, Netherlands, pp. 309–337.

Kellenberger, B., Marcos, D. and Tuia, D. (2018) 'Detecting mammals in UAV images: Best practices to address a substantially imbalanced dataset with deep learning', *Remote Sensing of Environment*, vol. 216, pp. 139–153.

Kim, K.S. and Beresford, R.M. (2009) 'Use of geographic information systems and satellite data for assessing climatic risk of establishment of plant pathogens', *New Zealand Plant Protection*, vol. 62, pp. 109–113.

Kozminski, K.G. (2015) 'Biosecurity in the age of big data: A conversation with the FBI', *Molecular Biology of the Cell*, vol. 26, pp. 3894–3897.

Lee, G. and Pak, S.I. (2017) 'A GIS-based mapping to identify locations at risk for highly pathogenic avian influenza virus outbreak in Korea', *Journal of Veterinary Clinics*, vol. 34, pp. 146–151.

Li, D. et al. (2018) 'DNA barcoding and real-time PCR detection of Bactrocera xanthodes (Tephritidae: Diptera) complex', *Bulletin of Entomological Research*, vol. 109, pp. 102–110.

Lopes, R., Kruse, A.B., Nielsen, L.R., Nunes, T.P. and Alban, L. (2019) 'Additive Bayesian Network analysis of associations between antimicrobial consumption, biosecurity, vaccination and productivity in Danish sow herds', *Preventive Veterinary Medicine*, vol. 169, p. 104702.

Maxwell, S.L., Rhodes, J.R., Runge, M.C., Possingham, H.P., Ng, C.F. and McDonald-Madden, E. (2015) 'How much is new information worth? Evaluating the financial benefit of resolving management uncertainty', *Journal of Applied Ecology*, vol. 52, pp. 12–20.

Maye, D., Dibdan, J., Higgins, V. and Potter, C. (2012) 'Governing biosecurity in a neoliberal world: Comparative perspectives from Australia and the United Kingdom', *Environment and Planning A: Economy and Space*, vol. 44, pp. 150–168.

Merrill, S.C. et al. (2019) 'Willingness to comply with biosecurity in livestock facilities: Evidence from experimental simulations', *Frontiers in Veterinary Science*, vol. 6.

Ministry for Primary Industries (2016) *Biosecurity 2025: Protecting to Grow New Zealand*. New Zealand Government, Wellington, New Zealand.

Miraballes, C. et al. (2019) 'Probability of Rhipicephalus microplus introduction into farms by cattle movement using a Bayesian Belief Network', *Ticks and Tick-Borne Diseases*, vol. 10, pp. 883–893.

Nalepa, C.A., Swink, W.G., Basham, J.P. and Merten, P. (2015) 'Comparison of Buprestidae collected by Cerceris fumipennis (Hymenoptera: Crabronidae) with those collected by purple prism traps', *Agricultural and Forest Entomology*, vol. 17, pp. 445–450.

Noordhuizen, J., Surborg, H. and Smulders, F.J. (2013) 'On the efficacy of current biosecurity measures at EU borders to prevent the transfer of zoonotic and livestock diseases by travellers', *The Veterinary Quarterly*, vol. 33, pp. 161–171.

Ogilvie, S., McCarthy, A., Allen, W., Grant, A., Mark-Shadbolt, M., Pawson, S., Richardson, B., Strand, T., Langer, E.L. and Marzano, M. (2019) 'Unmanned aerial vehicles and biosecurity: Enabling participatory-design to help address social licence to operate issues', *Forests*, vol. 10, p. 695.

Peccoud, J., Gallegos, J.E., Murch, R., Buchholz, W.G. and Raman, S. (2018) 'Cyberbiosecurity: From naive trust to risk awareness', *Trends in Biotechnology*, vol. 36, pp. 4–7.

Piper, A.M. et al. (2019) 'Prospects and challenges of implementing DNA metabarcoding for high-throughput insect surveillance', *GigaScience*, vol. 8.

Poland, T.M. and Rassati, D. (2019) 'Improved biosecurity surveillance of non-native forest insects: A review of current methods', *Journal of Pest Science*, vol. 92, pp. 37–49.

Rey, N., Volpi, M., Stephane, J. and Tuia, D. 'Detecting animals in African Savanna with UAVs and the crowds', *Remote Sensing of the Environment*, vol. 200, pp. 341–351.

Rock, F., Barsan, N. and Weimar, U. (2008) 'Electronic nose: Current status and future trends', *Chemical Reviews*, vol. 108, pp. 705–725.

Runge, M.C., Rout, T., Spring, D. and Walshe, T. (2017) *Value of Information Analysis as a Decision Support Tool for Biosecurity: Chapter 15*. Cambridge University Press, Cambridge, United Kingdom.

Rutledge, C.E., Hellman, W., Teerling, C. and Hellman, C.W. (2011) 'Two novel prey families for the Buprestid-hunting wasp Cerceris fumipennis Say (Hymenoptera: Crabronidae)', *Coleoptera Bulletin*, vol. 65, pp. 194–196.

Schaad, N.W., Frederick, R.D., Shaw, J., Schneider, W.L., Hickson, R., Petrillo, M.D. and Luster, D.G. (2003) 'Advances in molecular-based diagnostics in meeting crop biosecurity and phytosanitary issues', *Annual Review of Phytopathology*, vol. 41, pp. 305–324.

Scott, P., Bader, M.K.F., Burgess, T., Hardy, G. and Williams, N. (2019) 'Global biogeography and invasion risk of the plant pathogen genus Phytophthora', *Environmental Science and Policy*, vol. 101, pp. 175–182.

Seymour, A.C., Dale, J., Hammill, M., Halpin, P.N. and Johnston, D.W. (2017) 'Automated detection and enumeration of marine wildlife using unmanned aircraft systems (UAS) and thermal imagery', *Scientific Reports*, vol. 7, pp. 45127–45127.

Shi, Y. et al. (2016) 'Unmanned aerial vehicles for high-throughput phenotyping and agronomic research', *PLoS One*, vol. 11, p. e0159781.

Simpson, M. and Srinivasan, V. (2014) Australia's biosecurity future: Preparing for future biological challenges. Commonwealth Scientific and Industrial Research Organisation (CSIRO), Canberra, Australia.

Sims, N.C. et al. (2018) 'Spectral separability and mapping potential of cassava leaf damage symptoms caused by whiteflies (Bemisia tabaci)', *Pest Management Science*, vol. 74, pp. 246–255.

Stanaway, M.A., Mengersen, K.L. and Reeves, R. (2011) 'Hierarchical Bayesian modelling of early detection surveillance for plant pest invasions', *Environmental and Ecological Statistics*, vol. 18, pp. 569–591.

Stephenson, B.P., Gill, G.S.C., Randall, J.L. and Wilson, J.A. (2003) 'Biosecurity approaches to surveillance and response for new plant species', *New Zealand Plant Protection*, vol. 56, pp. 5–9.

Suma, P., La Pergola, A., Longo, S. and Soroker, V. (2014) 'The use of sniffing dogs for the detection of Rhynchophorus ferrugineus', *Phytoparasitica*, vol. 42, pp. 269–274.

Velusamy, V., Korostynka, O., Vaseashta, A. and Adley, C. (2012) 'Applications in food quality monitoring and biosecurity', in A. Vaseashta et al. (eds.) *Technological Innovations in Sensing and Detection of Chemical, Biological, Radiological, Nuclear Threats and Ecological Terrorism*. Springer, Netherlands, pp. 149–158.

Verma, S., Bhatia, A., Chug, A. and Singh, A.P. (2020) 'Recent advancements in multimedia big data computing for IoT applications in precision agriculture: Opportunities, issues, and challenges', *Intelligent Systems Reference Library*, vol. 163, pp. 391–416.

Vinatzer, B.A. et al. (2019) 'Cyberbiosecurity challenges of pathogen genome databases', *Frontiers in Bioengineering and Biotechnology*, vol. 7.

Welvaert, M., Al-Ghattas, O., Cameron, M. and Caley, P. (2017) 'Limits of use of social media for monitoring biosecurity events', *PLoS One*, vol. 12, p. e0172457.

Westphal, M.I., Browne, M., MacKinnon, K. and Noble, I. (2008) 'The link between international trade and the global distribution of invasive alien species', *Biological Invasions*, vol. 10, pp. 391–398.

Wilson, A.D. (2013) 'Diverse applications of electronic-nose technologies in agriculture and forestry', *Sensors*, vol. 13, pp. 2295–2348.

Wiltshire, S. et al. (2019) 'Network meta-metrics: Using evolutionary computation to identify effective indicators of epidemiological vulnerability in a livestock production system model', *JASSS*, vol. 22.

Witczuk, J., Pagacz, S., Zmarz, A. and Cypel, M. (2018) 'Exploring the feasibility of unmanned aerial vehicles and thermal imaging for ungulate surveys in forests: Preliminary results', *International Journal of Remote Sensing*, vol. 39, pp. 5504–5521.

Yokota, F. and Thompson, K.M. (2004) 'Value of information literature analysis: A review of applications in health risk management', *Medical Decision Making*, vol. 24(JOUR et al.), pp. 287–298.

Zuboff, S. (2019) *The Age of Surveillance Capitalism: The Fight for a Human Future at the New Frontier of Power*. Public Affairs, New York.

17

RISK ASSESSMENT FOR INVASIVE SPECIES

John D. Mumford and Mark A. Burgman

Introduction

Invasive species pose a particularly diverse challenge for risk assessment and management because of the vast numbers and different kinds of potentially invasive organisms, the diverse entry pathways, lack of data on pathway movements and the lack of information on the many susceptible environments. Invasive species cause impacts beyond the economic, affecting amenity and cultural values, social well-being, human health and biodiversity. The time scale of their impacts may be very long, potentially extending to hundreds of years, compared to the more immediate harm arising from introduced agricultural pests (Crooks, 2005; Waage and Mumford, 2008). Responsibility for management is also often unclear, landowners often are unaware of their management responsibilities and public authorities are often left to protect dispersed resources with limited opportunities to recover costs from stakeholders. This chapter describes the processes used in risk assessment for invasive species, particularly in the UK and the European Union. The geographical scope is motivated by the recent implementation of the Invasive Alien Species Regulation in Europe, which has focused attention on the need for risk assessment and the development of proportionate approaches (Roy et al., 2015). These developments represent a new phase in the evolution of procedures for biosecurity risk assessment.

Risk assessment processes for invasive species

ISPM 11 from the International Plant Protection Convention, or IPPC (FAO, 2017), is the international standard for risk analysis for agricultural pests. The approach described in this pest risk analysis standard is broadly relevant to a wider group of invasive species. It sets out a staged process of initiation, risk assessment (including entry, establishment, spread and impact), risk management and communication. Risk assessment is therefore a distinct part of this larger set of analytical steps.

Initiation of risk analysis, in the terms of the IPPC, can involve awareness or concern about a species or pathway, as expressed by authorities or wider stakeholders, or it can be the result of new policies or priorities for protection. Inherently, the risk related to invasive species primarily involves the properties of the harmful organism, or agent, itself, but conditions on the pathways and in the receptor environments should also be considered. The initiation points for risk analysis thus reflect species, pathways and policies related to the receptor environments.

The second step in risk analysis outlined in ISPM 11 is assessment of the four key components of risk arising from a new plant pest (or more broadly for an invasive species): entry, establishment, spread and impact. Risk combines likelihood and consequences of an event, such as the harm arising from an invasive species. Entry and establishment can be expressed as probabilities of occurrence in defined times and areas. Establishment is conditional on entry, so the two probabilities can be combined as a product.

The European Invasive Alien Species Regulation (EU Regulation 1143/2014) establishes a system for science-based risk assessment for invasive species of EU concern, although the method of risk assessment is not specified.[1] For the purposes of the regulation, a risk assessment should be conducted in relation to the current and potential range of invasive alien species, having regard to the following elements:

- a description of taxonomic identity, history, natural and potential range of the species;
- a description of reproduction and spread patterns and dynamics, including an assessment of whether the environmental conditions necessary for reproduction and spread exist;
- a description of the potential intentional and unintentional pathways of introduction and spread of the species, including any commodities with which the species is generally associated;
- a thorough assessment of the risk of entry, establishment and spread in relevant biogeographical regions in current conditions and in foreseeable climate conditions;
- a description of the current distribution of the species, including whether the species is already present in relevant areas, and a projection of its likely future distribution;
- a description of the adverse impact on biodiversity and related ecosystem services, including on native species, protected sites, endangered habitats, human health, safety and the economy, including an assessment of the potential future impact;
- an assessment of the potential costs of damage;
- a description of the known uses for the species and the social and economic benefits deriving from those uses.

While the regulation is focused on organisms that act as harmful agents, it also addresses pathways of introductions and the properties of vulnerable receptor environments (Mumford et al., 2017). The Invasive Alien Species Regulation differs from other legislation related to pest control, such as the Pesticides Directive or the GMO Directive, by including social and economic benefits associated with the selection of invasive species as considerations in assessing the priority of risks and management actions.

Spread is often difficult to predict and express. It can mean the rate of extension of the frontier of presence or the increase in density of populations within a range. The rate and pattern of spread depend on an organism's ability to move between adjacent units of suitable habitat or to pass beyond adjacent habitat barriers to other separated suitable habitats (With, 2002). Many terrestrial species spread radially through potential habits that may extend in all directions while riverine species may only be able to spread linearly along the watercourse. Others jump to new satellite locations or have habitats, such as ponds, that are not contiguous.

A unifying concept is to consider spread as a change in the proportion of a total potential habitat within a risk assessment area that has been occupied within a specified period of time. This defines spread as an extent relative to a potentially vulnerable resource. The product of spread and impact together describe the estimated consequence of an invasive species. So, if the organism is estimated to be capable of spreading into 10% of its potential habitat within a year and the scale of economic loss if all the vulnerable environment were occupied is £1 million/year, then the estimated annual impact for the risk assessment area would be £100,000.

Impact assessments for invasive species must include economic, social, environment and human health effects. The different dimensions of impacts usually have different measures. Some applications calibrate the dimensions to a common scale, such a monetary value. It is important for risk assessors to have guidelines of semantic descriptions for impact values, as shown in Table 17.1. Of course, the equivalence of impacts between dimensions involves value judgements, and such tables should be calibrated to reflect the preferences of stakeholders in areas of potential impact. Social dimensions of risk have been examined in detail in relation to the threats to tree health posed by a range of harmful organisms in Great Britain (Urquhart et al., 2017).

Uncertainty can be scaled for each score on an impact or probability dimension. Table 17.2 shows a set of definitions of levels of confidence applicable to a five point scoring scale (Holt et al., 2014). This description of confidence as a likelihood of being correct does not directly indicate by how much the score may be wrong. However, this is likely to be related to how

Table 17.1 Some definitions of scales in different dimensions of impact.

Score	Nominal £/year	Health	Environment	Social
Minimal	£10k	Local, mild, short-term reversible effects to individuals	Local, short-term population loss, no significant ecosystem effect	No social disruption
Minor	£100k	Mild, short-term reversible effects to identifiable groups, localised	Some ecosystem impact, reversible changes, but localised	Significant concern at local level
Moderate	£1m	Minor irreversible effects, or large numbers with reversible effects, localised	Measurable long-term damage to populations and ecosystems, but little spread, no extinction	Temporary changes to activity at a local level
Major	£10m	Significant irreversible local effects or reversible effects on a large area	Long-term irreversible ecosystem change, beyond local area	Some permanent change of activity locally, concern over larger area
Massive	£100m	Widespread, severe irreversible health effects	Widespread, long-term population loss, or extinction, affects several species, ecosystem effects	Long-term social change, significant loss of employment, migration from area

Table 17.2 Definition of terms for confidence.

Estimate	Definition of confidence
Very high	90% correct
High	80% correct
Moderate	50% correct
Low	35% correct

Source: Holt et al., 2014.

extreme the score for the risk factor is. So, for very high or very low scores, there is an implication that if there is an incorrect score given, the true value is more likely to be relatively near, and somewhat less extreme, than the score given. For a moderate score, the same level of uncertainty would be likely to include true values in both directions and with a wider distribution.

These confidence values reflect the assessor's expert judgement in how likely their choice of score is to be the correct score. Both the score and the confidence level expressed by a risk assessor can be partly validated through peer review of the rationale and documentation provided for the scores in the risk assessment. The invasive species risk assessment process used in Great Britain (Mumford et al., 2010) relies on an assessor and an independent reviewer for each specific risk assessment and a review panel that uses a consensus approach to calibrate scores and confidence values in line with estimates and documentation across the full range of risk assessments carried out within this scheme. This has provided a robust assessment scheme that has been applied to over 100 species since 2007.[2]

Likelihood and magnitude components of risk can be assigned values drawn from their respective ranges in a stochastic process (Baker et al., 2007) by assigning probability distributions to the scores for each risk factor, based on the confidence values, for each of the scores for entry, establishment, spread and impact (as used for pest risk assessment; FAO, 2017) or for the factors in the ENSARS scheme for assessing aquaculture risk (Copp et al., 2014b). Risk can then be represented as a cloud of points on a likelihood and magnitude graph, as shown in Figure 17.1. This helps illustrate the dimensions in which there is greater or lesser confidence in the scores and helps reviewers visualise the extent and acceptability of uncertainty in a risk assessment. Uncertainty could be reduced by gathering further information for the risk assessment or further research. It should be noted that collecting more information may be costly and may not always reduce uncertainty (Regan et al., 2002). Uncertainty is an important part of risk communication and an important consideration in predicting the range of returns from risk management (Milner-Gulland and Shea, 2017).

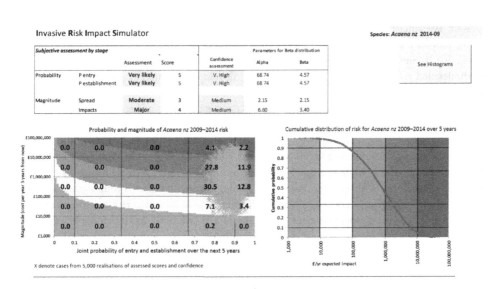

Figure 17.1   A risk simulation showing distribution of joint likelihood (entry × establishment) and magnitude (spread × impact) for pirri-pirri burr.

Source: Modified from Mumford et al., 2010.

Risk management is the final stage of risk analysis and will not be covered in detail in this chapter (see Mumford, 2013). Many authorities try to separate assessment and management of agricultural pests and invasive species into distinct operations with specialised staff. However, the two functions have many common features. Risk assessment is a systematic approach to the uncertain progress of the invasion process of an organism on a pathway affecting a receptor environment. Risk management is a systematic approach to the uncertain performance of mitigation options directed at the combination of agent, pathway and receptor. Both involve factors such as the fitness of the organism, the scale and potential control points on the pathway and the potential resilience of the environment.

There is considerable scope for similar approaches to the quantification of risk factors and the treatment of uncertainty. In the end, decisions on management actions are a product of the management utility and the level of risk (Booy et al., 2017). As a result, low risks with high management utility may be managed more actively than high risks with management options that score poorly on feasibility or other performance criteria. The European Invasive Alien Species Regulation 1143/2014[3] states that the European Commission 'should take into consideration the implementation cost for Member States, the cost of inaction, the cost-effectiveness and the socio-economic aspects' in proposing invasive species that may ultimately require management action by the member states.

The level of acceptable risk for invasive species should be related to that described in ISPM 11 (FAO, 2017), which describes acceptable level of risk in a number of ways, including: in reference to existing regulatory requirements; as indexed to estimated economic losses; as a level expressed on a scale of risk tolerance; and as compared with the level of risk accepted by other countries. Figure 17.2 shows an acceptability matrix for likelihood and magnitude scores for intentional releases of GM organisms in Australia which gives an example of 'a level expressed on a scale of risk tolerance' for a relatively similar risk. Where restrictions on international trade are applied, authorities may be required to show that similar levels of acceptable risk have been used to determine management for other similar risks to ensure a fair process.

Risk assessment for invasive species in Great Britain[4] (Baker et al., 2007; Mumford et al., 2010) follows a process similar to that for agricultural pests in Europe (EPPO, 2011; Schrader et al., 2012). Similar schemes have been proposed in Southeast Asia (Soliman et al., 2016), and more detailed risk assessment schemes have been developed for aquatic invasive species (Copp et al., 2014a, 2014b).

In plant health, the UK Plant Health Risk Register is an attempt to give a relatively rapid assessment of risks for a very large number of species affecting agriculture and forestry throughout the country.[5] It includes just over 1,000 species and gives a simple 1- to 5-point score for each of three attributes: likelihood, relative impact and the value of resource at risk. This gives a 125-point range of risk when the three scores are multiplied. This extensive listing was achieved through engagement with many industry stakeholder groups and gives a rough guide to risk as expressed by those groups and an indication of where more intensive risk assessments may be warranted. The highest scoring species has a value of 125, while the 90th percentile score is 60, so this scoring system gives very good discrimination across the range of higher relative risks. Such an extensive approach has not yet been taken for invasive species outside of agriculture and forestry.

Chapman et al. (2017) and Hulme (2009) describe factors that affect the likelihood of entry along trade pathways, such as the volume of trade (particularly from locations with similar climate) and the speed and frequency of movement along the pathway. They note the greatly expanded patterns of trade into Europe from China and South America in recent years.

	Negligible **Marginal**	Low **Minor**	Intermediate **Medium**	Major **High**
Negligible **Highly unlikely**	Negligible	Negligible	Low	Moderate
Low **Unlikely**	Negligible	Low	Moderate	Moderate
Medium **Likely**	Low	Low	Moderate	High
High **Highly likely**	Low	Moderate	High	High

Figure 17.2   An acceptability matrix for intentional releases of GM organisms in Australia; negligible risk is acceptable without management; low risk could be justified, sometimes without management; all risks can be justified if management can be shown to reduce risk to negligible or low level.

Source: See OGTR, 2013.

Chapman et al. (2017) propose a complex connectivity index as a way to represent strength of agent propagules, conducive transport linkages and the likelihood of introduction and establishment. Live plant movements pose a particular risk, both from the plants themselves as potential invasive species but also through hitchhikers on the plants and associated soil. Management that is focused on prevention addresses ways to reduce the likelihood component of risk, rather than the magnitude of risk.

Quantitative and qualitative risk assessment

Quantitative risk assessment, where it is possible, is likely to be more informative than qualitative or semi-quantitative risk assessment and is the preferred approach in ISPM 11. In practice, quantitative risk assessment can be difficult for invasive species risks where there is little information on the movement or potential establishment of the species. As a result, quantitative risk assessments may be substantially uncertain and difficult to interpret.

Qualitative risk assessments may be easier to carry out and interpret but will be no less uncertain than their quantitative counterparts and may not provide sufficient evidence to support difficult and expensive public decisions. Semi-quantitative approaches (Baker et al., 2007; Vanderhoeven et al., 2015; Soliman et al., 2016) provide a practical compromise in which scoring systems for the major components of likelihood and magnitude of risk are assigned to well-defined ranges of quantitative values with subjective confidence (or uncertainty) ratings. Some systems use five-point scales and include confidence ratings for the four key risk components in ISPM 11 (Baker et al., 2007). The Belgian system uses a three-point scale on four criteria: potential for spread; colonisation of natural habitats; adverse ecological impacts on native species; and adverse ecological impacts on ecosystems (Vanderhoeven et al., 2015). Scores are multiplied together with equal weight to give an index of risk that is not directly related to the likelihood and magnitude of risk but which gives a ranked priority of 'riskiness' in terms of establishment, spread and two types of impact. Speirs-Bridge et al. (2010) suggest a four-step approach to estimate values for likelihood or impact in which an assessor is asked to estimate a lower bound, an upper bound, a most likely estimate and the percentage likelihood that the true value is within the estimated limits. Combining the results from multiple independent assessors, where practical, improves the derived confidence in estimates.

The process of risk assessment for invasive species in Great Britain differs from agricultural pest risk assessments in that the latter is carried out by dedicated teams of professional risk analysts in the responsible ministry whereas the breadth of invasive species risks requires a wider range of experts with knowledge of specific taxa. A diverse range of assessors with limited experience of risk assessment (but substantial knowledge of the organisms) means the risk assessment process needs to be well formatted and closely prescribed to ensure a consistent calibration of scores across different species and assessors, as well as consistent application of confidence ratings and an appropriate level of documentation for the scores and confidence ratings (Baker et al., 2007). The individual risk assessments are reviewed by a second, independent assessor, and both the original assessment and the review go through a peer review panel to ensure greater consistency of scoring and documentation (Holt et al., 2012).

To preserve coherent scores, assessors for the Great Britain invasive species risk assessments are asked to give overall scores and confidence ratings for the four key risk components, based on their justifications for more detailed factors in each of the component categories. This avoids any explicit weighting of the many individual factors covered in the template and allows the relevant factors to be expressed in the overall scoring. Experience has shown that a standard weighting would not cover the range of different circumstances effectively. The review panel can refer to the individual factor scores and confidence as shown in Figure 17.3. Here the shaded bars show

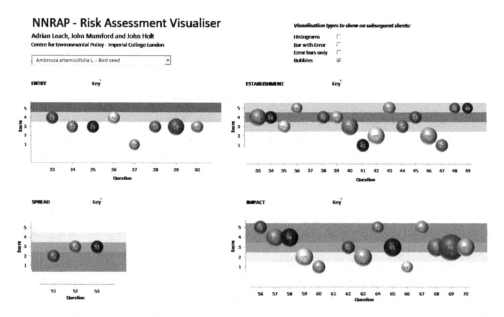

Figure 17.3   Visualisation of key risk components and individual risk factors.

Source: Modified from Holt et al., 2012.

the overall component scores and confidence distribution and the bubbles indicate the score and the level of confidence (larger bubbles have greater uncertainty) for each specific question in the risk assessment template (see Holt et al., 2012). The visualisation of individual scores and uncertainty that make up the overall values can be used in the peer review process to consider the implicit weights assessors are giving to specific information.

Risk summaries have been used in Britain to communicate the risks to relevant authorities and stakeholders. The summaries use a standard template to give basic descriptive information on the species, a brief history of concern in Britain, the native and current invasive distributions, the range of pathways for introduction, likely impacts in relevant dimensions and overall and component risk scores and confidence values.[6] The risk summary expresses likelihood and magnitude of risk in qualitative terms, with colour coding for emphasis.

Spatial and temporal scales of invasive species risk assessment

The spatial scale of risk assessments is determined by the extent of the vulnerable receptor environments or by the limits of the administrative responsibility of the relevant authority. The temporal scale for risk assessments is more subjective. Some invasive species establish and spread slowly but inexorably over decades or centuries (Crooks, 2005). This would imply taking a very long time period into account in risk assessments. Management actions and priorities, however, are generally considered and implemented over relatively short periods. Environmental managers place a greater weight on immediate impacts over long-term impacts (Freeman and Groom, 2013). For risk assessors, the level of uncertainty in assessments is greater over longer time frames. Risk managers need a common benchmark for the temporal dimension of risk to set priorities. This could be achieved by setting a common time scale for risk assessments and assessing present values for a stream of future benefits or costs (Mumford, 2001). Neither of these approaches is entirely satisfactory. The common time frame ignores the wide range of spread potential for different invasive species and gives greater weight to rapid spread over eventual impact. Setting a present value may undervalue the increasing uncertainty of estimates over time and may not allow for different social discount rates (Freeman and Groom, 2013), which may describe longer term societal values on future biodiversity that might apply for iconic or endangered environments.

The invasive species risk assessment process in Great Britain standardises the risk assessment time frame at five years, taking into the account the limited resources and the need for priorities based on short-term capacity. However, assessors are asked to indicate if a species poses a threat that would have high impact over longer time frames and where special consideration of long-term effects and management potential could suggest special consideration. In contrast, in Australian invasive species risk assessments, impacts consider the full extent of eventual potential spread, irrespective of the time frame over which these may occur. Species distribution models including climate matching models and maximum entropy models play a significant part in estimating potential extent (Elith, 2017).

Pathway vs. species risk assessment

Invasive species risks most commonly focus on species, but the risk is a combination of the attributes of the agent, the pathway and the receptor environment. Pathways are defined as any means by which species may enter or spread in a new environment (FAO, 2017). Often they are narrowly specified by area of origin, means of movement and association with particular commodities, trades or activity. The name of the European Invasive Alien Species Regulation 1143/2014 itself focuses on species, although it describes pathways and receptors as relevant factors.

Roy et al. (2015) reviewed both species and pathways in their assessment of priorities for invasive species risks into Europe. Escape from confinement (gardens, ponds, pets) was the main pathway for freshwater vertebrates and fish, plants and terrestrial vertebrates; transport contamination or associations were the main pathways for marine species and terrestrial invertebrates. Just over 40% of the species ranked come from Asia, and just under 40% come from the Western Hemisphere. Hulme (2009) highlights the importance of movement of pest species by air in recent times, citing 73% of US pest interceptions between 1984 and 2000 occurring at airports. Fresh produce in air freight is currently growing at 10% per year (Economist, 2017), so this pathway is likely to continue to increase in importance.

Exclusion or inclusion lists

Species may be regulated using lists that specify unwanted harmful species (exclusion lists) or, alternatively, acceptable species for entry (inclusion lists). Europe has adopted an exclusion list for invasive species under the current Invasive Alien Species Regulation. Member states are required to indicate potential harmful species of concern and must submit risk assessments in support of wider EU concerns. A species can be added to the list after technical review of the evidence for harm at a significant level over a large area of Europe. Member states are expected to take action against species on the list but may also take action under emergency measures against organisms of concern not yet on the list. Such a list can impose increasing responsibilities on member states under their obligations to control invasive species as a result of species added to the list through the concerns of other states. In the development of the current regulation, there was discussion about limiting the number of organisms that could be included because of concern that the scale of the obligations could become too great. The list is based on risk assessment criteria and not management feasibility or cost. As a result, species on the list may range from some that are easy to manage to others that are extremely difficult to manage.

Inclusion lists are used in some countries but are most practical as a reflection of past introductions or trade which have proved to be acceptable. An example in England is a list of fish species that are permitted for import which is periodically updated under the Import of Live Fish (England and Wales) Act 1980. Adding new species to an inclusion list requires evidence to indicate that they would be acceptable – for instance, that they would not establish, spread or have harmful impacts. This is often difficult to prove with sufficient certainty, and inclusion lists are therefore more restrictive than regulations based on exclusion lists.

Both types of lists are dependent on the concerns or interests of stakeholders, who have strong reason to exclude or permit introductions or use. Analysis in either case will involve structured evidence with uncertainty and needs a management regime to explicitly address the levels of acceptable risk involved.

Determining future invasiveness

Invasiveness is a function of the intrinsic ability of an organism to establish, spread and cause harm through competition with native species and the potential to enter a new environment through attachment to an entry pathway. Future invasiveness can be affected by changes in the agent itself through evolution and adaptation (Leger and Espeland, 2010); changes in the receptor environment that could make it more amenable to establishment or spread of a new entrant (Higgins and Richardson, 2014) or more vulnerable to damage (EEA, 2012); and pathways that may change in volume, speed, condition or route in a way that makes entry more probable for a species (Chapman et al., 2016).

Of these potential changes, pathways can change most quickly and have the most immediate effect. Over the past 30 years in Europe, trade from China and the Far East has grown at a very rapid rate, and the balance of entry of unwanted organisms has shifted away from a predominantly transatlantic route (Chapman et al., 2016). As a result, Asian species are now relatively common invaders in Europe. Entry probability often is a function of trade volume, so increases in volume directly increase risk, unless risk management is imposed on the pathway to reduce survival on the route.

Organisms entering a new environment undergo natural selection (Leger and Espeland, 2010). Where only a small proportion of a population is able to survive the transfer to a new location, the selection process may result in a subpopulation that is better adapted to new conditions than the wider population in the original range, leading to unexpected establishment in new habitats and conditions. In many circumstances, the receiving environment is entirely novel and the response of the invading population is inherently unpredictable. Natural selection on the pathway is difficult to anticipate and amplifies uncertainty about establishment and impact. The behaviour of ramorum blight in the UK is an example; it exhibits a very different host range and impact, particularly affecting larch rather than oak, compared to its expression as sudden oak death in the US (Harris and Webber, 2016).

Pathways for invasive species can change very rapidly as markets and travel patterns shift (Chapman et al., 2016; Waage and Mumford, 2008), quickly changing the expected probability of entry. Receptor environments can also change through disturbance and changes in land use, which can affect the likelihood of establishment (Pysek and Richardson, 2010). Where environments become more suitable to a particular invading species and the area of suitable environment expands, the probability of establishment will increase, but some environmental change may also inhibit establishment of invasive species (Bellard et al., 2018). While the rate of change in environmental vulnerability may be relatively slow, it may be more predictable than changes in pathways or the adaptability of organisms. Prediction may be complicated by the fact that climate change and other environmental processes may interact with pathway changes and the biology of invasive species (Bellard et al., 2018), precipitating unexpected events such as novel hybridisations and accelerated spread (Muhlfeld et al., 2014; Chapman et al., 2016).

Horizon scanning

Risk assessment is prompted by an expression and recognition of concern about particular species or pathways by stakeholders and authorities. In many cases this concern arises through the interception of organisms on an entry pathway or discovery of an organism that has successfully entered and begun to establish and spread. These are very immediate concerns that often stretch limited capacities for risk assessment and management. Despite these demands, longer term and less likely events should be monitored, as entry routes change and conditions for establishment or impact evolve, to detect and manage emerging problems as early and effectively as possible. Horizon scanning exercises (such as Roy et al., 2015) have been carried out in Europe to identify such potential risks. Roy et al. (2015) assigned scores on a five-point scale for the four key risk components, as in ISPM 11, and provided an overall score based on their product. This ranked 95 species drawn from extensive literature reviews and expert consultations. Roy et al. (2015) provide a complete description of the horizon scanning methodology in their report.

A similar large-scale exercise in ranking risks was carried out for the UK Plant Health Risk Register,[7] which also assigned five-point scores for the same key risk components for almost 1,000 potential agricultural, horticultural and forestry pests. This process involved an extensive series of workshops with industry and other stakeholders to provide the scores, facilitated by ministry officials. Because of the scale of the process, confidence levels were not included with the scores. The risk register includes two scores for each species, with and without anticipated management actions. This provides an opportunity to rank pests according to the risk they pose and to rank relative value of management for species. This could then be set against the costs and effectiveness of such management to indicate a return on investment for the species considered. A return-on-investment approach has been recommended as a general prescription for setting priorities for invasive species management internationally (Dodd et al., 2017).

Conclusion

Risk assessment for invasive species is carried out in response to concerns expressed by responsible authorities and other stakeholders to support risk management. It does this by setting priorities on management and enabling stakeholders and responsible authorities to discuss the potential outcomes of invasions within a recognised framework of risk. To set priorities, risk assessments must be conducted within a consistent framework that ensures that risks are calibrated according to clearly defined scales, whether quantitative, semi-quantitative or qualitative. The key risk components for both agricultural pests and invasive species include entry, establishment, spread and impact. Where establishment can be expressed as conditional on entry, their values may be combined, leading to the likelihood element of risk; the product of spread and impact is a measure of the magnitude element of risk. Management priorities depend on both the extent of the assessed risk from an invasive species and, very important, relevant criteria such as the opportunity, practicality and efficacy of management options. So, a low priority risk may be a high priority for management if the potential actions are effective and easy while high risks may need to be accepted without active management if there are no practical options. This view encapsulates a philosophy that embraces the maximisation of returns on investment in biosecurity, where returns may be in the form of reduced impacts on the economy, the environment, human health or cultural and social values. The acceptable level of risk for responsible authorities and other stakeholders, particularly those from whom there is an expectation of recovering costs of management, should be established in advance.

Notes

1 European Commission, Invasive Alien Species website, 2021 (see http://ec.europa.eu/environment/nature/invasivealien/index_en.htm).
2 Great Britain Non-native Species Secretariat, Risk Assessment website, 2021 (see www.nonnativespecies.org/index.cfm?pageid=143).
3 European Commission, Invasive Alien Species website, 2021 (see http://ec.europa.eu/environment/nature/invasivealien/index_en.htm).
4 Great Britain Non-native Species Secretariat website, 2021 (see www.nonnativespecies.org).
5 UK Department for Environment, Food and Rural Affairs, UK Plant Health Risk Register website, 2021 (see https://secure.fera.defra.gov.uk/phiw/riskRegister/).
6 Great Britain Non-native Species Secretariat website, 2021 (see www.nonnativespecies.org).
7 UK Department for Environment, Food and Rural Affairs, UK Plant Health Risk Register website, 2021 (see https://secure.fera.defra.gov.uk/phiw/riskRegister/).

References

Baker, R.H.A., Black, R., Copp, G.H., Haysom, K.A., Hulme, P.E., Thomas, M.B., Brown, A., Brown, M., Cannon, R.J.C., Ellis, J., Ellis, M., Ferris, R., Glaves, P., Gozlan, R.E., Holt, J., Howe, L., Knight, J.D., MacLeod, A., Moore, N.P., Mumford, J.D., Murphy, S.T., Parrott, D., Sansford, C.E., Smith,

G.C., St-Hilaire, S. and Ward, N.L. (2007) 'The UK risk assessment scheme for all non-native species', in W. Rabitsch, F. Essl and F. Klingenstein (eds.) *Biological Invasions: From Ecology To Conservation. NEO-BIOTA*, vol. 7, pp. 46–57.

Bellard, C., Jeschke, J.M., Leroy, B. and Mace, G.M. (2018) 'Insights from modeling studies on how climate change affects invasive alien species geography', *Ecology and Evolution*, vol. 8, pp. 5688–5700.

Booy, B., Mill, A.C., Roy, H.E., Hiley, A., Moore, N., Robertson, P., Baker, S., Brazier, M., Bue, M., Bullock, R., Campbell, S., Eyre, D., Foster, J., Hatton-Ellis, M., Long, J., Macadam, C., Morrison-Bell, C., Mumford, J., Newman, J., Parrot, D., Payne, R., Renals, T., Rodgers, E., Spencer, M., Stebbing, P., Sutton-Croft, M., Walker, K.J., Ward, A., Whittaker, S. and Wyn, G. (2017) 'Risk management to prioritise the eradication of new and emerging invasive non-native species', *Biological Invasions*, vol. 19, no. 8, pp. 2401–2417.

Chapman, D., Purse, B.V., Roy, H.E. and Bullock, J.M. (2017) 'Global trade networks determine the distribution of invasive non-native species', *Global Ecology and Biogeography*, vol. 26, pp. 907–917.

Chapman, D.S., Makra, L., Albertini, R., Bonini, M., Páldy, A., Rodinkova, V., Šikoparija, B., Weryszko-Chmielewska, E. and Bullock, J.M. (2016) 'Modelling the introduction and spread of non-native species: International trade and climate change drive ragweed invasion', *Global Change Biology*, vol. 22, pp. 3067–3079.

Copp, G., Godard, M., Russell, I., Peeler, E., Gherardi, F., Tricarico, E., Miossec, L., Goulletquer, P., Almeida, D., Britton, J., Vilizzi, L., Mumford, J., Williams, C., Reading, A., Rees, E.M. and Merino-Aguirre, R. (2014a) 'A preliminary evaluation of the European Non-native Species in Aquaculture Risk Assessment Scheme (ENSARS) applied to species listed on Annex IV of the EU Alien Species Regulation', *Fisheries Management and Ecology*, vol. 23, no. 1, pp. 12–20.

Copp, G., Russell, I., Peeler, E., Gherardi, F., Tricarico, E., MacLeod, A., Cowx, I.G., Nunn, A.D., Occhipinti-Ambrogi, A., Savini, D., Mumford, J. and Britton, J.R. (2014b) 'European Non-native Species in Aquaculture Risk Analysis Scheme (ENSARS): A summary of assessment protocols and decision-support tools for use of alien species in aquaculture and stock enhancement', *Fisheries Management and Ecology*, vol. 23, no. 1, pp. 1–11.

Crooks, J.A. (2005) 'Lag times and exotic species: The ecology and management of biological invasions in slow-motion', *Écoscience*, vol. 12, pp. 316–329.

Dodd, A.J., Ainsworth, N., Hauser, C.E., Burgman, M.A. and McCarthy, M.A. (2017) 'Prioritizing plant eradication targets by re-framing the project prioritization protocol (PPP) for use in biosecurity applications', *Biological Invasions*, vol. 19, pp. 859–873.

Economist (2017) 'Trends in the air-freight business', *The Economist*, April 14, www.economist.com/business/2017/04/14/trends-in-the-air-freight-business, accessed 26 August 2019.

EEA (2012) 'The impacts of invasive alien species in Europe', European Environment Agency, Brussels, Belgium, 114pp.

Elith, J. (2017) 'Predicting distributions of invasive species', in A.P. Robinson, T. Walshe, M.A. Burgman and M. Nunn (eds.) *Invasive Species: Risk Assessment and Management.* Cambridge University Press, Cambridge, United Kingdom.

EPPO (2011) 'Guidelines on pest risk analysis', PM 5/3(5) European and Mediterranean Plant Protection Organisation, Paris, France, 44pp.

FAO (2017) 'Pest risk analysis for quarantine pests', ISPM 11, International Plant Protection Convention, FAO, Rome, Italy, 40pp.

Freeman, M. and Groom, B. (2013) 'Biodiversity valuation and the discount rate problem', *Accounting, Auditing and Accountability Journal*, vol. 26, pp. 715–745.

Harris, A.R. and Webber, J.F. (2016) 'Sporulation potential, symptom expression and detection of *Phytophthora ramorum* on larch needles and other foliar hosts', *Plant Pathology*, vol. 65, pp. 1441–1451.

Higgins, S.I. and Richardson, D.M. (2014) 'Invasive plants have broader physiological niches', *Proceedings of the National Academy of Sciences*, vol. 111, pp. 10610–10614.

Holt, J., Leach, A.W., Knight, J.D., Griessinger, D., MacLeod, A., van der Gaag, D.J., Schrader, G. and Mumford, J.D. (2012) 'Tools for visualising and integrating pest risk assessment ratings and uncertainties', *EPPO Bulletin*, vol. 42, pp. 35–41.

Holt, J., Leach, A.W., Schrader, G., Petter, F., MacLeod, A., van der Gaag, D.J., Baker, R.H.A. and Mumford, J.D. (2014) 'Eliciting and combining decision criteria using a limited palette of utility functions and uncertainty distributions: Illustrated by application to Pest Risk Analysis', *Risk Analysis*, vol. 34, pp. 4–16.

Hulme, P.E. (2009) 'Trade, transport and trouble: Managing invasive species pathways in an era of globalization', *Journal of Applied Ecology*, vol. 46, pp. 10–18.

Leger, E.A. and Espeland, E.K. (2010) 'Coevolution between native and invasive plant competitors: Implications for invasive species management', *Evolutionary Applications*, vol. 3, pp. 169–178.

Milner-Gulland, E.J. and Shea, K. (2017) 'Embracing uncertainty in applied ecology', *Journal of Applied Ecology*, vol. 54, pp. 2063–2068.

Muhlfeld, C.C., Kovach, R.P., Jones, L.A., Al-Chokhachy, R., Boyer, M.C., Leary, R.F., Lowe, W.H., Luikart, G. and Allendorf, F.W. (2014) 'Invasive hybridization in a threatened species is accelerated by climate change', *Nature Climate Change*, vol. 4, pp. 620–624.

Mumford, J.D. (2001) 'Environmental risk evaluation in quarantine decision making', in K. Anderson, C. McRae and D. Wilson (eds.) *The Economics of Quarantine and the SPS Agreement*. Centre for International Economic Studies, Adelaide, and AFFA Biosecurity Australia, Canberra, Australia.

Mumford, J.D. (2013) 'Biosecurity management practices: Determining and delivering a response', in A. Dobson, K Barker and S. Taylor (eds.) *Biosecurity: The Socio-Politics of Invasive Species and Infectious Diseases*. Earthscan, London, United Kingdom.

Mumford, J.D., Booy, O., Baker, R.H.A., Rees, M., Copp, G.H., Black, K., Holt, J., Leach, A.W. and Hartley, M. (2010) 'Invasive species risk assessment in Great Britain', *Aspects of Applied Biology*, vol. 104, pp. 49–54.

Mumford, J.D., Leach, A., Holt, J., Suffert, F., Moignot, B. and Hamilton, R.A. (2017) 'Integrating crop bioterrorism hazards into pest risk assessment tools', in M.L. Gullino, J. Stack, J. Fletcher and J.D. Mumford (eds.) *Practical Tools for Plant and Food Biosecurity*. Springer, Cham, Switzerland.

OGTR (2013) *Risk Analysis Framework*. Office of the Gene Technology Regulator, Canberra, Australia.

Pysek, P. and Richardson, D.M. (2010) 'Invasive species, environmental change and management, and health', *Annual Review of Environment and Resources*, vol. 35, pp. 25–55.

Regan, H.M., Colyvan, M. and Burgman, M. (2002) 'A taxonomy and treatment of uncertainty for ecology and conservation biology', *Ecological Applications*, vol. 12, pp. 618–628.

Roy, H.E., Adriaens, T., Aldridge, D.C., Bacher, S., Bishop, J.D.D., Blackburn, T.M., Branquart, E., Brodie, J., Carboneras, C., Cook, E.J., Copp, G.H., Dean, H.J., Eilenberg, J., Essl, F., Gallardo, B., Garcia, M., García-Berthou, E., Genovesi, P., Hulme, P.E., Kenis, M., Kerckhof, F., Kettunen, M., Minchin, D., Nentwig, W., Nieto, A., Pergl, J., Pescott, O., Peyton, J., Preda, C., Rabitsch, W., Roques, A., Rorke, S., Scalera, R., Schindler, S., Schönrogge, K., Sewell, J., Solarz, W., Stewart, A., Tricarico, E., Vanderhoeven, S., van der Velde, G., Vilà, M., Wood, C.A. and Zenetos, A. (2015) 'Invasive alien species: Prioritising prevention efforts through horizon scanning', ENV.B.2/ETU/2014/0016. European Commission, Brussels, Belgium.

Schrader, G., MacLeod, A., Petter, F., Baker, R.H.A., Brunel, S., Holt, J., Leach, A.W. and Mumford, J.D. (2012) 'Consistency in pest risk analysis: How can it be achieved and what are the benefits?', *EPPO Bulletin*, vol. 42, pp. 3–12.

Soliman, T., MacLeod, A., Mumford, J.D., Nghiem, T.P.L., Tan, H.T.W., Papworth, S.K., Corlett, R.T. and Carrasco, L.R. (2016) 'A regional decision support scheme for pest risk analysis in Southeast Asia', *Risk Analysis*, vol. 36, no. 5, pp. 904–913.

Speirs-Bridge, A., Fidler, F., McBride, M., Flander, L., Cumming, G. and Burgman, M. (2010) 'Reducing overconfidence in the interval judgments of experts', *Risk Analysis*, vol. 30, pp. 512–523.

Urquhart, J., Potter, C., Barnett, J., Fellenor, J., Mumford, J. and Quine, C.P. (2017) 'Expert risk perceptions and the social amplification of risk: A case study in invasive tree pests and diseases', *Environmental Science and Policy*, vol. 77, pp. 172–178.

Vanderhoeven, S., Adriaens, T., D'hondt, B., Van Gossum, H., Vandegehuchte, M., Verreycken, H., Cigar, J. and Branquart, E. (2015) 'A science-based approach to tackle invasive alien species in Belgium: The role of the ISEIA protocol and the Harmonia information system as decision support tools', *Management of Biological Invasions*, vol. 6, pp. 197–208.

Waage, J.K. and Mumford, J.D. (2008) 'Agricultural biosecurity', *Philosophical Transactions of the Royal Society B*, vol. 363, pp. 863–876.

With, K. (2002) 'The landscape ecology of invasive spread', *Conservation Biology*, vol. 16, pp. 1192–1203.

18

THE EMERGENCY MODALITY

From the use of figures to the mobilisation of affects

Francisco Tirado, Enrique Baleriola and Sebastián Moya

Introduction

So-called global health has meant, among other things, opening a field of reflection that has problematised, in a novel way, both the notion of health and the notion of health intervention (Lakoff and Collier, 2008). Thus, it has been posed, first, that the processes of globalisation constitute a new source of risk since the accelerated movement of people, animals, technologies and merchandise of all kinds contributes to the rapid dissemination of new biological threats (Peckham and Sinha, 2017). Second, the aforementioned phenomenon puts on the table the problem of competencies in the regulation and attention of new threats. What institutions or governments are responsible for acting in the context of a global threat? Who is responsible for coordinating the intervention? Questions such as these are regularly formulated by experts and laypeople and have not yet received clear and convincing answers. Finally, and this is the element we will interrogate in this chapter, global threats share an interpretative approach called the 'emergency modality of intervention'.

This modality does not suppose a long-term intervention in the cultural, social, political or economic dimensions associated with the appearance of outbreaks of infectious diseases. Instead, it is committed to fast-acting medical practices, the use of standardised protocols for each biological emergency and the massive dissemination of surveillance and data recording systems that are easy to implement and maintain. Mobility is the essential characteristic of such techniques. They can be deployed quickly, with low economic cost, anywhere in the world regardless of their idiosyncratic local characteristics (Collier and Lakoff, 2008).

Several factors are used to justify the application of an emergency modality in the health field. First, emergency measures quickly attract the attention of public opinion and generate broad political acceptance. Among other characteristics, they are easy to describe and demonstrate in the media with some assiduity. Second, emergency implementation is relatively easy to perform in contrast to major measures that seek to transform cultural or population habits. Finally, all these measures avoid the complex webs of significances and local uses that can be found in each specific territory by imposing a set of standardised protocols that have general application.

Our work aims to deepen this line of analysis. Starting from postulates inherited from philosopher Michel Foucault and actor-network theory (Latour, 2005), we make a nominalist approach to the notion of 'emergency modality'. We argue that there is no univocal meaning to the concept of 'emergency', the meaning of which depends directly on technology and the set of practices that accompany the use of the word. In this way, we argue that an emergency modality is not the same when the threat is treated with the old technology of risk calculation than when modern protocols or scenarios are used. In each case, the notion of emergency acquires a concrete and differentiated meaning that redefines the biological threat in a very specific sense. We support this analysis by examining three concrete sets of practices that have been put into operation in well-known historical biothreats: *risk calculation*, the use of *protocols* and *action through scenarios*.

We begin by reviewing the meaning of the word 'emergency' in the field of global health. Second, we consider what the notion means when working with classical *risk calculation* in epidemiology, using the example of SARS (severe acute respiratory syndrome) and the influenza A (H1N1) epidemic. Third, using the example of Zika, we examine the meaning of 'emergency' when *protocols* and action guides are used. Finally, drawing on the example of the Ebola epidemic in 2014, we pay attention to the meaning acquired by the notion of biological emergency as produced through *scenarios*.

What is a biological emergency?

The notion of an epidemic has always been considered an example of what Dupuy (1999) calls a 'panic phenomenon'. When the news of a possible mass contagion breaks out, the masses become individualised and the social order is fragmented. Irrationality, fear and finally panic appear. This reading has two great foundations. One is etymological. The Greek word 'epidemic' (*epi-demos*) admits the translation of 'against the people'. That is to say, the epidemic is an event that hits human life as a collective life or aggregate of individuals and destroys it. Therefore, we must fear it. In that sense, it is interesting to note that for Hippocrates, father of medicine, an epidemic did not exactly refer to the phenomenon of an infection in itself but rather to the condition in which such an infection begins to be suffered by a collective, affecting not only individuals but the whole village (*demos*). The second foundation is historical. Along these lines, Michel Foucault (2003) puts forward the epidemic is something more than a particular form of disease. For instance, in the bilious fever outbreaks of Marseilles in 1721, Bicêtre in 1780 and Rouen in 1769, what occurred was the emergence of epidemiology as the 'autonomous, coherent, and adequate evaluation of disease' (Foucault, 2003, 23).

Furthermore, Foucault emphasised that a medicalisation of epidemics could exist only if supplemented by the police: to supervise the location of mines and cemeteries; to get as many corpses as possible cremated instead of buried; to control the sale of bread, wine and meat; to supervise the running of abattoirs and dye works; and to prohibit unhealthy housing (Foucault, 2003, 25). Intervention to manage them quickly became an official matter and contributed to the construction of what we know as the modern state (Maureira et al., 2018).

This link between epidemics and panic continues today and has been generalised to every biological threat. In a broad sense of the term, 'emergency' is defined as a serious, unexpected and often dangerous situation requiring immediate action (Oxford Dictionary, 2018). In epidemiological and technical terms, a 'public health emergency' is defined as an occurrence or imminent threat of an illness or health condition – caused by bioterrorism, an epidemic or pandemic disease or a novel and highly fatal infectious agent or biological toxin – which poses a substantial risk of a significant number of human fatalities or incidents of permanent or long-term disability. This includes, but is not limited to, an illness or health condition resulting from a natural disaster (CDC, 2001).

There are several global institutions that participate in the management of these emergencies. The most well known is the World Health Organization (WHO), which has developed an Emergency Response Framework (ERF) to establish its functions and responsibilities in these

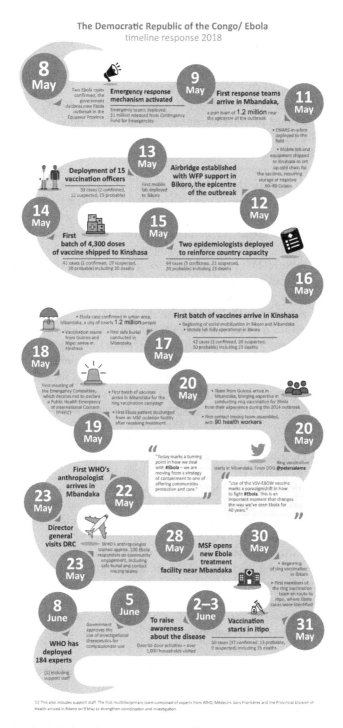

Figure 18.1 Outbreak of Ebola in the Democratic Republic of Congo

Source: WHO, 2018b.

EWARS – Early Warning, Alert and Response System; WFP – World Food Programme; MSF – Médecins Sans Frontières

cases and provide a common approach to the work it must accomplish during an emergency, as well as its obligation to act with urgency and predictability (WHO, 2013). This framework incorporates various levels of action conforming to the degrees of emergency (1, 2, and 3), which are defined according to their consequences to public health and the demand of the World Customs Organization (WCO) and the WHO response.

Let us take one example of an emergency response. In 2018 the WHO and the Democratic Republic of Congo worked together to contain an outbreak of Ebola virus disease (EVD) in Bikoro, Equateur. There were 34 reported cases in five weeks between October and November that year, including 2 confirmed cases, 18 probable cases (deceased) and 14 suspected cases (WHO, 2018a). When the emergency response mechanism was activated, these institutions were able to generate a response that included multiple and diverse actors (Figure 18.1; WHO, 2018b). That is, a rapid response entails mobilising different actors affected by the outbreak and, what is more, ensuring these actors align as a single entity.

This pattern of action or logics can be identified from the beginning of the 21st century. Thus, we could add to the aforementioned examples of global health emergencies SARS in 2003 and the influenza A (H1N1) pandemic in 2009. The first appeared in Guangdong, China, and affected 26 countries with more than 8,000 cases (WHO, 2003), and the second was first detected in the US, and it was estimated that between 151,700 and 575,400 people worldwide died during the first year (Figure 18.2; WHO, 2009b).

An emergency, therefore, is a situation of danger that demands a quasi-immediate response. In the case of biological emergencies, the risk of their extension to the global scale is undoubtedly assumed, and it is argued that an adequate response must be a coordinated action that brings together multiple actors. Therefore, the notion of emergency recalls what was affirmed by Foucault (2003): an emergency is much more than a situation of risk or danger since it involves the creation of a playing field in which: a) a scientific (medical) authority appears that is not restricted to one authority of knowledge or one expert – on the contrary, it refers to a social institution that makes decisions that directly affect people, cities and even countries; b) a field of intervention is defined that usually goes beyond the mere biological problem of air, water, cultivated lands, infrastructures, etc. and becomes the focus of regulation and intervention by the aforementioned authority; c) medical administration mechanisms are introduced, such as data recording, comparison of statistics, obtaining samples, etc.; and d) an analytical economy of health and care that will be applied to the affected areas is determined and will operate as an internationally legitimated form of foreign intervention.

These general characteristics define the notion of emergency that governments and international agencies, experts and media use. However, despite these common denominators, the 'emergency modality' that operates through interventions into biological threats articulates different universes of meaning depending on the technical devices that are deployed and their associated practices. Let us consider, first of all, what happens when the biological emergency utilises the tools of classical risk assessment.

SARS, H1N1 and classical risk assessment: we are not satisfied with less than eradication

The SARS crisis occurred in an interconnected world with an integrated economy and was considered a global threat of a type not seen before (Tan and Enderwick, 2006). It had a range of impacts, from affecting emergency services with a reduction in patient care (Huang et al., 2005) to affecting the economies of China and Hong Kong mainly due to the impact of the disease on the behaviour of people (Lee and McKibbin, 2004).

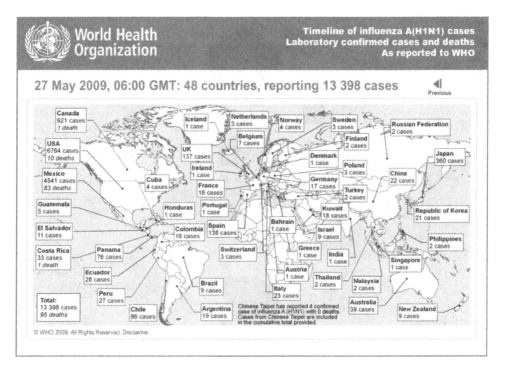

Figure 18.2   Timeline of influenza A (H1N1) cases. Laboratory-confirmed cases and deaths as reported to WHO.

Source: WHO, 2009b.

SARS was contained in less than four months due to unprecedented international cooperation, from the WHO and the Global Outbreak Alert and Response Network (GOARN) together with 115 national health services to academic institutions, technical institutions and individuals. This response started with the notification of a case of atypical pneumonia in Guangdong (November 2002). The WHO then issued a global alert and activated the chain of steps provided by the Emergency Operations Center (EOC) of the CDC, publishing provisional guidelines for state and local health departments on SARS. Subsequently, a 'health warning notice' and associated precautions were issued for the spraying of patients suspected of SARS in hospitals and especially in crossing borders (CDC, 2013). In respect to the notification of the first case, there is a theory that transmission between humans did not occur until February 21, when a doctor (case A; Figure 18.3), whose illness began on 15 February, stayed at the Metropole Hotel in Hong Kong, generating the spread of SARS (CDC, 2003).

After the SARS crisis, different models about SARS transmission patterns were published in order to identify patterns that could be used in the future for comparison and employed in the event of a new SARS crisis. But these models had limited impact because no reliable projections were provided to the authorities that incorporated the impact of overestimated management, which possibly contributed to media misunderstanding. Following this logic, previous models about SARS spreading were based on the modelling of smallpox (2002) and anthrax (2003), and their forecast was not met mainly because SARS was a new and not well-known human disease (Glasser et al., 2011). Since then, China has built effective public health emergency management systems (PHEMS) based on protocols and drills about previous epidemics and has made

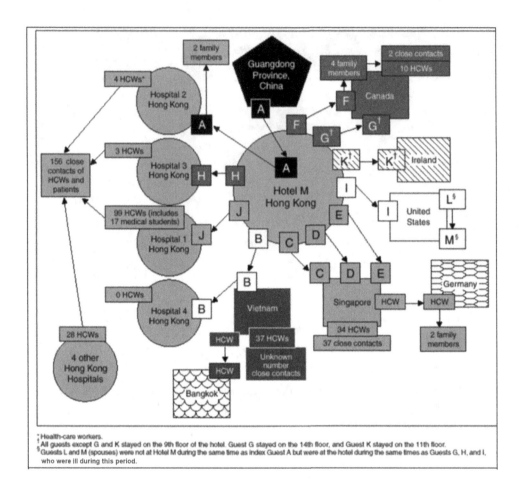

Figure 18.3 Chain of transmission from Hotel M.

Source: CDC, 2003.

comprehensive progress and improvements in preparation, readiness, response and recovery (Sun et al., 2018).

The SARS emergency was immediately followed by the arrival of the influenza A (H1N1) pandemic, which generated responses to contain the emergency with marked similarities to those of the previous SARS outbreak. These operations are described in the WHO document *Human Infection with Pandemic (H1N1) 2009 Virus: Updated Interim WHO Guidance on Global Surveillance*, whose main objective was to guide countries in what should be done to address this particular threat. This document was presented in four sections that have characteristics common to a wide range of epidemic surveillance protocols of recent decades and are also explained in more detail in the classic manuals on emergency management (e.g. Lindgren and Bandhold, 2003): what we term the emergency modality of classical risk assessment. Following this document, the classic emergency modality can be summed up as follows: a) it is a process focused on taking *prophylactic measures* against emergent threats that are already occurring; b) it is based on the employment of statistical data and decision trees or chains in order to manage risk *on the basis of what was previously known*; c) it is addressed to calculate the consequences after an outbreak through *the use of statistics* from countries' reports; and d) it is a process managed through its representation – that is, by *bringing to the present past statistical data* and related knowledge and displaying it in graphs, diagrams or maps.

According to classical risk assessment, the systems modelling SARS and H1N1 were based on the transformation of causal maps into mathematical representations that can be simulated (Lindgren and Bandhold, 2003, 119), dragging (bio)risk from the sphere of uncertainty to the realm of objective reality. The result of this modelling is a diagram or graph built through data collected from different sources, producing a piece of knowledge that is a copy of a parcel of reality (Figures 18.2 and 18.3). This operation has important consequences. First, it means that it is possible to substitute a fragment of reality with its corresponding graph. Second, the substitution allows elements of reality to be managed and for current knowledge to be produced and stored such that, when a new risk arrives, it can be managed with the previously accumulated knowledge. Thus, representation is considered a precondition to intervene and to change reality, but, at the same time, representation is the final justification of the knowledge produced. So, investing large amounts of funding in representational technologies is considered by experts, politicians and stakeholders as crucial to prevent new outbreaks. This fragment of a report from the WHO illustrates the essence of the emergency modality of classic risk assessment:

> WHO will use the information provided to inform global risk assessments, including mathematical modelling of the epidemic, to better understand the spread of the pandemic and the effectiveness of mitigation measures. Scientists from countries providing data will be invited to participate in the development of, and be co-authors of, publications that draw on their country specific data. Countries will always be consulted in the development of any articles in which their data has been used. WHO will report and visualize the surveillance data provided. Reports will include alerts, situational summaries, tables, charts and maps of the evolving pandemic situation.
>
> *(WHO, 2009a, 7)*

Summing up, the use of classical risk assessment gives the notion of emergency a very clear and simple definition. It alerts us to a threat that can be located with more or less effort and, more important, that must be eradicated. The deployment of human actors, technologies and intervention practices is driven by a risk calculation where the objective is to detect and eliminate the factor that has triggered the danger based on prior knowledge. In this logic, notions such as case zero, origin, chain of transmission, etc. have major relevance. However, it is not always possible to eradicate the factor of origin, it is often very expensive to invest in detection technologies and monitor cases and, as we have previously seen, the calculations made on the intensity and speed of transmission are often unreliable. In addition, classical risk assessment offers a series of operations that face a biological threat based on its objective reality. In other words: the scientific knowledge of what is being discovered about the concrete threat is articulated with the knowledge of what has been done previously with similar risks.

Zika and protocols: the promise of containment

The Zika virus was identified in humans in 1952 in Uganda and the United Republic of Tanzania. Its first major outbreak was on the Island of Yap in 2007, its second in French Polynesia in 2013 and its third in Brazil in 2015. In this outbreak, the virus was associated with Guillain-Barré syndrome and microcephaly. Currently 86 countries have reported infections by Zika transmitted by mosquitoes (species *Aedes Aegypti*; WHO, 2018c).

The WHO has given support to various countries to properly manage the new threat through a Zika Strategic Response Framework. The support has consisted of means of detection, prevention, care and research (Figure 18.4). To undertake these actions, the WHO has involved different types of protocols or guidelines addressed to different agents. In fact, there are currently more than 20 documents produced by the WHO dealing with different topics around Zika, such as those related to diagnoses by analytics (e.g. *Guidelines for the Serological Diagnosis of Zika Virus Infection*), transmission (e.g. *Prevention of Sexual Transmission of Zika Virus Interim Guidance*), risk groups (e.g. *Pregnancy Management in the Context of Zika Virus Infection*) and scientific studies (e.g. *Prospective Longitudinal Cohort Study of Newborns and Infants Born to Mothers Exposed to Zika Virus during Pregnancy*). All these involve agents and actors corresponding to different levels and scales, from local and international institutions (e.g. Zika Strategic Response Plan, for coordination and collaboration among WHO and its partners) to laypeople and small towns.

There are several clinical guidelines specially designed for the use of laypeople: pregnant women, infants and children or women of reproductive age. For instance, in the first case, the guide outlines why it is not recommended to travel to areas with severe risk of Zika and, if the trip is inevitable, important steps to avoid mosquito bites (CDC, 2017a, 2017b) and protocols for after travelling (Figure 18.5). In addition to the WHO's guidance, private associations and NGOs such as MotherToBaby offer information (web details, FAQs, phone numbers, etc.) that can be consulted in case of concerns. As Rosa et al. (2017) point out, these protocols proved useful in the Zika emergency in several ways. First, they offered information about prevention – through geographical data of risk localities – and advice to avoid mosquitos. Second, they created a unified protocol allowing the channelling of suspected patients to a single pathway. Third, they allowed coordinated actions among different countries. Finally, they showed unified and standardised information that helped eliminate confusion and panic.

During this emergency, the mass media played an important role in the dissemination of the aforementioned information. There were various collaborations between national institutions (e.g. ministries of Brazil) and agencies (e.g. prefectures and universities) that were disseminated through newspapers and digital resources, which focused on detecting and controlling mosquitos (see Figure 18.5). In a similar way, the information movement among various professionals and experts working in institutions was also relevant, as Baker (2016) pointed out:

> In the last 2 months, I have received at least one email a week that includes information about the Zika virus; over the past several weeks, that has increased to at least one or more per day. The titles of these stories included 'Fears Over Spread of Zika Virus Grow in the Caribbean,' 'CDC Guidelines on Preventing Sexual Transmission of Zika Virus Issued,' 'CDC Updates Interim Zika Virus Guidelines for Pregnant Women and Women of Reproductive Age,' 'CDC Issues Updated Zika Guidelines for Health Care Providers,' 'EU Drugs Agency Sets Up Zika Task Force to Speed Vaccine Work,' 'Zika Virus a Global Health Emergency, W.H.O. Says,' 'Zika Virus Isn't the Only Concern for Rio Olympics' and 'Public Health Agencies, Hospitals Prepare for Potential Zika Spread.'

And, unlike in other emergencies of this type, digital social networks also played a key role in spreading information among the population (Chandrasekaran et al., 2017).

The notion of emergency deployed through the use of protocols and guidelines is different from the one we found in the case of risk calculation. It is assumed that the cause of a threat may not be eradicated or that such an operation would require considerable time and effort. Instead, the universe of containment unfolds. Before the most negative effects of a biological threat are spread, the mobilisation of the necessary resources for stopping them is considered. In

Figure 18.4 Zika strategic response.

Source: WHO, 2016a.

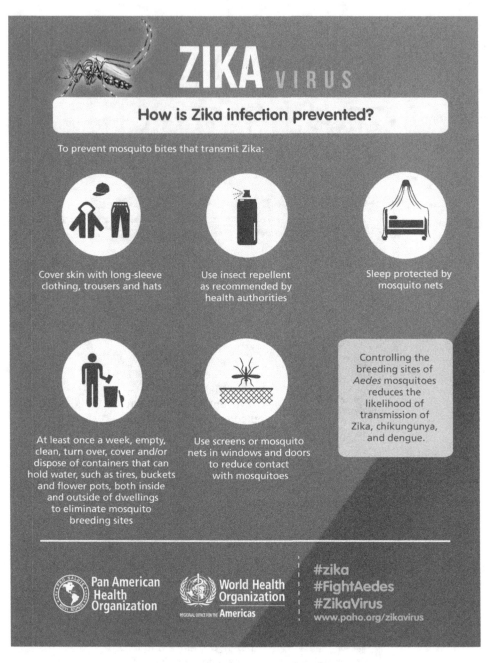

Figure 18.5 Zika virus.

Source: PAHO, 2016.

its maximum aspiration, prevention before the appearance of such a threat in the future would be the ideal objective of any protocol or action guide. In this new logic, the calculation of risk has not disappeared. On the contrary, it continues to exist and plays an important role. However, now it is subsumed in the parameters of performance and interpretation that mark these protocols. This new logic has been called 'preparedness' and has been well described by various authors (Anderson, 2010; Collier, 2008; Fearnley, 2005).

The development of protocols and guidelines for action has become so important in the last two decades that their format has changed. Their structure has gone from being a mere form that established a small chain of relationships in which the steps of an intervention were ordered and hierarchised to a compendium of instructions (sometimes with hundreds of pages) that establish true grammars of life (Tirado et al., 2012). This logic of complexity has led to the evolution of preparedness, and the massive use of scenarios has been added to its complicated protocols. As we argue in the following sections, such incorporation supposes the establishment of a new logic to think about biological emergencies. In fact, we consider that this new tool converts so-called preparedness into a different situation that demands its own detailed description and analysis.

Ebola scenarios: mobilising affects

Scenario building has been used in many fields to produce knowledge about very different future uncertainties (such as pandemics, environmental disasters and biological accidents). In a very naive sense, a scenario is simply a play or performance that is focused on a possible future event and used to be better prepared in case it actually occurs. Scenarios have been implemented across a range of realms such as war, terrorism, economics, decision-making, organisation, psychology, sociology, physics and health. For instance, in the opinion of several authors, scenarios have become the most important way to obtain information about future health issues (Anderson and Adey, 2011; De Goede, 2008; Kaufmann, 2016; Krasmann, 2015), displacing other classical means such as statistical calculation (Anderson, 2010).

Let us introduce a couple of quotations from documents about scenario planning in communicable diseases, where the aim is to prevent and prepare for epidemics, and a graph predicting cases of Ebola in 2014 (Figure 18.6):

> The objective of this preparedness work is to perform a prospective assessment of different vaccine efficacy designs under different scenarios. Experts agreed that it is

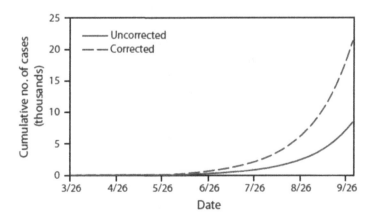

Figure 18.6 Estimated number of Ebola cases, Liberia and Sierra Leone combined, 2014.

Source: CDC, 2014.

important to provide researchers with a framework and a trial simulator to guide quantitative and qualitative assessments of the pertinence of trial designs in view of various epidemic scenarios for each priority pathogens.[1]

Ebola is a new risk in our country. . . . So, a lot of preparations for a potential Ebola patient involved simulation which then fall on my hands.[2]

These samples are interesting because they show some common features of scenario-planning logics. First, we find semantics about the future. This is always the topic of every scenario we can imagine. Second, the data and statements that appear in the samples are based on and endorsed by institutions and experts. These have the function of providing a burden of veridiction to the aforementioned semantics about the future. Third, scenarios join together different scales in a unique and logical claim (global institutions, citizens, countries, local hospitals, airports, physicians, nurses, families, etc.). Finally, they share an attempt to represent a very specific situation. 'Represent' must be understood in a common sense – that is, displaying knowledge either gathered or created by experts in their performance with charts, graphics or decision trees (Lewnard et al., 2014). In all cases they bring to the present some future information to prevent or be prepared against new epidemics, thereby representing or simulating the future. Thus, Lindgren and Bandhold (2003) posed that system modelling is a key element defining scenarios because this method is based on the transformation of causal maps into mathematical representations that can be simulated.

This is classical logic as attributed to scenarios: to build a representation. However, scenarios diverge from other representational forms. That is, a graph is produced through data collection from different sources, and a piece of knowledge is generated and considered a copy of a fragment of reality. It is possible to substitute the graph for this reality and manage and change elements of the reality thanks to the capacity of the graph to enable the further production of knowledge about reality. However, scenarios are much more.

Scenarios are existential territories

The French historian Patrick Zylberman (2013) considers that scenarios are tools that operate beyond mere representation. In his opinion, the great novelty they contribute is that they manage to convey emotions that are offered to their users. No matter what content we give to a scenario, its significance lies in delimiting an emotion and exploring how it develops and is lived by the user. This analysis rejects reducing scenarios to a form of representation. However, Zylberman's work responds to the coordinates of a historical analysis and cannot go beyond affirming the inclusion of certain emotions and affects in the scenarios he has diachronically analysed. If we combine their perspective with a synchronic one – that is, an analysis of social and cultural elements (obtained through individual and group interviews, image examinations, etc.) – we can identify something further and altogether more exciting: that scenarios go beyond the delimitation of affections as they build what we have called, borrowing a concept from Guattari (2008), *existential territories*. That is, the current scenarios open spaces for doubt, reflexivity, tension and conflict; offer clues to define emotions; and, what is more, establish coordinates so that the actors re-signify their relationship with the scenario itself and with the other actors, services or institutions included within it. In the following section, we look at these claims in more detail.

Scenarios open reflexivity

The statements and quotes in the following sections are based on the results of research carried out in the Universitat Autònoma de Barcelona, Spain. The research gathered the following empirical data: a) documentary materials from institutions like the Centres for Disease Control and Prevention, the World Health Organization and the Food and Agriculture Organization; b) hospital, laboratory and veterinary protocols; c) laws and regulations from the European Union and different Spanish cities; d) 12 focus groups with different society cohorts from Barcelona, Bilbao and Almería such as data experts, veterinary researchers, university students, elderly people, feminist groups and activist groups; e) a great variety of images and photographs related to epidemics from diverse sources: newspapers, international health institutions, hospitals and laboratories; and f) interviews with researchers and lab technicians.

The first point we found in the analysis of our results was that current scenarios frequently contained images, audio-visual elements accompanied and interacting with text, a video or even infographics that were not mere graphs based on statistical calculation (examples included a global map, a person with a containment suit, a naked body opened up to show the different infected tissues, virus on a microscopic scale, etc.). The second was that these scenarios were not addressed to experts but to laypeople (they are published on YouTube, they appear in newspapers and some others are fostered through public participation).[3] So, scenarios pursue forms of engagement and an interpretative relation between the scenario and everyday people. For instance, we encountered statements like these:

> [The scenario] makes me think that diseases are invented by laboratories; they are in fact, science-fiction, because they seem to be diseases made to control the world. It makes me think there are some powers that are trying to control us.
>
> *(young woman in the activist group from Barcelona)*

In a certain scenario, any citizen interprets its origin and its motive and positions themselves before it. Conspiracy theories are frequent, but they are not the only theories. Speculations about its usefulness and possible social application, about who has created them and why, are also common. The important thing, however, is not so much the content of such an interpretation as the emergence and realisation of it. At the moment when a citizen is confronted by a scenario and must make the effort to interpret it, s/he has been captured by it. Immediately, the interpretation gives way to the unfolding of a set of emotions that accompany this exercise of interpretation.

Emotions

RESEARCHER: In any moment, do you feel you could be infected, either by Ebola last year or H1N1 in 2009?

GROUP IN GENERAL: No . . .

FOCUS GROUP PARTICIPANT: Yes, I do, because I was in Argentina and one person I know was infected and I was going to her house, and I was living with her. That was the moment I was afraid. . . . I remember their flatmates were washing their hands all day with alcohol.

(feminist group in Bilbao)

Fear, then, was a clear emotion deployed in scenarios about Ebola, influenza or any kind of virus. And with fear a complex social and symbolic universe appears:

INTERVIEWER: How would you react if the media announces that this unknown virus has arrived in Europe?

FOCUS GROUP PARTICIPANT 1: There is not so much scope for action . . . if the case is very close maybe I would become paranoid and I would put a mask on.

FOCUS GROUP PARTICIPANT 2: Sincerely, if there are some alerts . . . it would scare me. I don't know if you are like 'Action Man' or made of ice. If I am told that Ebola arrives in Spain, it would scare me because it is to realise that: 'oh, it is here!'

(students from the University of Almeria)

The aforementioned emotions can be interpreted as affects because they realised what Rancière (2007) calls a redistribution of sensibility. That is, they determined what behaviour had to be carried out and how to approach the threat, how to define it, how to manage relationships with other actors and what to demand from institutional actors. They offered a concrete experience of reality. Therefore, the fiction created by the scenario is able to redefine how we engage with the world around us: cleaning and hygiene, self-care, fear, washing our hands, wearing masks, worrying about our peers, dealing with proximity to the virus, claiming social services and even thinking about our death.

Engagement with actors: from global to local and vice versa

A third important result is related to the role of institutions and actors. Throughout focus groups, participants talked about very different institutions and actors with which they would engage if they were in the epicentre of an epidemic: from hospitals to the World Health Organization, police, nurses, laboratories, governments, universities or social psychologists, among others. Let us see several examples:

INTERVIEWER: From the moment the virus is detected, what do you think is the pathway in order to contain it?

GROUP: In the case of animal outbreaks, a farmer detects something suspicious and calls the vet. The vet then reports to the veterinary official systems. Next, the official vet goes and gather samples that are analysed in a laboratory. . . . In a severe case, the Generalitat [the Catalan government] and the Spanish Ministry of Health are the institutions that may decide how to act.

(Veterinary group, Barcelona)

FOCUS GROUP PARTICIPANT 1: I don't know if the army would act how they are expected to. I think if a serious pandemic arrives, soldiers could not repress the population. If I were a nurse in Africa and I was infected . . . I think I would run away.

FOCUS GROUP PARTICIPANT 2: I think when you are infected, you are isolated from the rest of humankind . . . because people interact with you through a special suit . . . this thing that a physician touches you and can comfort you . . . is lost. You interact with a body from which you don't want anything.

(focus group of data experts from Barcelona)

What stands out extensively in the quoted fragments is the production of scales of activity and the relationship among them. The actors are able to define local scales and scales that link these with global ones. The conformation of scales is more relevant than it seems at first since it determines what our subsequent courses of action will be and how they will be articulated with those of other actors. For example, it is not the same to think about how to survive in a city besieged by a virus in which there are certain social services as it is to imagine the same from a small locality that requires you to move to get help.

Scenarios as existential territories

The French philosopher Jacques Rancière insists that images do not represent, that they are not manifestations of the properties of a technical meaning (Rancière, 2007). Rather, they are operations: relations between the whole and the parts; between a visibility and both a potential signification and an affect associated with them. They are operations between the expectations and that which executes them. In this sense, understanding scenarios as images means to put forward that these are mediators, engagements of effects, meanings and expectations. In fact, as Rancière (2005) showed, an image is not the creation of an imaginary world in opposition to the real world. It is an action that creates dissension, which changes the modes of the distribution of sensibility, and the forms of enunciation, by changing frames, scales or rhythms and building new relationships and ways of seeing and feeling. Thus, we can sustain that scenarios are tools to produce existential territories.

Nevertheless, they do not create any kind of territory; they work with risk (risk of infection, risk of an outbreak, risk of danger for the country, etc.). Thus, the sense of risk in scenarios is precisely what surrounds everything (emotions, behaviours, ideas, etc.). Risk is not a probability or number, but a frame of details where action and preparedness are blurred. Scenarios teach or show us new visibilities, sensitivities and forms of life that become possible and thinkable, and a new existential option emerges in its definition.

It must be clear that the use of mathematical models in strategic decision-making has not been excluded, they are simply subordinated to a broader objective: to 'think the unthinkable' (Gosselin and Tindemans, 2016). This means, first, to broach sensitive topics that could not have been discussed before by approaching abstract concepts; second, to become an open tool for risk work; and finally, to use a powerful and research-based method for approaching unpredictable and complex environments with a capacity to involve actors and knowledge from a range of domains by 'resocializing disciplines' towards a global health equity (Farmer et al., 2013). That is, with the scenarios we move beyond the definition of the emergency as a containment or prevention exercise. Now, it appears defined as an existential territory that affects all facets of our daily activity.

It could be argued that scenarios are a tool at the service of the logic of preparedness. This is correct, but it does not include all the effects that the widespread use of scenarios is producing. This supposes the opening of a new intelligence or logic to think about health. In this logic we find the following features: a) The scenario uses fiction to create a kind of play or role play. However, that fiction is loaded with data and elements from science that gives it a strong burden of veracity. b) Each scenario can be developed from free interpretation or, as is usually the case, presented as a kind of graphic script in which both experts and citizens can improvise. c) The scenarios are represented in different sets: one will be a scenario where the best of the possible cases will happen, another will be the worst-case scenario and others will be a mixture of both situations. d) The scenarios provide diagrams and equations that aim to 'capture' human behaviour – the rational or emotional variables that determine individual and collective behaviour.

Conclusions: the reality of fiction

Classical risk assessment, as we have seen, puts the focus on the reality of the (bio)risk: what we know about it, what we have done coping with similar epidemics and, on the basis of both, what we can do now. Certainty, the specific aim of 'dragging' the risk to the light of objectiveness – and the sphere of what we already know (and already know how to handle) – is the key feature of this first emergency modality. In order to achieve this goal, different techniques are employed: diagrams, decision trees or chains and statistics, among others. Next, we explored how protocols, as they articulate courses of action in a hypothetical future, opened, in some way, the epistemic reflection to certain fictional content. Finally, fiction was shown to be key to scenarios.

Improvising in scenarios, the actors immerse themselves in an imaginary world and through that mimesis become familiar with the attitudes, knowledge, techniques and points of reference that will take them to the centre of a situation like a tense and difficult epidemic crisis, an environmental disaster or a planetary contingency. Therefore, through fiction, experience and learning are generated. However, this fiction oscillates continuously between parameters that clearly show that we are not dealing with real facts and data or elements that give verisimilitude or veracity to the lived experience. In the case of scenarios, the threat is perceived as a different universe, where the danger is transcendent and unpredictable. In this it is not possible to perform an internal injury of a determined socio-technical system. On the contrary, risk is always open: we know what will happen but not how or when. It is the ineffable and always external to the system.

Doubtless, scenario planning is a practice that is building a new regime of conceptualisation of the future. This is no longer understood as an anticipation, a forecast or a figure that indicates a trend. Now, the future becomes life experience of wide uncertainty that can and should be felt in the present. The future opens as an experience in all the breadth of sensations and meanings it can offer. The future as scenario is based on a specific risk (Ebola, terrorist attack, water shortage, etc.) but is deployed as if it were a plant, avid for light in the form of the integration of emotions, relationships, interpretations and, ultimately, close experiences that make up our day-to-day. That is, fiction becomes the main organiser of our present.

CODA The emergency modality and COVID-19: real-time and biosurveillance citizenship

Practically since the moment it emerged, the COVID-19 pandemic has sparked analysis and interpretation from leading authors in the sphere of Human and Social Sciences. A notable example was to be found in the columns published by Giorgio Agamben (2020), in which he argued that the state of emergency declared in places like Italy and Spain was simply a logical development of the way industrialised countries had been going about the exercise of governance for decades. According to this particular author, their approach had been legitimised by the fear of indiscriminate terrorism, but once this threat was neutralised, these governments sought legitimacy for their actions in the panic generated by biological emergencies. The writings of the philosopher Slavoj Zizek (2020) are also worth citing. He works with two basic ideas; the first relates to how, thanks to this new virus, humanity has remembered that it is totally interconnected, while the second is that overcoming this crisis can only lead to a world that will take a more united and collectivized approach to problems and their solutions. Finally, we might also mention Byung-Chul Han's (2020) analysis of the relative success of different countries in the medical fight against the virus, and the influence of the type of culture and traditions dominant in their historical contexts, be they Asian or European.

Though unquestionably interesting, these analyses all apply old frameworks of reflection to a present phenomenon that supposedly corroborates them. Thus, the COVID-19 pandemic is simply an event that serves to illustrate and give substance to warnings, ideas, cognitive frameworks, hunches and so on that we were already aware of. None of these interpretations, nor indeed the many other interpretations currently being made, invoking notions such as biopolitics, climate change or voracious capitalism, provides any analysis of the novel and idiosyncratic aspects that have been thrown up by the pandemic. In this sense, we are witnessing the curious paradox of a situation which is claimed to be entirely new and unknown, unparalleled in history, being described with old instruments, and the empirical data produced by the pandemic being used to endorse or justify the continued use and reliability of these concepts.

We believe that this paradox should be ended. As an event, the so-called COVID-19 pandemic has its own genuine and completely new and unexpected dimensions and features, which necessitate fresh analyses and conceptualisations of biosecurity. In this respect, the current pandemic offers us elements to rethink the issues that we have raised regarding the notion of 'emergency modality' in our chapter, but the particular aspect that clearly sets it apart from the emergencies experienced in the cases of the Ebola virus, SARS and H1N1 flu is that it has been followed, broadcast and assessed live, practically in real time. Day after day, we have seen images of our hospitals at saturation point, data on the numbers of people infected and deaths, growth charts tracking the expansion of the virus and comparisons of countries and continents. The press, meanwhile, have offered computer applications that trace the advance of the virus both locally and worldwide, comparing it and making projections of future scenarios. Our local and national leaders have felt compelled to make almost daily public appearances to provide updates on all the latest developments. In short, this is the first pandemic to be broadcast live and with real-time tracking. Obviously, this has made possible by the infrastructure now available to us, and the phenomenon points to a future that will see the appearance of categories of reflection that will rely on this temporal immediacy to conceptualise the meaning of a situation of emergency.

Hence, the gist of this opuscule is that, rather more than the mere result of technological deployment, this instantaneousness has ontological effects because it shapes our understanding of a biohealth emergency and our response to it. Moreover, such instantaneousness speaks to two different phenomena that we must distinguish between. As Erik Sadin (2013) has pointed out, the phenomenon of real time is often confused with that of live transmission, but they are not the same thing. When we refer to something that is happening live, we are alluding to a technical configuration that allows events to be relayed in the form of images and sounds at the very moment they occur, a typical example being the live broadcast of a sporting event. The relevance of live broadcasting in everyday life lies in its ability to transcend the limitations of the body, geography and the isolation of populations and collectives. Real time, however, goes far beyond live transmission. The meaning of this phenomenon stems originally from the technical configuration that consists of sending commands to a computer and processing the results of these commands at the moment they occur. Thus, real time implies the absence of any appreciable time lag between a user's action and the demand being met. Real time, therefore, broadly refers to an operation to capture a number of phenomena from 'the real' at the very moment they occur. Thus, real time denotes the capacity to follow a large number of events and activities without delay, and in their dimension of continuous development. The main technology on which real time is based is the large-scale deployment of different types of sensors in our immediate surroundings. As the connected surfaces in our everyday lives continue to multiply and the power of systems just goes on growing, our contact with real time is becoming a rather everyday experience. These connected objects that intercept information at source imply increasingly comprehensive visibility in every aspect of life. Indeed, they will bring about the entry of real time in our very anthropological condition, enabling us to immediately tap into permanently fluctuating information flows from the world, and then act accordingly.

And so we come to a question thrown up by the COVID-19 pandemic which is highly relevant to the notion of 'emergency situation', but which none of the aforementioned interpretations has thematised. The pandemic has operated as if we were faced with two completely different emergency situations. For that matter, at times it has felt like contemplating the spread and evolution of two completely different pandemics. On the one hand, we have witnessed the handling and approach of countries like China, South Korea, Singapore, Japan or New Zealand. Going way beyond simply tracking and mapping the infection, these places have deployed an infrastructure of sensors and procedures to literally pursue and monitor it in the quasi–real time of its appearance. The virus has been tracked as it has spread, and attempts have been made to anticipate fresh breakouts. On the other, we have observed how some European countries, in particular Spain and Italy, and at certain times France, have used conventional epidemiological methods, limiting their efforts to dealing with the evolution of the virus live, but not in real time. In other words, these countries have provided constant updates on the pandemic, but in tackling the emergency they have waited for outbreaks to occur before taking action, they have failed to track the movement of the virus (in fact, the lack of tracing and tracking is seen, at the time of writing in September 2020, as one of the main reasons why Spain has one of the highest contagion rates per 100,000 inhabitants in the world) and they have turned to general lockdowns as a last resort to control community spread.

The countries in the first group that we mentioned have defined their emergency situations on the basis of this notion of real time, whereas those in the second group have restricted themselves to a live commentary of the catastrophe. As can be seen, both the treatment and the consequences of the vector of infection have been different. We would like to add that the real-time phenomenon is not simply a question of capturing a whole range of emerging phenomena in an endlessly unfolding present but involves a different relationship with the immediate environment. The distance separating bodies from their environment gradually melts away, until we see a snapshot of what was previously opaque now becoming visible, a vector of infection, for instance. A good example of this are the apps on a Smartphone, which allow us to identify elements to be found in our immediate location. Such a feature generates an augmented reality effect that allows us to go beyond the boundaries of the classical science approach to the natural state of things, merely describing and monitoring the course of their evolution. We are now able to alter their course, intervening and, where we can, adjusting things according to our requirements. It is worth mentioning here that Spain somewhat belatedly created a good app to track people infected with SARS-CoV-2, but for various organisational and administrative reasons has been unable to implement it nationwide.

To conclude this opuscule, we would like to mention that the phenomenon of real-time action has given us a glimpse of an interesting new sociological and anthropological development that we shall call 'biosurveillance citizenship'. It is, like others which will need to be determined and thematised, the child of real time, and it is characterised by the capacity to monitor things at the very moment they occur. It is a sort of control mechanism which not only feeds off past or present archives but also the state of 'the real' in the very instant of its conception, eliminating any opacity from experience, its uncertainty and randomness. This biosurveillance citizenship has four features:

(1) Life is watched and watchful at the same time. The handling of the pandemic has required that citizens be both watched and watchful. They have become an active and crucial part of the monitoring and surveillance circuits deployed to manage biohealth emergency situations.

(2) The quality of citizenship with full rights vacillates and depends on its exercise of surveillance in real time. Controversial tools in European countries, like the health passport or medical tests before or after travelling, show that the condition of citizenship might be shifting from being an absolute status held everywhere and at every moment being an intermittent situation, requiring some mechanism to activate or shut it down. Biological emergencies would provide the framework for this transition.

(3) The citizen is a biowatcher. The prerequisite for leading a normal life in the aforementioned emergencies appears to be that the citizen must acquire skills and abilities in the observation and surveillance of the actors involved in these emergencies.

(4) The biosurveillance citizen is the producer of information over which she/he has no control regarding ultimate use.

Contrary to what some interpretations of the current pandemic seem to indicate, we are not heading towards a general panopticon that would observe and register even the most trivial aspects of our daily lives. We believe that the sociology and anthropology that has focused on action taken in response to SARS-CoV-2 points to the birth of a device or governance which will have a kinepolitical (cinepolitical) feel to it, in which the management of movement in real time will shape our everyday lives.

Notes

1 From the document *An R&D Blueprint for Action to Prevent Epidemics. Plan of Actions May 2016* (WHO, 2016b). Available at: www.who.int/csr/research-and-development/WHO-R_D-Final10.pdf.
2 From the video scenario 'Ebola Simulation Drill | The Little Couple' made by TLC television and performed by experts at the Texas Children's Hospital. Available at: www.youtube.com/watch?v=x50SzazSjoY.
3 For instance, see this one about Ebola that appeared in the *Daily Telegraph*: www.youtube.com/watch?v=-tCTh4AsZ9Y&t=3s or the GleamViz project (www.gleamviz.org/case-study/).

References

Agamben, G. (2020) 'The enemy is not outside, it is within us', http://bookhaven.stanford.edu/2020/03/giorgio-agamben-on-coronavirus-the-enemy-is-not-outside-it-is-within-us/

Anderson, B. (2010) 'Pre-emption, precaution preparedness: Anticipatory action and future geographies', *Progress in Human Geography*, vol. 34, pp. 777–798.

Anderson, B. and Adey, P. (2011) 'Affect and security: Exercising emergency in "UK civil contingencies"', *Environment and Planning D: Society and Space*, vol. 29, pp. 1092–1109.

Baker, D. (2016) 'Zika virus and the media', *Hospital Pharmacy*, vol. 51, no. 4, pp. 275–276.

Centers for Disease Control and Prevention (CDC) (2001) *The Model State Emergency Health Powers Act.* Atlanta, www.aapsonline.org/legis/msehpa.pdf

Centers for Disease Control and Prevention (CDC) (2003) 'Update: Outbreak of severe acute respiratory syndrome-worldwide, 2003', *Morbidity and Mortality Weekly Report*, vol. 52, no. 12, pp. 241–248.

Centers for Disease Control and Prevention (CDC) (2013) 'CDC SARS response timeline', www.cdc.gov/about/history/sars/timeline.htm

Centers for Disease Control and Prevention (CDC) (2014) 'Estimating the future number of cases in the Ebola epidemic – Liberia and Sierra Leone, 2014–2015', *Morbidity and Mortality Weekly Report Supplements*, vol. 63, no. 3, pp. 1–14.

Centers for Disease Control and Prevention (CDC) (2017a) 'Caring for pregnant women', www.cdc.gov/pregnancy/zika/testing-follow-up/pregnant-woman.html

Centers for Disease Control and Prevention (CDC) (2017b) 'After your trip', https://wwwnc.cdc.gov/travel/page/after-trip

Chandrasekaran, N., Gressick, K., Singh, V., Kwal, J., Cap, N., Koru-Sengul, T. and Curry, C. (2017) 'The utility of social media in providing information on Zika virus', *Cureus*, vol. 9, no. 10, p. e1792.

Collier, S. and Lakoff, A. (2008) 'The problem of securing health', in A. Lakoff and S.J. Collier (eds.) *Biosecurity Interventions: Global Health and Security in Question*. Columbia University Press, New York, pp. 7–32.

Collier, S.J. (2008) 'Enacting catastrophe: Preparedness, insurance, budgetary rationalization', *Economy and Society*, vol. 37, no. 2, pp. 225–250.

De Goede, M. (2008) 'Beyond risk: Premediation and the post-9/11 security imagination', *Security Dialogue*, vol. 39, pp. 155–176.

Dupuy, J.P. (1999) *El Pánico*. Barcelona, Gedisa (buscar si hay versión en inglés).

Farmer, P., Kleinman, A., Kim, J. and Basilico, M. (2013) *Reimagining Global Health*. California University Press, California.

Fearnley, L. (2005) 'Pathogens and the strategy preparedness', http://anthropos-lab.net/wp/publications/2007/01/fearn_pathogens.pdf

Foucault, M. (2003) *The Birth of the Clinic: An Archaeology of Medical Perception*. Taylor and Francis e-Library, Abingdon.

Glasser, J., Hupert, N., McCauley, M. and Hatchett, R. (2011) 'Modeling and public health emergency responses: Lessons from SARS', *Epidemics*, vol. 3, pp. 32–37.

Gosselin, D. and Tindemans, B. (2016) *Thinking Futures: Strategy at the Edge of Complexity and Uncertainty*. LannooCampus Publishers, Leuven.

Guattari, F. (2008) *The Three Ecologies*. Continuum, London.

Han, B. (2020) 'Why Asia is better at beating the Pandemic than Europe: The key lies in civility', *El País*, https://english.elpais.com/opinion/2020-10-30/why-asia-is-better-at-beating-the-pandemic-than-europe-the-key-lies-in-civility.html

Huang, C., Hung-Tsang, D., Kao, W., Wang, L., Huang, C. and Lee, C. (2005) 'Impact of Severe Acute Respiratory Syndrome (SARS) outbreaks on the use of emergency department medical resources', *Journal of the Chinese Medical Association*, vol. 68, no. 6, pp. 254–259.

Kaufmann, M. (2016) 'Exercising emergencies: Resilience, affect and acting out security', *Security Dialogue*, vol. 47, pp. 99–116.

Krasmann, S. (2015) 'On the boundaries of knowledge: Security, the sensible, and the law', *InterDisciplines, Journal of History and Sociology*, vol. 6, pp. 187–213.

Lakoff, A. and Collier, S.J. (2008) *Biosecurity Interventions: Global Health and Security in Question*. Columbia University Press, pp. 7–32, www.jstor.org/stable/10.7312/lako14606.3

Latour, B. (2005) *Reassembling the Social*. Oxford University Press, Oxford.

Lee, J. and McKibbin, W. (2004) *Estimating the Global Economic Costs of SARS*. In: Institute of Medicine (US) Forum on Microbial Threats; Knobler S, Mahmoud A, Lemon S, et al., editors. Learning from SARS: Preparing for the Next Disease Outbreak: Workshop Summary. Washington (DC): National Academies Press (US); 2004. Available from: https://www.ncbi.nlm.nih.gov/books/NBK92473/

Lewnard, J., Ndeffo, M., Alfaro-Murillo, J., Frederick, A., Bawo, L., Nyenswah, T. and Galvani, A. (2014) 'Dynamics and control of Ebola virus transmission in Montserrado, Liberia: A mathematical modelling analysis', *The Lancet Infectious Diseases*, vol. 14, pp. 1189–1195.

Lindgren, M. and Bandhold, H. (2003) *Scenario Planning: The Link between Future and Strategy*. Palgrave Macmillan, New York.

Maureira, M., Tirado, F., Torrejón, P. and Baleriola, E. (2018) 'The epidemiological factor: A genealogy of the link between medicine and politics', *International Journal of Cultural Studies*, vol. 21, no. 5, pp. 505–519.

Oxford Dictionary (2018) Oxford University Press, Oxford.

Pan American Health Organization (PAHO) (2016) 'Zika virus', https://www.paho.org/hq/dmdocuments/2016/8x11preventionENG.pdf

Peckham, R. and Sinha, R. (2017) 'Satellites and the new war on infection: Tracking Ebola in West Africa', *Geoforum*, vol. 80, pp. 24–38.

Rancière, J. (2005) *The Politics of Aesthetics: The Distribution of the Sensible*. Continuum, New York.

Rancière, J. (2007) *The Future of the Image*. Verso, London.

Rosa, R., Abbo, L., Kapur, G., Paige, P., Jean, R., Rico, E. and Memon, A. (2017) 'Development and implementation of a Zika virus disease response protocol at a large academic medical center', *Disaster Medicine and Public Health Preparedness*, vol. 11, no. 2, pp. 256–258.

Sadin, E. (2013) *L'humanité augmentee. L'administration numérique du monde*. Éditions L'échappée, Paris.

Sun, M., Xu, N., Li, C., Wu, D., Zou, J., Wang, Y., Luo, L., Yu, M., Zhang, Y., Wang, H., Shi, P., Chen, Z., Wang, J., Lu, Y., Li, Q., Wang, X., Bi, Z., Fan, M., Fu, L., Yu, J. and Hao, M. (2018) 'The public health emergency management system in China: Trends from 2002 to 2012', *BMC Public Health*, vol. 18, p. 474.

Tan, W. and Enderwick, P. (2006) 'Managing threats in the Global Era: The impact and response to SARS', *Thunderbird International Business Review*, vol. 48, no. 4, pp. 515–536.

Tirado, F., Castillo, J. and Galvez, A. (2012) 'Movimiento y regímenes de vitalidad. La nueva organización de la vida en la biomedicina', *Politica Y Sociedad*, vol. 49, no. 3, pp. 571–590.

World Health Organization (WHO) (2003) 'SARS (Severe Acute Respiratory Syndrome)', www.who.int/ith/diseases/sars/en/

World Health Organization (WHO) (2009a) 'Human infection with pandemic (H1N1) 2009 virus: Updated interim WHO guidance on global surveillance', World Health Organization, Geneva, www.who.int/csr/disease/swineflu/guidance/surveillance/WHO_case_definition_swine_flu_2009_04_29.pdf?ua=1

World Health Organization (WHO) (2009b) 'Timeline of Influenza A (H1N1) cases', https://www.who.int/csr/disease/swineflu/interactive_map/en/

World Health Organization (WHO) (2013) 'Emergency response framework', www.who.int/hac/about/erf/es/

World Health Organization (WHO) (2016a) 'Zika virus outbreak global response', www.who.int/emergencies/zika-virus/response/en/

World Health Organization (WHO) (2016b) 'An R&D blueprint for action to prevent epidemics. Plan of Actions May 2016', www.who.int/csr/research-and-development/WHO-R_D-Final10.pdf

World Health Organization (WHO) (2018a) 'WHO and partners working with national health authorities to contain new Ebola outbreak in the Democratic Republic of the Congo', www.who.int/news-room/detail/11-05-2018-who-and-partners-working-with-national-health-authorities-to-contain-new-ebola-outbreak-in-the-democratic-republic-of-the-congo

World Health Organization (WHO) (2018b) 'The Democratic Republic of the Congo/Ebola timeline response 2018', www.who.int/ebola/ebola-response-first-month-june-2018.pdf?ua=1https://www.who.int/ebola/ebola-response-first-month-june-2018.pdf?ua=1

World Health Organization (WHO) (2018c) 'Zika virus', www.who.int/news-room/fact-sheets/detail/zika-virus

Zizek, S. (2020) *Pandemic!* Polity Press, New York.

Zylberman, P. (2013) *Tempêtes microbiennes: Essai sur la politique de sécurité sanitaire dans le monde transatlantique.* Gallimard, Paris.

19

BIOSECURITY IN THE LIFE SCIENCES

Limor Samimian-Darash and Ori Lev

Introduction

In March 2017, US-based pharmaceutical company Tonix announced that its attempts to develop a safer smallpox vaccine in collaboration with scientists from the University of Alberta, Canada – that is, a less risky vaccine for a disease that was eradicated over 40 years ago – had been successful. The vaccine, made of synthesised horsepox virus, showed protective activity when tested on mice (Tonix Pharmaceuticals, 2017). This was the first *de novo* synthesis (a gene alteration) of an orthopoxvirus, a term that refers to a closely related group of viruses that includes horsepox, cowpox and the variola virus that causes smallpox. In response to this development, some argued that by demonstrating that it was possible to synthesise orthopoxviruses, the scientists were possibly encouraging further experimentation with the creation of infectious viruses from synthetic DNA – a development that could potentially risk the intentional or unintentional release of a virus (Koblentz, 2017). Others argued, however, that orthopox synthesis experiments were technically difficult to conduct, at least for the time being, and therefore that potential misuse of this kind remained highly unlikely (DiEuliis et al., 2017).

This type of debate on biosecurity in the life sciences is rather new. Indeed, it was only after the events of 9/11 and the publication of the Fink Report (2004) that the problem of biosecurity began to diffuse into the life sciences and the particular problem of dual use was raised. And although several events of concern in which the problem of dual use was raised took place after the establishment of the National Science Advisory Board for Biosecurity (NSABB) in 2004, no explicit US biosecurity policy was initially developed in response to these. It was thus not until the 2012 controversy surrounding research on the H5N1 virus (discussed in the following) that a new focus emerged on biosecurity in the context of dual-use research of concern (DURC) – and, in consequence, new policies and regulations on scientific work were established.

This chapter focuses on the developments in the biosecurity apparatus and approach as they manifest in the US. This is because the US took a leading role in this regard and seems to have gone the furthest in its actions to address biosecurity risks, especially as they pertain to dual-use research. However, other countries have also been discussing policy measures. Germany, Australia, Canada, Netherlands, the UK and Israel have been either developing or using existing regulations to address biosecurity risks in general and dual-use research in particular (Smith and Kamradt, 2014; William-Jones et al., 2014; Knesset of Israel, 2008).[1] Yet, and as mentioned previously, the US appears to have taken the most noteworthy actions in this regard, and thus we chose to focus on its actions.

Biosecurity: WMDs, bioterrorism and the life sciences threat

Although a number of states (including France, Japan, the UK and the US) had sponsored programmes to study and create biological weapons in the early part of the 20th century, by the end of World War II and the early days of the Cold War, it was mainly the arms race between the US and the Soviet Union that sustained such programmes. Increasing investment in research and development to tackle the possible use of biological weapons during this period led not only to the creation of weapons but also to scientific discoveries and the development of vaccines. Although most of these biological weapons research programmes had been dissolved by the end of the 20th century, at the beginning of the 21st century, the prospect that biological weapons might be used had once more become a substantial threat. Events such as the anthrax postal attacks in the US, which took place shortly after the events of 11 September 2001 (see the following), and the proliferation of WMDs prompted the establishment of the US Department of Homeland Security and the initiation of preparedness programmes. Biological threats thus became discursively entangled with the rise of global terrorist groups and rogue states, leading the US to respond with solutions such as a smallpox vaccination programme in 2002–2003 (Rose, 2008). Furthermore, state and science relations in the context of biological threats grew increasingly close as, on the one hand, US government resources were mobilised to advance research and development of biomedical measures (e.g. vaccines) to combat such threats and, on the other, new scientific knowledge and biotechnologies heightened the potential for accidental or intentional release of viruses (Guillemin, 2005).

According to Joseph Masco (2014), the US post-9/11 counterterror state represents a transformation of the Cold War nuclear state, whereby the events of 9/11 were harnessed by the existing US security apparatus to refashion itself for the 21st century. In Masco's view, both the expansion of the US national security state in the 1950s and the rise of the counterterror state in the 21st century were enabled and justified through reference to events that had caused major shockwaves around the world – the nuclear tests carried out by the Soviet Union and the communist revolution in China in the case of the nuclear security state and the events of 9/11 in the context of the modern counterterror state. The rise of the latter required the existing US security apparatus to develop ways of anticipating and preempting potential terror threats and create new infrastructure to facilitate such activities.

Between September and October 2001, not long after the 9/11 attacks on New York and Washington, DC, the US underwent another terror attack, this time biological. Letters containing anthrax bacteria had been sent to various media organisations and public officials, leading to the mobilisation of resources and efforts towards tackling what seemed to be a growing threat posed by bioterrorism. Soon enough, the issues of WMDs, biological threats and terrorism were grouped together under the same logic of anticipatory preemption (Masco, 2014, 145–150). Thus, while the issue of the threat posed by biological weapons had been attached to the heightened nuclear threat during the Cold War, the project of biosecurity became an important site for the 21st-century US counterterror state. Subsequent efforts to anticipate and preempt future terror threats, however, were the onset for the creation of new dangers. As Masco (2014, 43) put it, 'Biosecurity installs capacities in the future that are unknown, setting the stage not only for the potential preemption of a deadly virus but also for creating new kinds of dangers'.

The end of the Cold War era thus saw the emergence of bioterrorism threats as a political and technical problem for US national security. During the 1990s, the uncovering of

Soviet and Iraqi bioweapon programmes led to growing concerns over possible linkages among rogue states, global terrorism and WMD proliferation. These concerns increased following the 2001 anthrax letters incident (and the 1995 Aum Shinrikyo gas attack on the Tokyo subway). Moreover, concerns related to biodefence against bioterrorism were joined by increasing worries over new developments in the life sciences that could make it easier than ever to create or synthesise lethal viruses. Together with the domains of 'emerging infectious disease' and 'food safety', these worries and concerns catalysed the formation of a biosecurity assemblage composed of scientists, experts, policymakers, activists and officials from state institutions, NGOs and fields such as the life sciences, public health and security (Lakoff and Collier, 2008, 7–12).

Within the US, the question of biosecurity underwent a major transformation in the decades following the end of the Cold War, especially after the post-9/11 anthrax attacks – as it went from being a concern embedded within a more general Cold War nuclear threat strategy to a specific threat to be managed within the national security framework. Moreover, the growing concern over bioterrorism threats was further enhanced by particular developments in the life sciences. This, in turn, led to an expansion in the field of biosecurity (Samimian-Darash et al., 2016).

Biosecurity and US life sciences

The discourse on biosecurity that has emerged over the past two decades in the US has included a concern with threats in four interconnected domains: emerging infectious diseases, bioterrorism, cutting-edge life sciences and food safety (Lakoff, 2008, 8). In the US, and especially since the anthrax attacks that closely followed 9/11, there has been growing concern about the threat of bioterrorism and how particular developments in the life sciences might contribute to that threat.

Joseph Masco (2013, 206–207) divides the chronology of US biosecurity concerns since the end of the Cold War into three epochs:

1991–2001: food production, with a focus on the isolation and containment of livestock.
2001–2005: terrorism, with a focus on intentional attacks on the US population and domestic territory.
2005–today: all the prior plus infectious disease, with a focus on the global preemption of emerging dangers.

In our analysis of biosecurity in the life sciences, we propose an analytical distinction rather than a historical one. We define the threat of food safety as 'accidental biosecurity', the threat of bioterrorism as 'intentional biosecurity' and the problem of biosecurity caused by scientists as 'unintentional biosecurity'. Our definitions are not normative but descriptive, based on the main problems observed and defined at each period/stage identified by Masco.

After the Bush administration's declaration of a 'war on terror', the importance of the biosecurity discourse within the US grew to such an extent that it became a form of governance (Masco, 2013) and part of the preparedness apparatus (Lakoff and Collier, 2008). This preparedness rationale led to a proliferation of laboratories devoted to research on dangerous pathogens. The number of laboratories devoted to BSL-4 pathogens (e.g. smallpox, Lassa fever, Ebola) in the US grew from 5 before 2001 to 15 in 2019. BSL-3 laboratories (researching pathogens such as SARS, West Nile virus, tuberculosis and anthrax) have also multiplied, and there are now 1,356 registered BSL-3 laboratories across the US (Masco, 2013, 211).

Furthermore, 'the total U.S. government spending on civilian biodefence research between 2001 and 2005 [increased] from $294.8 million to $7.6 billion' (Lakoff, 2008, 10). Most of this funding was directed to life sciences research to improve defence against biothreats. Ironically, however, the proliferation of defence research also led to increased concern about possible misuse of the biological material and information being produced in the name of preparedness and defence. Thus, although the anthrax attacks were not the result of any new capacities on the part of biology or biologists – that is, unintentional biosecurity – they were perceived as such and opened 'strategic spaces' for a new biosecurity problematisation and possible interventions (Rabinow, 1996, 126).

In 2004, the National Academy of Sciences, which was tasked with investigating the problem of biosecurity in US life sciences, issued a report entitled *Biotechnology Research in an Age of Terrorism*, also known as the Fink Report. The report included a listing of select dangerous agents that had been identified as early as 2001 as requiring control to prevent their possible use in bioterrorism:

> In the United States, the US PATRIOT Act of 2001 and the Bioterrorism Preparedness and Response Act of 2002 already establish the statutory and regulatory basis for protecting biological materials from inadvertent misuse. Once fully implemented, the mandated registration for possession of certain pathogens (the "select agents"), designation of restricted individuals who may not possess select agents, and a regulatory system for the physical security of the most dangerous pathogens within the United States will provide a useful accounting of domestic laboratories engaged in legitimate research and some reduction in the risk of pathogens acquired from designated facilities falling into the hands of terrorists.
>
> *(Fink Report, 2004, 2)*

The report reframed the problem of biosecurity in the life sciences, shifting the focus away from other aspects of biosecurity – such as the creation and use of secure spaces for research activity – to what was referred to as the 'dual-use dilemma'. This term was borrowed from the language of arms control and disarmament – where 'Dual Use refers to technologies intended for civilian application that can also be used for military purposes' (Fink Report, 2004, 18) – and applied to a case in which 'the same technologies can be used legitimately for human betterment and misused for bioterrorism' (Fink Report, 2004, 1). In the light of this definition, the Fink Report presented the following recommendations:

- educate scientists regarding the dual-use dilemma;
- expand the existing biosafety oversight system before research is conducted;
- review research plans according to seven criteria for review of experiments 'involving microbial agents that raise concerns about their potential for misuse' (Fink Report, 2004, 130);
- conduct a self-review of scientists and scientific journals before publication; and
- establish the National Science Advisory Board for Biosecurity.

From biosecurity to dual use and dual-use research of concern

In 2004, US Department of Health and Human Services Secretary Tommy G. Thompson announced the establishment of the National Science Advisory Board for Biosecurity, stating that 'the new board will advise all Federal departments and agencies that conduct or support Life Sciences research that could fall into the "Dual Use" category. The NSABB will be managed by the National Institutes of Health (NIH)' (DHHS, 2004).

With the establishment of the NSABB, then, the problem of biosecurity was reframed as the problem of dual use – that is, how to prevent misuse of the results of scientific research and technologies. However, dealing with biosecurity in general, and specifically with dual use, raised another problem: how to identify potentially dangerous studies or agents requiring oversight.

In March 2006, the NSABB held a meeting in which the term 'dual-use research of concern' (DURC) was presented. This term derived specifically from the assertion that '*most* if not *all* Life Sciences research *could* be considered Dual Use'; therefore, 'the Working Group wanted to focus [the criteria of dual-use research] to identify specific Life Sciences research that could be of greatest concern for misuse' (Dual Use Criteria Working Group, 2006, emphasis added). In other words, only certain kinds of life sciences research were considered to be 'of concern' and therefore subjected to oversight.

In June 2007, the NSABB published its proposed framework for the 'Oversight of Dual-Use Life Sciences Research: Strategies for Minimizing the Potential Misuse of Research Information'. This document defined dual-use research of concern as 'a subset of Dual Use Research that has the highest potential for generating information that could be misused' (NSABB, 2007). The definition of DURC narrowed the broader problem of dual use to only a few studies that could be marked in their early stages as being 'of concern'. It would be the NSABB's role to initiate measurements and practices of oversight that were appropriate for the 'dual use' and 'dual-use of concern' problem. However, no criteria were set down regarding how to identify activities of concern.

Here, it is important to note that vexing questions related to biosecurity and dual-use research in the life sciences have long-standing parallels in other fields: physics, particularly nuclear physics, has given rise to similar concerns (Evens, 2013), for example, and the problem is also relevant within the social sciences (see Hausman, 2007; Kitcher, 2001). But let us first be clear what the problem is.

Dual-use research is commonly understood as research in which scientists are pursuing a socially valuable study (e.g. developing a better vaccine, as mentioned earlier) with the intention of generating valuable knowledge but where the same study and the knowledge and methods it generates could enable malevolent actors to cause harm (e.g. by providing a blueprint for the synthesis of dangerous viruses). It is here that the dilemma of dual use arises and from here that the biosecurity risks emanate in the case of the life sciences. We should be clear what this means, at least from the perspective of the researchers pursuing such studies. In order to recognise that a dual-use dilemma arises in a particular case – that is, that a scientist is aware that her well-intended research could be abused – she has to be able to foresee the possibility of such abuse. Some scientists might not be aware of such a possibility, either because they lack sufficient knowledge about how their study could be abused or because such a possibility is not on their horizons for other reasons. In other words, the question of the foreseeability of the potential of abuse could be regarded as a subjective matter. However, as in other domains, the relevant question is not whether a particular scientist could actually foresee the danger but rather whether the danger is *reasonably foreseeable* (Douglas, 2003). We should ask whether scientists ought to foresee the potential for misuse of their work. In this way, a more objective standard is adopted. If misuse were reasonably foreseeable and a scientist did not take that into consideration, we could then accuse her of negligence – or possibly worse if she had foreseen the possibility of misuse and did not act accordingly. On the other hand, according to this approach, if the misuse could not have been reasonably foreseen but somehow materialised, the scientist would still be regarded as having acted appropriately.

In most well-known cases – including the one mentioned earlier of synthesising horsepox – there is very little discussion among scientists about the foreseeability of the potential for misuse,[2] even though such potential is usually relatively easy to discern. The questions raised are usually about the balance between the risks and benefits of conducing the research and subsequently publishing the findings.

In the aforementioned, we have examined the nature of the dilemma surrounding dual-use research from the point of view of a particular scientist, but this is not the only perspective that might be adopted. The dilemma could arise for funding institutions, journals and governments. Regardless of the perspective adopted, however, the dilemma remains the same, as do the criteria to determine whether it manifests – namely, the nature and value of the study, its potential to be misused by third parties and the reasonable foreseeability of such a potential. When these transpire, an appropriate course of action is required. Multiple options for dealing with such a situation are available – for instance, ignoring the risks, stopping the research, delaying it, adding safety features, modifying the research design and/or not publishing details of the study.

Having clarified what dual-use research is, we can now look at how it is legally defined. The definition found in the 'US Government Policy for Oversight of Life Sciences Dual Use Research of Concern' states:

> For the purpose of this Policy, DURC is life sciences research that, based on current understanding, can be reasonably anticipated to provide knowledge, information, products, or technologies that could be directly misapplied to pose a significant threat with broad potential consequences to public health and safety, agricultural crops and other plants, animals, the environment, materiel, or national security.
>
> *(United States Government, 2012, 1–2)*

This definition is focused exclusively on life sciences research narrowly defined as biological sciences (United States Government, 2014, 7). The definition presented here, however, is of 'dual-use research of concern', not of 'dual-use' research more generally. This is because the category of dual-use research has been recognised by many as being too broad since almost any study could be abused to cause some harm. The methods described in most biological studies could be abused, as could the information gleaned from them or the materials used in such studies. It was recognised early on in the debate over dual-use research and associated biosecurity risks that the concern should not be of all science (NSABB, 2007). A much narrower category needed to be identified so that biosecurity risks could be addressed more effectively without hampering scientific endeavours in general.

The definition should be such that it does not leave out studies that ought to concern us. But it should exclude the vast majority of studies that raise little to no concern. To achieve a balance between these two objectives is not easy, but the aforementioned definition attempts to do so by suggesting that the risks associated with a study identified as being of concern should be 'significant', with 'broad potential consequences'. In other words, the likelihood and magnitude of the risks are such that we should be alarmed by them and therefore assess them carefully. Most studies do not pose significant risks with wide-ranging consequences and therefore remain outside the definition. Importantly, the definition directs our attention to a category of risks that are qualitatively different from the 'everyday' risks of possible misuse of scientific studies. However, the only way to ascertain whether a study poses a significant risk is to conduct an empirically driven risk-benefit analysis.

It is worth noting that the definition does not single out human populations as the only ones that could be harmed by dual-use research. Indeed, it clearly states that the threats of dual-use research are much wider; they apply to crops, livestock and more. Dual-use research of concern can generate information, technologies and methods that could enable harm to animals and the environment at large. Since such harms could potentially have significant consequences, studies in those fields should be closely examined in order to ascertain their dual-use potential.

The definition set out in the 'US Government Policy for Oversight of Life Sciences Dual Use Research of Concern' was adapted, with a few revisions, from an earlier definition that the NSABB had articulated in its report 'Proposed Framework for the Oversight of Dual Use Life Sciences Research: Strategies for Minimizing the Potential Misuse of Research Information':

> Research that, based on current understanding, can be reasonably anticipated to provide knowledge, products, or technologies that could be directly misapplied by others to pose a threat to public health and safety, agricultural crops and other plants, animals, the environment, or materiel.
>
> *(NSABB, 2007)*

The two definitions are similar, but there is at least one key difference worth noting: in the new definition, the phrase 'by others' had been excised. This phrase was originally intended to express the idea that scientists conducting research have good intentions and the risk is simply that 'others' – malevolent actors – might misapply their research. The elimination of the phrase 'by others' could be understood as suggesting that scientists themselves could possibly misuse research findings – a shift that was most likely influenced by the 2001 US anthrax attacks mentioned previously.

Having clarified the various features of dual-use research and how it is defined, we can list its risks, of which there are three kinds. First, there are biosecurity risks stemming from *knowledge and information* generated by well-intended research. The risk here is that malevolent agents could use published or presented information, or unauthorised access to information that had not been published, to cause serious harm. In other words, information and knowledge gained from socially beneficial research could enable bad actors to inflict harm. Second, there are biosecurity risks generated by the *'products'* of well-intended research. The risk here is that malevolent actors might gain access to these products – for example, dangerous pathogens – and use them to cause harm. Third, there are biosecurity risks generated by the *technologies* created in beneficial research – as exemplified by the recent horsepox study discussed at the start of this chapter. Here, again, beneficial research could generate valuable technologies, yet if bad actors have access to such technologies, they could employ them to cause harm (Lev and Rager-Zisman, 2014).

Key biosecurity events in the life sciences

The issue of dual use has been raised several times in the past three decades in the context of life science research. In each case it was in relation to studies in which scientists had developed dangerous pathogens or methods that seemed to present the kinds of biosecurity risks described above.

Around 1998–1999, a group of Australian scientists experimented with clinical mousepox (ectromelia virus) to develop an artificial strain that would cause infertility in mice. However, by adding a mouse gene involved in immune response to the mousepox virus, the scientists created a new violent strain that killed most of the mice, even those genetically immune to or vaccinated against mousepox. The scientists first informed the Australian government and military of their discovery yet soon afterwards went on to publish the results of their study in *Journal of Virology*. The mousepox virus itself, it was noted, poses no threat to humans. However, as humans possess an immune-system gene similar to the one used in the study, scientists expressed worries about the possibility of unethical use of these findings, mainly in the sense of turning human viruses into weapons (Broad, 2001; Jackson et al., 2001).

In 2001–2002, the US government initiated its smallpox vaccination programme. This attempt to address a possible future re-emergence of a disease that had not been present in human populations for several decades came in the light of calls by public health and national security experts to tackle the growing threat posed by potential malicious use of biological agents (Rose, 2008).

In another dual-use case, using internet-published mapping of the polio virus RNA genome, researchers followed instructions for a written sequence and managed to artificially synthesise a polio virus by combining two corresponding strands of DNA they purchased via mail order (Cello et al., 2002).

In 2005, scientists conducted a study on the properties of the virulent 1918 influenza virus. The virus's coding sequence allowed them to use reverse genetics to first produce this influenza virus and then compare it with the H1N1 influenza virus (Tumpey et al., 2005).

Also in 2005, researchers Lawrence M. Wein and Yifan Liu (2005) published a mathematical model that indicated that a single deliberate release of botulinum toxin in a milk-processing facility could lead to several hundred thousand people being poisoned as a result of the rapid distribution and consumption of products if detection from early symptomatics was not timely. Accordingly, they suggested using an in-process testing mechanism that could likely eliminate the threat.

In 2012, the H5N1 controversy erupted when two groups of scientists reported that they had conferred airborne mammalian transmissibility to the extremely dangerous highly pathogenic avian influenza virus (H5N1) (Herfst et al., 2012; Imai et al., 2012). Concerns regarding the implications of the H5N1 studies were raised after they were submitted for publication, one to the journal *Nature* and the other to *Science*. The editors of those journals sought the NSABB's advice on whether to approve publication. Their concern was that publication would enable malevolent agents to use the information reported in harmful ways, thereby endangering public health and safety (Lev and Samimian-Darash, 2014; Samimian-Darash, 2016; Samimian-Darash et al., 2016).

Although previous cases had been brought to the attention of the NSABB for consideration – for example, the research on H1N1 and the 1918 flu virus in 2005 – the NSABB and the scientific community appeared to have little interest in amending US policy on biosecurity. For example, a letter by the editor of *Science* in 2005 was still framed according to the discourse of biosafety, rather than biosecurity (Enserink and Malakoff, 2012). In addition, the NSABB has consistently maintained that the best way to address the issue of DURC is through self-regulation and awareness-raising activities, not through government action (Kaiser and Moreno, 2012).

The controversy surrounding research on the H5N1 flu virus, however, was a situation that challenged the existing framing of dual-use research of concern and related US policy and will be discussed in more detail in the following section. Before the occurrence of that event, the main biosecurity risk within US life sciences was generally framed as unintentional, and the main practices adopted to address such risk were those of biosafety – such as the use of designated secure spaces for research. With the H5N1 controversy, however, questions arose as to whether the risk was indeed unintentional and whether biosafety measures were enough.

Constructed H5N1: 'an engineered doomsday'

In 2011, working independently, researchers Ron Fouchier and Yoshihiro Kawaoka created strains of H5N1 viruses that spread via the air among ferrets. Since ferrets are generally considered to provide the best model for predicting how a flu virus might behave in humans, it was assumed that these strains exemplified human pandemic strains as well.

On 13 September 2011, Fouchier presented his study at a meeting in Malta and announced that his lab had 'discovered that only 1–3 substitutions are sufficient to cause large changes in antigenic drift . . . [and that] large antigenic differences between and within H5N1 clades could affect vaccine efficiency and even result in vaccine failure' (*Influenza Conference Newspaper*, 2011). Fouchier and his team reportedly

> introduced mutations, by reverse genetics into laboratory ferrets. They then collected a nasal wash from each infected ferret and inoculated another ferret after a few days. They repeated this process ten times. The result? H5N1 had been transmitted to three out of four ferrets.
>
> *(Influenza Conference Newspaper, 2011)*

Fouchier's study caused turmoil in the scientific community, especially among virologists and microbiologists: On 19 September, *Scientific American* published an article entitled 'What Will the Next Influenza Pandemic Look Like?' (Harmon, 2011), which described a possible H5N1 pandemic as 'topping the worst-case scenario list for most flu experts', and raised the question of whether 'the dreaded H5N1' would become transmissible in humans. The answer to the article's title question lay in the study by Fouchier and his team, who had 'mutated the hell out of H5N1' and found that within 'as few as five single mutations it gained the ability to latch onto cells in the nasal and tracheal passageways' (Harmon, 2011). On 26 September, the *New Scientist* reported on Fouchier's work:

> H5N1 bird flu can kill humans, but has not gone pandemic because it cannot spread easily among us. That might change: five mutations in just two genes have allowed the virus to spread between mammals in the lab. What's more, the virus is just as lethal despite the mutations.
>
> *(MacKenzie, 2011)*

Because of the concerns raised by the research, in October 2011 the NSABB was called on to review the two studies, which by then had been submitted for publication. The board was to examine the question of their publication in relation to DURC with regard to the potential that the information presented might be misused. As explained by a board member (in an interview with one of the authors):

> The work provides information about [properties of infectious agents] . . . it's the information that allows someone else to create these things with these properties. Two of the most serious properties to give an infectious agent are *high virulence* and . . . *aerosol transmission between people*, and this was the case with the new H5N1 strain.

According to *Science* editor Bruce Albert, when Fouchier's paper arrived at the journal, 'it was obvious' that it needed special review. The journal

> quickly recruited outside specialists, including biosecurity experts who serve on the NSABB. NSABB itself was first alerted to the studies by NIAID (National Institute of Allergy and Infectious Diseases) in late [summer 2011] and received copies of the papers in mid-October.
>
> *(Enserink and Malakoff, 2012)*

The research was presented by all actors involved as an internal security threat: a new lethal virus had been introduced into the world, and scientists were responsible for its creation. Many spoke of it as 'a man-made flu virus that could change world history if it were ever set free' (Enserink, 2011).

Paul Keim, NSABB chair at the time, said he could not 'think of another pathogenic organism that is as scary as this one . . . . Anthrax is [not] scary at all compared to this' (Enserink, 2011). Laurie Garrett, a Pulitzer Prize–winning science journalist, wrote a piece entitled 'The Bioterrorist Next Door' (Garrett, 2011). The perception that science had created the next global pandemic or the next bioterror event began to spread.

At the heart of the debate, the critical question was raised as to whether Fouchier and Kawaoka had acted responsibly in conducting their studies in the first place. Thomas Inglesby and colleagues criticised the H5N1 research and asked, 'Should we purposely engineer avian flu strains to become highly transmissible in humans? In our view, no'. They cited three reasons for their position: first, the deadly strain could 'escape accidentally from the laboratory'; second, the idea that the engineered strain could help scientists identify similar characteristics in currently circulating strains of H5N1 'is a speculative hope but not worth the potential risk'; and, third, the assertion that the creation of the strain would motivate scientists to search for H5N1 vaccines was also speculative (Inglesby et al., 2012).

The position of the *New York Times* on the matter was decidedly negative, as expressed in a 2012 editorial entitled 'An Engineered Doomsday':

> Defenders of the research in Rotterdam . . . say the findings could prove helpful in monitoring virus samples from infected birds and animals. . . . But it is highly uncertain, even improbable, that the virus would mutate in nature along the pathways prodded in a laboratory environment, so [any such] benefit . . . seems marginal.
>
> *(New York Times, 2012)*

A very different position, however, was expressed by NIH officials Anthony Fauci, Gary Nabel and Francis Collins, who supported the publication of the H5N1 studies, in a commentary in the *Washington Post*:

> Understanding the biology of influenza virus transmission has implications for outbreak prediction, prevention and treatment. . . . The question is whether benefits of such research outweigh risks. . . . [Ne]w data provide valuable insights that can inform influenza preparedness.
>
> *(Fauci et al., 2011)*

This statement reflects a new approach to the H5N1 research – the application of risk-benefit analysis. In the opinion of these officials, the public health benefits of the H5N1 research outweighed its risks. In their eyes, the research constituted 'a flu virus risk worth taking'.

In December 2011, at the height of the controversy, the NSABB published its recommendations following its review of the articles. It declared:

> While the public health benefits of such research can be important, certain information obtained through such studies has the potential to be misused for harmful purposes. . . . Due to the importance of the findings to the public health and research communities, the *NSABB recommends that the general conclusions highlighting the novel outcome be published, but that the manuscripts not include the methodological and other*

details that could enable replication of the experiments by those who would seek to do harm.

(National Institutes of Health, 2011, emphasis added)

Thus, the H5N1 research was to be assessed according to its perceived risks and benefits, and such assessment would determine what biosecurity measures (if any) should be taken in relation to it. To minimise the risk of the experiments being replicated by those not authorised to do so, the NSABB had recommended that the articles not be published in full. At the same time, emphasising the important contributions of the studies to public health research, the board recommended that full details be provided to a closed network of scientists 'authorised' to use the information to continue to conduct 'responsible' research on the topic, thus continuing the long-standing practice of differentiating between good and bad users of scientific products (although in this case between scientists in the 'safe' US and those in the 'unsafe' third world).

On 30 December 2011, the World Health Organization (WHO) released a statement expressing concern that limiting dissemination of information from Fouchier's and Kawaoka's work would undermine the international Pandemic Influenza Preparedness (PIP) framework. On 16–17 February 2012, the WHO held a meeting of members from around the globe. Participants reached consensus on two related issues: redaction of the studies was not a viable option because of urgent public health needs, and a mechanism to limit access was not practical at that moment (WHO, 2012).

Some of those who assessed the H5N1 research in terms of its risks and benefits ended up arguing forcefully against publication of the results. Michael Osterholm, a public health expert and member of the NSABB, and physician and biosecurity expert Donald Henderson explained in *ScienceExpress* why the risks associated with publication of the two H5N1 papers outweighed the proposed benefits and challenged the rationales offered in defence of the work:

> The current circulating strains of influenza A/H5N1, with their human case fatality rate of 30 to 80%, place this pathogen in the category of causing one of the most virulent known human infectious diseases. . . . We can't unring a bell; should a highly transmissible and virulent H5N1 influenza virus that is of human making cause a catastrophic pandemic, whether as the result of intentional or unintentional release, the world will hold Life Sciences accountable for what it did or did not do to minimize that risk.

(Osterholm and Henderson, 2012)

This claim, however, was also refuted by numerous researchers, who noted that publishing the results of previous research involving engineered viral strains had not generated any controversy:

> No one can guarantee that the ferret-passaged H5N1 virus would not be lethal and transmissible in humans. However, the same could be said about many laboratory-modified viruses, none of which have attracted the attention of the NSABB or the press.

(Racaniello, 2012)

Policy after H5N1

Following this public debate, the NSABB was assembled to re-evaluate its previous decision. Shortly afterwards, the NSABB's formerly open-to-the-public meetings became closed, top-secret sessions. Attendees were subjected to FBI clearance, no documents could be taken out of the meeting rooms and NSABB members were asked not to discuss matters considered in the meetings with outsiders. It was at this point that risk assessment experts were called in to help create a tool to assess the level of risk associated with publication of the H5N1 studies. It took the NSABB members approximately four months to reassess the H5N1 manuscripts (from November 2011 to March 2012). At different points in this process, they reached opposing conclusions: that the research did and did not present a significant risk to the public.

In February 2012, the American Society of Microbiology hosted a meeting on 'Biodefense and Emerging Diseases' during which an ad hoc session took up the H5N1 work. In this session, Fouchier defended his work and provided a fuller explanation of the issue of pathogenicity. At the same meeting, NIAID director Anthony Fauci announced that he had asked the two researchers to revise their papers and the NSABB to review the revised manuscripts. That same month, a gathering of NSABB members and more than a dozen observers, including NIH director Francis Collins and WHO member Keiji Fukuda, took place at the NIH campus. At the gathering, the participants read both the original and revised reports and subsequently voted to allow full publication of the revised studies (NSABB, 2012).

Following the controversy surrounding publication of the H5N1 studies, a new 'United States Government Policy for Oversight of Life Sciences Dual Use Research of Concern' was issued (on 29 March 2012). Taking effect immediately, it defined such research as follows:

> For the purpose of this Policy, DURC is life sciences research that, based on current understanding, can be reasonably anticipated to provide knowledge, information, products, or technologies that could be directly misapplied to pose a significant threat with broad potential consequences to public health and safety, agricultural crops and other plants, animals, the environment, materiel, or national security.
>
> *(United States Government, 2012)*

The new policy established a review process for studies that are conducted either in US government labs or in institutions funded by the federal government. The policy lists 15 pathogens and toxins, including H5N1, that are subject to its terms. It then lists seven kinds of experiments that should be reviewed. If a study meets these two criteria, a review is then carried out to determine whether it meets the DURC definition provided previously. If that is the case, a risk-benefit assessment is then conducted to determine whether the study should be conducted, under which conditions and the manner of its publication.

This review is to be conducted by the US agencies that are funding the study. However, the institution in which the study is conducted can challenge the determination reached by the agency, in which case the conditions under which the study will be conducted are jointly agreed upon. The policy states that such reviews should be conducted twice a year and reported to the assistant to the president for homeland security and counterterrorism (United States Government, 2012).

The policy thus places at the centre the worry that studies carried out in an academic setting might be misused by malevolent actors and represents a shift away from the notion that academic work should enjoy vast freedom. Indeed, the policy implies that third-party risks can override academic freedom even if researchers' intentions are benign. Moreover, the policy closes a regulatory gap that had existed for many years with regard to biosecurity. The regulations in the US previously focused mainly on biosafety – for example, through the Federal Select Agent Program (E-CFR, 2005). Such regulations focused on protecting the public from accidental release of dangerous agents and the protection of lab workers and their immediate environments. However, the existing policies did not fully address the risk that work carried out under the best biosafety conditions could be misused by third parties not involved in those studies. The new regulation closes this biosecurity gap and at the same time enhances the biosafety conditions.

On 21 February 2013, less than a year after the publication of the new US policy statement, the US Department of Health and Human Services announced a new 'Framework for Guiding Funding Decisions About Research Proposals with the Potential for Generating Highly Pathogenic Avian Influenza H5N1 Viruses That Are Transmissible Among Mammals by Respiratory Droplets To Guide Funding Decisions on Proposals for Research Anticipated To Generate HPAI H5N1 Viruses' (DHHS, 2013). This framework applies only to Department of Health and Human Services (DHHS) funding decisions and, as the title indicates, only to research that might create airborne H5N1. This step seeks to make clear that the federal oversight policy was not enough to reduce the risks of the ongoing H5N1 research agenda.

Within the space of another year, another policy was announced: On 24 September 2014, the White House issued a document entitled 'United States Government Policy for Institutional Oversight of Life Sciences Dual Use Research of Concern', and the policy it set out took effect the following year. The new policy is similar to the federal one in terms of the methods to be used to identify research projects that pose biosecurity risks, the main difference being that it requires federally funded researchers and research institutions to conduct their own reviews and report their findings to the federal government. In order to conduct such reviews, funded institutions are required to create local committees – Institutional Review Entities (IREs) – composed of researchers and experts in biosecurity policies (Science and Technology Policy Office, 2013).

On 17 October 2014, just a few weeks after the announcement of the government's institutional oversight policy, the White House Office of Science and Technology Policy and the Department of Health and Human Services instituted a moratorium on funding for gain-of-function research that involves influenza, Middle East respiratory syndrome (MERS) and severe acute respiratory syndrome (SARS) viruses while officials deliberated the risks and benefits associated with such research (Science and Technology Policy Office and the Department of Health and Human Services, 2014). The deliberative process behind this decision was conducted by the NSABB and the National Research Council, whose recommendations were submitted to the DHHS secretary (NSABB, 2016). Three years later, in December 2017, the funding moratorium was lifted. The end of the funding moratorium was enabled by the publication of a new review framework for studies that involve lab-enhanced pathogens that are highly virulent and transmissible in human populations (DHSS, 2017). The publication of this framework, together with the other policies created since 2013, seems to have closed most of the regulatory gaps that existed before the H5N1 controversy. However, it remains to be seen whether new cases, such as the artificially created horsepox virus mentioned at the beginning of this chapter, might highlight new gaps or alternatively show that the currently existing biosecurity policies are sufficient.

Conclusions

In this chapter, we have provided an overview of the developments in the fields of biosecurity and the life sciences. The chapter has focused on the most recent development in the intersection of these fields – namely dual-use research of concern. In the past biosecurity concerns pertained to at least two main possibilities: 1) intentional use of pathogens to inflict harm and 2) fears that nature will introduce dangerous pathogens. As a result, much life sciences research was devoted to developing countermeasures to both concerns. In the last 20 years, a new biosecurity concern has emerged: the worry that life sciences research itself, unintentionally, will enable malevolent agents to cause harm. We described the H5N1 case as a paradigm example of these new biosecurity risks. We have also shown that the US has taken a leading role in addressing these risks, yet it is quite clear that the US alone cannot completely mitigate them. Since science is global, such risks are likely to arise elsewhere. Given the global impact of US regulations, its actions are likely to motivate other countries to follow its lead. Nonetheless, its impact is not likely to generate a global mechanism for addressing biosecurity risks stemming from life sciences research; accordingly, a global approach is warranted.

Notes

1 The Netherlands has established a Biosecurity Office with the purpose of developing a biosecurity policy (see www.bureaubiosecurity.nl/en/Policy).
2 See, for example, Cello et al. (2002), Herfst et al. (2012), Imai et al. (2012), Jackson et al. (2001), Tumpey et al. (2005), Watanabe et al. (2014).

References

Broad, W.J. (2001) 'Australians create a deadly mouse virus', *New York Times*, January 23, p. A6.

Cello, J., Paul, A.V. and Wimmer, E. (2002) 'Chemical synthesis of poliovirus cDNA: Generation of infectious virus in the absence of natural template', *Science*, vol. 297, no. 5583, pp. 1016–1018.

Department of Health and Human Services (DHHS) (2004) 'HHS will lead government-wide effort to enhance biosecurity in "dual use" research', https://fas.org/sgp/news/2004/03/hhs030404.html, accessed 11 June 2018.

Department of Health and Human Services (DHHS) (2013) 'A framework for guiding US Department of Health and Human Services funding decisions about research proposals with the potential for generating highly pathogenic avian influenza H5N1 viruses that are transmissible among mammals by respiratory droplets', www.phe.gov/s3/dualuse/Documents/funding-hpai-h5n1.pdf, accessed 11 June 2018.

Department of Health and Human Services (DHHS) (2017) 'Framework for guiding funding decisions about proposed research involving enhanced potential pandemic pathogens', www.phe.gov/s3/dualuse/Documents/p3co.pdf, accessed 11 June 2018.

DiEuliis, D., Berger, K. and Gronvall, G. (2017) 'Biosecurity implications for the synthesis of horsepox, an orthopoxvirus', *Health Security*, vol. 15, no. 6, pp. 629–637.

Douglas, H.E. (2003) 'The moral responsibilities of scientists (tensions between autonomy and responsibility)', *American Philosophical Quarterly*, vol. 40, no. 1, pp. 59–68.

Dual Use Criteria Working Group (2006) 'Oversight of dual-use life sciences research: Strategies for minimizing the potential misuse of research information', http://oba.od.nih.gov/biosecurity/biosecurity_documents.html, accessed 16 February 2013.

Electronic Code of Federal Regulations (E-CFR) (2005) 'Title 42: Public health. Part 73 – selected agents and toxins', www.ecfr.gov/cgi-bin/retrieveECFR?gp=&SID=8a4be60456973b5ec6bef5dfeaffd49a&r=PART&n=42y1.0.1.6.61, accessed 11 June 2018.

Enserink, M. (2011) 'Scientists brace for media storm around controversial flu studies', *Science Insider*, November 23, http://news.sciencemag.org/scienceinsider/2011/11/scientists-brace-for-media-storm.html, accessed 20 February 2013.

Enserink, M. and Malakoff, D. (2012) 'Will flu papers lead to new research oversight?', *Science*, vol. 335, no. 6064, pp. 20–22.

Evens, N.G. (2013) 'Contrasting dual-use issues in biology and nuclear science', in B. Rappert and M.J. Selgelid (eds.) *On the Dual Uses of Science and Ethics: Principles, Practices, and Prospects.* Australian National University Press, Canberra.

Fauci, A.S., Nabel, G.J. and Collins, F.S. (2011) 'A flu virus risk worth taking', *Washington Post – Opinions*, December 30, http://articles.washingtonpost.com/2011-12-30/opinions/35285482_1_influenza-virus-public-health-, accessed 20 January 2013.

Fink Report (2004) *Biotechnology Research in an Age of Terrorism.* National Academies Press, Washington, DC, http://books.nap.edu/openbook.php?record_id=10827&page=1, accessed 3 November 2012.

Garrett, L. (2011) 'The bioterrorist next door', *Foreign Policy*, December 15, www.foreignpolicy.com/articles/2011/12/14/the_bioterrorist_next_door, accessed 20 February 2013.

Guillemin, J. (2005) *Biological Weapons: From the Invention of State-Sponsored Programs to Contemporary Bioterrorism*. Columbia University Press, New York.

Harmon, K. (2011) 'What will the next pandemic look like?', *Scientific American*, www.scientificamerican.com/article.cfm?id=next-influenza-pandemic, accessed 19 February 2013.

Hausman, D.M. (2007) 'Group risks, risks to groups, and group engagement in genetics research', *Kennedy Institute of Ethics Journal*, vol. 17, no. 4, pp. 351–369.

Herfst, S., Schrauwen, E.J.A., Linster, M., Chutinimitkul, S., de Wit, E., Munster, V.J., Sorrell, E.M., Bestebroer, T.M., Burke, D.F., Smith, D.J., Rimmelzwaan, G.F., Osterhaus, A.D.M.E. and Fouchier, R.A.M. (2012) 'Airborne transmission of influenza A/H5N1 virus between ferrets', *Science*, vol. 336, no. 6088, pp. 1534–1541.

Imai, M., Watanabe, T., Hatta, M., Das, S.C., Ozawa, M., Shinya, K., Zhong, G., Hanson, A., Katsura, H., Watanabe, S., Li, C., Kawakami, E., Yamada, S., Kiso, M., Suzuki, Y., Maher, E.A., Neumann, G. and Kawaoka, Y. (2012) 'Experimental adaptation of an influenza H5 HA confers respiratory droplet transmission to a reassortant H5 HA/H1N1 virus in ferrets', *Nature*, vol. 486, no. 7403, pp. 420–428.

Influenza Conference Newspaper (2011) 'Scientists provide strong evidence for pandemic threat', *The Influenza Times – Conference Newspaper, Fourth ESWI Influenza Conference*, www.eswiconference.org/Downloads/FEIC_news_1.aspx, accessed 26 February 2013.

Inglesby, T.V., Cicero, A. and Henderson, D.A. (2012) 'The risk of engineering a highly transmissible H5N1 virus', *Biosecurity and Bioterrorism: Biodefense Strategy, Practice, and Science*, vol. 10, no. 1, pp. 151–152.

Jackson, R.J., Ramsay, A.J., Christensen, C.D., Beaton, S., Hall, D.F. and Ramshaw, I.A. (2001) 'Expression of mouse interleukin-4 by a recombinant ectromelia virus suppresses cytolytic lymphocyte responses and overcomes genetic resistance to mousepox', *Journal of Virology*, vol. 75, no. 3, pp. 1205–1210.

Kaiser, D. and Moreno, J. (2012) 'Dual use research: Self-censorship is not enough', *Nature*, vol. 492, no. 7429, pp. 345–347.

Kitcher, P. (2001) *Science, Truth, and Democracy*. Oxford University Press, Oxford.

Knesset of Israel (2008) *Regulation of Research into Biological Disease Agents Law*. Jerusalem, www.health.gov.il/LegislationLibrary/Shonot03.pdf, accessed 2 August 2019.

Koblentz, G.D. (2017) 'The de novo synthesis of horsepox virus: Implications for biosecurity and recommendations for preventing the reemergence of smallpox', *Health Security*, vol. 15, no. 6, pp. 620–628.

Lakoff, A. (2008) 'The generic biothreat, or, how we became unprepared', *Cultural Anthropology*, vol. 23, no. 3, pp. 399–428.

Lakoff, A. and Collier, S.J. (eds.) (2008) *Biosecurity Interventions: Global Health and Security in Question*. Columbia University Press, New York.

Lev, O. and Rager-Zisman, B. (2014) 'Protecting public health in the age of emerging infections', *The Israel Medical Association Journal: IMAJ*, vol. 16, no. 11, pp. 677–682.

Lev, O. and Samimian-Darash, L. (2014) 'Biosecurity policy in the US: A critical assessment', *Frontiers in Public Health*, vol. 2, article 110 (1–3).

MacKenzie, D. (2011) 'Five easy mutations to make bird flu a lethal pandemic', *New Scientist*, September 21, www.newscientist.com/article/mg21128314.600-five-easy-mutations-to-make-bird-flu-a-lethal-pandemic.html, accessed 30 May 2015.

Masco, J. (2013) *The Nuclear Borderlands: The Manhattan Project in Post-Cold War New Mexico*. Princeton University Press, Princeton.

Masco, J. (2014) *The Theater of Operations: National Security Affect from the Cold War to the War on Terror*. Duke University Press, Durham.

National Institutes of Health (NIH) (2011) 'Press statement on the NSABB review of H5N1 research', *NIH NEWS*, December 20, www.nih.gov/news/health/dec2011/od-20.htm, accessed 20 February 2013.

National Science Advisory Board for Biosecurity (NSABB) (2007) 'Proposed framework for the oversight of dual use life sciences research', https://osp.od.nih.gov/wp-content/uploads/Proposed-Oversight-Framework-for-Dual-Use-Research.pdf, accessed 10 June 2018.

National Science Advisory Board for Biosecurity (NSABB) (2012) 'March 29–30, 2012 meeting of the National Science Advisory Board for Biosecurity to review revised manuscripts on transmissibility of A/H5N1 influenza virus', http://oba.od.nih.gov/oba/biosecurity/PDF/NSABB_Statement_March_2012_Meeting.pdf, accessed 16 February 2013.

National Science Advisory Board for Biosecurity (NSABB) (2016) 'Recommendations for the evaluation and oversight of proposed gain-of-function research', https://osp.od.nih.gov/wp-content/uploads/2016/06/NSABB_Final_Report_Recommendations_Evaluation_Oversight_Proposed_Gain_of_Function_Research.pdf, accessed 11 June 2018.

New York Times (2012) 'An engineered doomsday', January 7, www.nytimes.com/2012/01/08/opinion/sunday/an-engineered-doomsday.html?_r=0, accessed 20 February 2013.

Osterholm, M.T. and Henderson, D.A. (2012) 'Life sciences at a crossroads: Respiratory transmissible H5N1', *Science*, vol. 335, no. 6070, pp. 801–802.

Rabinow, P. (1996) *Essays on the Anthropology of Reason*. Princeton University Press, Princeton.

Racaniello, V.R. (2012) 'Science should be in the public domain', *MBio*, vol. 3, no. 1, pp. 4–12, http://mbio.asm.org/content/3/1/e00004-12.full, accessed 20 February 2013.

Rose, D. (2008) 'How did the smallpox vaccination program come about? Tracing the emergence of recent smallpox vaccination thinking', in A. Lakoff and S.J. Collier (eds.) *Biosecurity Interventions: Global Health and Security in Question*. Columbia University Press, New York.

Samimian-Darash, L. (2016) 'Biosecurity in the US: "The Scientific" and "the American" in critical perspective', In V.R. Dominguez and J. Habib (eds.) *America Observed: On an International Anthropology of the United States*. Berghahn Books, New York.

Samimian-Darash, L., Henner-Shapira, H. and Daviko, T. (2016) 'Biosecurity as a boundary object: Science, society, and the state', *Security Dialogue*, vol. 47, no. 4, pp. 329–347.

Science and Technology Policy Office (2013) 'United States government policy for institutional oversight of life sciences dual use research of concern', https://federalregister.gov/a/2013-04127, accessed 31 May 2018.

Science and Technology Policy Office and the Department of Health and Human Services (2014) 'US government gain-of-function deliberative process and research funding pause on selected gain-of-function research involving influenza, MERS, and SARS viruses', www.phe.gov/s3/dualuse/Documents/gain-of-function.pdf, accessed 10 June 2018.

Smith, F.L. and Kamradt-Scott, A. (2014) 'Antipodal biosecurity? Oversight of dual use research in the United States and Australia', *Frontiers in Public Health*, vol. 2, article 142.

Tonix Pharmaceuticals (2017) 'Tonix Pharmaceuticals announces demonstrated vaccine activity in first-ever synthesized chimeric horsepox virus', www.tonixpharma.com/news-events/press-releases/detail/1052/tonix-pharmaceuticals-announces-demonstrated-vaccine, accessed 26 April 2018.

Tumpey, T.M., Basler, C.F., Aguilar, P.V., Zeng, H., Solórzano, A., Swayne, D.E., Cox, N.J., Katz, J.M., Taubenberger, J.K., Palese, P. and García-Sastre, A. (2005) 'Characterization of the reconstructed 1918 Spanish influenza pandemic virus', *Science*, vol. 310, no. 5745, pp. 77–80.

United States Government (2012) 'United States government policy for oversight of life sciences dual use research of concern', www.phe.gov/s3/dualuse/Documents/us-policy-durc-032812.pdf, accessed 31 May 2015.

United States Government (2014) 'United States government policy for institutional oversight of life sciences dual use research of concern', www.phe.gov/s3/dualuse/Documents/durc-policy.pdf, accessed 31 May 2015.

Watanabe, T., Zhong, G., Russell, C.A., Nakajima, N., Hatta, M., Hanson, A., McBride, R., Burke, D.F., Takahashi, K., Fukuyama, S., Tomita, Y., Maher, E.A., Watanabe, S., Imai, M., Neumann, G., Hasegawa, H., Paulson, J.C., Smith, D.J. and Kawaoka, Y. (2014) 'Circulating avian influenza viruses closely related to the 1918 virus have pandemic potential', *Cell Host & Microbe*, vol. 15, no. 6, pp. 692–705.

Wein, L.M. and Liu, Y. (2005) 'Analyzing a bioterror attack on the food supply: The case of botulinum toxin in milk', *Proceedings of the National Academy of Sciences of the United States of America*, vol. 102, no. 28, pp. 9984–9989.

Williams-Jones, B., Olivier, C. and Smith, E. (2014) 'Governing "dual-use" research in Canada: A policy review', *Science and Public Policy*, vol. 41, no. 1, pp. 76–93.

World Health Organization (WHO) (2012) 'Report on technical consultation on H5N1 research issues', www.who.int/influenza/human_animal_interface/mtg_report_h5n1.pdf, accessed 10 February 2013.

20

REWILDING AND INVASION

Timothy Hodgetts and Jamie Lorimer

Introduction

Rewilding is an increasingly influential approach in wildlife conservation. Although it has historical roots in practices of ecological restoration and wilderness preservation, it emerged in its contemporary incarnation in North America in the 1990s. Since then, rewilding has proliferated in geographic reach, conceptual scope and varieties of practice. It tends to involve the (re) introduction of species of large carnivore or herbivore into areas of their former range or into new areas where a similar (but now extinct) species used to roam. The term has been applied to conservation schemes that cross continents as well as those within single fields and to projects that expand existing habitats as well as those seeking to create novel ecologies, sometimes even with new forms of life. But among the diversity of rewildings that are currently enacted and discussed, there is a shared core commitment to promoting or restoring ecological processes rather than simply 'managing extinctions' (Donlan et al., 2005, 913).

In this chapter, we outline the historical emergence of contemporary forms of rewilding and then introduce five axes of difference through which the various approaches might be categorised. These are: i) the objectives of the project, particularly with respect to ecological baselines; ii) the intervention type, from the active reintroduction of megafauna to passive inaction in abandoned agricultural landscapes; iii) the spatialities of the project (scale and boundedness); iv) its legal status (official or unsanctioned); and (v) the political issues and approaches involved. We then identify the major tensions that exist between rewilding and regimes of biosecurity. The first tension derives from rewilding's commitment to a functionalist approach in ecological management (i.e. targeting ecological processes), rather than the compositional logics that underpin traditional biodiversity conservation and invasive species management (that target particular species and species communities as contextually desirable or not). The second emerges from the open-ended and experimental management practices that characterise rewilding, in contrast to the fixed goals that often underpin the logics of biosecurity and traditional species-based conservation. The third comes from the emphasis on connectivity and circulation in rewilding, as compared to the emphasis on borders and control in regimes of biosecurity. These are caricatured positions compared to the nuanced practices through which rewilding and biosecurity are enacted; nevertheless, we suggest they are informative. We conclude with some targeted suggestions as to how the tensions can be addressed in both theory and practice.

Throughout the chapter, we introduce a wide range of contemporary rewilding projects to give a flavour of the diversity in this movement and the diverse implications it has for biosecurity and invasive species management.

A history and geography of rewilding in theory and practice

The genesis of the term 'rewilding' is usually traced to back to the early 1990s and the Wildlands Project in North America (Jørgensen, 2015). Founded by a stellar cast of conservation luminaries including conservation biologist Michael Soule, Earth First! activist Dave Foreman and the environmental philanthropist Doug Tompkins, the aim was to re-imagine wildlife conservation at a continental scale (WildlandsNetwork, 2018). The project emphasised the importance of connectivity between protected areas and proposed a series of vast landscape linkages to connect these core areas. They called it 'rewilding': conservation action that protected threatened ecosystems by facilitating necessary ecological processes – specifically the movement of large apex predators whose behaviour regulated trophic interactions throughout the system (Soulé and Noss, 1998). This combination of 'cores, corridors and carnivores' became known as the '3 Cs' and continues to underpin what is now known as the Wildlands Network (Sandom et al., 2013). Representing one of the most ambitious and wide-ranging rewilding programmes in the world, the Wildlands Network brings together a range of collaborative projects that collectively contribute towards creating 'wildways' across North America. As the project website explains,

> Rewilding means making our landscapes whole again. Today's national parks and other protected areas, although critical to conservation, are too small and isolated from one another to support wildlife migrations and dispersals, native plant communities, and services provided by nature – like pollination, carbon storage, and clean water. Wildlands Network is helping to rewild North America by protecting core reserves, reconnecting them via vast corridors of habitat, and restoring apex predators.
>
> *(Wildlands Network, 2018)*

The rewilding concept was further developed by an expanded team of biologists and conservationists in the mid-2000s through a series of influential interventions that proposed a new approach to conservation: Pleistocene rewilding (Donlan et al., 2006). This vision went further in asserting the means by which to achieve the restoration of 'lost' or 'missing' ecological processes (Donlan et al., 2005). Rather than simply (re)connecting habitats for extant apex predators or mega-herbivores (whose ecosystem-shaping roles were increasingly being incorporated into rewilding logics by this time) or reintroducing such animals to parts of their historical range, Donlan and colleagues proposed introducing 'proxy species' in North America to enact ecological processes that had been lost with the extinction of various megafauna species. They suggested, for example, introducing Bactrian camels to re-establish ecological processes (such as browsing on woody plants) that were lost when *Camelops*, a North American camelid, went extinct. Similarly, African elephants (*Loxodonta africana*) might be used in the US to maintain grasslands, fulfilling a role 'probably' played in the Pleistocene by now extinct mammoths and mastodons (ibid., 914).

More controversially, the Pleistocene rewilding approach also advocated the release of predator proxies to regulate prey species, such as African lions and cheetahs (*Panthera leo* and *Acinonyx jubatus*), into the American plains. The logic was that species historically co-evolved in communities and that important ecological processes supporting such communities were lost when megafauna were removed from the system. While wild-living large animals were once ubiquitous in the earth's ecosystems, in the Anthropocene they are increasingly rare (Dirzo et al., 2014). Advocates suggest that additional conservation benefits accrue from creating new spaces for species that are critically endangered in their home ranges – as wild Bactrian camels most certainly are (Hare, 2008). Of course, the effects of introducing non-native species into new ecosystems are hard to predict in advance, especially for megafauna species whose behaviours often lead to significant trophic reverberations through an ecological community. As such, Pleistocene rewilding is controversial within the conservation community (Caro, 2007). We discuss these uncertainties, and their implications for invasive species management, in the final section.

Rewilding ideas and projects have not been limited to North America but have been developed over the past two decades across diverse geographical contexts. In Europe, for example, high-profile rewilding schemes have tended to focus more on the introduction of herbivores to enact trophic cascades, compared to the North American emphasis on carnivores. The European schemes often seek to apply an understanding of the role of grazing behaviour in the past in an attempt to promote more resilient, biodiverse 'wild' landscapes for the future (RewildingEurope, 2015). Furthermore, the grazers utilised are often 'back-bred' domesticated species (Lorimer and Driessen, 2016) rather than translocated wild species from other continents (as per the Pleistocene model). Many European projects are part of the 'European Rewilding Network', an information-sharing and support network led by the NGO Rewilding Europe to foster collaboration among the numerous (and diverse) rewilding projects active across the continent. By the end of 2017, the network included 61 different projects, in 26 countries, covering 6 million hectares of land (RewildingEurope, 2018). Examples include the introduction of Heck cattle and Konik ponies to the Oostvaardersplassen in the Netherlands and Garrano horses and Maronesa cattle to the Faia Brava nature reserve in Portugal and the (re)introduction of free-ranging bison in Germany (ibid.). Another high-profile approach to rewilding that also emphasises the role of herbivory as an ecological process has been the use of non-native tortoises to restore processes of grazing and seed dispersal on oceanic islands where the historical chelonian residents have been extirpated. For example, 'proxy' species of tortoises have been introduced to several islands in the Galapagos, and also in Mauritius, in an effort to restore historic plant ecologies (Griffiths et al., 2010; Hunter et al., 2013).

Ideas about rewilding have proliferated in recent years, as have conservation projects that claim to follow a rewilding approach. As well as the introduction of locally extirpated apex predators or proxy non-native species, rewilding has come to encompass a range of strategies for restoring ecological processes in contemporary ecosystems and is sometimes even used as a promotional label for more traditional ecological restoration activities. The term has become, as the environmental historian Dolly Jørgensen puts it, somewhat 'plastic': so diffuse as to be increasingly imprecise, sometimes contradictory and perhaps therefore unhelpful (Jørgensen, 2015). Yet these variant strategies for rewilding tend to be linked by certain shared commonalities: i) a focus on ecological processes, ii) the use of fauna (often megafauna) to instigate or restore such processes and iii) a hands-off management approach. The first shared aim, as discussed prior, is to restore ecological processes as an alternative way to protect biodiversity compared to the usual focus on particular species. The second commonality is a tendency to utilise (mega)fauna to this end – often having first identified which keystone species are 'missing' through historical research. Missing megafauna are reintroduced to instigate trophic cascades, with these cascades understood as occurring 'where apex consumers shape ecosystems via effects on prey and other resources, as well as competitors, and their multidirectional propagation through food webs' (Svenning et al., 2016, 899).

The classic example usually given of 'trophic rewilding' is the reintroduction of wolves to Yellowstone National Park. Overgrazing by dense elk populations had limited the recruitment of aspen and willow species, but the return of wolves to the system has allowed these species to flourish – both through direct predation (less elk) and through affecting elk behaviour (predator avoidance; Ripple and Beschta, 2012). But trophic logics also underpin the (re)introduction of megafaunal herbivores, whose grazing behaviours drive trophic interactions, as well as smaller scale reintroductions of locally extirpated species, such as beavers in parts of the UK and Europe. The third similarity is that rewilding strategies tend to rely on the self-regulating properties of rewilded ecosystems to minimise the need for subsequent human interventions. Forms of intensive management may be used to instigate rewilding projects, such as species reintroductions or invasive species removal programmes. However, once initial conditions have been created, the rewilding approach is best characterised as a 'hands-off' form of adaptive management, where ecological processes are allowed to unfold, with intervention only necessary if the ecosystem shifts towards an undesirable state (Sandom and Wynne-Jones, 2018). To make the contrast clear: traditional conservation management specifies what ecosystem state is desirable (with all other states as undesirable by implication) whereas rewilding does not specify desirable outcomes but can specify what is undesirable.

These various forms of contemporary rewilding have longer roots within conservation's ecological and social history. Most obvious are significant links with restoration ecology, particularly when rewilding approaches draw explicitly on historical ecological baselines such as that found in the Pleistocene approach (we discuss baselines in the next section). The influence of ecological resilience theory (Gunderson and Pritchard, 2002) is also apparent in the focus on maintaining ecological processes and avoiding undesirable (eco)systemic states. For example, there are similarities between resilience-inspired approaches to coastal management that work through re-wetting previously reclaimed saltmarshes (often termed 'managed realignment of the coastline' in the UK context) and forms of rewilding that rest on similar logics (allowing ecological processes to unfold and thus reducing the need for human intervention). Rewilding with beavers, whose dams slow the release of flood waters and thus reduce the risk of flooding downstream, is another example. And there are also relevant conservation and social histories that influence rewilding practices and theories, such as the role of wilderness ideals in conservation (Cronon, 1996) and the related colonial (and post-colonial) practices of exclusion to create wilderness (Neumann, 1998), which we discuss later in this chapter.

A typology of rewildings

In this section, we set out a typology for comparing these various rewildings. We identify five axes of difference which can be used to map rewilding practices and theories, relating to their objectives, intervention type, spatialities, legal status and politics (Table 20.1). Other typologies are of course possible (see, for instance, Lorimer and Driessen, 2016), but we suggest that these axes of difference are of particular relevance when considering the implications of rewilding for biosecurity and invasive species management.

Objectives

Rewilding schemes all promote ecological processes. However, there are choices to be made concerning exactly which ones to promote and why. Rewilding can involve the *restoration* of previously existing processes, such as the roles in seed dispersal historically played by a locally (or globally) extinct large herbivore. One example is the introduction of Aldabra tortoises to the Mauritian islands to disperse the seeds of the endemic palm species, an ecological process that had been lost with the extinction of the native tortoises in the 19th century (Griffiths et al.,

Table 20.1 Characteristics of rewilding schemes.

Axis of difference	Forms of rewilding in theory or practice		
Objectives	Specific historical baselines	←→	Open-ended novel ecologies
Intervention type	Active	←→	Passive
	(e.g. megafauna reintroduction)		(e.g. agricultural abandonment)
Spatiality	Bounded	←→	Unbounded
	Intensive	←→	Extensive
	Purified	←→	Hybrid
Legal status	Official	←→	Unofficial
	(e.g. government approved)		(e.g. 'beaver bombing')
Political issues	Imposition	←→	Collaboration
(Who decides? Who benefits?)	Inequality	←→	Equality

2010). Another is the use of large numbers of bison, horses, moose, musk ox and reindeer – and their grazing and defecating behaviours – to try and recreate a mammoth steppe ecosystem in a 16,000 hectare Siberian reserve called the Pleistocene Park (Zimov, 2005). Rewilding can also involve the *conservation* of existing but threatened ecological processes, such as the wide-ranging movement of large carnivores that, as apex predators, shape trophic relations throughout eco-systems (as promoted, for example, in the Wildlands Network). Or it may involve the *creation* of new ecological relations and processes through experimental, open-ended regimes of spe-cies introduction and hands-off management that do not seek to recreate a lost past but a more resilient future.

Perhaps the most common difference between rewilding schemes is, therefore, their choice of ecological baseline. The concept of baselines comes from restoration ecology and refers to the historical species community that is to be restored. Since different species communities have existed in the same location over time, restoration ecology always involves a choice. For example, much of North American conservation implicitly presumes a 'pre-Columbian' base-line wherein management aims at restoring the species communities found across the continent before the arrival of European settlers. In more recent decades, paleo-ecologists and environ-mental historians have done much to challenge the notion of pre-Columbian ecologies as 'pris-tine wildernesses', and the role of Native American communities in shaping these ecosystems is becoming better appreciated within Western knowledge systems (Cronon, 1996). An alternative has been to target restoration towards recreating ecosystems that existed before the widespread impacts from human agriculture began to shape ecologies or – as in Pleistocene rewilding – yet further into pre-history, to the period before human actions had caused widespread megafaunal extinctions. Indeed, much recent research by paleo-ecologists has sought to specify the role of human hunting in the megafaunal extinctions of the late Pleistocene (Sandom et al., 2014), and such work is often used as part of the ethical justification for rewilding projects.

Many European rewilding projects also operate with an explicit or assumed baseline in deep history, in their case often influenced by the work of Dutch ecologist Franz Vera and his 'open-woodland' hypothesis (Vera, 2000). Vera proposed that the ecology of Europe before widespread human agriculture would not, contrary to widespread belief, have been dominated by closed canopy forests. Instead, he suggests that mega-herbivores like aurochs (progenitors to domes-ticated cattle) would have shaped, through grazing, the vegetational community into a mosaic of woodland and open patches of grassland. For example, one of the earliest and most influential European rewilding projects was initiated in the 1980s by a team, including Vera, at the Oostvaardersplassen in the Netherlands. This is a nature reserve close to the city of Amsterdam, where a regime of 'naturalistic grazing' was instigated using herds of de-domesticated Heck cattle and Konik horses (bred and utilised as proxies of the now extinct aurochs and tarpan, their wild-living predecessors; Lorimer and Driessen, 2014). However, as noted previously, rewilding also tends to promote a hands-off, open-ended form of adaptive management. Thus, even in those projects that have an explicit or implicit historical baseline underpinning their objectives, there remains a tension between strict adherence to that historical objective (as would be the goal in classic restoration ecology) and the commitment to fostering processes rather than particular community compositions. Indeed, in many forms of rewilding, it would be more accurate to say that the goal is not the restoration of a lost ecological past; instead, the objective is to *learn from that past* (especially with respect to the role played by large animals) in order to create functioning, biodiverse ecological communities for the future (Moorhouse and Sandom, 2015).

Finally, conservation schemes tend to have multiple ecological objectives, and rewilding schemes are no different. While the functional roles of large animals as predators, browsers, grazers or seed dispersers is emphasised within rewilding logics, such schemes also promote the conservation of these species (or their required habitats) for their own sakes. For example, the Pleistocene rewilding schemes proposed by Donlan et al. (2006, 2005) emphasise that translocations of African megafauna to North American habitats could also help conserve these species. Wolves, by contrast, are not threatened with extinction in a global sense, but their extirpation from most of North America due to deliberate human actions was part of the ethical justification for their reintroduction. Similar logics apply to the reintroduction of beavers in some western European countries (Crowley et al., 2017a) and the translocation of pine martens from Scotland to Wales (Hodgetts, 2017).

Intervention types

Rewilding schemes can be implemented through a range of active interventions. Introducing or reintroducing large animals is the most common form, but there is variety even here: from former native species to exotic proxies and from endangered species to potential future genetically engineered 'de-extinct' species. An example of the latter, reminiscent of the book and film *Jurassic Park*, is the possibility of recreating living mammoths and reintroducing them to the Pleistocene Park in Siberia using frozen mammoth DNA from the tundra and elephant eggs (Jørgensen, 2015). A less extreme example is the attempt to back-breed domesticated cattle to recreate something approximating aurochs, for use as naturalistic grazers in European rewilding schemes past and present (Lorimer and Driessen, 2016). Other forms of active intervention are also possible. For example, in the original Wildlands Project, the progenitor of much rewilding practice, much of the intervention relates to creating and protecting habitat corridors to enable the movements of large carnivores.

All these examples are active interventions. However, there is an important form of rewilding that is created through inaction. For instance, the abandonment of agricultural land, especially in Europe and North America, has created opportunities for multiple species to repopulate former (or populate new) areas as forests have begun to regenerate (Navarro and Pereira, 2015; Schnitzler, 2014). Indeed, large carnivores in Europe, such as wolves and brown bears, have benefitted significantly from this trend (Buller, 2008; Chapron et al., 2014). Furthermore, there are also forms of rewilding that are unintentional but are instigated through active interventions of other types. For example, Svenning et al. (2016) suggest that the introduction of feral horses to North

America might be understood as a form of trophic rewilding, but one that was unintentional in two ways: it wasn't done with rewilding objectives as motivation, and it wasn't known at the time that wild horses were part of the continent's historic (but extinct) megafauna. The subsequent protection of ecologies that have been created through inaction and/or unintentionally are often justified in terms of rewilding theory. Furthermore, unintentional rewildings can act as a source of empirical information that can be used to inform more activist rewilding projects, particularly given the experimental character of such schemes (Svenning et al., 2016). Of course, the rewilding benefits of agricultural land abandonment need not be unintentional. For instance, one approach to rewilding is to deliberately leave agricultural land to 'wild' itself (Taylor, 2005). However, other forms of unintentional passive rewilding are less amenable to imitation. For example, the radioactive aftermath of the Chernobyl nuclear meltdown has created an unintentional form of habitat rewilding wherein many large-bodied non-human animal species appear to be flourishing (Deryabina et al., 2015).

Spatialities

Rewilding projects and proposals also involve a range of spatial scales and degrees of boundedness. They can be spatially limited in extent, such as the scheme launched in England to reintroduce beavers to a 6.5 hectare enclosure in the Forest of Dean. Despite its small size, the project, led by England's Forestry Commission and backed by Environment Secretary Michael Gove (DEFRA, 2017), has been widely discussed and reported as a form of rewilding. The flagship rewilding project at the Oostvaardersplassen discussed earlier is bigger, covering around 56 square kilometres of wetland. But rewilding schemes can also be applied across vast areas, such as the Wildlands Project, with its 'wildways' criss-crossing the entire North American continent. These differences in scale are notable for the forms of ecology and land management they respectively envisage and the different social effects they may imply.

The second notable form of spatiality concerns the degree of boundedness enacted within a rewilding scheme. Both the English beaver ponds and the Dutch rewilded polder at the Oostvaardersplassen rely on fences to keep certain (but not all) animals enclosed. But spatial bounding can also operate across much bigger areas, reflecting a long-standing and mainstream approach in protected area management that is sometimes labelled (critically) as 'fortress conservation' (Brockington, 2002). For example, in setting out their ideas for Pleistocene rewilding, Donlan and colleagues proposed:

> The third stage in our vision for Pleistocene re-wilding would entail one or more 'ecological history parks', covering vast areas of economically depressed parts of the Great Plains. As is the case today in Africa, perimeter fencing would limit the movements of otherwise free-living ungulates, elephants and large carnivores.
>
> *(Donlan et al., 2005, 914)*

Again, there is a contrasting approach to rewilding that enacts a less bounded form of spatiality, where species (including reintroduced species) are not bound by fences or restricted to designated areas. In between, there is a wide range of more or less restrictive practices and spatialities. For example, the habitat corridors envisaged in the continent-crossing Wildlands Network utilise different forms of bounding depending on the local contexts – fences may be necessary next to roads but less relevant in the mountains, for example. Rewilding can therefore imply very different management practices and social relations within ecologies depending on the extent to which particular schemes rely on enclosing 'wild' lands.

Different degrees of bounding also reflect a third form of spatiality, relating to the extent to which rewilding projects 'purify' spaces by removing human activities or create more hybrid spaces in which rewilded ecologies incorporate human activities. Fenced nature reserves utilising rewilding principles are extreme examples of the former approach, with the Oostvaarders-plassen being an archetypical example. While animals are kept in by the fences, at the same time people are kept out, with access by humans being strictly limited and controlled. In contrast, rewilding principles have also been applied within agricultural landscapes to create wildlife-rich ecologies that can exist alongside food production – here, the rewilding project at Knepp Castle in England is exemplary (Tree, 2018). There are degrees of hybridity, with different levels of human activity being tolerated in different projects – for example, low impact eco-tourism enacts a more 'purified' form of spatiality than food production. The differences between 'purified' and 'hybrid' spatialities reflect a wider debate within conservation between *land-sparing* approaches, that seek to separate 'wild' lands from those inhabited or industrialised by humans, and *land-sharing* approaches, which advocate conservation actions within a matrix of more or less human-influenced habitats – and rewilding can adopt either approach, as well as a hybrid combination of both in larger areas. As the geographer Henry Buller notes, 're-wilding is really all about boundaries, both their presence and their transgression' (Buller, 2013, 189).

Legal status

The diverse conservation situations and schemes united under the banner of rewilding can also be categorised by their legal status and by the types of additional legal and regulatory challenges they create. Active-intervention rewilding schemes tend to be 'official', agreed in advance with relevant regulatory agencies and sanctioned accordingly. Indeed, the release or reintroduction of large animals without the proper permissions is illegal in many countries, on the grounds of biosecurity as well as public safety (see later discussion). Furthermore, although rewilding approaches tend to emphasise the reduced role of human management implied in such schemes, there is often an acceptance that reintroduced large animals will need occasional management if they become a threat to human safety, property or other valued forms of domesticated or wild life. If national or other official wildlife agencies are to take on the responsibility of managing, say, the impacts of reintroduced wolves in agricultural communities surrounding a 'rewilded' area, it is important that such releases are officially sanctioned to begin with.

However, there are examples of active-intervention forms of rewilding that are unsanctioned and in some cases illegal; the unofficial release of beavers in southern England is an example. In this and similar cases, the legal status of the released animals has been ambiguous – although beavers are a formerly native species, they are not currently protected by conservation legislation and can be culled without permit (albeit subject to existing animal welfare laws). In practice, however, wider societal interests and concerns have complicated any actions to remove them once in place (Crowley et al., 2017a). Feral animals are another category of actively introduced megafauna whose trophic interactions may shape ecosystems and whose legal status is often complicated – feral boar in the UK, feral camels in Australia and feral horses in North America are some examples. The difference in their case is that the intervention was active (i.e. humans deliberately moved these animals to new locations) but often unintentional (they may have become feral through escaping enclosures). Nevertheless, and despite the underlying motivation, the ecological effect of releasing large animals can often be consonant with official rewilding projects, and if established, the reintroduced animals may require management and/or protection. Passive rewilding often raises a different set of questions and challenges around the legal status of animals that enter new jurisdictions. The spread of large carnivores from east to west in

Europe, following the path of agricultural abandonment, is an example (Chapron et al., 2014; Navarro and Pereira, 2015). Some of these animals are protected by European Union conservation legislation and may come into conflict with agricultural communities as they move into new areas (Buller, 2008).

A different set of questions around legal status relate to the welfare of animals managed according to rewilding principles. The issues involved are perhaps best exemplified through the situation of horses and cattle at the Oostvaardersplassen. Enclosed by a fence, the herbivores are imagined by the reserve's managers as wild grazers rather than domesticated equine companions. In harsh winters, as in 2018, when there is little food to be found, they go hungry. This has led to significant controversy, with debates on the welfare of these animals reaching the national parliament and supplementary winter feeding and population management being instigated by the government (Lorimer and Driessen, 2014). Another related difference of note is that between the capture and translocation of wild-living animals and the release into the 'wild' of captive-bred animals. The legal status of reintroduced species in a rewilding scheme can thus help characterise the type of wildness actually enacted.

Political issues

As these sections suggest, rewilding can be a politically contentious activity. It raises issues around the desirability of particular ecologies, about access and exclusion and concerning animal and human welfare. Who gets to decide about such issues, who benefits and who bears the costs are all relevant political questions. For example, the reintroduction of wolves to an area with agriculture might be popular with conservationists and those in the tourist industry but might be a less attractive proposition for livestock farmers. Compensation schemes can be developed to address the economic impacts of livestock losses, but the funding and management of such schemes involves political choices, as does the initial decision to allow or impose reintroductions of large carnivores. Given that conservation has a sometimes-fraught social history involving forced exclusions of marginalised people from lands and livelihoods (Dowie, 2009), such concerns have heightened salience. Introducing large herbivores is also not without social impacts, which might relate to the purification of spaces, biosecurity for agriculture or even risk to human and animal health – for example, releasing elephants in North America, as envisaged in some 'Pleistocene rewilding' proposals, must be viewed in the context of elephant-linked human mortality in areas of the world where these animals are free roaming. The various issues raised by rewilding projects are addressed through forms of politics that stretch from outright imposition by various actors (e.g. state wildlife agencies, international conservation NGOs) to collaborative, locally instigated and inclusive forms of organisation. The degree to which relevant stakeholders are engaged in rewilding projects, and the stage at which they are consulted (e.g. before, during or after activities begin), can be considered as another axis of difference. Indeed, critical social science also highlights the role that these different types of politics and rewilding activities can play in producing, and reproducing, wider social relations and economic arrangements (Hintz, 2007; Jørgensen, 2015; Lorimer and Driessen, 2016).

Summary

The term 'rewilding' covers a wide range of management approaches, united by their focus on ecological processes, the role of (mega)fauna in regulating trophic systems and a style of minimal interventions after initial conditions have been set. These various approaches can be characterised by their objectives (historical baselines/open-ended future wilds), intervention style (active/

passive), spatialities (scale and boundedness), legal status (official/unofficial) and politics (inclusive/exclusive). In the next section, we consider the intersections with, and frictions between, these various rewildings and concerns about invasive species and biosecurity.

Key philosophical tensions between rewilding and invasive species/biosecurity

Biosecurity at its simplest refers to the protection of life, often human or livestock. Rewilding, when done through the (re)introduction of apex predators, or the encouragement of their wider movement, can put human and livestock lives at risk. This comparison suggests a tension between biosecurity as promoting life and rewilding as facilitating death. But that is too simple. Biosecurity, and the biopolitics through which it is enacted, tends to work through managing the circulations of various organisms and in making some die so that others might live (Dobson et al., 2013; Hinchliffe et al., 2013; Hinchliffe and Bingham, 2008). Similarly, rewilding operates according to a logic suggesting that through restoring or creating trophic relations and related actions (be they grazing or predatory behaviours), a more resilient ecosystem will emerge (Buller, 2013). In both cases, lives are secured through facilitating processes of death. There are also some notable similarities in methods. For example, similar to the use of predators to enact ecological processes in rewilding, there is a history of utilising non-native predators as biological control agents in invasive species management, albeit a fraught history full of unexpected consequences (Thompson, 2015). Likewise, although rewilding tends to emphasise the self-managing properties of wilded landscapes, in practice (and especially within 'bounded' schemes like the Oostvaardersplassen), the culling of animals – a staple of invasive species management – is practised to control population sizes or remove 'problem' animals. Nevertheless, there do exist some significant philosophical and practical tensions between rewilding approaches and regimes of biosecurity (Table 20.2).

Functionalism vs. compositionalism

Invasive species management rests on what are often described as compositional logics. This is the idea that different locations each have ecological communities 'composed' of a particular set of species and that the aim of conservation should be to protect or restore that set of species if human action has led to local extirpations or degradation. It is the dominant framework for conservation theory and practice in much of the world, and it rests on an important assumption that species are linked to particular places. As a result, location-based lists of species and habitat types (found presently or historically) have been created and serve as baselines for location-specific conservation action. This is the basis of the categories 'native' (to a place), 'non-native' and thus also 'invasive'. Rewilding, by contrast, is less concerned about conserving or restoring the full list of species to a place. Instead, as we have seen, the emphasis is on restoring or facilitating

Table 20.2 Tensions between rewilding and biosecurity/invasive species management.

	Rewilding	*Biosecurity/invasive species management*
Theoretical basis	Functionalism	Compositionalism
	– ecological processes	*– located species*
Management style	Open-ended/experimental	Fixed objectives
Spatial practices	Facilitating connectivity	Controlling movement
	(e.g. habitat corridors)	*(e.g. borders/boundaries)*

ecological processes. The lack of concern with species lists in rewilding inspires serious questions about the status of non-functionally important species. Rewilders tend to emphasise the importance of megafauna for regulating the species at lower trophic levels. Compositional logics, by contrast, often spread the emphasis more widely. Indeed, the famous Ehrlich rivet hypothesis (which argues that like the rivets in an aeroplane wing, you don't know which species are crucial for the resilience of the ecosystem until it fails) is an important theoretical underpinning to composition-based conservation.

Furthermore, the tension between compositional and processural logics is also apparent with respect to the introduction of non-native species to an area, as envisaged in Pleistocene rewilding, for example, with the call to translocate African megafauna to North America. This move is in strong contradiction with logics of composition. African elephants are not on the North American list. Nor are they on the Australian list, where their introduction has been proposed in an effort to control invasive grasses (Bowman, 2012). The concerns are related not so much to the aesthetics of species being out of place (although this does play a role in the politics of conservation in many situations) but to the possibility of unexpected ecological consequences that may occur from the introduction of non-native species to a new area. Such species might become invasive. They might drive some native species towards local (or even global) extinction through the trophic cascades they initiate. For the rewilder, if the ecological processes generated make the wider ecosystem more resilient, this may be an acceptable price; for the compositionalist, the loss of some native species, even locally, would be regarded as failure. Thus, and given the dominance of compositional logics in contemporary Western conservation, some rewilding schemes have been directly blocked because of concerns about the nativity of the proposed reintroductions. In certain contexts, this can become an argument about ecological baselines and the question of whether a species is native or not if it went locally extinct centuries ago. Indeed, one of the concerns expressed about Pleistocene rewilding is that ecosystems have changed in many ways in the last several thousand years and that even historical nativity is not a good guide to the likely consequences of a species reintroduction.

The tension concerning nativity and invasion is magnified because of the specific history of invasive species management as a field. Rewilders were not the first to think about using non-native species to enact ecological processes. For example, non-native species have been introduced to new locations in attempts to control agricultural pests. The unintended consequences of these deliberate introductions became one of the central and founding problems of invasive species ecology and management, with the disastrous introduction of non-native cane toads to Australia in an attempt to control cane beetles being the classic example (Buller, 2013).

Open-ended vs. fixed goal management

The tensions between compositional and process-facilitating approaches to conservation are exacerbated by the emphasis on open-ended management in rewilding. In invasive species management, and in regimes of biosecurity, objectives are often specified clearly and unambiguously. Certain system states are categorised as desirable, and threats to them are identified and countered. In rewilding, by contrast, although certain system states might be deemed as undesirable, there is much less prescription about what a resilient ecology should consist of, beyond that it should be biodiverse and resilient. Indeed, in a recent articulation of the principles of rewilding, there is an explicit disavowal of the kinds of fixed goals relevant to invasive species management; as the authors write, 'Under these guidelines rewilding cannot be . . . used to manage for a particular target species, habitat or ecosystem outcome' (Sandom and Wynne-Jones, 2018, 242).

In practical terms, this means that management styles in rewilding and invasive species management/biosecurity can differ enormously. Invasive species and biosecurity management both have tendencies towards intensive forms of practical action: resource-heavy programmes of culling or removing invasives where possible, developing monitoring capabilities for pre-emptive action or creating effective borders and barriers (such as customs controls between countries that regulate the movement of non-native species). In contrast, rewilding is based on the notion that self-willed ecosystems will, by and large and if helped to begin with, self-manage – a characteristic that makes the approach attractive within neoliberal modes of government that emphasise reduced costs and outsourcing. Rewilding also tends to emphasise 'experimental' forms of management, where the term is not always, or even often, used in the sense of replicated treatments, but more often to indicate interventions where the end point is deliberately unpredictable. The rewilder's ethos of 'let's just see what happens' could easily be anathema to the biosecurity manager.

Nevertheless, it is important not to overplay this tension. Both biosecurity and rewilding are more nuanced than their extreme caricatures may suggest. Rewilding may utilise less intensive management techniques when a scheme is operational, but the initial intervention – introducing megafauna, or removing invasives, to create a favourable starting point – is often intensive in character and also does much to limit the possible outcomes of a supposedly open-ended scheme. Furthermore, biosecurity regimes can specify high-level fixed goals but can attain these through experimental and non-intensive forms of management. If a rewilded ecosystem is more resilient than before an intervention, it may also be better able to withstand potential invasive species and may be more biosecure in terms resistance to disease and perturbation. Both approaches also share some technologies, particularly selective fences.

Connectivity vs. control

The third area of philosophical tension relates to connectivity and control. The notion of connectivity is fundamental to rewilding, from the early incarnations of 'cores, corridors and carnivores' in the Wildlands Project to present day concerns with resilient rewilded ecosystems able to adapt to climate change. The mobility of large animals is crucial to the efficacy of many ecological processes, including seed dispersal, resilience to disease, population regulation through predation and metapopulation survival. Rewilding schemes thus often focus on improving connectivity between habitats and across landscapes. While this is beneficial for many species, however, it also raises a series of concerns with respect to biosecurity and invasive species management. Connectivity schemes, and corridors in particular, often favour the movement of invasive species as well as, and in some cases more than, native species (Hilty et al., 2006). Enhancing ecological mobilities can also aid the spread of diseases, including those affecting plants, animals and humans. Regimes of biosecurity tend to work through controlling and regulating the circulations of organisms (Dobson et al., 2013; Hodgetts, 2017) rather than promoting a mobility free-for-all. When coupled with rewilding's non-interventionist, experimental management style and process-based objectives, the promotion of unfettered connectivity without adequate safeguards in place may easily conflict with the imperatives of biosecurity.

However, as detailed in the previous section, rewilding is a broad term, covering a range of approaches with some shared characteristics but also a range of key differences. When it comes to connectivity and its implications for biosecurity or invasive species management, the form of spatiality assumed in a rewilding approach is central to understanding the likelihood and substance of any friction. Rewilding schemes that create and maintain boundaries (especially those using fences) will sit more easily within a biosecurity regime based on borders, separation and control. Nevertheless, the possibility of escapes or disease transmission through fence lines will likely generate some concerns. By contrast, rewilding schemes that enact or envisage unbounded movement may raise different, and perhaps more pressing, worries. Context and particularity matter. Rewilding will not be appropriate everywhere, as its architects and advocates explicitly acknowledge (Donlan et al., 2005; Sandom and Wynne-Jones, 2018). Biosecurity and invasive species concerns are thus an important factor in deciding where rewilding can be appropriate. Rather than conflict, such regimes can coexist.

Resolving tensions between rewilding and invasive species management

> Challenging objections to Pleistocene re-wilding include the possibility of disease transmission, the fact that habitats have not remained static over millennia, and the likelihood of unexpected ecological and social consequences of reintroductions. These issues must be addressed by sound research, prescient management plans and unbiased public discourse on a case-by-case and locality-by-locality basis. Well-designed, hypothesis-driven experiments will be needed to assess the impacts of potential introductions before releases take place.
>
> *(Donlan et al., 2005, 914)*

In this final section, we suggest three key interventions through which to address the possible tensions between rewilding schemes and biosecurity or invasive species management: i) improving stakeholder engagement, ii) emphasising local context and iii) applying best practice.

The first move to address the tensions outlined prior is to encourage a more frank and explicit recognition that, although rewilding envisages allowing ecological processes to unfold in an open-ended way, there is nevertheless a value hierarchy attached to possible ecosystem states. This hierarchy is discernible from the allowance made for intensive initial interventions within rewilding schemes, before retreating to less intensive forms of management. Indeed, since these initial interventions can involve the removal of invasive species, such actions can act to resolve some tensions early on in a practical project or theoretical discussion. The hierarchy is also apparent from the willingness of rewilders to intervene when 'experiments' lead to undesirable outcomes. Rewilding is thus in some important ways more about adaptive management than a lack of management. These frank recognitions, especially in the context of practical projects, are best made in real discussions among relevant and involved stakeholders. It is thus imperative for rewilders to engage with multiple interests when proposing or designing projects, and those concerned with the possible impacts of invasive species and biosecurity should be part of such exercises (Crowley et al., 2017b). Doing so would allow rewilders to incorporate such concerns into their management plans, especially through pre-defining what an unfavourable ecosystem state might incorporate.

Indeed, the call for stakeholder engagement within rewilding is more than an empty gesture. Certain articulations of the rewilding approach have been much critiqued in some quarters for their neocolonial tendencies, such as imposing large-scale landscape planning regimes from above or promoting the removal of human communities from newly designated wild areas (Jørgensen, 2015). The echoes of 'fortress conservation' are real and of concern (Brockington, 2002), as are the sometimes reified ideas of wilderness as a place without people (Cronon, 1996) that have been given renewed vigour through some rewilding narratives (Jørgensen, 2015). Furthermore, analysing carnivore (re)introductions from the standpoint of political ecology illuminates how the costs of such schemes may hit the poorest, marginalised communities (such as upland livestock farmers or rural dwellers newly exposed to large carnivores), with the benefits mostly accruing to wealthy elites (urban-living tourists). Given these concerns, rewilding schemes should only proceed with the genuine support of local communities (Sandom and Wynne-Jones, 2018).

Stakeholder engagement is thus a means by which rewilding projects can engage with the particularities of local contexts. It is incumbent on such schemes to carefully consider, among other things, the historical and contemporary ecology of an area, including its vulnerability to invasive species and the likely pathways through which ecologically problematic species may migrate. Stakeholder engagement can also provide information as to the cultural importance of particular species and species communities that might otherwise be neglected (and harmed) through the instigation of an open-ended rewilding scheme. Not all shepherds view their sheep as purely economic assets whose deaths in the jaws of reintroduced wolves can be compensated for with cash. Local people may have conflicting ideas about the nativity, and cultural desirability, of particular extant or potentially introducible species – both between themselves and compared to invasion biologists and rewilding advocates. Indeed, similar calls for genuine participation in management, and appreciation of local socio-political and ecological contexts, have been made in the context of invasive species management more broadly (Crowley et al., 2017b). As rewilding advocates repeatedly stress, the approach will not be appropriate everywhere.

The inherent tensions between rewilding and invasion can also be addressed through learning from 'best practice'. Although perhaps reminiscent of unhelpful management jargon, the notion of best practice can be very helpful in conservation, especially given the long-standing difficulties in assembling conservation evidence (relating to the tendency not to publish about 'failures'; Sutherland et al., 2004). In particular, selecting the most appropriate species to (re)introduce or encourage can be challenging, given existing ecologies and likely future ecological and bioclimatic changes (Svenning et al., 2016). Rewilding can draw on three sources in particular. First, given that species introductions are a common feature of rewilding schemes, projects can gain from engaging with the various sets of guidelines that have been published in academic and practitioner literatures pertaining to such actions; see, for example, the IUCN Guidelines for Reintroductions and Other Conservation Translocation (IUCN/SSC, 2013). Nevertheless, such guidelines may conflict with rewilding schemes that propose introductions of species beyond their historical range (Buller, 2013). Second, rewilding projects have much to gain from a closer cooperation with invasion biology (with respect to both the unintended consequences of species reintroductions *and* the trophic cascades that can occur when disruptive species enter a new habitat), as well as from social science analyses of invasion biology (especially with respect to the forms of politics enacted through regimes of nativity and metaphors of invasion). And third, given the explicit emphasis on experiments, rewilding projects have much to learn about best practice from their own and similar endeavours.

References

Bowman, D. (2012) 'Conservation: Bring elephants to Australia?', *Nature*, vol. 482, p. 30.

Brockington, D. (2002) *Fortress Conservation: The Preservation of the Mkomazi Game Reserve, Tanzania*. The International African Institute in association with James Currey, Mkuki Na Nyota and Indiana University Press, Bloomington.

Buller, H. (2008) 'Safe from the wolf: Biosecurity, biodiversity, and competing philosophies of nature', *Environment and Planning A*, vol. 40, pp. 1583–1597.

Buller, H. (2013) 'Introducing Aliens, reintroducing natives: A conflict of interest for biosecurity?', in A. Dobson, K. Barker and S.L. Taylor (eds.) *Biosecurity: The Socio-Politics of Invasive Species and Infectious Diseases*. Routledge, Oxford.

Caro, T. (2007) 'The Pleistocene re-wilding gambit', *Trends in Ecology & Evolution*, vol. 22, pp. 281–283.

Chapron, G., Kaczensky, P., Linnell, J.D., von Arx, M., Huber, D., Andrén, H., López-Bao, J.V., Adamec, M., Álvares, F., Anders, O., Balčiauskas, L., Balys, V., Bedő, P., Bego, F., Blanco, J.C., Breitenmoser, U., Brøseth, H., Bufka, L., Bunikyte, R., Ciucci, P., Dutsov, A., Engleder, T., Fuxjäger, C., Groff, C., Holmala, K., Hoxha, B., Iliopoulos, Y., Ionescu, O., Jeremić, J., Jerina, K., Kluth, G., Knauer, F., Kojola, I., Kos, I., Krofel, M., Kubala, J., Kunovac, S., Kusak, J., Kutal, M., Liberg, O., Majić, A., Männil, P., Manz, R., Marboutin, E., Marucco, F., Melovski, D., Mersini, K., Mertzanis, Y., Mysłajek, R.W., Nowak, S., Odden, J., Ozolins, J., Palomero, G., Paunović, M., Persson, J., Potočnik, H., Quenette, P.-Y., Rauer, G., Reinhardt, I., Rigg, R., Ryser, A., Salvatori, V., Skrbinšek, T., Stojanov, A., Swenson, J.E., Szemethy, L., Trajçe, A., Tsingarska-Sedefcheva, E., Váňa, M., Veeroja, R., Wabakken, P., Wölfl, M., Wölfl, S., Zimmermann, F., Zlatanova, D. and Boitani, L. (2014) 'Recovery of large carnivores in Europe's modern human-dominated landscapes', *Science*, vol. 346, pp. 1517–1519.

Cronon, W. (1996) 'The trouble with wilderness: Or, getting back to the wrong nature', *Environmental History*, vol. 1, pp. 7–28.

Crowley, S.L., Hinchliffe, S. and McDonald, R. (2017a) 'Nonhuman citizens on trial: The ecological politics of a beaver reintroduction', *Environment and Planning A: Economy and Space*, vol. 49, pp. 1846–1866.

Crowley, S.L., Hinchliffe, S. and McDonald, R.A. (2017b) 'Conflict in invasive species management', *Frontiers in Ecology and the Environment*, vol. 15, no. 3, pp. 133–141.

DEFRA (2017) 'Environment secretary backs release of Beavers in Forest of Dean', www.gov.uk/government/news/environment-secretary-backs-release-of-beavers-in-forest-of-dean, accessed 3 September 2019.

Deryabina, T., Kuchmel, S., Nagorskaya, L., Hinton, T., Beasley, J., Lerebours, A. and Smith, J. (2015) 'Long-term census data reveal abundant wildlife populations at Chernobyl', *Current Biology*, vol. 25, pp. R824–R826.

Dirzo, R., Young, H.S., Galetti, M., Ceballos, G., Isaac, N.J. and Collen, B. (2014) 'Defaunation in the Anthropocene', *Science*, vol. 345, pp. 401–406.

Dobson, A., Barker, K. and Taylor, S.L. (2013) *Biosecurity: The Socio-Politics of Invasive Species and Infectious Diseases*. Routledge, Oxford.

Donlan, C.J., Berger, J., Bock, C.E., Bock, J.H., Burney, D.A., Estes, J.A., Foreman, D., Martin, P.S., Roemer, G.W., Smith, F.A., Soulé, M.E. and Greene, H.W. (2006) 'Pleistocene rewilding: An optimistic agenda for Twenty-First Century conservation', *The American Naturalist*, vol. 168, pp. 660–681.

Donlan, J., Greene, H.W., Berger, J., Bock, C.E., Bock, J.H., Burney, D.A., Estes, J.A., Foreman, D., Martin, P.S., Roemer, G.W., Smith, F.A. and Soulé, M.E. (2005) 'Re-wilding north America', *Nature*, vol. 436, pp. 913–914.

Dowie, M. (2009) *Conservation Refugees: The Hundred-Year Conflict between Global Conservation and Native Peoples*. MIT Press, Cambridge.

Griffiths, C.J., Jones, C.G., Hansen, D.M., Puttoo, M., Tatayah, R.V., Müller, C.B. and Harris, S. (2010) 'The use of extant non-indigenous tortoises as a restoration tool to replace extinct ecosystem engineers', *Restoration Ecology*, vol. 18, pp. 1–7.

Gunderson, L.H. and Pritchard, L. (2002) *Resilience and the Behavior of Large-Scale Systems*. Island Press, New York.

Hare, J. (2008) 'Camelus ferus', *The IUCN Red List of Threatened Species*, www.iucnredlist.org/species/63543/12689285, accessed 5 February 2018.

Hilty, J.A., Lidicker, W.Z. and Merenlender, A.M. (2006) *Corridor Ecology: The Science and Practice of Linking Landscapes for Biodiversity Conservation*. Island Press, New York.

Hinchliffe, S., Allen, J., Lavau, S., Bingham, N. and Carter, S. (2013) 'Biosecurity and the topologies of infected life: From borderlines to borderlands', *Transactions of the Institute of British Geographers*, vol. 38, pp. 531–543.

Hinchliffe, S. and Bingham, N. (2008) 'Securing life: The emerging practices of biosecurity', *Environment and Planning A*, vol. 40, pp. 1534–1551.

Hintz, J. (2007) 'Some political problems for rewilding nature', *Ethics, Place and Environment*, vol. 10, no. 2, pp. 177–216.

Hodgetts, T. (2017) 'Wildlife conservation, multiple biopolitics and animal subjectification: Three mammals' tales', *Geoforum*, vol. 79, pp. 17–25.

Hunter, E.A., Gibbs, J.P., Cayot, L.J. and Tapia, W. (2013) 'Equivalency of Galápagos giant tortoises used as ecological replacement species to restore ecosystem functions', *Conservation Biology*, vol. 27, pp. 701–709.

IUCN/SSC (2013) 'Guidelines for reintroductions and other conservation translocations version 1.0. Gland, Switzerland: IUCN Species Survival Commission', https://portals.iucn.org/library/efiles/documents/2013-009.pdf, accessed 3 September 2019.

Jørgensen, D. (2015) 'Rethinking rewilding', *Geoforum*, vol. 65, pp. 482–488.

Lorimer, J. and Driessen, C. (2014) 'Wild experiments at the Oostvaardersplassen: Rethinking environmentalism in the Anthropocene', *Transactions of the Institute of British Geographers*, vol. 39, pp. 169–181.

Lorimer, J. and Driessen, C. (2016) 'From "Nazi cows" to cosmopolitan "ecological engineers": Specifying rewilding through a history of Heck cattle', *Annals of the American Association of Geographers*, vol. 106, pp. 631–652.

Moorhouse, T.P. and Sandom, C.J. (2015) 'Conservation and the problem with "natural"-does rewilding hold the answer?', *Geography*, vol. 100, p. 45.

Navarro, L.M. and Pereira, H.M. (2015) 'Rewilding abandoned landscapes in Europe', in H.M. Pereira and L.M. Navarro (eds.) *Rewilding European Landscapes*. Springer, London.

Neumann, R.P. (1998) *Imposing Wilderness: Struggles Over Livelihood and Nature Preservation in Africa*. University of California Press, Oakland.

RewildingEurope (2015) 'What is rewilding?', https://rewildingeurope.com, accessed 20 April 2015.

RewildingEurope (2018) 'European rewilding network', https://rewildingeurope.com/european-rewilding-network/, accessed 23 May 2018.

Ripple, W.J. and Beschta, R.L. (2012) 'Trophic cascades in Yellowstone: The first 15 years after wolf reintroduction', *Biological Conservation*, vol. 145, pp. 205–213.

Sandom, C., Donlan, C.J., Svenning, J.C. and Hansen, D. (2013) 'Rewilding', in D.W. Macdonald and K.J. Willis (eds.) *Key Topics in Conservation Biology 2*. John Wiley & Sons, Oxford.

Sandom, C., Faurby, S., Sandel, B. and Svenning, J.-C. (2014) 'Global late quaternary megafauna extinctions linked to humans, not climate change', *Proceedings of the Royal Society B*, vol. 281, no. 1787, p. 20133254.

Sandom, C. and Wynne-Jones, S. (2018) 'Rewilding a country: Britain as a study case', in N. Pettorelli, S.M. Durant and J. Dutoit (eds.) *Rewilding*. Cambridge University Press, Cambridge.

Schnitzler, A. (2014) 'Towards a new European wilderness: Embracing unmanaged forest growth and the decolonisation of nature', *Landscape and Urban Planning*, vol. 126, pp. 74–80.

Soulé, M.E. and Noss, R. (1998) 'Rewilding and biodiversity: Complementary goals for continental conservation', *Wild Earth*, vol. 8, pp. 18–28.

Sutherland, W.J., Pullin, A.S., Dolman, P.M. and Knight, T.M. (2004) 'The need for evidence-based conservation', *Trends in Ecology & Evolution*, vol. 19, pp. 305–308.

Svenning, J.-C., Pedersen, P.B., Donlan, C.J., Ejrnæs, R., Faurby, S., Galetti, M., Hansen, D.M., Sandel, B., Sandom, C.J. and Terborgh, J.W. (2016) 'Science for a wilder Anthropocene: Synthesis and future directions for trophic rewilding research', *Proceedings of the National Academy of Sciences*, vol. 113, pp. 898–906.

Taylor, P. (2005) *Beyond Conservation: A Wildland Strategy*. Earthscan, London.

Thompson, K. (2015) *Where Do Camels Belong? The Story and Science of Invasive Species*. Profile Books, London.

Tree, I. (2018) *Wilding: The Return of Nature to an English Farm*. Picador, London.

Vera, F.W.M. (2000) *Grazing Ecology and Forest History*. CABI, Wallingford.

Wildlands Network (2018) 'History', https://wildlandsnetwork.org/history/, accessed 8 February 2018.

Zimov, S.A. (2005) 'Pleistocene park: Return of the mammoth's ecosystem', *Science*, vol. 308, pp. 796–798.

INDEX

Note: Page numbers in *italics* indicate figures; page numbers in **bold** indicate tables.

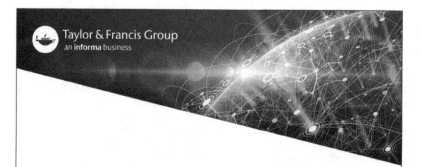

Printed in the United States
by Baker & Taylor Publisher Services